CONCEPTUAL DYNAMICS
An interactive approach

Kirstie Plantenberg
Richard Hill

SDC
Publications

SDC Publications

P.O. Box 1334
Mission KS 66222
(913) 262-2664
www.SDCpublications.com

Publisher: Stephen Schroff

Examination Copies:
Books received as examination copies are for review purposes only and may not be made available for student use. Resale of examination copies is prohibited.

Electronic Files:
Any electronic files associated with this book are licensed to the original user only. These files may not be transferred to any other party.

CONCEPTUAL DYNAMICS

PREFACE

Difficulties teaching dynamics

The subject of dynamics is often a difficult one for undergraduate students to master. One source of this difficulty is the great number of permutations of problems that exist. Students must have a firm grasp of the underlying principles; they cannot simply memorize a cookbook approach to solving problems. It is also difficult to solve real-world problems because of their complexity. Obtaining the right level of abstraction and knowing what can be neglected and what can't are difficult tasks for even the most seasoned engineer.

Conceptual dynamics approach

In order to overcome these difficulties, the authors propose placing a greater emphasis on concepts and involving students more actively in the learning process so that they may build physical intuition for themselves. With a greater understanding of the underlying concepts, the students will then be able to solve more traditional calculation-based problems with greater confidence and success. This theory of instruction is supported by a growing body of pedagogical research. This emphasis on concepts is also consistent with the types of problems typically seen on the Fundamentals of Engineering (FE) examination. Problems within the book that are particularly representative of FE-type problems are identified by the letters, fe.

The problems and activities included in this book will assist instructors with adopting this more active approach to teaching. The approach of this book is becoming particularly relevant as an increasing number of faculty at world-class universities make their lectures available online. This book may be used as a standalone text or in conjunction with these on-line lectures. This book, can effectively assist an instructor in "inverting the classroom,". In this sense, students can watch or read lectures on their own, leaving class time as an opportunity for the students to actively interface with the material. The instructor is then able to act as a moderator and can focus on the specific areas where students are struggling. The use of simple quizzes can be employed to encourage students to watch or read any assigned materials. This text allows the instructor to choose how much instruction is provided within or outside of the classroom.

Conceptual problems

Throughout this book, sets of "conceptual" problems are included that are meant to test understanding of the fundamental ideas presented in the text without requiring significant calculation. These problems can be assigned as homework or can be employed in class as exercises that more actively involve the students in lecture. When employed in class, these problems can provide the instructor with real-time feedback on how well the students are grasping the presented material. For example, following a ten-minute presentation of some new material, either inside or outside of class, a conceptual example could be presented to the students. The students could then be given a couple of minutes to attempt to answer the problem on their own. A couple of additional minutes could also be allowed for students to discuss their answers with a neighboring student so that they may correct or solidify their reasoning. After this time has been taken, the instructor can then poll the class either through the use of a personal response system ("clickers") or more informally by a show of hands, in order to assess how well the class has understood the concept. This process of polling is facilitated by the multiple-choice format of the problems. If the class as a whole scored well on the problem, then the lecture can proceed to the next topic or onto a more complex problem, perhaps involving more calculation. If the class seemed to not understand some concept, then the instructor can present additional material on the topic to try to address the misunderstandings.

The time taken to think about and discuss the examples help the students to construct meaning which helps their understanding and retention of the material. The active nature of these examples also helps the student maintain their focus and attention during class. When class consists of an instructor simply talking at the students, their attention is more likely to drift, especially during a long class period.

Supplemental instructional materials

In order to assist the instructor, PowerPoint lecture slides are provided to accompany the book. Boxes are included throughout the text leaving places where students can record important definitions and the correct responses to the conceptual questions presented within the PowerPoint slides. Students are encouraged to write in their books. In this sense, the book is meant to be "used." It is not necessarily the most comprehensive text on dynamics meant to be employed as a reference for some expert in the field. Rather, it is meant to be used as a tool by which undergraduate students can come to learn and appreciate the subject of dynamics.

These lecture slides also include animations that are useful for students to visualize the sometimes complex motions of bodies. Furthermore, interactive learning modules are available to provide students an alternative means for being introduced to the material. These animations and learning modules are made available via the book's website.

www.ConceptualDynamics.com

Problem solutions

Students have access, through the website, to complete solutions of all of the conceptual and example problems given in the book. Instructors will be provided a solution manual that contains worked solutions of the end-of-chapter problems.

Activities

Students are further encouraged to be active participants in their learning through activities presented at the end of each chapter. These activities can be performed in class involving the students or as demonstrations, or can be assigned to the students to perform outside of class. These activities help the students build physical intuition for the sometimes abstract theoretical concepts presented in the book and in lecture.

Computer and design problems

Along with the standard dynamics problems that are assigned as part of a student's homework, this book also includes computer and design problems. The proliferation of powerful software packages have greatly modified the way in which the subject of dynamics is applied in practice. Traditional dynamics homework problems typically involve the algebraic solution for a particular instant of time or set of conditions. The advent of software tools allows an engineer to model a dynamic system by a general equation of motion that can then be solved (often numerically) for many situations. In this book, problems are provided that require the student to derive the equation of motion and to sometimes solve the resulting differential equation. The computer problems range from problems that may be completed using a spreadsheet to problems that require coding or a specialized software package (e.g. Mathematica®, Maple®, or MATLAB/Simulink®).

Design problems are included in each chapter in order to motivate the importance of the material for students, as well as to get the students to think about real-world considerations. For example, practicing engineers must consider what simplifying assumptions are appropriate, as well as how the required information will be obtained. The application of the fundamental subject material to various design problems helps students see the material from a different perspective. It will also help them solidify their understanding of the material.

Contacting the authors

It is an exciting time to be an engineering educator with many new ideas available for improving student learning outcomes. It is our hope that this book will help you adopt some of these new pedagogical techniques. If you have any questions or comments regarding the book, we highly encourage you to contact us.

Kirstie Plantenberg
Richard C. Hill
(dynamics@engineeringessentials.com)

University of Detroit Mercy
Detroit, Michigan

NOTES

CONCEPTUAL DYNAMICS

TABLE OF CONTENTS

NOTES

PART I: INTRODUCTION

CHAPTER 1: BASIC CONCEPTS AND UNITS

CHAPTER OUTLINE

CHAPTER SUMMARY

The goal of this chapter is to introduce you to the subject of *dynamics*, explain what will be learned throughout the text, illustrate the importance of units, cover gravitation and present some core concepts in the study of dynamics. Dynamics is the study of motion. The study of dynamics is broken up into two major categories: *kinematics* and *kinetics*. In general, kinematics relates the position, velocity and acceleration of a body through the use of defining relations and calculus. Kinetics relates the motion of a body to the forces that are applied to it. Dynamics traditionally analyzes two body types: *particles* and *rigid bodies*. Particles have mass but no size and rigid bodies have both mass and size. Bodies may also be flexible, but these types of objects are beyond the scope of this book. It is often the goal of engineering analysis to quantify (i.e. model) a particular attribute of a system with an equation or numerical value. This numerical value has no meaning without units attached to it. This book will use both *Metric* and *US Customary* units. Before we dive in and start analyzing dynamic systems, there are a few core concepts that should be reviewed and/or introduced. Information learned in courses on *statics* and *calculus* will be the basis from which we build our knowledge of *dynamics*. From there, we will learn about *position*, *velocity*, *acceleration* and *force* along with other important topics that relate to the analysis of moving bodies.

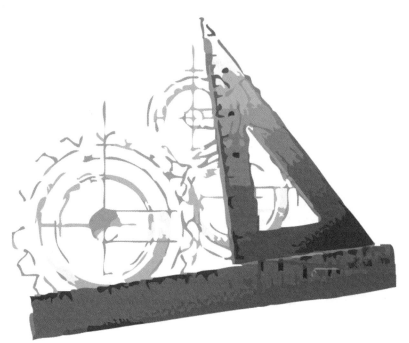

1.1) DYNAMICS

1.1.1) WHAT IS DYNAMICS?

Statics is the study of non-moving bodies in equilibrium. In statics we generally analyze the forces and reactions acting on these non-moving bodies. **Dynamics**, on the other hand, is the study of bodies in motion. These bodies may be in equilibrium or non-equilibrium, but they are always in motion. Given that the bodies under consideration are moving, dynamics is an inherently more complex subject than statics and involves more math, especially calculus. However, don't let the complexity of dynamics, as a subject, deter you from its study. It is very important. The majority of machines and devices that engineers need to analyze and design are in motion in some way. This includes everything from the very simple (e.g. a ball) to the very complex (e.g. a car). We need to know how objects move and how their motion will affect the objects around them. Thus, it is imperative that every engineer have a working knowledge of dynamics.

What is *dynamics*?

Why should we study *dynamics*?

1.1.2) WHAT WILL WE LEARN?

When studying the subject of statics and analyzing non-moving bodies, one method reigned supreme, Newtonian mechanics. The Newtonian equations of equilibrium given in Equations 1.1-1, state that the forces and moments applied to a body in equilibrium sum to zero.

Equations of equilibrium: 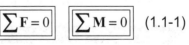 $\boxed{\sum \mathbf{F} = 0} \quad \boxed{\sum \mathbf{M} = 0}$ (1.1-1)

In the subject of dynamics, there are several methods of analysis that are commonly applied. Which one is employed depends on the type of body being analyzed, the information given and what you need to calculate. Many problems may be solved using multiple methods. This book does not cover all the methods available, but it does cover all the methods that are appropriate for an undergraduate course in dynamics.

When analyzing a moving object, the first decision you have to make is whether or not that object can be assumed to be a *particle* or if it needs to be analyzed as a *rigid body*. That is why, if you look at the table of contents, you will see that all the analysis methods are broken up into two parts; one where we use a method to analyze rigid bodies and another where we use a simplified version of the method to analyze particles.

In real life, all objects have both size and mass. Some bodies, if they move in a specific way, may be modeled as having no size (i.e. as a particle). Therefore, we divide dynamic analysis into two body types: *particles* and *rigid bodies*. **Particles** have mass but negligible size. They are treated as points and all forces act on the body at a single point. Particles are simpler to analyze than rigid bodies because without size they can only translate and not rotate. **Rigid Bodies** are non-deformable bodies that have both mass and size. They cannot be treated as a point; their rotation must be considered. Both of these types of bodies are a simplification of the real world. It may sometimes be the case that a body cannot be considered as a particle or as rigid. It is in general, the goal of a good engineer to employ as simple a model as possible that still allows the relevant questions to be answered. For example, if we wish to determine the time it takes a car to travel from Chicago to Detroit, a particle model makes sense. The particle model would be valid because the curvature of the road is large compared to the length of the car. If we are trying to determine a car's load transfer between the front and rear axles during braking, then a rigid body model would be necessary.

What is the difference between a *particle* and a *rigid body* model?

Additionally, dynamic analysis methods are divided into two different categories: *kinematics* and *kinetics*. **Kinematic analysis** involves the study of the geometric and time aspects of motion. There is no reference to the forces that cause the motion. **Kinetic analysis** involves the study of how forces and energies relate to the ensuing motion. Kinetic analysis is broken up into three major approaches (1) Newtonian mechanics, (2) work-energy and (3) impulse-momentum. One of the primary goals of this book is to help the student understand which method is the most appropriate to use based on the given situation. Depending on the information available, we may need to use kinematics, kinetics or both to analyze a given problem.

What is the difference between *kinematic* and *kinetic* analysis?

Conceptual Example 1.1-1

What method of analysis is being used to solve for the block's acceleration in the two cases shown in the figure? Kinematics or kinetics?

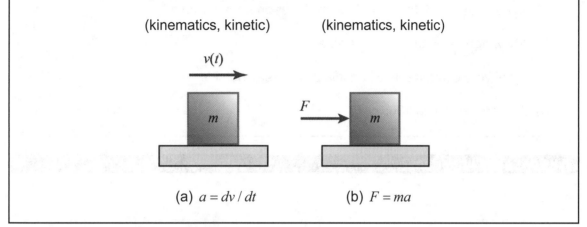

(kinematics, kinetic)　　　　(kinematics, kinetic)

(a) $a = dv / dt$　　　　(b) $F = ma$

1.1.3) PREREQUISITE KNOWLEDGE

A prerequisite for the study of dynamics is a good understanding of the subject of statics. Your course in statics gave you a good foundation in manipulating and using vectors, Newton's first law (i.e. the equations of equilibrium), understanding support reaction forces and drawing a proper free-body diagram. All of these topics will be used and expanded upon in this course. In addition, learning the subject of dynamics will rely on a good differential and integral calculus foundation. Many of the foundational subjects are either reviewed within the text or in the appendices.

1.2) UNITS

1.2.1) WHY UNITS ARE IMPORTANT

Some people have the mistaken impression that units are meaningless and a waste of time. Engineers cannot afford to be among this group. One of the most famous unit blunders occurred with NASA's Mars orbiter in 1999. The following is a composite summary from CNN and space.com.

1999 - NASA lost a $125 million Mars orbiter because a Lockheed Martin engineering team used English units of measurement while the agency's team used the more conventional metric system for a key spacecraft operation. According to a NASA official, trying to explain the loss of the craft, Lockheed Martin engineers sent NASA key maneuvering data in non-standard units. Miscalculations due to the use of English units instead of metric units apparently sent the craft slowly off course -- 60 miles in all -- leading it on a suicide course through the Martian atmosphere.

Solving a dynamics problem and obtaining the correct numerical value is a challenge, but you are not finished until you have attached units to your answer. A number does not make sense unless it has units attached to it.

Conceptual Example 1.2-1

What is missing from the following sentences?

   a) My daughter is 6 today and I started her on baby formula.

   b) My weight is 55 and I wear a size 8.

   c) I feel very comfortable when it is 28 degrees outside.

   d) The speed limit is 100.

Why are *units* important?

1.2.2) UNIT OF MEASURE

A **unit of measure** is used to define the physical quantity or magnitude of something. For example, the *unit of measure* of length could be either miles or kilometers depending on what measurement system in which you are working. The measurement systems that people in the United States are most familiar with are the *US Customary* system and the *Metric* system. There are many other measurement systems, but these are the two systems that this book will focus on. However, units from other systems may be mentioned periodically as points of interest.

Many engineering texts have switched to using only the *Metric* system. Using a single measurement system, especially the Metric system, makes calculations easier and less confusing for the students. However, the United States is still a dual unit society. For example, look at the power and handling specifications for the 2012 Mustang GT® given below. Notice that the description of the Mustang's power and handling specifications use dual systems of units. Specifically, horsepower (hp), pound-foot (lb-ft) and inches (") are *US Customary,* while Liters (L) is *Metric*.

Power and Handling - 2012 Mustang GT®

- 550 hp @ 6,200 rpm
- 510 lb-ft of torque @4,250 rpm
- 5.4L supercharged 4V Ti-VCT V8 engine
- 15 city/23 hwy mpg
- 19" front and 20" rear forged aluminum wheels
- 6-speed manual transmission

Until the United States has completely converted to a single unit society, it is important that students understand both systems, know the differences between the two and know how to convert from one to the other.

1.2.3) SI UNITS

All measurement systems, metric and non-metric, are linked through a network of international agreement supporting the *International System of Units (SI)*. The **SI system** is made up of 7 base units, 22 derived units with special names and many other derived units without special names and some units that are outside the SI that are accepted for use. The seven SI system base units are given in Table 1.2-1. The base units are consistent with the Metric system. Other SI units are defined algebraically in terms of the base units. For example, the SI unit of force, the Newton (N), is defined as the force that accelerates a mass of one kilogram at the rate of one meter per second per second ($kg\text{-}m/s^2 = N$).

The SI does allow the use of some metric and non-metric units that are traditional in various fields such as *degrees* for angle measurement, *liter* for volume measurement and *minute* for time measurement. See Appendix C or the References listed at the end of this chapter for more detailed coverage of the SI system.

1.2.4) METRIC SYSTEM

The *Metric* system got its start in 1799 with the *meter* and *kilogram*. In 1832 a unit for time and the decimal based aspect of the Metric system were incorporated. More and more units were added through the years until 1971 when the SI units became complete and reached the form that is known today. As was shown above, the SI base units have very precise definitions, however, the *Metric* system was not always based on those definitions. As our technological abilities have increased, so has the accuracy of our units of measure. Historically, the *meter* was intended to be equal to one ten-millionth of the length of the meridian through Paris from pole to equator, a *kilogram* was the mass of one cubic decimeter of water and a *second* was defined as the fraction 1/86,400 of the mean solar day.

The **Metric system** is a subset of the SI units. Its base units and derived units that are relevant to the subject of dynamics are given in Table 1.2-2. A complete list of units and their conversions are given in Appendix C.

In the Metric system, designations of any derived unit within the same measurement class may be arrived at by adding the prefixes deka, hecto and kilo meaning, respectively, 10, 100 and 1000 and deci, centi and milli, meaning, respectively, one-tenth, one-hundredth and one-thousandth. In certain cases, it becomes convenient to provide for multiples larger than 1000 and for subdivisions smaller than one-thousandth. A subset of all prefixes is given in Table 1.2-3 and a complete list of prefixes is given in Appendix C.

SI Base Unit	Symbol	What it measures	Basis of measurement
Meter	m	Distance	The distance light travels in a vacuum during a time interval of 1/299,792,458 of a second.
Kilogram	kg	Mass	Equal to the mass of the international prototype of the kilogram.
Second	s	Time	The duration of 9,192,631,770 periods of the radiation corresponding to the transition between the two hyperfine levels of the ground state of the cesium 133 atom.
Ampere	A	Electric current	The constant current which, if maintained in two straight parallel conductors of infinite length, of negligible circular cross-section and placed 1 meter apart in vacuum, would produce a force equal to 2×10^{-7} Newton per meter of length.
Kelvin	K	Temperature	The fraction 1/273.16 of the temperature of the triple point of water.
Mole	mol	Amount of substance	The amount of substance of a system which contains as many elementary entities as there are atoms in 0.012 kg of carbon 12.
Candela	cd	Intensity of light	The intensity in a given direction of a source that emits monochromatic radiation of frequency 540×10^{12} hertz and that has a radiant intensity in that direction of 1/683 watt per steradian.

Table 1.2-1: SI base units

Metric Unit	Symbol	What it measures
Meter	m	Distance
Kilogram	kg	Mass
Second	s	Time
Newton	$N = kg\text{-}m/s^2$	Force
Joule	$J = N\text{-}m$	Energy
Watt	$W = J/s$	Power

Table 1.2-2: Metric units

Prefix	Symbol	Meaning	Prefix	Symbol	Meaning
tera	T	10^{12}	deci	d	10^{-1}
giga	G	10^{9}	centi	c	10^{-2}
mega	M	10^{6}	milli	m	10^{-3}
kilo	k	10^{3}	micro	μ	10^{-6}
hecto	h	10^{2}	nano	n	10^{-9}
deka	da	10^{1}	pico	p	10^{-12}

Table 1.2-3: Metric prefixes

1.2.5) US CUSTOMARY SYSTEM

The *US Customary* system finds its roots in the *British Imperial* system. Traditionally this system was based on the measurement of the human body. The *inch* represented the width of a thumb. The *foot* (12 in) was originally the length of a human foot. The *yard* (3 ft) was the distance from the tip of the nose to the end of the middle finger of the outstretched hand. And, a *mile* was the length of 1000 paces. There are many other "natural units" such as the *palm* (3 in) and the *hand* (4 in) that we do not commonly use today. You could imagine that given the person, region or reigning king at the time, this system of measure could vary quite significantly. Today the **US Customary** units are based on SI units and are consistent and unchanging.

The US Customary system base units and derived units that are relevant to the subject of dynamics are given in Table 1.2-4. A complete list of units and their conversions are given in Appendix C.

US Customary Unit	Symbol	What it measures
Foot	ft	Distance
Slug	slug = lb_f-s^2/ft	Mass
Pound-mass (avoirdupois pound)	lb_m	Mass
Second	s	Time
Pound	lb_f = slug-ft/s^2	Force
British Thermal Units	BTU	Energy
Horsepower	hp	Power

Table 1.2-4: US customary units

The units for mass seem to be the most confusing unit class for students to grasp in the US Customary system. One major factor contributing to this confusion is that the US Customary system is a force-based system. This means that one of its base units is weight or force. We are all used to giving our weight in pounds, but do we know what our mass is? If we start with weight, then how do we get the mass? The weight of an object is defined as mass times gravity as shown in Equation 1.2-1. Therefore, if you want to calculate the mass of an object, you need to divide the weight of the object by the acceleration due to gravity. The mass unit calculated in this manner is called the *slug*. The name "slug", derives its name from the concept of inertia or "sluggishness." The relationship between pounds and slugs is not in itself very confusing other than the fact that we have a good physical sense of what 100 lbs of weight feels like, but we generally have no physical sense of what 3 slugs feels like. It turns out that 3 slugs is approximately 100 lbs of weight. What becomes really confusing is that the US customary system has another unit for mass that is not calculated using Equation 1.2-1.

Weight / Mass relationship: $\boxed{\mathbf{W} = m\mathbf{g}}$ (1.2-1)

$\mathbf{W}$ = weight
m = mass
$\mathbf{g}$ = acceleration due to gravity = 32.1740 ft/s^2

The *avoirdupois pound* and the *slug* are both units of mass and are both derived from the *pound-force*, but they have different values. One slug equals 32.1740 avoirdupois pounds. At standard conditions (sea level at a latitude of 45°), a body weighing 1 lb$_f$ has a mass of 1/32.1740 slugs or 1 lb$_m$. The confusion is twofold. One, the avoirdupois pound and the pound-force both have pound in their name and both are usually just referred to as pound. Two, the pound-mass and the pound-force seem to have the same value. Because of the confusion between the avoirdupois pound and the pound-force, this book will focus on the use of the slug mass unit. If the pound-mass unit is used, it will always carry the "m" subscript. Therefore, anywhere you observe a number with units of lbs, you may assume that it is a force unless otherwise specified.

Conceptual Example 1.2-2

Without calculating, which one in each of the following sets of objects has a different weight than the others?

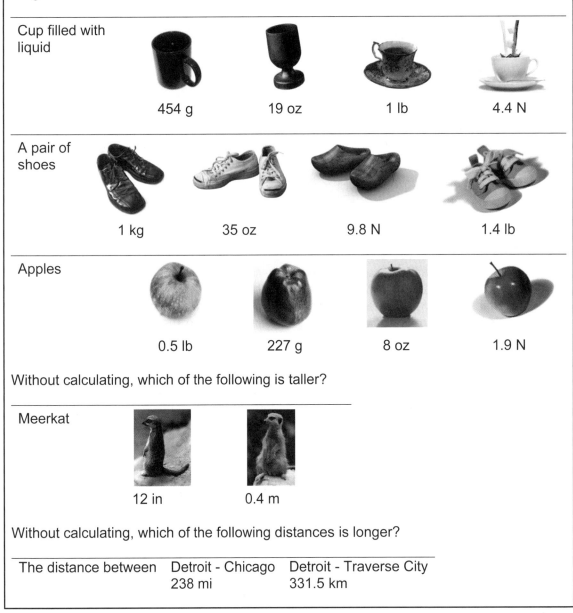

Cup filled with liquid			
454 g	19 oz	1 lb	4.4 N

A pair of shoes			
1 kg	35 oz	9.8 N	1.4 lb

Apples			
0.5 lb	227 g	8 oz	1.9 N

Without calculating, which of the following is taller?

Meerkat	
12 in	0.4 m

Without calculating, which of the following distances is longer?

The distance between	Detroit - Chicago	Detroit - Traverse City
	238 mi	331.5 km

Conceptual Example 1.2-3

Convert the following units. Use the conversion tables in Appendix C to complete this example.

Original unit	Equals	Converted unit
5 cm	=	in
1200 ft	=	mi
15°	=	rad
125 lb	=	slugs
85 kg	=	lb
3,000 J	=	BTU
450 hp	=	kWatts

1.3) BASIC CONCEPTS

There are several basic concepts that we need to review, or learn maybe for the first time, before diving into the study of dynamics. Most of these concepts were covered in previous courses such as physics, statics and calculus. This section is meant to refresh the memory, illustrate the important foundations of the coming chapters and introduce some new concepts that will be built upon throughout your journey through dynamics. The coverage of the topics in this section is meant to be brief and is here only to illustrate their importance and to show how they fit within the subject of dynamics. More comprehensive reviews will be given within the book on a need-to-know basis, while further information on the math components are given in Appendix B.

1.3.1) VECTORS

In dynamics, we study objects that move. When an object moves, it not only moves with a certain speed, but it also moves in a particular direction. This makes *vectors* a perfect vehicle to describe the position, velocity and acceleration of a moving body. A **vector** not only has a size or magnitude, but it also has a direction as shown in Figure 1.3-1. **Scalars** on the other hand will be used to describe quantities that only have size. *Mass* (m) is an example of a scalar quantity and *Weight* (**W**) is an example of a vector quantity.

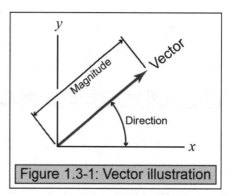

Figure 1.3-1: Vector illustration

Scalars and vectors will be used quite often throughout this book. The variable notation for each will be slightly different. A scalar variable will be represented by italicized Times New Roman font (e.g. mass = *m*), while a vector variable will be represented by bold faced Times New Roman font (e.g. weight = **W**).

What is the difference between a *scalar* and a *vector*?

Conceptual Example 1.3-1

Which is a scalar and which is a vector?

    a) I walked 10 miles. (vector, scalar)
    b) I walked 10 miles west. (vector, scalar)

1.3.2) SPACE

Dynamics studies bodies that, in reality, have mass and size and are located somewhere. This "somewhere", where the bodies live and interact, is defined as **space**. Not specifically "outer space", although that is a space, but the space around you. Let's use a car as an example. When a car exists in space, or more specifically on a road, we would like to be able to locate where exactly that car starts from and after it drives down the road, where it ends up. To do this we need to define our space. This space and how and where the bodies occupy it are usually described in terms of a **coordinate system**. There are several coordinate

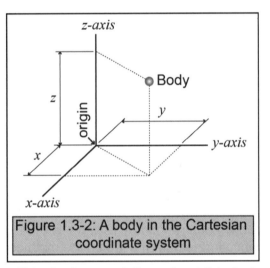

Figure 1.3-2: A body in the Cartesian coordinate system

systems that we can use to define space. We will look at several throughout this text; however, the one that you are probably most familiar with is referred to as the *Cartesian* or the *rectangular* coordinate system. This coordinate system defines space in terms of a set of three orthogonal linear directions, the *x*-, *y*- and *z*-coordinate axes that intersect at a defined origin as shown in Figure 1.3-2. It should be noted that any coordinate system can be defined as fixed or moving, but we will get into that more later.

What is *space*?

How do we usually define *space*?

1.3.3) POSITION

Position gives a body's location in space. Position is usually a vector quantity because, relative to a coordinate system, a body has both a distance from the origin and a direction.

What is *position*?

What are the units of *position*?

1.3.4) TIME

Once we know an object's position in space we are ready for it to move. How fast or slow an object moves depends on time. The quantity of **time** allows us to define when things occur, the duration of events and the rate at which events occur such as the motion of a body.

Why do we need *time* in dynamics?

What are the units of *time*?

1.3.5) VELOCITY

When a body moves, it can move fast or slow. This is represented by its *speed*. The greater its speed the faster the body is moving; that is, the faster its position is changing. *Velocity* on the other hand is more than that. It encompasses both how fast or slow a body is moving and in what direction. Therefore, velocity is a vector quantity and **speed** is the magnitude or size of the velocity vector. In technical terms, **velocity** is the time rate of change of position. Therefore, the units of velocity are length per time.

What is *velocity* and is it a vector?

What is *speed* and is it a vector?

What are the units of *velocity*?

1.3.6) ACCELERATION

Acceleration comes into play when we want to know if a moving body is picking up or losing speed. Acceleration tells us whether velocity is increasing or decreasing. More precisely, **acceleration** is the time rate of change of velocity making its units length per time per time or length per time squared. Since velocity is a vector, it makes sense that acceleration is also a vector because it has both magnitude and direction. This may make more sense if you imagine physically undergoing a change in velocity. For example, if you have ever been strapped into a roller coaster car going quickly around a curve, you have felt the side of the car pushing you around the curve. In addition, you have felt the seat of the car pushing into your back causing your speed to increase. Therefore, in this situation, your acceleration has both a sideways and a forward component to it.

What is *acceleration* and is it a vector?

What are the units of *acceleration*?

1.3.7) FORCE

Forces are important in dynamics in two ways. First, a body does not accelerate unless there is an imbalance of forces acting on it. Second, a moving body has the ability to apply dynamic forces to other bodies that it comes into contact with. A **force** is a push or pull that causes the velocity of an object, with mass, to change direction and/or magnitude. We know that velocity and acceleration are vectors; therefore, the force that causes this change in velocity (i.e. acceleration) must act in the direction of the acceleration. Hence, force is also a vector.

What is *force* and is it a vector?

What are the units of *force*?

1.3.8) MASS AND WEIGHT

It is easy to confuse *mass* and *weight*, but they are different. **Mass** is the quantity of matter in an object and **weight** is the force gravity applies to mass. Mass is the measure of how much a body resists acceleration (i.e. how much inertia it has). It determines an object's acceleration in the presence of an applied force. A body's mass stays the same no matter where it is located, whether it is here on earth or on the moon. However, a body's weight depends on its location. A body will have a different weight here on earth than it does on the moon. This is because the gravitational force in the two locations is different. The relationship between mass and weight is given in Equation 1.3-1.

Mass has no direction and therefore is a scalar. The weight of a body is determined by multiplying its mass by the acceleration due to gravity. The acceleration due to gravity acts towards the center of the earth or whatever body is being considered (e.g. the moon). That makes the acceleration due to gravity (**g**) a vector which, in turn, makes weight (**W**) a vector.

Weight / Mass relationship: $$\mathbf{W} = m\mathbf{g}$$ (1.3-1)

$\mathbf{W}$ = weight
m = mass
$\mathbf{g}$ = acceleration due to gravity = 9.807 m/s^2 = 32.1740 ft/s^2

What is *weight*? Is it the same as *mass*? If not, what is the difference between *weight* and *mass*?

Name two things that could make you weigh less.

Is *weight* a vector or a scalar? Is *mass* a vector or a scalar?

What are the units of *mass*? What are the units of *weight*?

1.3.9) CALCULUS AND DYNAMICS

Calculus can be helpful in solving many engineering problems, particularly those that involve quantities that change with respect to time, such as in dynamics. As was introduced in the previous sections, velocity is the time rate of change of position and acceleration is the time rate of change of velocity. Velocity and acceleration are two very important concepts in dynamics. Many times these quantities are continuous functions of time.

You may recall from calculus that a **derivative** represents a rate of change. Therefore, taking derivatives of continuous functions is an essential part of dynamics. For example, if you know a particle's velocity as a function of time ($v(t)$), you can determine its acceleration by taking the derivative of $v(t)$ with respect to time. The position (x), velocity (v) and acceleration (a) of a particle are related through Equations 1.3-2.

$$v = \frac{dx}{dt} = \dot{x} \qquad a = \frac{dv}{dt} = \ddot{x} \qquad (1.3\text{-}2)$$

x = position
v = velocity
a = acceleration
t = time

The relationships shown in Equations 1.3-2 will be explored further in the kinematics chapters. They are just placed here for illustrative purposes. A particle's position, velocity and acceleration may be represented by functions taking a variety of forms. Therefore, it is necessary to know how to differentiate many different types of functions. A short review of differential calculus is given in Appendix B.

The majority of the derivatives performed in dynamic analysis are with respect to time. You may be familiar with the shorthand notation for a derivative $y' = dy/dx$. When the derivative is specifically with respect to time, then it has its own special shorthand notation. A single dot is used if a first derivative is with respect to time ($\dot{x} = dx/dt$) and a double dot is used for a second derivative with respect to time ($\ddot{x} = d^2x/dt^2$). This notation can be seen in Equations 1.3-2.

Integration is the inverse operation of differentiation. Integrals are important in dynamics if you need to calculate the velocity of an object given its acceleration, or you need to calculate the position of a body given its velocity. The position (x), velocity (v)

and acceleration (a) of a particle are related through Equations 1.3-2. These equations may be rearranged to give the integral forms shown in Equations 1.3-3. More intuitively, an integral represents a summation. Therefore, integrating velocity over time represents how much distance has been accumulated. This is because velocity represents how quickly position changes with respect to time. A short review of integral calculus is given in Appendix B.

$$\boxed{\int v\,dt = \int dx}\quad \boxed{\int a\,dt = \int dv}\ (1.3\text{-}3)$$

x = position
v = velocity
a = acceleration
t = time

Why, when studying dynamics, is it important to understand derivatives and integrals?

1.3.10) SOLVING DYNAMICS PROBLEMS

Throughout this book a clear systematic approach to problem solving will be consistently used. A systematic approach to problem solving has the benefit of enabling the problem to move forward while minimizing simple mistakes. The approach laid out in this section is meant to help guide you through what may sometimes seem like complex and confusing problems. It will also help others interpret your work. For most problems, the following steps should be used when solving and preparing a problem for submission.

Step 1: Read the problem carefully. You will probably end up reading the problem several times while preparing the solution.

Step 2: Create a section on your paper labeled PROBLEM. This section is used to write down details of the problem statement that you think will help you wrap your mind around the problem.

Step 3: Identify what is given to you in the problem. Assign a variable to each given quantity and write this down next to the heading GIVEN.

Step 4: Identify what needs to be found in the problem. Assign a variable to each unknown quantity and write this down next to the heading FIND.

Step 5: Create a section labeled SOLUTION. This is where you will draw figures and perform calculations that will lead you to the solution of the problem.

    a) Choose an approach (i.e. a technique/equations that connect the givens to the unknowns). We will learn several approaches throughout this book; it is your job to choose the most appropriate technique to use.

    b) Draw a picture/schematic that represents the body/system described in the problem statement. This drawing will be used and referred to in the analysis of the problem; therefore, it should be simple, neat and the variables should be clearly identified. This drawing should/may include: coordinate axes. (always a must), a free-body diagram, variables and vectors that represent forces and/or displacements, velocities, accelerations and dimensions.

Sometimes it may be desirable to draw two different pictures to capture different aspects of the given situation(s).

c) Perform the solution clearly stating any simplifying assumptions along the way. Many times it is much easier, cleaner and more accurate to perform all calculations symbolically. Ideally, the answer should be given in both variable form and numerical form. The answer is never complete without units attached to the numerical value. The final answer for the quantities identified in the FIND section should be highlighted in some way. A simple way of doing this is by drawing a box around the answer.

Step 6: Sanity check. Ask yourself the following questions.

a) Do the signs, magnitudes and units of the answers make sense? This is always a must.

b) Can you perform the solution in an alternate manner and still get the same results?

1.4) UNIVERSAL GRAVITATION

Gravity is a concept that most everyone is familiar with, it is what causes things to fall. It conjures pictures of Isaac Newton being hit on the head by an apple, but the funny thing is no one really knows what causes it. In spite of not knowing for certain what causes gravity, scientists and engineers are able to observe it and describe it mathematically. Experiments to derive and validate a mathematical model for gravitational forces have been performed countless times. Thus, there is great confidence in our ability to quantify the magnitudes and directions of gravitational forces. These experiments were first performed by observing the orbits of celestial bodies (e.g. planets). As experimental techniques have improved, the theory has been extended to smaller bodies.

Various experiments have proven that, under most conditions, a gravitational force of attraction exists between all bodies in direct proportion to the product of their masses and in inverse proportion to the square of the distance separating them. Therefore, the bigger the mass of two objects and the closer together they are, the larger the force of gravity between them. These same experiments have also derived a constant of proportionality simply called the *gravitational constant* (G). In summary, the *Law of Universal Gravitation* is expressed mathematically by Equation 1.4-1.

$$\text{Law of universal gravitation:} \quad \boxed{F = G\frac{m_A m_B}{d^2}} \quad (1.4\text{-}1)$$

F = force of gravity
G = gravitational constant
    = 6.67 x 10^{-11} N m^2/kg^2

m_A, m_B = mass of body A and body B respectively
d = distance between the mass centers of body A
    and body B

Newton's Law of Universal Gravitation starts to break down when particularly massive bodies are considered or when bodies are moving at very high velocities. In this case, Einstein's General Theory of Relativity is able to more accurately describe the gravitational forces. For most practical application, Newton's Law of Gravitation is sufficiently accurate.

Recalling that forces are vectors, the gravitational force between two bodies acts on the line connecting their two mass centers as shown in Figure 1.4-1. The gravitational attraction between two bodies is an example of Newton's Third Law which is widely expressed as "every force of action has an equal and opposite reaction." This law will

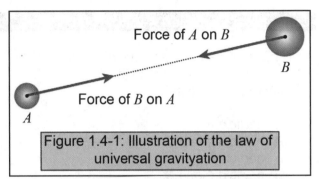

Figure 1.4-1: Illustration of the law of universal gravityation

be examined in further detail in the Newtonian mechanics chapters. In this case, it means that the force of gravitational attraction experienced by body A from body B in an interaction pair is equal in magnitude and opposite in direction to the gravitational force experienced by body B from body A. This means that the force of gravity you experience standing on the surface of the earth is also experienced by the earth, but in the opposite direction. In practice, however, the mass of the earth is so great that this gravitational force has little effect on its motion.

The Universal Law of Gravitation can be used to derive the constant $\mathbf{g}$ representing the acceleration due to gravity. Often a few simplifying assumptions are made that can cause the actual constant $\mathbf{g}$ to deviate slightly from the derived value. Specifically, it is often assumed that the earth's mass center is located at its geometric center when in fact it is slightly off center. This is because the earth does not have uniform density. Furthermore, it is usually assumed that the earth is a sphere of constant radius, when in reality the earth tends to bulge at the equator. Its surface is also covered with mountain ranges, valleys and other geographic features. Finally, the rotation of the earth is also often neglected. For engineering purposes, it is usually the case that these details can be neglected without greatly affecting the resulting calculations and conclusions.

Conceptual Example 1.4-1

Two pairs of particles are shown below. Which particle(s) experience the greatest gravitational attraction?

a) particle A
b) particle A and B
c) particle C
d) particle C and D

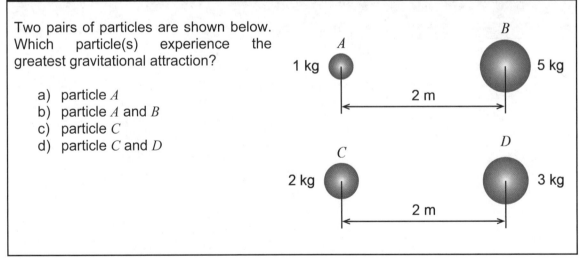

Conceptual Example 1.4-2

Two pairs of particles are shown below. Which particle(s) experience the greatest gravitational attraction?

a) particle A
b) particle A and B
c) particle C
d) particle C and D

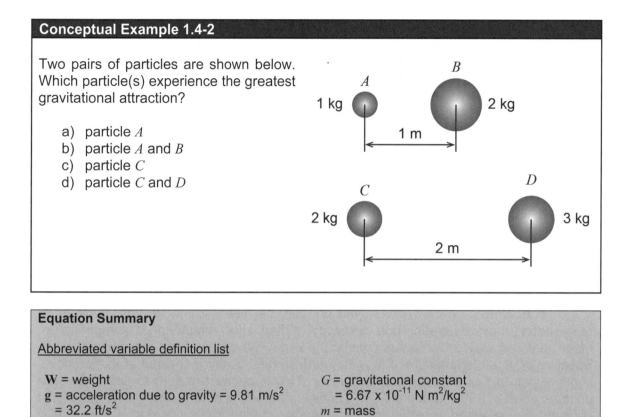

Equation Summary

Abbreviated variable definition list

W = weight
g = acceleration due to gravity = 9.81 m/s^2
   = 32.2 ft/s^2
F = force of gravity

G = gravitational constant
   = 6.67 x 10^{-11} N m^2/kg^2
m = mass
d = distance between the mass centers of body A and body B

Weight

$$W = mg$$

Law of universal gravitation

$$F = G\frac{m_A m_B}{d^2}$$

Example Problem 1.4-3

Considering that the earth has a mass of 5.97 x 10^{24} kg and an average radius of 6371 km, derive the average acceleration due to gravity experienced at the earth's surface.

Given:

Find:

Solution:

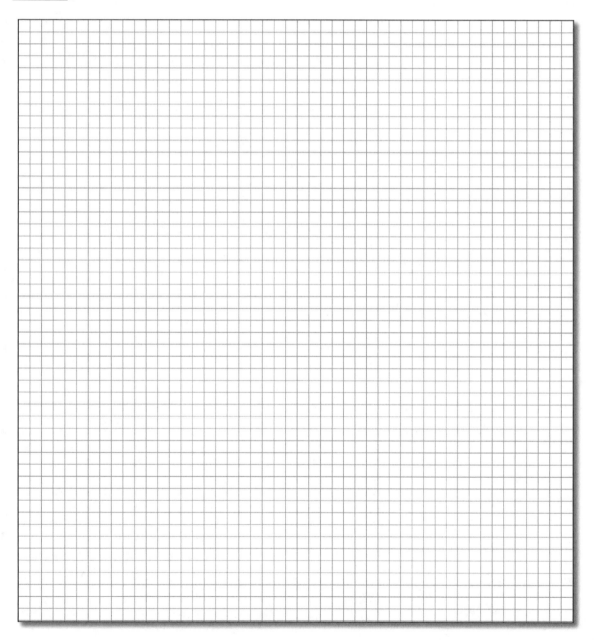

Example Problem 1.4-4

A 75-kg astronaut standing on a scale on the moon's surface determines that she weighs 123 N. If the moon is known to have a mass of 7.347 x 10^{22} kg, determine the moon's radius at the astronaut's current location.

Given:

Find:

Solution:

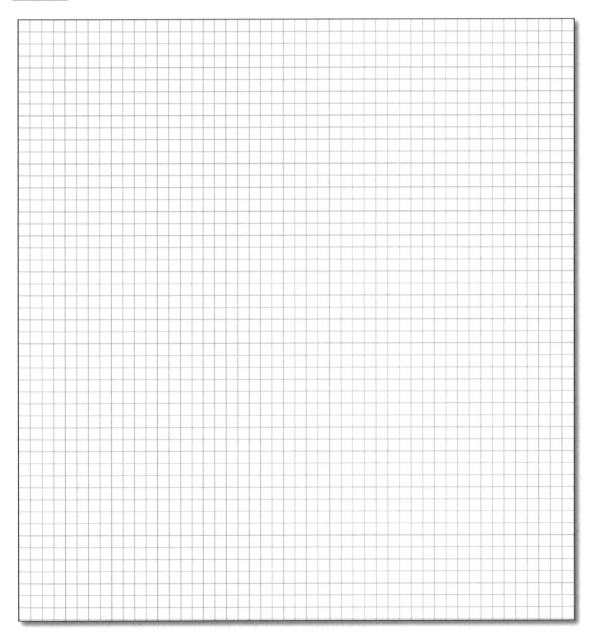

Solved Problem 1.4-5

At what point between the Earth and the Sun does an object have equal pull by both celestial bodies?

Solution

Getting familiar with the problem

The problem statement doesn't give much concrete information. It would be a good idea to draw a schematic of the problem. We will also have to look up a lot of information.

Law of universal gravitation

Let's write down the law of universal gravitation to see what information we need to look up.

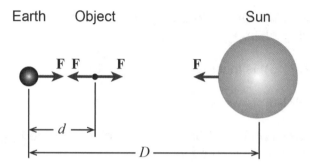

$$F = G\frac{m_A m_B}{d^2}$$

The masses of the Earth and Sun and the distance between the Earth and Sun may be looked up in the Appendix.

m_{Earth} = 5.97 x 10^{24} kg
m_{Sun} = 1.98892 x 10^{30} kg
D = 147098074 to 152097701 km

The force of gravity (**F**) between the Earth and the object and the force of gravity (**F**) between the Sun and the object are equal.

$$G\frac{m_{Earth}m_{object}}{d^2} = G\frac{m_{Sun}m_{object}}{(D-d)^2} \qquad \frac{m_{Earth}}{d^2} = \frac{m_{Sun}}{(D-d)^2} \qquad m_{Earth}(D-d)^2 = m_{Sun}d^2$$

$$\left(\frac{m_{Sun}}{m_{Earth}}-1\right)d^2 + 2Dd - D^2 = 0$$

We will use the quadratic formula to solve the above equation.

$$d = \frac{-b \pm \sqrt{b^2 - 4ac}}{2a}$$

$$\boxed{d = 254410 - 263057 \text{ km}}$$

CHAPTER 1 REVIEW PROBLEMS

RP1-1) What is the difference between a *particle* and a *rigid body*?

RP1-2) What is the difference between *kinematics* and *kinetics*?

RP1-3) Of the following units, which may be used to describe weight? (kg, lb$_f$, slug, N, lb$_m$)

RP1-4) Of the following units, which may be used to describe mass? (kg, lb$_f$, slug, N, lb$_m$)

RP1-5) Is mass a scalar or a vector?

RP1-6) Is weight a scalar or a vector?

RP1-7) (True, False) The acceleration due to gravity is constant no matter where you are standing on earth.

CHAPTER 1 PROBLEMS

P1.1) BASIC LEVEL UNITS PROBLEMS

P1.1-1)^{fe} Determine the mass in slugs of a person who weighs 150 pounds on earth.

a) m = 4.658 slug   b) m = 4.658 lb$_m$ c) m = 15.3 slug   d) m = 15.3 lb$_m$

P1.1-2)^{fe} Assuming that the acceleration due to gravity some distance from the earth is $g_1 = g / 3600$, where g is the acceleration due to gravity at the surface of the earth. Express g_1 in terms of kilometers and hours.

a) g_1 = 0.278 km/hr^2 b) g_1 = 1000 km/hr^2 c) g_1 = 0.001 km/hr^2 d) g_1 = 35.3 km/hr^2

P1.1-3)^{fe} Which of the following quantities is equal to a metric ton?

a) 500 N   b) 1000 kg   c) 2000 kg   d) 2000 N

P1.1-4)^{fe} Which of the following quantities is equal to an English ton?

a) 500 lb   b) 1000 lb   c) 2000 slug   d) 2000 lb

P1.1-5) What is 70 mph in ft/s and kph?

P1.2) INTERMEDIATE LEVEL UNITS PROBLEMS

P1.2-1) The weight of 4 grapefruits is 3 lb. Determine the average mass of one grapefruit in both SI units and US Customary units. What is the average weight of one grapefruit in SI units?

P1.2-2) What are the simplest units of the following quantities given their equations? The tables shown give the units for each quantity and the relationship between the base units and derived units.

a) $T = \dfrac{1}{2}mv^2$

b) $V = mgh$

c) $U = Fd$

d) $G = mv$

e) $I = Ft$

f) $H = mvr$

Quantity	Units
m	kg
t	s
h, d, r	m
v	m/s
g	m/s²
F	N

Derived unit	Base units
Force (N)	N = kg-m/s²
Energy (J)	J = N-m
	= kg-m²/s²

P1.2-3) Check the following equations for dimensional consistency. The tables shown give the units for each quantity and the relationship between the base units and derived units.

a) $F = ma$

b) $\dfrac{1}{2}mv_1^2 + mgh_1 + U_{1-2} = \dfrac{1}{2}mv_2^2 + mgh_2$

c) $Ft_{1-2} = mv_2 - mv_1$

Quantity	Units
m	kg
t	s
h	m
v	m/s
g, a	m/s²
F	N
U	J

Derived unit	Base units
Force (N)	N = kg-m/s²
Energy (J)	J = N-m
	= kg-m²/s²

P1.3) BASIC LEVEL GRAVITATION PROBLEMS

P1.3-1)[fe] What is the force of gravitational attraction between two 500-mm diameter spheres 1.5 m apart? The mass of one sphere is 3 slugs and the mass of the other is 5 slugs.

a) $F = 3.4 \times 10^{-7}$ lb   b) $F = 2.3 \times 10^{-7}$ lb   c) $F = 0.7 \times 10^{-7}$ lb   d) $F = 2.1 \times 10^{-7}$ lb

P1.3-2) A satellite in geosynchronous orbit with the earth and is positioned at a height of 1200 km, measured from the earth's surface, and has a mass of 2500 kg. Determine its weight on earth and its weight in orbit.

P1.3-3) Determine your mass in kg and slugs and your weight on the moon in pounds and in Newtons if you weigh W in pounds on earth.

P1.3-4) Calculate the range of the force of attraction between the earth and the moon.

P1.3-5) Calculate the range of the force of attraction between the earth and the sun.

P1.4) INTERMEDIATE LEVEL GRAVITATION PROBLEMS

P1.4-1) Calculate the weight in lb_f of a 5-lb_m and a 5-slug object on the moon.

P1.4-2) What is the force of gravitational attraction between two solid 200-mm diameter spheres that are 3 meters apart? One sphere is made of copper and the other is made from lead.

P1.4-3) At what altitude above the earth would you weigh half of what you do on the earth's surface?

P1.4-4) At what point between the Earth and the Sun does an object have twice the pull from the Sun as from the Earth? Use the average distance between the Sun and Earth.

CHAPTER 1 COMPUTER PROBLEMS

C1-1) Plot the force of gravitational attraction between two 100-kg spheres from touching to 1 km apart. Both spheres have a diameter of 100 mm.

CHAPTER 1 REFERENCES

[1] "NIST Handbook 44, Appendix C, Specifications, Tolerances and Other Technical Requirements for Weighing and Measuring Devices, General Tables of Units of Measurement" http://ts.nist.gov/weightsandmeasures/publications/appxc.cfm

[2] "The NIST Reference on Constants, Units and Uncertainty" http://physics.nist.gov/cuu/index.html

[3] "A Dictionary of Units of Measure" http://www.unc.edu/~rowlett/units/index.html

PART II: KINEMATICS

CHAPTER 2: KINEMATICS OF PARTICLES - RECTILINEAR MOTION

CHAPTER OUTLINE

CHAPTER SUMMARY

In this chapter, we will study kinematics of particles. **Kinematics** involves the study of a body's motion without regard to the forces that generate that motion. In particular, kinematics involves studying the relationship between displacement, velocity and acceleration. This chapter will focus on analyzing simple one-dimensional motion. The next chapter will move on to more complex two-dimensional motion.

The treatment of particles precedes rigid bodies because they are simpler to analyze. A **particle** may be treated as a point; it has mass but no size. Therefore, we will only need to consider translational motion and not worry about rotation.

2.1) RECTILINEAR MOTION

2.1.1) RECTILINEAR MOTION

The motion of a real object with size and mass is very complex. Take, for example, the stunt airplane shown in Figure 2.1-1. A plane can move in the forward direction, turn around and move in the opposite direction. It can gain altitude, lose altitude, and also turn left and right. It becomes even more complex when we consider that the

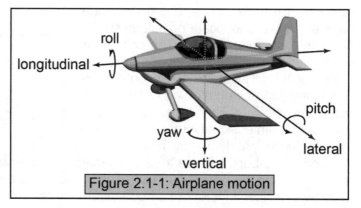

Figure 2.1-1: Airplane motion

plane can yaw, pitch and roll. These are the various rotations the plane can undergo. In total, there are six variables, or degrees of freedom, that are needed to describe the position and orientation of this plane. If we include velocities and accelerations, the number of variables needed increases to eighteen.

As students who are just learning dynamics, you don't want to jump in and learn how to analyze the most complex body and system first. Therefore, we are going to start with a simple body and constrain this body to move in a particular way. First, we will start by analyzing a particle. A **particle** is a body that has mass but no size. This means that it can translate but not rotate. Using a particle to describe the body of interest, we simplify our calculations by removing all of the rotational degrees of freedom. Second, we will constrain our body to move along a straight line. This means that the body can only move left and right, up and down or forwards and backwards, depending on the situation. This simplifies the calculations even further because now we only need three variables or degrees of freedom to describe how the body is moving. One variable to describe where the body is on the line, one variable to describe how fast it is moving along the line and one variable to describe how its speed is changing.

You may be thinking that such a simple analysis method can't have any practical use in the real world. However, there are many situations where this type of analysis can give a good approximation of the actual motion. Consider a car traveling down a straight road. If the road is relatively smooth, under normal driving conditions we can often neglect the rotational motion of the car and assume that it is a particle. This method is perfect for the analysis of simple vehicle acceleration problems. Another example is a ball that is thrown straight up in the air. These problems may be analyzed using what we will refer to as rectilinear analysis. There are many other examples that you will discover throughout this chapter.

2.1.2) RECTILINEAR COORDINATE AXIS

Rectilinear motion of a particle is motion along a straight line. The particle is allowed to move in a one-dimensional space. In order to locate objects in space, we use a coordinate axis. The coordinate axis used in rectilinear motion is shown in Figure 2.1-2. The particle can only move back and forth along the line or s-axis and not up or down. Therefore, we consider this a one-dimensional problem. A **one-dimensional** problem only needs one parameter to describe where the particle is

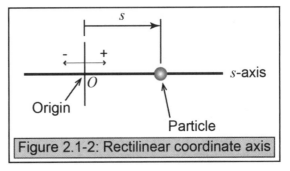

Figure 2.1-2: Rectilinear coordinate axis

located. This book will use the s coordinate to describe the position of a particle relative to the origin of the coordinate axis. The variable x is also commonly used to represent a particle's position in a rectilinear coordinate frame.

What is a *particle*?

What is *rectilinear motion*?

2.1.3) POSITION

Position (s) is where a particle is located on the line or s-axis. If you look at the rectilinear coordinate axis shown in Figure 2.1-2, you can see that it has an origin. The particle can be located s units to the right of the origin or it could be s units to the left of the origin. We need some way of differentiating the two. This can be done using vectors. Technically speaking, position is a vector because it has both a magnitude and a direction. However, because the direction is either to the right

Units of Position (length)
Metric Units:
- meters (m)
- centimeter (cm = 0.01 m)
- millimeters (mm = 0.001 m)
- kilometers (km = 1000 m)
US Customary Units:
- feet (ft = 12 in)
- inch (in)
- mile (mi = 5280 ft)

or left of the origin, we will forgo vector notation for the moment and just use signs to indicate direction. For example, a particle that is 10 meters to the left of the origin has a position of s = -10 meters and a particle that is 5 meters to the right of the origin has a position of s = 5 meters.

What is *position*?

Position has both magnitude and direction. Is *position* a vector or a scalar?

2.1.4) DISPLACEMENT

Displacement (Δs) is the change in the particle's position and has the same units as *position*. It is simply the difference between the particle's final and initial positions. Displacement is visually illustrated in Figure 2.1-3 and can be calculated using Equation 2.1-1.

Displacement also has a direction. In rectilinear motion, displacement can occur to the right or left. In this case, we will indicate the direction of the displacement with a positive or negative sign. For example, if a particle starts at 5 meters and ends at 12 meters, then its displacement is $\Delta s = 12 - 5 = 7$ meters. The positive sign indicates that the particle moved to the right. However, if the particle starts at 5 meters and ends at 2 meters, then its displacement is -3 meters. The negative sign indicates that the particle moved to the left.

Displacement: $$\boxed{\Delta s = s_{final} - s_{initial}}$$ (2.1-1)

Δs = displacement

$s_{final}, s_{initial}$ = final position / initial position

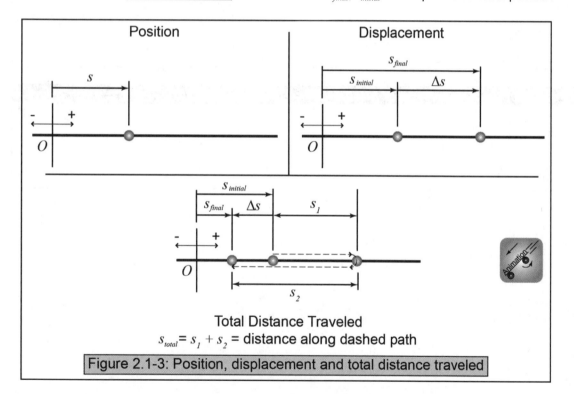

Total Distance Traveled

$s_{total} = s_1 + s_2$ = distance along dashed path

Figure 2.1-3: Position, displacement and total distance traveled

The **total distance traveled** (s_{total}) by a particle is equal to the length of the path taken by the particle. Distance is different from displacement in that it depends on the route between the initial and final positions of the particle, not just the endpoints themselves. The total distance traveled is the same as the displacement if the particle travels in a straight line without making any turns. However, if the particle changes direction and proceeds in the opposite direction, these two quantities will be different. In the case of rectilinear motion, the total distance traveled is calculated by adding the lengths of the legs between changes in direction of the particle's motion. The total distance traveled is visually illustrated in Figure 2.1-3.

The *total distance traveled* is a scalar in that it disregards direction. For example, consider a particle starting at the origin that first goes 5 meters in the positive s-direction before turning and proceeding 15 meters in the opposite direction. The particle made a turn producing two segments of travel that need to be added together. The total distance traveled ends up being s_{total} = 5 + 15 = 20 meters. Distance is, therefore, a scalar that does not include whether the travel was in the positive or negative direction. By contrast, the displacement of the particle in this case is -10 meters since the particle began at s = 0 meters and ended at s = -10 meters.

What is the difference between *position* and *displacement*?

What is the difference between *displacement* and *total distance traveled*? When are they the same?

Is *displacement* a vector? Is *total distance traveled* a vector?

Conceptual Example 2.1-1

The given figure shows several s-t graphs where s is the position of the particle and t is time. Rank the graphs from greatest to least amount of particle displacement over the time interval from 0 to 3 seconds. Repeat the ranking for total distance traveled.

Displacement
Greatest: _____ Next: _____ Next: _____ Next: _____ Next: _____ Least: _____

Total distance traveled
Greatest: _____ Next: _____ Next: _____ Next: _____ Next: _____ Least: _____

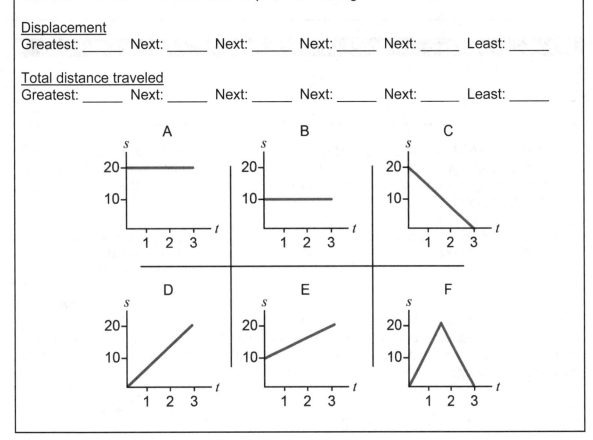

2.1.5) VELOCITY

We now know that *position* is where a particle is located on a line, for example, where a car is located on a road. We also know that dynamics is the study of particles in motion. So, how do we describe the motion component? We do this with the

> Units of Velocity (length/time)
> Metric Units:
> - meters per second (m/s)
> - kilometers per hour (kph = km/h)
> US Customary Units:
> - feet per second (ft/s)
> - miles per hour (mph = mi/h = 1.4667 ft/s)

concepts of *velocity* and *acceleration*. Velocity is how fast the particle or car is moving. You may be going 60 mph, 70 mph, or if you suddenly find that a policeman is following you, your velocity may be way too fast. To get technical, **velocity** (v) is the rate at which your position (s) is changing with respect to time. Did you cover 1 mile in a short period of time (i.e. high velocity) or are you riding a bicycle and covered the same mile in a long period of time (i.e. low velocity)?

Consider a car driving along I-80 going west. We can indicate the car's position at a particular time by mile markers, but is the car driving on I-80 West or I-80 East? Technically velocity is a vector. But again, just like with position, we will use sign to tell us whether velocity is proceeding to the right (i.e. positive) or to the left (i.e. negative).

What is *velocity*?

Velocity has both magnitude and direction. Is _velocity_ a vector or a scalar?

Velocity (v) is the time rate of change of position (s). The larger a body's velocity, the quicker it accumulates displacement. Averaged over a finite interval of time, the average velocity is equal to the change in position divided by the change in time as shown in Equation 2.1-2.

$$\text{Average Velocity:} \quad \boxed{v_{ave} = \frac{\Delta s}{\Delta t}} \quad (2.1\text{-}2)$$

v_{ave} = average velocity
Δs = displacement
Δt = change in time

As the change in time in Equation 2.1-2 becomes infinitesimally small, the average rate of change becomes an instantaneous rate of change also known as a derivative. Therefore, velocity is the instantaneous rate of change of position with respect to time as shown in Equation 2.1-3. Note that the dot notation ($\dot{s}$) represents a derivative with respect to time.

$$\text{Velocity:} \quad \boxed{v = \frac{ds}{dt} = \dot{s}} \quad (2.1\text{-}3)$$

$v = \dot{s}$ = velocity
s = position
t = time

Velocity is the derivative of position s with respect to time t (Equation 2.1-3). Graphically, a derivative is the slope of a line tangent to the function at a particular point. Velocity at a given instant of time is the slope of the line tangent to the $s\text{-}t$ curve at that instant as shown in Figure 2.1-4. If a particle changes direction, the slope of the $s\text{-}t$ curve will be momentarily zero, hence, the velocity at the turn will be zero.

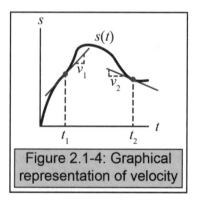

Figure 2.1-4: Graphical representation of velocity

Conceptual Example 2.1-2

The following illustrations show time-lapse pictures of rolling balls. The time between each successive time-lapsed image is equal. List each case in order from highest to lowest ball velocity for the following situations.

a) Starting velocity

Greatest: _____   Next: _____   Next: _____   Least: _____

b) Ending velocity

Greatest: _____   Next: _____   Next: _____   Least: _____

c) Also, list which cases have constant velocity.

Constant velocity: _____

Conceptual Example 2.1-3

The figure shows several s-t graphs where s is the position of the particle and t is time. Rank the graphs from greatest to least achieved instantaneous velocity. Then rank the graphs from greatest to least average velocity over the time interval 0 to 3 seconds.

Instantaneous
Greatest: _____   Next: _____   Next: _____   Next: _____   Next: _____   Least: _____

Average
Greatest: _____   Next: _____   Next: _____   Next: _____   Next: _____   Least: _____

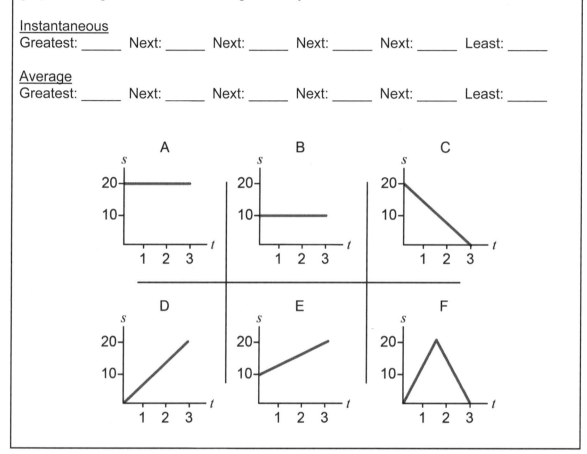

Conceptual Example 2.1-4

A car drives along a straight road. The car's position changes with time as shown in the graph. The graph indicates that ...

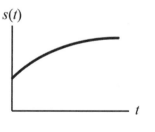

   a) the car speeds up with time.
   b) the car slows down with time.
   c) the car initially speeds up and then slows down.
   d) the car moves with a constant velocity.

2.1.6) ACCELERATION

Acceleration also describes a particle's motion. Let's say that you are behind a slow vehicle and you want to pass it. What do you do? You press on the gas pedal accelerating your car and increasing your velocity. **Acceleration** is the rate at which your velocity changes with respect to time. As with position and velocity, acceleration is technically a vector quantity. However, in rectilinear analysis, we indicate direction with sign. If the particle is speeding up, the acceleration is positive. If it is slowing down, the acceleration is negative.

> Units of Acceleration (length/time^2)
> Metric Units:
> - meters per second squared (m/s^2)
>
> US Customary Units:
> - feet per second squared (ft/s^2)

What is *acceleration*?

Conceptual Example 2.1-5

You are driving a car on a city street.

The acceleration of your car is _____ when you start moving after being stopped at a red light.

The acceleration of your car is _____ when approaching a yellow light.

The acceleration of your car is _____ when your velocity is constant.

Acceleration (a) is the time rate of change of velocity (v). Over a finite time interval, the average acceleration is given by Equation 2.1-4.

Average Acceleration: 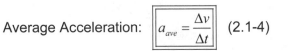 (2.1-4)

$$a_{ave} = \frac{\Delta v}{\Delta t}$$

a_{ave} = average acceleration
Δv = change in velocity
Δt = change in time

At a specific point in time, acceleration is the instantaneous rate of change of the velocity with respect to time. This is represented by the derivatives shown in Equation 2.1-5. Note that the double-dot notation ($\ddot{s}$) represents the second derivative with respect to time. Graphically, acceleration is the slope of the line tangent to the *v-t* curve at a given point in time as shown in Figure 2.1-5.

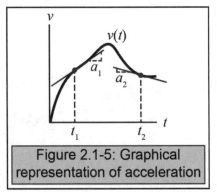

Figure 2.1-5: Graphical representation of acceleration

Acceleration: $\boxed{a = \dfrac{dv}{dt} = \dot{v} = \dfrac{d^2 s}{dt^2} = \ddot{s}}$ (2.1-5)

$a = \dot{v} = \ddot{s}$ = acceleration
v = velocity
s = position
t = time

Conceptual Example 2.1-6

Five balls are thrown up in the air from the same level with the initial velocity indicated in the figure. How high will each ball go relative to each other? List each case from highest height to lowest.

Highest: _____ Next: _____ Next: _____ Next: _____ Lowest: _____

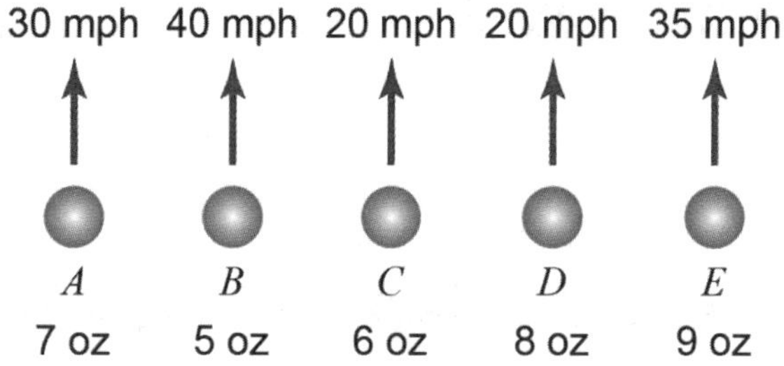

30 mph	40 mph	20 mph	20 mph	35 mph
↑	↑	↑	↑	↑
A	B	C	D	E
7 oz	5 oz	6 oz	8 oz	9 oz

Conceptual Example 2.1-7

The following illustrations show time-lapse pictures of rolling balls. The time between each successive time-lapsed image is equal. Determine which cases have positive, negative or zero acceleration.

Positive: _____ Negative: _____ Zero: _____

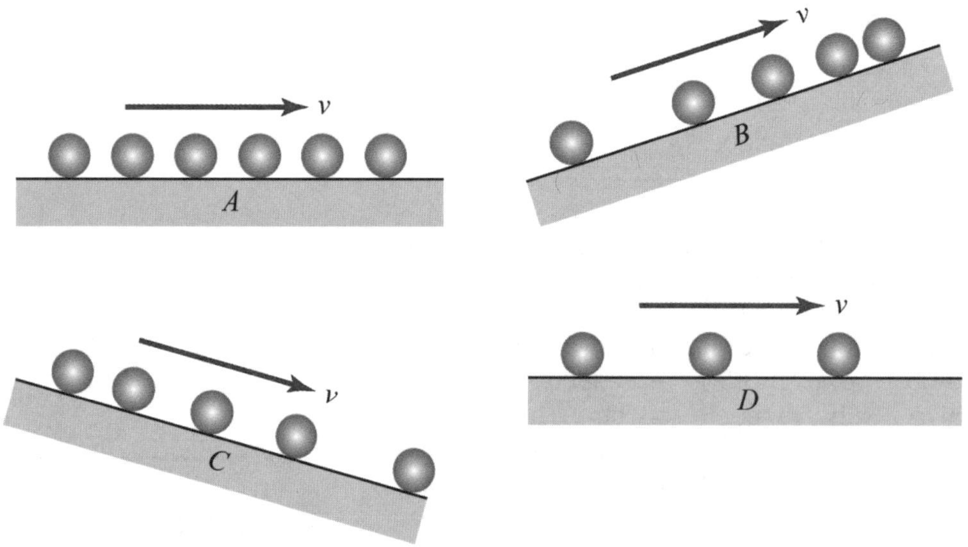

Conceptual Example 2.1-8

The position of two rolling balls are captured at 0.1 second intervals in the figure below. The balls are moving towards the right. What can be said about the balls' accelerations?

a) Ball A's acceleration is greater than ball B's acceleration.
b) Ball B's acceleration is greater than ball A's acceleration.
c) Both accelerations are equal and greater than zero.
d) Both accelerations are equal to zero.
e) Not enough information is given to answer the question.

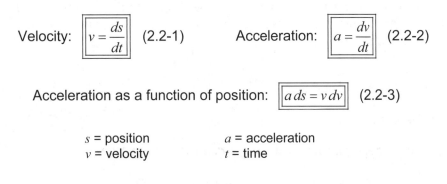

2.2) SOLVING RECTILINEAR PROBLEMS

2.2.1) THE BASIC EQUATIONS

Almost every particle rectilinear kinematic problem can be solved by manipulating the following three equations: Equation 2.2-1, 2.2-2 and 2.2-3. Equation 2.2-3 is arrived at by algebraically eliminating the quantity dt between Equation 2.2-1 and 2.2-3.

Velocity: $v = \dfrac{ds}{dt}$ (2.2-1)   Acceleration: $a = \dfrac{dv}{dt}$ (2.2-2)

Acceleration as a function of position: $a\,ds = v\,dv$ (2.2-3)

s = position a = acceleration
v = velocity t = time

2.2.2) TIME-DEPENDENT EQUATIONS

In general, the progression of derivatives with respect to time can be thought of as $s \rightarrow v \rightarrow a$. Proceeding from left to right, the derivative of displacement with respect to time is velocity and the derivative of velocity with respect to time is acceleration. This progression gives some indication of which equations to use when you are given information about a particle's motion as a function of time. For example, given a particle's position as a function of time $s(t)$, its velocity as a function of time $v(t)$ can be found using Equation 2.2-1. Subsequently, the particle's acceleration as a function of time can be found using Equation 2.2-2.

We can also go in the other direction $a{\rightarrow}v{\rightarrow}s$ using the mathematical inverse of the derivative, the integral. The integral of velocity with respect to time is equal to the change in position over that time as shown in Equation 2.2-4. Equation 2.2-4 can be arrived at by recognizing that differential elements like dt can be thought of as very small differences (i.e. $\Delta t \rightarrow 0$). Therefore, differential elements can be multiplied and divided just like any number. Referring to Equation 2.2-1, the differential element dt can be multiplied on both sides and then each side of the equation can be integrated leading to Equation 2.2-4. Likewise, the integral of acceleration with respect to time is equal to the change in velocity over that time as shown in Equation 2.2-5. It should be noted that the integral equations require limits of integration to solve. Therefore, we need to have some information about the particle's initial and final states.

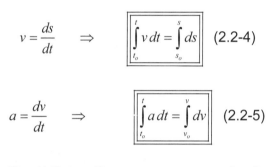

$$v = \frac{ds}{dt} \quad \Rightarrow \quad \boxed{\int_{t_o}^{t} v\,dt = \int_{s_o}^{s} ds} \quad (2.2\text{-}4)$$

$$a = \frac{dv}{dt} \quad \Rightarrow \quad \boxed{\int_{t_o}^{t} a\,dt = \int_{v_o}^{v} dv} \quad (2.2\text{-}5)$$

s, s_o = position / initial position a = acceleration
v, v_o = velocity / initial velocity t, t_o = time / initial time

2.2.3) POSITION-DEPENDENT EQUATIONS

Equation 2.2-3 is usually integrated to give Equation 2.2-6. This equation is useful if the particle's acceleration is a function of its position $a(s)$. Since this is an integral equation, it is again necessary that you have some knowledge of the limits of integration in order to solve the equation. For example, to find $v(s)$ using Equation 2.2-6, knowledge of the initial displacement s_o and initial velocity v_o are needed in addition to the acceleration function $a(s)$.

$$\boxed{\int_{s_o}^{s} a\,ds = \int_{v_o}^{v} v\,dv} \quad (2.2\text{-}6) \qquad\qquad \begin{array}{c} s,\, s_o = \text{position / initial position} \\ v,\, v_o = \text{velocity / initial velocity} \\ a = \text{acceleration} \end{array}$$

2.2.4) ACCELERATION AS A FUNCTION OF VELOCITY

So far we have addressed the situation where it is desired to find a particle's velocity and acceleration given as either a function of time (Equation 2.2-5) or as a function of position (Equation 2.2-6). Equation 2.2-5 and Equation 2.2-6 can be rearranged to provide an equation that may be employed to solve for velocity when the acceleration is given as a function of velocity as shown in Equation 2.2-7 and Equation 2.2-8. More specifically, Equation 2.2-7 is employed if the velocity at a specific instant of time is desired and Equation 2.2-8 is employed if the velocity at a specific position is desired. These equations are of practical significance because the acceleration of a body often does depend on velocity, such as when the body experiences wind drag or certain types of friction forces.

$$\int_{t_o}^{t} dt = \int_{v_o}^{v} \frac{1}{a} dv \quad (2.2\text{-}7) \qquad \int_{s_o}^{s} ds = \int_{v_o}^{v} \frac{v}{a} dv \quad (2.2\text{-}8)$$

s, s_o = position / initial position a = acceleration
v, v_o = velocity / initial velocity t, t_o = time / initial time

2.2.5) CONSTANT ACCELERATION EQUATIONS

If the acceleration of a particle is constant, then the defining relationships (Equations 2.2-1 through 2.2-4) can be integrated to give Equations 2.2-9 through 2.2-11.

Velocity if acceleration is constant: $\boxed{v = v_o + a_o(t - t_o)}$ (2.2-9)

Position if acceleration is constant: $\boxed{s = s_o + v_o(t - t_o) + \dfrac{a_o(t - t_o)^2}{2}}$ (2.2-10)

Velocity if acceleration is constant: $\boxed{v^2 = 2a_o(s - s_o) + v_o^2}$ (2.2-11)

s, s_o = position / initial position a_o = acceleration
v, v_o = velocity / initial velocity t, t_o = time / initial time

Equation Derivation

We will start by integrating Equation 2.2-2 and setting the acceleration to a constant value of a_o.

$$a_o = \frac{dv}{dt} \quad \Rightarrow \quad \int_{t_o}^{t} a_o \, dt = \int_{v_o}^{v} dv \quad \Rightarrow \quad a_o(t - t_o) = v - v_o$$

Rearranging the above equation gives us the expression for the velocity given below.

$$v = v_o + a_o(t - t_o)$$

Next we will integrate Equation 2.2-1 using the above equation as the expression for the velocity $v(t)$.

$$v = \frac{ds}{dt} \quad \Rightarrow \quad \int_{t_o}^{t} v\, dt = \int_{s_o}^{s} ds \quad \Rightarrow \quad \int_{t_o}^{t} (v_o + a_o(t-t_o))dt = \int_{s_o}^{s} ds$$

$$v_o(t-t_o) + a_o \frac{t^2 - t_o^2}{2} - (a_o t_o)(t-t_o) = s - s_o$$

Rearranging the above equation gives us the expression for the position given in Equation 2.2-10.

$$s = s_o + v_o(t-t_o) + \frac{a_o(t-t_o)^2}{2}$$

Next we will integrate Equation 2.2-3 to obtain an alternative expression for the velocity given below.

$$\int_{s_o}^{s} a_o\, ds = \int_{v_o}^{v} v\, dv \quad \Rightarrow \quad a_o(s - s_o) = \frac{v^2 - v_o^2}{2} \quad \Rightarrow \quad v^2 = 2a_o(s - s_o) + v_o^2$$

2.2.6) GENERAL NOTES

The set of equations 2.2-1, 2.2-2, and 2.2-3 can be rearranged and evaluated to solve a vast array of particle kinematics problems. We have given several examples on the preceding pages, but we have not exhausted all possible arrangements. The choice of which equations to use and how to rearrange them depends on what you need to find and the information that is given. Specifically, whether the given information is a function of time, displacement or velocity.

It is recommended that when you are using any of the above equations, a symbolic equation (e.g. $v(t)$) be found before substituting any specific numerical values. There are many instances where knowledge of the function may be of use for other parts of the problem.

Equation Summary

Abbreviated variable definition list

s = position

Δs = displacement

v = velocity

a = acceleration

t = time

Displacement

$$\Delta s = s_{final} - s_{initial}$$

Velocity

$$v = \frac{ds}{dt} = \dot{s} \qquad v_{ave} = \frac{\Delta s}{\Delta t}$$

Acceleration

$$a = \frac{dv}{dt} = \dot{v} = \frac{d^2 s}{dt^2} = \ddot{s} \qquad a_{ave} = \frac{\Delta v}{\Delta t} \qquad a\,ds = v\,dv$$

Non-constant acceleration

$$\int_{t_o}^{t} v(t)\,dt = \int_{s_o}^{s} ds \qquad \int_{t_o}^{t} a(t)\,dt = \int_{v_o}^{v} dv \qquad \int_{s_o}^{s} a(s)\,ds = \int_{v_o}^{v} v\,dv \qquad \int_{t_o}^{t} dt = \int_{v_o}^{v} \frac{1}{a(v)}\,dv$$

$$\int_{s_o}^{s} ds = \int_{v_o}^{v} \frac{v}{a(v)}\,dv$$

Constant acceleration

$$v = v_o + a_o(t - t_o) \qquad s = s_o + v_o(t - t_o) + \frac{a_o(t - t_o)^2}{2} \qquad v^2 = 2a_o(s - s_o) + v_o^2$$

Example 2.2-1

Determine the appropriate equation that should be used to find the answer to the problems shown below and then solve the problem. Any unknown initial conditions may be assumed to be zero.

Given	Find	Equation(s)	Answer
Position $s(t)$ $s = t^3 - 2t + 5$	Velocity $v(t)$		
Velocity $v(t)$ $v = t - 4$	Position $s(t)$		
Acceleration $a(t)$ $a = \cos(\omega t)$	Velocity $v(t)$		
Position $s(t)$ $s = \cos(\theta(t))$	Acceleration $a(t)$		
Acceleration $a(s)$ $a = s^2 + 1$	Velocity $v(s)$		
Velocity $v(s) = \dfrac{1}{s+1}$	Position $s(t)$		

Example Problem 2.2-2

A particle travels along a straight line with a velocity given by $v = (3 + 12t - 3t^2)$ m/s. When $t = 1$ s, the particle is located 50 m to the left of the origin. Determine the acceleration when $t = 4$ s, the displacement from $t = 0$ to $t = 10$ s, and the total distance the particle travels during this time period.

Given:

Find:

Solution:

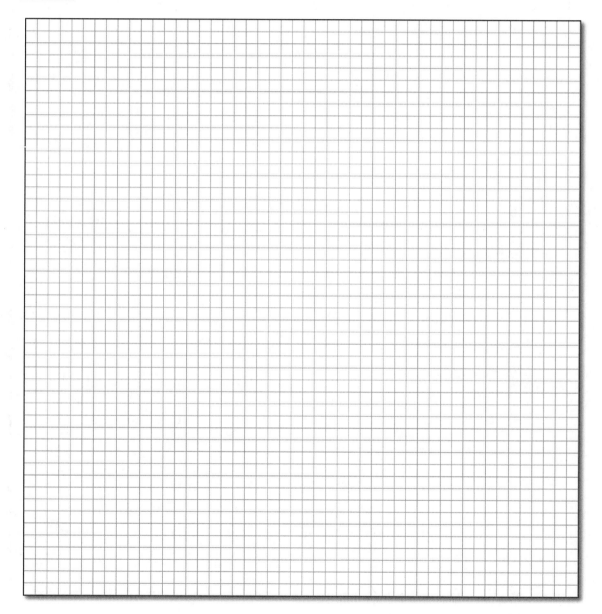

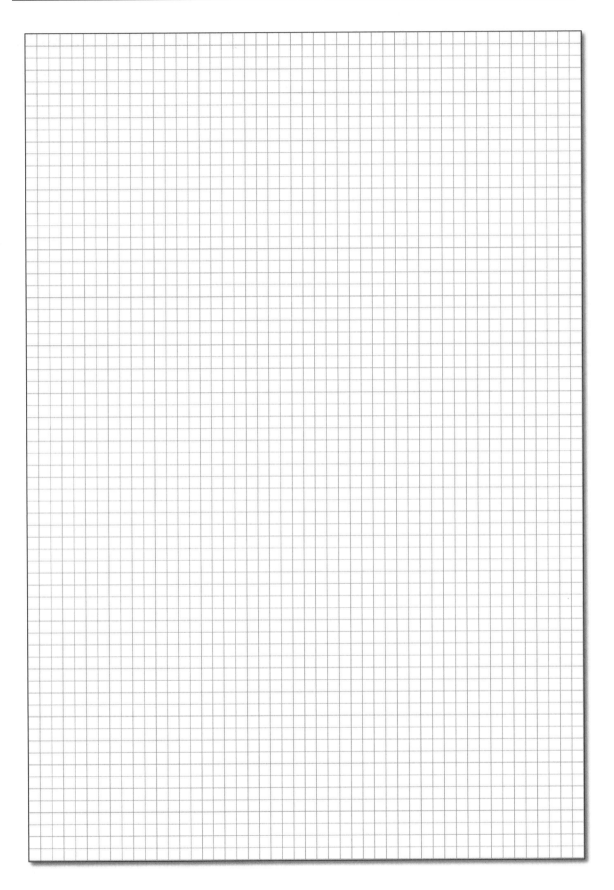

Example Problem 2.2-3

A car is driving along a straight road at 60 mph. The driver suddenly applies the brakes decelerating the car at a constant rate of 10 ft/s^2. What is the stopping distance of the car? If the deceleration of the car is a function of the distance traveled and given by $a = -\sqrt{s}$ ft/s^2, where s is in ft, what is the stopping distance of the car in this case?

Given:

Find:

Solution:

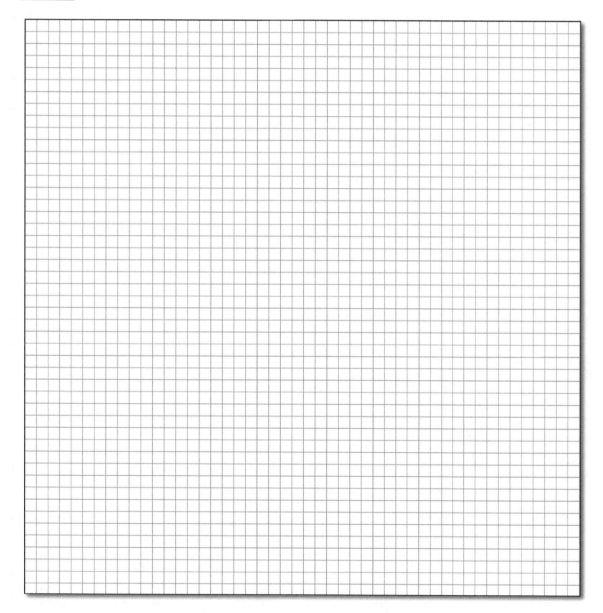

Example Problem 2.2-4

A rocket is traveling straight up at 80 m/s when it runs out of fuel at a height of 7,000 m above the ground. Determine the maximum height reached by the rocket and the time it takes before it falls back down to the ground. Neglect the effect of air resistance.

Given:

Find:

Solution:

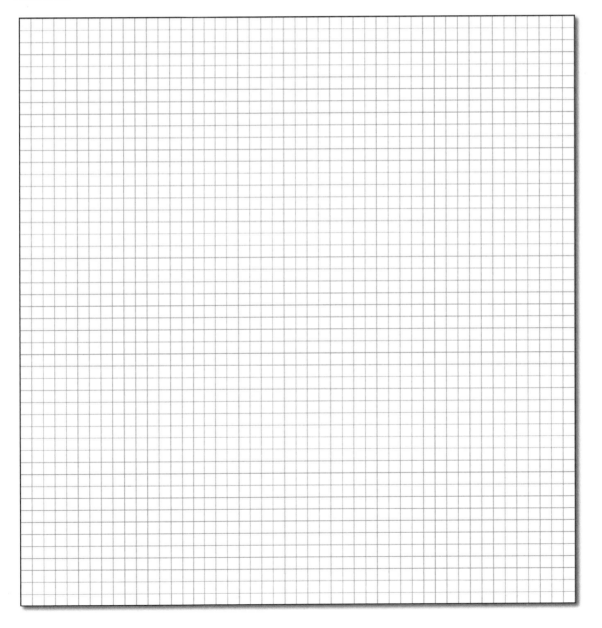

Solved Problem 2.2-5

A particle travels in a straight line with a velocity of $v(t) = t^2 - 3$ ft/s, where t is in seconds. At $t_o = 0$, the particle's position is $s_o = 2$ ft. Determine the particle's position, velocity and acceleration after 5 seconds.   Also, determine the particle's displacement and total distance traveled between 0 and 5 seconds.

<u>Given:</u> $v(t) = t^2 - 3$ ft/s
$\qquad\quad s_o = 2$ ft @ $t_o = 0$
$\qquad\quad t_f = 5$ s

<u>Find:</u> $s_f = s(5)$, $v_f = v(5)$, $a_f = a(5)$
$\qquad\quad \Delta s$, s_{total} between 0 and 5 seconds

Solution:

Velocity

We are given the function for the velocity of the particle, therefore, it is a simple task to find the velocity at 5 seconds.

$$v(t) = t^2 - 3 \text{ ft/s} \qquad\qquad v_f = 5^2 - 3 \qquad\qquad \boxed{v_f = 22 \frac{\text{ft}}{\text{s}}}$$

Acceleration

Knowing the velocity as a function of time, we can determine the acceleration of the particle through differentiation.

$$a = \frac{dv}{dt} = \frac{d(t^2 - 3)}{dt} = 2t \frac{\text{ft}}{\text{s}^2} \qquad\qquad a_f = 2(5) \frac{\text{ft}}{\text{s}^2} \qquad\qquad \boxed{a_f = 10 \frac{\text{ft}}{\text{s}^2}}$$

Position

Knowing the velocity as a function of time, we can determine the position of the particle through integration. Note that the lower limits on the integrals correspond to the known initial conditions, while variables are employed for the upper limits (i.e. s and t). If we had used the final conditions as the upper limits, we would have obtained $s(5)$ and that's all. Leaving the upper limits as variables allows us to find the more general equation for s as a function of time. Often finding such a general equation can be helpful for other parts of our analysis. In this problem, knowledge of $s(t)$ will be helpful for determining the total distance traveled by the particle.

$$v = \frac{ds}{dt} \qquad \rightarrow \qquad \int v\,dt = \int ds$$

$$\int_0^t (t^2 - 3)\,dt = \int_2^s ds \qquad\qquad \frac{t^3}{3} - 3t \Big|_0^t = s \Big|_2^s \qquad\qquad \frac{t^3}{3} - 3t = s - 2$$

$$s(t) = \frac{t^3}{3} - 3t + 2 \qquad\qquad s_f = \frac{5^3}{3} - 3(5) + 2 \text{ ft} \qquad\qquad \boxed{s_f = 28.67 \text{ ft}}$$

Displacement

Now that we have a general equation for the particle's position, we can calculate the displacement.

$$\Delta s = s_{final} - s_{initial} \qquad\qquad \Delta s_{t=0\to5} = s_f - s_o = 28.67 - 2 \text{ ft} \qquad\qquad \boxed{\Delta s_{t=0\to5} = 26.67 \text{ ft}}$$

Total distance traveled

The total distance traveled by the particle and the displacement of the particle are only the same if the particle does not turn. Therefore, our first task in calculating s_{total} is to determine if the particle made a turn in the specified time interval. At a turn, the velocity of the particle will momentarily be zero.

$$v_{turn} = t_{turn}^2 - 3 = 0 \qquad\qquad t_{turn} = 1.732 \text{ s}$$

$$s_{turn} = \frac{1.732^3}{3} - 3(1.732) + 2 \text{ ft} \qquad\qquad s_{turn} = -1.46 \text{ ft}$$

Based on the above calculations, we know that the particle turns. Before you calculate s_{total}, it is a good idea to plot the particle's position at various times of interest in order to help you visualize what is happening in the problem.

Position plot

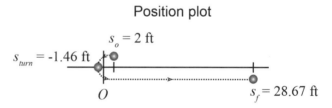

The total distance traveled is the distance between s_o and s_{turn} plus the distance between s_{turn} and s_f.

$$s_{total} = |s_{turn} - s_o| + |s_f - s_{turn}| = 3.46 + 30.1 \text{ ft} \qquad\qquad \boxed{s_{total} = 33.6 \text{ ft}}$$

Solved Problem 2.2-6

A car drives along a straight and flat highway. Starting from rest, the car accelerates to a speed of 100 mph in a span of 700 ft. If the acceleration of the car is approximately constant, determine the car's acceleration and the time it takes for the car to travel 700 ft.

Given: $v_o = 0$ $v_f = 100$ mph
        $\Delta s = 700$ ft $a = $ constant

Find: $a, \Delta t$

Solution:

Acceleration

The acceleration of the car is constant, therefore, we can use the constant acceleration equations. If we know the velocity and distance and want to find the acceleration, we need to use the equation that has all three of those variables.

$$v^2 = 2a_o(s - s_o) + v_o^2$$

It is important that when you use the constant acceleration equations, you understand how each of the values relates to the variables in your particular situation.

$$v_f^2 = 2a\Delta s + v_o^2 \qquad \left(100 \text{ mph}\frac{5280 \text{ ft}}{\text{mile}}\frac{\text{hr}}{3600 \text{ s}}\right)^2 = 2a(700 \text{ ft}) + 0 \qquad \boxed{a = 15.37 \frac{\text{ft}}{\text{s}^2}}$$

Time

If we know the velocity and acceleration and want to find the time, we need to use the equation that has all three of those variables.

$$v = v_o + a_o(t - t_o) \qquad\qquad v_f = v_o + a\Delta t \qquad\qquad 146.67 \frac{\text{ft}}{\text{s}} = 0 + 15.37 \frac{\text{ft}}{\text{s}^2}\Delta t$$

$$\boxed{\Delta t = 9.5 \text{ s}}$$

2.3) ERRATIC RECTILINEAR MOTION

At the beginning of this chapter, it was mentioned that rectilinear analysis works well as a tool for analyzing the acceleration and deceleration of a car driving on a straight flat road. In the last section, we considered particles that had a position, velocity and acceleration given as single continuous functions. That means that for a given time period, these functions had no jumps or sharp bends. Consider, however, the following question: "How often have you driven down a road in your car with a constant acceleration or with a velocity that didn't change abruptly?" Unless you're in the country, probably never. In your car, you are always braking for lights, speeding up and slowing down. The point is, your car's position, velocity and acceleration are erratic (more so for some drivers than others). This means that the functions describing these variables are not smooth and they are sometimes discontinuous or **erratic**. Figure 2.3-1 shows examples of erratic functions. As you can see, portion A of each motion is defined by a different equation than is portion B. In other words, these functions must be defined in a piecewise manner.

Erratic motion problems can still be rectilinear. This means that the same equations used to solve rectilinear problems can be used to solve erratic motion problems. The trick to solving erratic rectilinear problems is to look at each segment separately. Also, the end condition of a given motion segment will be the initial condition for the subsequent segment.

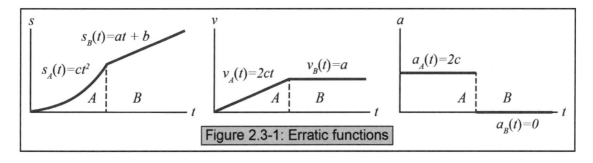

Figure 2.3-1: Erratic functions

Example 2.3-1

Consider a jogger that is running on a marked line as shown in the figure. Draw the jogger's $s(t)$ curve for the following situation. The jogger starts at position 3 and stays there for a while. He then jogs to position 1 and stays there a moment. Next, he runs very quickly to position 2 and stays a moment before walking back to position 3 and taking a rest.

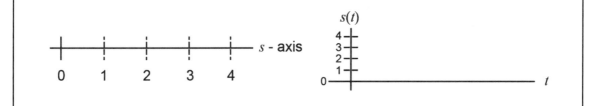

Example Problem 2.3-2

A car is driving down a straight flat road. The acceleration of the car follows the a-t graph shown. The car starts from rest at $t_o = 0$ seconds, reaches its maximum velocity of 45 m/s, and drives at that velocity for 5 seconds. The driver then applies the brakes slowing the car to an eventual stop. At what time does the car stop?

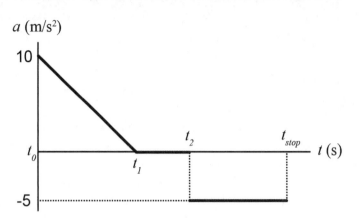

Given:

Find:

Solution:

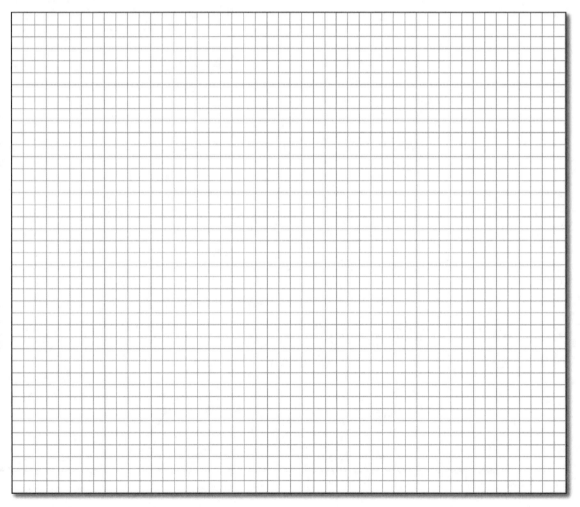

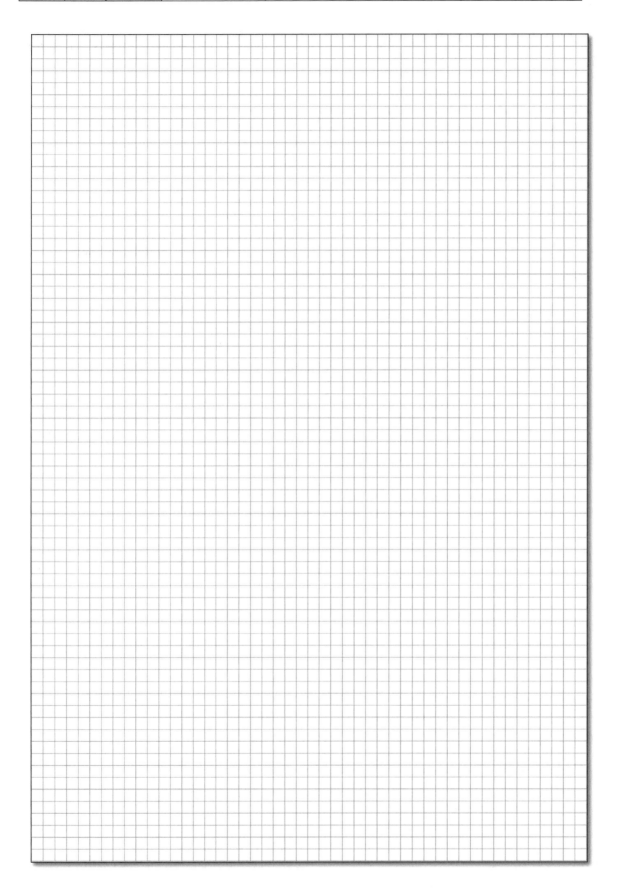

Example Problem 2.3-3

A speeder in a car passes a parked police car. The speeder is driving at a constant velocity of 100 mph. It takes the policeman 3 seconds to react, at which time, he accelerates the police car at 20 ft/s^2 until he reaches a maximum velocity of 120 mph. The police car continues to travel at 120 mph until it overtakes the speeding car. Starting at the point when the speeder passes the police car, what distance does the police car have to travel before it overtakes the speeding car?

Given:

Find:

Solution:

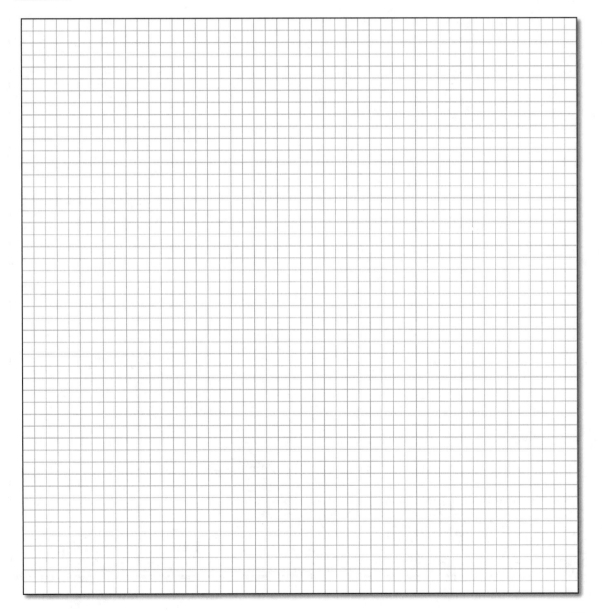

A car, traveling in a straight line, accelerates from rest according to the velocity profile $v(t) = 30\ln(t+1)$ mph, where t is in seconds. After 10 seconds, the driver applies the brakes and decelerates at a constant rate of 20 ft/s^2. Determine the time at which the car comes to a complete stop and the total distance traveled by the car.

<u>Given:</u> $v_A(t) = 30\ln(t+1)$ mph from $t_o = 0$ to $t_{Af} = 10$ s
$\qquad a_B = $ -20 ft/s^2

<u>Find:</u> t_{stop}, s_{stop}

<u>Solution:</u>

Setting up the problem

It is very important, when dealing with erratic rectilinear problems, that you understand the problem statement. The best way to do this is to draw graphs that depict the particle's motion. In this case, we will draw the v-t graph. It is clear that the velocity is an erratic function. Therefore, we will define two regions (i.e. region A, region B) where the velocity is continuous within the region.

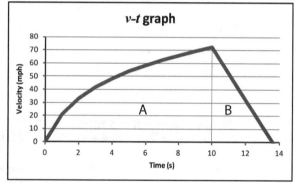

Time

To solve for the stopping time we only need to look at region B. In region B, the acceleration is a constant -20 ft/s^2, therefore, we can use the constant acceleration equations.

$$v = v_o + a_o(t - t_o)$$

Region B starts with a car velocity of v_A evaluated at 10 seconds and ends when the car stops.

$v_A = 30\ln(t+1)$ $\qquad\qquad v_{Af} = 30\ln(11) = 71.94 \text{ mph} = 105.51 \dfrac{\text{ft}}{\text{s}}$

$v_{stop} = v_{Af} + a_B(t_{stop} - t_{Af})$ $\qquad 0 = 105.51 - 20(t_{stop} - 10)$ $\qquad \boxed{t_{stop} = 15.3 \text{ s}}$

Distance traveled

To determine the total distance traveled, we need to consider both region A and B. In region A, the acceleration is not constant so we need to integrate the velocity to determine the position.

$$v = \frac{ds}{dt} \quad \rightarrow \quad \int v \, dt = \int ds$$

$$\int_0^t (30\ln(t+1)) \, dt = \int_0^{s_A} ds \qquad 30((t+1)\ln(t+1)-t)\Big|_0^t = s\Big|_0^{s_A}$$

$$s_A = 30((t+1)\ln(t+1)-t) \qquad s_{Af} = 491.31 \text{ ft}$$

The acceleration in region B is constant.

$$v^2 = 2a_o(s-s_o)+v_o^2 \qquad\qquad v_{stop}^2 = v_{Af}^2 + 2a_B(s_{stop}-s_{Af})$$

$$0 = 105.51^2 + 2(-20)(s_{stop}-491.31) \qquad \boxed{s_{stop} = 769.62 \text{ ft}}$$

2.4) SOLVING RECTILINEAR PROBLEMS GRAPHICALLY

So far, we have learned how to solve rectilinear problems mainly through the use of calculus. If we want to know the displacement of a particle $s(t)$ given the velocity $v(t)$, we have learned that we should integrate the function $v(t)$ to get $s(t)$ as shown below.

$$v = \frac{ds}{dt} \quad \Rightarrow \quad s(t) = \int_{t_o}^t v \, dt + s_o$$

If we take the above equation and integrate it between times t_1 and t_2, we see that the displacement $\Delta s = s_2 - s_1$ is equal to the integral of v with respect to t between t_1 and t_2 as shown in the derivation below. From calculus, we know that an integral is really calculating the area under a curve. Therefore, we can conclude that the displacement between t_1 and t_2 is equal to the area under the v-t curve during that time interval as shown in Figure 2.4-1.

$$s_2 = \int_{t_1}^{t_2} v \, dt + s_1 \quad \rightarrow \quad s_2 - s_1 = \int_{t_1}^{t_2} v \, dt \quad \rightarrow \quad \Delta s = \int_{t_1}^{t_2} v \, dt$$

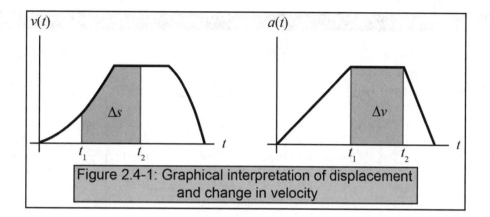

Figure 2.4-1: Graphical interpretation of displacement and change in velocity

The above reasoning holds true for the change in velocity, Δv, as shown in the derivation below. The area under the a-t curve between times t_1 and t_2 is equal to the change in velocity during that time interval as shown in Figure 2.4-1.

$$a = \frac{dv}{dt} \quad \Rightarrow \quad v(t) = \int_{t_o}^{t} a\,dt + v_o$$

$$v_2 = \int_{t_1}^{t_2} a\,dt + v_1 \quad \rightarrow \quad v_2 - v_1 = \int_{t_1}^{t_2} a\,dt \quad \rightarrow \quad \Delta v = \int_{t_1}^{t_2} a\,dt$$

Solving problems using the graphical method is useful if the area under the curve is easily calculated. It is also useful for situations where the graph cannot be expressed as a function. For example, it is a very efficient method to use for solving erratic motion problems. However, there is a caveat. The graphical method gives numerical answers and not functions. If you need a function for subsequent steps of a problem, the graphical method may not be preferred. It may be of interest to note that computer simulation tools, in essence, use a graphical approach to numerically approximate solutions to problems that do not have a closed-form solution or that are too complicated to solve by hand.

Conceptual Example 2.4-1

Given the following v-t graphs, rank them from largest to smallest particle displacement.

Largest: _____
Next: _____
Next: _____
Next: _____
Next: _____
Smallest: _____

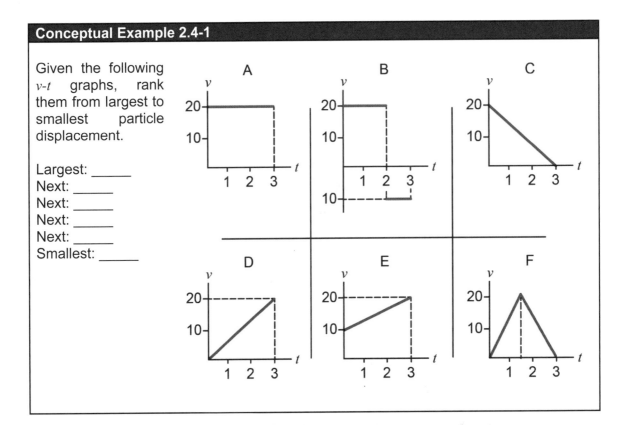

Example 2.4-2

A particle moves along a straight line. Its velocity profile is given in the graph below. Determine the following quantities related to the motion of the particle.

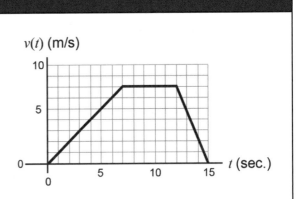

a) Distance traveled by the particle between 0 and 15 seconds.

b) The average velocity of the particle in the first 10 seconds.

c) Acceleration of the particle at t = 5 seconds.

d) Acceleration of the particle at t = 10 seconds.

Example Problem 2.4-3

A motorcyclist drives his motorcycle in a straight line. The motorcycle starts from rest and moves with the acceleration profile shown where the motorcycle is brought to rest at $t = 25$ seconds. Draw the velocity profile and determine the motorcycle's total distance traveled.

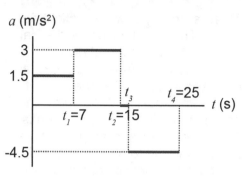

Given:

Find:

Solution:

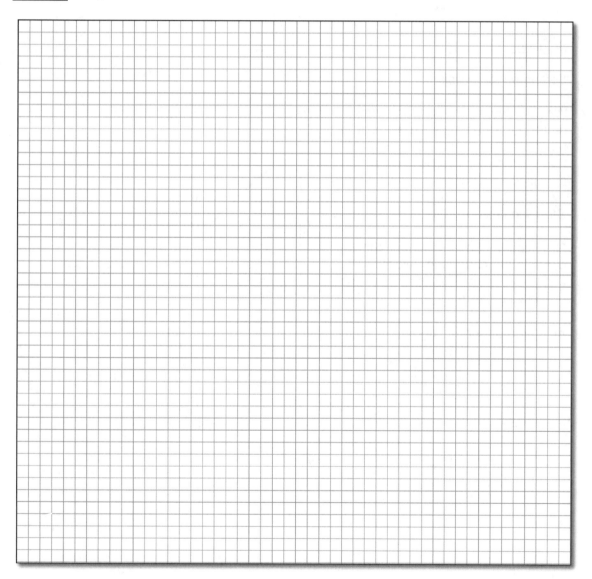

Solved Problem 2.4-4

A car is driving down a straight flat road. The acceleration of the car follows the a-t graph shown. The car starts from rest at $t_o = 0$ seconds, reaches its maximum velocity of 45 m/s, and drives at that velocity for 5 seconds. The driver then applies the brakes slowing the car to an eventual stop. At what time does the car stop?

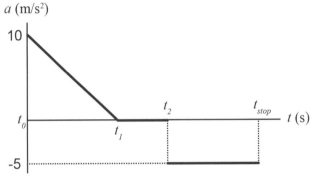

Given: $v_o = 0$ $v_1 = 45$ m/s
       $a_o = 10$ m/s^2 $t_2 - t_1 = 5$ s

Find: t_{stop}

Solution:

Setting up the problem

When dealing with an erratic rectilinear kinematics problem, it is good practice to label the segments that are continuous within a specified time period.

Stopping time

Graphically, the area under the a-t curve is the change in velocity. We can use this fact to determine the unknown times.

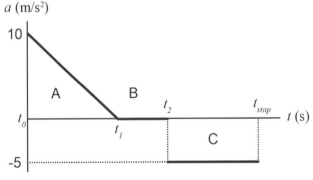

Δv = area under a-t curve

$$v_1 - v_o = Area_A = \frac{1}{2}(t_1 - t_o)a_o \qquad 45 - 0 = \frac{1}{2}(t_1 - 0)10 \qquad t_1 = 9 \text{ s}$$

$$t_2 = t_1 + 5 = 14 \text{ s}$$

$$v_{stop} - v_2 = Area_C = (t_{stop} - t_2)a_C \qquad 0 - 45 = (t_{stop} - 14)(-5) \qquad \boxed{t_{stop} = 23 \text{ s}}$$

RP2-1) Consider a particle moving in 1-D space. When the particle changes direction, what is its velocity?

RP2-2) The positions of two rolling balls are captured at 0.1 second intervals in the figure below. The balls are moving towards the right. Do the balls have the same speed at any point during their motion?

    a) No.
    b) Yes, at instant 2.
    c) Yes, at instant 5.
    d) Yes, between instant 3 and 4.
    e) Yes, between instant 1 and 2.

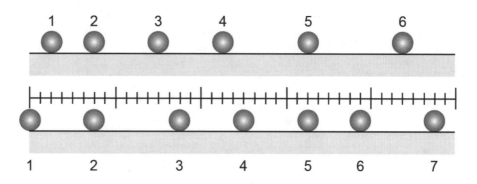

RP2-3) Is it possible for a body to have a positive velocity and a negative acceleration at the same instant of time? (Yes, No)

RP2-4) The slope of a line tangent to the position curve $s(t)$ for a particle's motion at a given instant of time gives which of the following.

    a) the particle's displacement
    b) the particle's average velocity
    c) the particle's instantaneous velocity
    d) the particle's instantaneous acceleration

RP2-5) The slope of a line tangent to the velocity curve $v(t)$ for a particle's motion at a given instant of time gives which of the following.

    a) the particle's displacement
    b) the particle's average velocity
    c) the particle's instantaneous velocity
    d) The particle's instantaneous acceleration

RP2-6) The area under a velocity-time curve is equal to the ...

    a) position
    b) displacement
    c) total distance traveled

RP2-7) The area under an acceleration-time curve is equal to the ...

    a) velocity
    b) average velocity
    c) change in velocity

RP2-8) If a body has a positive velocity and a negative acceleration at a particular instant of time, what would the v-t curve look like at that instant?

RP2-9) A particle moves along a straight line with a velocity of $v = (3t^2 - 5t)$ m/s, where t is the time in seconds. If it is initially located at the origin O, determine the particle's acceleration at $t = 2$ s and its total distance traveled between $t = 0$ and 2 s.

Given: $v = (3t^2 - 5t)$ m/s, $s = 0$ when $t = 0$

Find: $a_{t=2 \text{ s}}$, s_{total}

Solution:

Acceleration

Find the acceleration $a(t)$.
Circle the equation that should be used.

$$v = \frac{ds}{dt} \qquad a = \frac{dv}{dt} \qquad a\, ds = v\, dv$$

$a(t) =$ _____

$a_{t=2 \text{ s}} =$ _____

Position

Find the position $s(t)$.
Circle the equation that should be used.

$$v = \frac{ds}{dt} \qquad a = \frac{dv}{dt} \qquad a\, ds = v\, dv$$

Total Distance traveled

Does the particle turn?
If the particle turns, at what time does it turn?

Yes   No

$t_{turn} =$ _____

Calculate the total distance traveled between $t = 0$ and $t = 2$ s.

$s(t) =$ _____

$s_{total} =$ _____

RP2-10)   A boat going down river is travels at 25 knots (1 knot = 1.151 mph) relative to the shore for 1 minute before cutting its engines.   The river current flows at 2 knots causing the boat's velocity relative to the shore to follow the function $v = 42.2 - 6\ln(t - 59)$ ft/s after the engines are cut, where t is in seconds. Determine the distance traveled by the boat in 5 minutes.

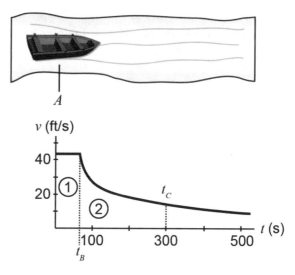

Note: $\int \ln(x - a)dx = (x - a)\ln(x - a) - x + C$

Given: $v_1 = 25$ knots $= 42.2$ ft/s
$\qquad v_2 = 42.2 - 6\ln(t - 59)$ ft/s
$\qquad t_B = 1$ min $= 60$ s $\qquad\qquad t_C = 5$ min $= 300$ s

Find: s_C

Solution:

Find s_B.

Find s_C.

$s_B =$ _____

$s_C =$ _____

P2.1) BASIC LEVEL RECTILINEAR PROBLEMS

P2.1-1)[fe]   The position of a particle moving along a straight line is given by $s = (1/2)(bt^2 + 2ct + d)$, where t is time and b, c and d are constants. Determine the particle's acceleration.

a) $a = b$ b) $a = c$ c) $a = d$ d) $a = bt + c$

P2.1-2)   The position of a particle moving along a straight line is given by $s = b\cos(dt + c)$, where t is time and b, c and d are constants. Determine the particle's velocity and acceleration as functions of time and the constants b, c and d. Also, find the maximum velocity of the particle.

Ans: $v = -db\sin(dt + c)$, $a = -d^2 b\cos(dt + c)$, $v_{max} = db$

P2.1-3) The velocity of a particle moving in a straight line is given by $v = 5 + 7t^{5/2}$ meters per second, where t is in seconds. Determine the position, velocity and acceleration of the particle when t is equal to 5 seconds. The particle is located 10 meters to the right of the origin at $t = 0$ seconds.

Ans: $s_{t=5}$ = 594 m, $v_{t=5}$ = 396.3 m/s, $a_{t=5}$ = 195.7 m/s^2

P2.1-4)[fe] The acceleration of a particle is given by $a = 2s^2$ meters per second squared, where s is the particle's position in meters. Find the velocity of the particle as a function of position. The particle is at rest at the origin when time is equal to zero.

a) $v = 4s$ m/s   b) $v = 4s^3/3$ m/s   c) $v = \sqrt{4s^3}$ m/s   d) $v = \sqrt{4s^3/3}$ m/s

P2.1-5)[fe] The acceleration of a particle is given by $a = -n^2 s$, where s is the particle's position. Find the velocity of the particle as a function of position if the particle starts at time t_o from a position of s_o with a velocity of v_o.

a) $v = \sqrt{-n^2 s^2}$ b) $v = \sqrt{v_o^2 - n^2(s^2 - s_o^2)}$ c) $v = -2n$ d) $v = -n^2$

P2.1-6) A car comes to a complete stop from an initial speed of 60 mph in a distance of 100 ft. With the same constant acceleration, what would the stopping distance (s_{stop}) be from an initial speed of 100 mph?

Ans: s_{stop} = 277.8 ft

P2.1-7)[fe] The position of point P on the head of the piston shown in the figure is given by geometry to be $s = 0.6\cos\theta$, where s is in meters and θ is in radians. If at the instant when $\theta = 0.5$ rad the angular velocity is $d\theta/dt = 1.5$ rad/sec, determine the velocity of point P.

a) $v = -0.29$ m/s   b) $v = 0.29$ m/s

c) $v = -0.43$ m/s   d) $v = 0.43$ m/s

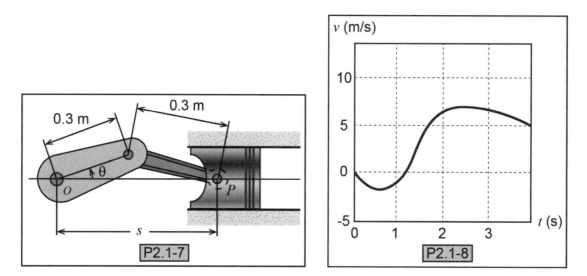

P2.1-8) Consider the given graph that displays a particle's velocity in m/s versus time in seconds. Assuming the particle moves along a straight line, use the given graph to estimate the following quantities. Remember units.
   a) particle's speed at $t = 3$ second
   b) particle's speed at $t = 1$ second
   c) particle's acceleration at $t = 3$ seconds
   d) particle's displacement between $t = 0$ seconds and $t = 3$ seconds
   e) distance traveled by particle between $t = 0$ seconds and $t = 3$ seconds

Ans: a) $v \cong 6$ m/s, b) $v \cong -1$ m/s, c) $a \cong -4/3$ m/s^2, d) $\Delta s \cong 7$ m, e) $s_{total} \cong 9.5$ m

P2.1-9) A particle moving along a straight line is initially traveling at 24 m/s directed to the right when it is subjected to a constant acceleration of 8 m/s^2 directed to the left. Determine the particle's total distance traveled for the first 5 seconds after the acceleration is applied.

Ans: $s_{total} = 52$ m

P2.2) INTERMEDIATE LEVEL RECTILINEAR PROBLEMS

P2.2-1)   The position of a car driving along a straight flat road is given by $s = e^{0.2t}(-0.001t^4 + 0.2t^2 - 1) + t + 1$, where s is in meters and t is in seconds. Plot the position, velocity, and acceleration versus time for the first 10 seconds of motion. Determine the maximum velocity and acceleration of the car and the times at which both occur. Also determine when the car has zero acceleration.

Ans: t = 8.72 s, v_{max} = 10.84 m/s, t = 6.36 s, a_{max} = 2.06 m/s^2, t = 8.72 s

P2.2-2)   The velocity of a particle is given by $v(t) = at - b$, where v is in m/s, t is in seconds, a = 2 and b = 8. Plot the position, velocity and acceleration of the particle between t = 0 and t = 5 seconds. Determine the position, velocity and acceleration of the particle at t = 2 seconds if the position of the particle is at the origin when t = 0. Also, determine the displacement of the particle and total distance traveled by the particle between t = 0 seconds and 5 seconds.

Ans: $s(t) = t^2 - 8t$ m, a = 2 m/s^2, Δs = -15 m, s_{total} = 17 m

P2.2-3) A particle travels along a straight line with a velocity of $v = 15 - 6t^2$ m/s. When t = 1 second, the particle is located 5 meters to the left of the origin. Determine the acceleration when t = 2 seconds. Also, determine the displacement and the total distance traveled between t = 0 and 2 seconds.

Ans: Δs = -14 m, s_{total} = 17.6 m, a = -24 m/s^2

P2.2-4) The position of a particle is given by $s = 2t^3 - 30t^2 + 150t - 10$ meters, where t is in seconds. Plot the position (s), velocity (v) and acceleration (a) as functions of time from t = 0 to 10 seconds. Determine at what time the velocity is zero. Also, determine the displacement and the total distance traveled between t = 0 and 10 seconds.

Ans: $v(t) = 6t^2 - 60t + 150$ m/s, t = 5 s, Δs = s_{total} = 510 m

P2.2-5) The position of a particle is given by $s = 5t^3 - 60t^2 + 150t - 20$ meters, where t is in seconds. Plot the position (s), velocity (v) and acceleration (a) as functions of time between t = 0 and 6 seconds. Determine at what time the velocity is zero. Also, determine the displacement and the total distance traveled between t = 0 and 6 seconds.

Ans: Δs = -180 m, $v(t) = 15t^2 - 120t + 150$ m/s, t_{stop} = 1.55 s, s_{total} = 394 m,
    $a(t) = 30t - 120$ m/s^2

2.2-6) The acceleration of a particle traveling in a straight line is $a = A\cos(nt)$, where t is time and A and n are constants. If the particle starts at time t_o at a position of s_o with a velocity of v_o, determine the position of the particle as a function of time and in terms of the constants A and n.

Ans: $s(t) = (A/n^2)(\cos(nt_o) - \cos(nt)) + (t - t_o)(v_o - (A/n)\sin(nt_o)) + s_o$

P2.2-7) Upon landing a jetliner, the pilot extends the craft's wing flaps and applies reverse thrust to slow the plane down. During this coast-down, the jetliner has an acceleration given by $a = -0.003\ v^2$ m/s^2 where the velocity v is in m/s.
   a) Determine the time it takes the plane's velocity to decrease from 100 m/s to 20 m/s.
   b) Determine the distance traveled by the plane during that time.

Ans: t = 13.33 s, Δs = 536.5 m

P2.3) ADVANCED LEVEL RECTILINEAR PROBLEMS

2.3-1) A particle moves along a straight line with a velocity $v = \sqrt{bs^2 + 2cs + d}$, where s is the position of the particle and b, c and d are constants. Find the acceleration of the particle as a function of position and in terms of the constants b and c. Then, prove that $v = \sqrt{bs^2 + 2cs + d}$ using the $a(s)$ calculated above.

Ans: $a(s) = bs + c$

P2.3-2) A ball is thrown vertically upward from the ground next to a house with an initial speed of 45 ft/s. The height of the house is 20 feet. Determine the maximum height the ball attains, the time it takes for the ball to land in the house's gutter and the impact velocity of the ball as it hits the gutter. Neglect air resistance.

Ans: h_B = 31.44 ft, t_C = 2.24 s,
     v_C = -27.14 ft/s

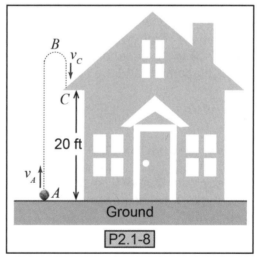

P2.3-3) A ball is thrown vertically upwards, from ground level, with an initial velocity of v_o. It takes the ball t_1 = 10 seconds to reach a vertical height of d, a total time of t_2 seconds to reach its apex (highest point), and a total time of t_3 = 20 seconds to come back down to a vertical height of d again. Determine the ball's initial velocity v_o.

a) v_o = 98.1 m/s   b) v_o = 147.15 m/s   c) v_o = 294.3 m/s   d) v_o = 483 m/s

P2.4) BASIC LEVEL ERRATIC MOTION PROBLEMS

P2.4-1) Consider the given graph that displays a particle's velocity in m/s versus time in seconds. Assuming the particle moves along a straight line, use the given graph to answer the following questions.
- a) At what time(s) does the particle change direction?
- b) At what time(s) does the particle return to its starting position?

Ans: a) $t = 2.75, 6.5$ s, b) $t = 3.67$ s

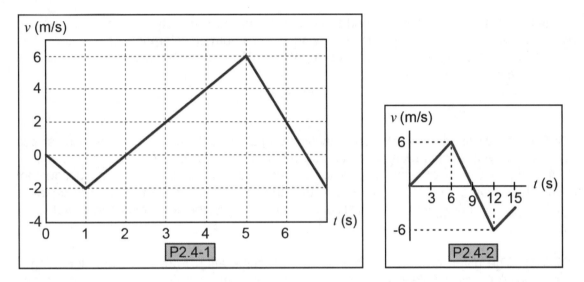

P2.4-2)[fe] A particle travels in a straight line with the velocity profile shown. Determine the particle's total distance traveled between $t = 6$ s and $t = 12$ s.

a) $s_{total} = 18$ m   b) $s_{total} = 0$ m

c) $s_{total} = 9$ m   d) $s_{total} = 12$ m

P2.4-3)   A racecar starts from rest and accelerates for 15 seconds before the driver applies the brakes and brings the car to a stop at $t = 25$ s. A graph of the car's velocity as a function of time is given. Use the given graph to estimate the following quantities. Remember units and show work.
- (a) car's velocity at $t = 10$ seconds
- (b) car's acceleration at $t = 10$ seconds
- (c) distance traveled by the car while it is accelerating
- (d) distance traveled by the car while it is braking

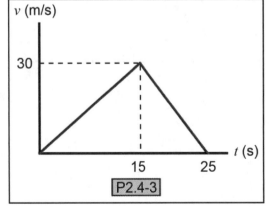

Ans: a) $v = 20$ m/s, b) $a = 2$ m/s^2, c) $\Delta s = 225$ m, d) $\Delta s = 150$ m

P2.4-4)[fe] A car initially traveling at 60 mph coasts for 5 seconds. Due to friction and wind resistance, the car decelerates while coasting at a rate of 1 ft/s^2. After the 5 seconds of coasting, the driver applies the brakes and decelerates at a constant rate eventually brining the car to a stop. If the total time that the car travels is 8 seconds, determine the car's braking acceleration.

a) a = -27.7 ft/s^2 b) a = -18.3 ft/s^2 c) a = -31.5 ft/s^2 d) a = -6.8 ft/s^2

P2.5) INTERMEDIATE LEVEL ERRATIC MOTION PROBLEMS

P2.5-1) A train starts from rest and travels along its track in a straight line. The engine has the capability of moving the train with a maximum acceleration of 3.5 ft/s. The maximum safe speed of the train is 65 mph. What is the minimum time needed for the train to travel 15 miles?

Ans: t_{min} = 14.1 min

P2.5-2) A train starts from rest and travels along a straight track. The train accelerates from v_o = 0 with a constant acceleration a_A until it reaches a maximum velocity of v_{max}. It then continues with constant speed for a time and then decelerates with a constant deceleration of a_C until it stops. Prove that the total time the train travels is equal to $t_{total} = (s_{total} / v_{max}) + (v_{max} / 2)((1 / a_C) + (1 / a_A))$, where s_{total} is the total distance traveled by the train.

P2.5-3) The given graph describes a truck's acceleration from rest after departing a metering light at a freeway onramp. Determine the truck's velocity as it passes s = 500 feet.

Ans: v = 68.9 mph

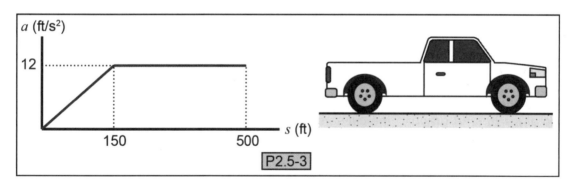

P2.5-4) A delivery truck travels along a straight road while accelerating and decelerating as described in the given graph. The truck starts from rest at time $t = 0$ s and returns to rest at the unknown time t'. Draw the v-t and s-t graphs for the truck during this time period. Include scales on the axes and determine the time t' at which the truck comes to a stop.

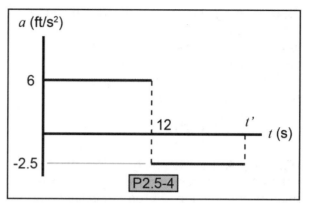

Ans: $t' = 40.8$ s

P2.6) ADVANCED LEVEL ERRATIC MOTION PROBLEMS

P2.6-1) A car is traveling down a straight flat road at a constant velocity of 70 mph as it passes point A. Some distance down the road, the driver applies the brakes for a total of 5 seconds causing the car to uniformly decelerate to a final speed of 50 mph at which point he releases the brakes. The car maintains the speed of 50 mph until it passes point D. Determine the distance the car travels before the brakes are applied if the total distance the car travels between point A and D is 2 miles and the time traveled before the driver hits the brakes is the same as the time traveled after the driver releases the brakes.

Ans: $\Delta s_{A\text{-}B} = 5904.55$ ft

P2.6-2) You are designing a material handling process where you wish automated carts to move material a distance of 100 meters in as short a time as possible. Furthermore, you wish to move the material "smoothly" so as to not cause any damage. It is proposed that your carts follow the velocity profile shown where the carts are accelerated from rest to a velocity of 10 m/s in the span of 1 second. This initial velocity profile is described by the equation $v(t) = at^3+bt^4+ct^5$ m/s where t is in seconds. Once a

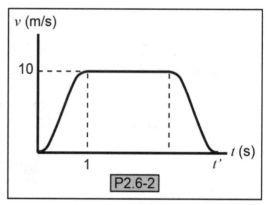

cart reaches its maximum speed, it maintains that level until it approaches its destination, at which point it begins to decelerate following a profile symmetric to the one employed during the acceleration phase of motion. Determine the coefficients a, b and c that will accelerate the carts as prescribed, while achieving a continuous acceleration profile. Furthermore, it is required that at $t = 1$ second the derivative of acceleration (jerk) be zero. Once you have determined the coefficients, determine the total time t' required for the carts to traverse the required 100-m distance.

Ans: $a = 100$, $b = -150$, $c = 60$, $t' = 11$ s

P2.6-3) A bicycle travels along a straight road where its velocity changes as a function of its position as shown in the given graph. Determine the acceleration of the bicycle at $s = 25$ m and $s = 75$ m. Also, draw the a-s graph.

Ans: $a_{s=25} = 1$ m/s^2, $a_{s=75} = -1$ m/s^2

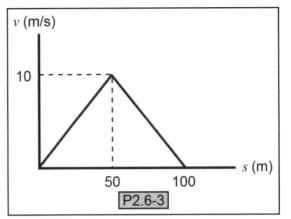

P2.6-4) A boy learning to fly an RC helicopter is practicing only the rise, hover and descent motions. From an initial point on the ground, the boy controls the helicopter to rise to an eventual height of h. At this height, he hovers the helicopter for 5 seconds before descending the helicopter back down to the ground. The velocity profile of the helicopter is shown in the figure. Determine the total flight time and the maximum height attained by the helicopter.

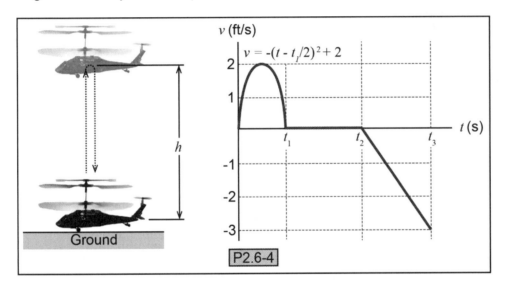

CHAPTER 2 COMPUTER PROBLEMS

C2-1) A rocket is traveling straight up at 80 m/s when it runs out of fuel at a height of 7,000 m above the ground. Taking into account air resistance, determine the maximum height reached by the rocket and the time it takes to reach this height. Also, plot the velocity of the rocket from the time it leaves the ground to the time it reaches its apex. Assume that the rocket has an acceleration of $a = (-0.01v^2 - 9.81)$ m/s^2 while it is ascending. The determination of a closed-form solution to this problem can be rather challenging due to the inclusion of air resistance. Therefore, employ a numerical approach to the solution of this problem. Compare your results to the solution where air resistance is neglected. In this case, do you think that omitting air resistance is a good modeling decision? Explain.

C2-2) The U.S. Environmental Protection Agency prescribes a few different drive cycles that are employed for testing vehicle emissions and fuel economy. Drive cycles specify a vehicle's velocity as a function of time. Automobile manufacturers also use these standard drive cycles for testing in order to enable comparison between various designs. Download the file for the Urban Dynamometer Drive Schedule (uddscol.txt) from http://www.epa.gov/nvfel/testing/dynamometer.htm. The first column represents time in seconds while the second column is speed in mph. This drive cycle is meant to represent city driving conditions and is sometimes referred to as "the city test." Employing this file, numerically estimate the corresponding position and acceleration of the vehicle as a function of time. The position data can be used, for instance, to determine the vehicle's fuel economy since it is necessary to know how far the vehicle travelled, in addition to how much fuel it consumed. The acceleration data can be used to help size various vehicle components, such as the engine (based on maximum acceleration), and the brakes (based on maximum deceleration).

CHAPTER 2 DESIGN PROBLEMS

D2-1) You are designing a material handling process where you wish automated carts to move material a given distance quickly. You wish to move the material "smoothly" so as to not cause any damage. The carts should be accelerated from rest to the cart's maximum attainable velocity of 10 m/s in a duration not less than 1 second, otherwise there is a possibility that the contents will be damaged. Once a cart reaches its maximum speed, it maintains that velocity until it approaches its destination. Design a velocity profile that will accelerate the carts as prescribed while also achieving a continuous acceleration profile. Once you have determined the velocity profile, calculate the distance travelled by the cart in the process of accelerating it from rest to its maximum velocity.

CHAPTER 2 ACTIVITIES

A2-1) This activity gives students practice calculating average velocity and provides physical intuition.

Group size: Minimum of 2 students

Supplies

- Masking tape
- Stop watch

Procedure

1) Mark two positions on the floor with masking tape.
2) Measure the distance between the two marked positions with a tape measure.
3) Have one student walk from one marked position to the other in a straight line while another student times how long it takes with a stop watch.

Calculations and Discussion

- Use the collected data to calculate the average velocity of each student.
- List all of the assumptions that were applied in your calculations.
- Discuss your results
- List possible sources of experimental error.

A2-2) This activity gives students practice calculating average velocity, average acceleration.

Group size: Minimum of 2 students

Supplies

- Masking tape
- Stop watch that can time multiple laps

Procedure

1) Mark five positions, in series, on the floor with masking tape.
2) Measure the distance between the marked positions with a tape measure.
3) Have one student walk from one marked position to the other in a straight line. The student should increase his/her velocity after passing each marked position. Another student will time how long it takes the first student to get from one marked position to another with a stop watch.

Calculations and Discussion

- Use the collected data to calculate the average velocity of each student for each individual segment.
- Use the collected data to calculate the overall average velocity of each student.
- Use the collected data to determine the average acceleration between segments 1 - 2, 2 - 3 and 3 - 4.
- Use the collected data to determine the overall average acceleration.
- Create a velocity versus time graph for each student. Draw a horizontal line on the graph that represents the overall average velocity.
- Create an acceleration versus time graph for each student. Draw a horizontal line on the graph that represents the overall average acceleration.
- List all of the assumptions that were applied in your calculations.
- Discuss your results
- List possible sources of experimental error.

PART II: KINEMATICS

CHAPTER 3: KINEMATICS OF PARTICLES - CURVILINEAR MOTION

CHAPTER OUTLINE

CHAPTER SUMMARY

In this chapter, we will continue to study kinematics of particles. Remember, **Kinematics** involves the study of a body's motion without regard to the forces that generate that motion. This chapter will focus on studying the relationship between position, velocity, and acceleration in two-dimensions. We will also study how to describe the motion of one particle relative to another moving particle, as well as what happens to the motion of a particle when it is constrained in some way.

Since this chapter deals with two-dimensions, we will need vectors to describe a body's motion. In this book, variables representing vector quantities will be denoted in bold face (e.g. $\mathbf{r}$, $\mathbf{v}$, $\mathbf{a}$) and scalar variables will be italicized (e.g. t, s, v, a).

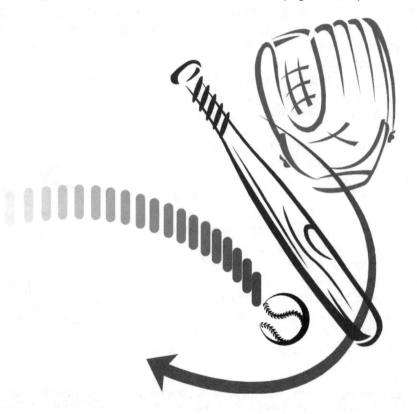

3.1) PLANE CURVILINEAR MOTION: *x-y* COORDINATES

3.1.1) PLANE CURVILINEAR MOTION

Two-dimensional **curvilinear motion** occurs when a particle moves along a curved path that lies in a plane. For example, the tip of your pencil writing out your homework moves in the plane of your paper. A particle moving unconstrained in a plane has two degrees of freedom and requires two coordinates to describe its motion. Unlike rectilinear motion where the motion of a particle can be described by a scalar, vectors are needed to completely describe the curvilinear motion of a particle. This is because the direction of the particle is no longer constrained to just move left or right, it is now free to move left, right, up, and down. Therefore, a simple negative sign is no longer sufficient to describe the direction of motion of the particle because the motion is more complex. It then follows that the position (**r**), velocity (**v**), and acceleration (**a**) are now all considered vectors since they will all have a magnitude and a direction. In order to proceed through this section, it is important that you know basic vector terminology and simple vector operations (e.g. vector addition, computing a dot product, calculating the magnitude of a vector). For information on vectors and vector operations, see Appendix B.

In the discussion of plane curvilinear motion we will examine several coordinate frames. Each coordinate frame may be used to describe the location and motion of a particle in a plane.   The choice of which coordinate frame to use will depend on the kind of motion the particle is experiencing and the particular information we are looking for in the problem.   One thing is always true no matter what coordinate system is chosen; the velocity vector is always directed tangent to the particle's path of motion as shown in

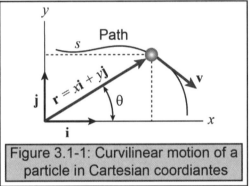

Figure 3.1-1: Curvilinear motion of a particle in Cartesian coordiantes

Figure 3.1-1. Even though velocity is always tangent to the path curve, in general, acceleration is not.

What is the difference between *rectilinear motion* and *curvilinear motion*?

3.1.2) PLANE CURVILINEAR MOTION: CARTESIAN COORDINATES

A set of coordinate axes that is frequently used to describe the motion of a particle in a plane is *rectangular* or *Cartesian* coordinates. The **Cartesian coordinate frame** consists of two perpendicular axes: the *x*-axis and the *y*-axis as shown in Figure 3.1-2. The Cartesian coordinate system that we will consider in this section is an inertial coordinate frame. This means that the coordinate system does not accelerate. But for our purposes, we can think of an inertial reference frame as one where the origin is fixed and does not move and the axes do not rotate. Some coordinate systems, such as the *n-t* coordinate system, are body-fixed coordinate frames. This means that the coordinate axes are fixed to and move with the body and the axes may rotate.

A Cartesian coordinate system is commonly used to solve problems in which the motion proceeds in a straight line, is expressed in terms of the *x*- and *y*-components or is non-circular in nature.   More generally, Cartesian coordinates are useful when the *x*- and *y*-components of a body's acceleration are independent, as in the case of projectile motion which will be discussed later in this chapter.

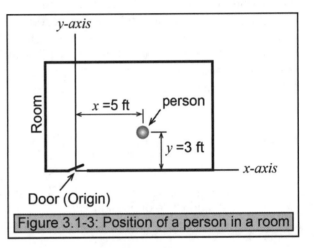

Figure 3.1-2: Cartesian coordinate system

3.1.3) POSITION

If I am a person standing in a room, how would I identify my position in that room? One way to do this would be to measure my distance relative to some object in the room. Let's say we choose the door. So, I could say that my position inside the room is 5 ft east of the door and 3 ft north of the door (see Figure 3.1-3). This is exactly how we locate objects in the Cartesian coordinate system. First, we choose an origin, in this case the door. Second, we measure two distances that are mutually orthogonal or perpendicular to each other to locate the particle's position within the space.

Figure 3.1-3: Position of a person in a room

The location of an object in space is given by its **position vector r**. The position vector **r** (given in Equation 3.1-1) has units of length and is drawn as an arrow from the origin of the coordinate frame to the particle as shown in Figure 3.1-4. The length of the arrow indicates the magnitude of the vector and the arrowhead indicates the direction of the vector.   In Figure 3.1-4, the particle's current position can be reached by traveling *x* units in the *x*-direction and *y* units in the *y*-direction.   Therefore, the vector **r** can be considered as the summation of a vector in the *x*-direction and a vector in the *y*-direction. This fact can be represented mathematically using unit direction vectors **i** and **j**. Both vectors have a magnitude of one, but **i** points in the positive *x*-direction while **j** points in the positive *y*-direction. If considering vectors in three dimensions, the unit vector **k** corresponding to the *z*-direction is needed.

Units of Position (length)
Metric Units:
• meters (m)
• centimeter (cm = 0.01 m)
• millimeters (mm = 0.001 m)
• kilometers (km = 1000 m)
US Customary Units:
• feet (ft = 12 in)
• inch (in)
• mile (mi = 5280 ft)

Position: $r = x\mathbf{i} + y\mathbf{j}$ (3.1-1)

r = position vector
x = distance of the particle from the origin in the x-direction
y = distance of the particle from the origin in the y-direction

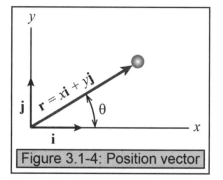

Figure 3.1-4: Position vector

Given the components of a vector, its magnitude and direction can be determined using geometry. In particular, the vector itself can be considered the hypotenuse of a right triangle with the two component vectors being the legs. Therefore, the magnitude of the position vector (Equation 3.1-2) can be determined using the Pythagorean Theorem (i.e. $a^2 + b^2 = c^2$). Additionally, the angle that the position vector makes with the x-axis (Equation 3.1-3) can be determined using the tangent function* (i.e. $\tan\theta = $ opposite/adjacent).

*Technical Note: The tangent function in Equation 3.1-3 is multi-valued so care must be taken when evaluating its inverse. For example, $y/x = -1$ could correspond to ($y = -1$, $x = 1$) or ($y = 1$, $x = -1$). The convention that is usually employed is that the function $\tan^{-1}$ will return an angle between -180 and 180 degrees, even when that is not where the body is actually located.

Magnitude of **r**: $r = |\mathbf{r}| = \sqrt{x^2 + y^2}$ (3.1-2)   Direction of **r**: $\theta = \tan^{-1}\left(\dfrac{y}{x}\right)$ (3.1-3)

r = position vector
r = magnitude of position

x = x-axis position of the particle
y = y-axis position of the particle
θ = angle between **r** and the x-axis

The components of the position vector can also be determined by decomposing it into its components (Equation 3.1-4 and 3.1-5) when the magnitude and direction are known using the sine (i.e. $\sin\theta = $ opposite/hypotenuse) and cosine functions (i.e. $\cos\theta = $ adjacent/hypotenuse).

Vector components of position: $x = |\mathbf{r}|\cos\theta$ (3.1-4) $y = |\mathbf{r}|\sin\theta$ (3.1-5)

r = position vector
x = x-axis position of the particle

y = y-axis position of the particle
θ = angle between **r** and the x-axis

3.1.4) DISPLACEMENT AND TOTAL DISTANCE TRAVELED

Displacement is the change in position of a body and has units of length just like position. It is the difference between the final and initial position and can be calculated using Equation 3.1-6. Displacement ($\Delta\mathbf{r}$) is represented by a vector as shown in Figure 3.1-5.

Just like in rectilinear motion, displacement is not necessarily equal to the total distance traveled. The **total distance traveled** (s_{total}) is the length of the path that the particle traverses (see Figure 3.1-5).

Displacement: $$\Delta \mathbf{r} = \mathbf{r}_{final} - \mathbf{r}_{initial}$$ (3.1-6)

$\Delta \mathbf{r}$ = displacement
$\mathbf{r}_{final}$, $\mathbf{r}_{initial}$ = final and initial position respectively

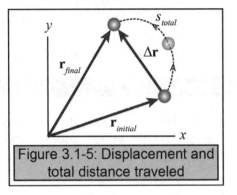

Figure 3.1-5: Displacement and total distance traveled

3.1.5) VELOCITY

The **velocity** of a particle is the time rate of change of position. Velocity of a particle moving in a plane has the same definition as the velocity of a particle moving in a straight line. The difference is now position and velocity are vectors. Recall that velocity has units of length per time.

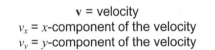
Units of Velocity (length/time)
Metric Units:
- meters per second (m/s)
- kilometers per hour (kph = km/h)
US Customary Units:
- feet per second (ft/s)
- miles per hour (mph = mi/h = 1.4667 ft/s)

Consider a runner running on flat ground. In plain language, the faster the runner covers ground, the higher her velocity is going to be. The difference between rectilinear and curvilinear motion is that for rectilinear motion the runner has to run in a straight line. With curvilinear motion, the runner can run in circles or whatever path she cares to take. Because the runner is able to change direction and move in two dimensions, her velocity is now a vector. The velocity has both an x- and y-component as shown in Equation 3.1-7.

Velocity vector: $$\mathbf{v} = v_x \mathbf{i} + v_y \mathbf{j}$$ (3.1-7)

$\mathbf{v}$ = velocity
v_x = x-component of the velocity
v_y = y-component of the velocity

Velocity has a very interesting and useful property. Consider a runner running on flat ground, as discussed earlier. She runs on the path shown in Figure 3.1-6. Her velocity will always be tangent to the path on which she is running. This is true for any particle and any path. Velocity is always tangent to the path curve.

What is the difference

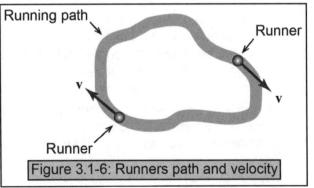

Figure 3.1-6: Runners path and velocity

between velocity and speed? Many people use these two terms interchangeably. Let's take a look at a car driving down a road. Is the velocity of the car 50 mph or is the speed

50 mph. Does it matter? If we say that a car is going 50 mph, are we giving all the pertinent information? No. What we are missing is the direction of the car's motion. The car may be driving northeast or heading south. If we say that the car is going 50 mph in the northeast direction, this is a vector quantity because it has both magnitude and direction. If we say that the car is going 50 mph, this is a scalar quantity because it has only size and no direction. So, what is the difference between velocity and speed? Velocity is a vector and **speed** is its magnitude.

Example 3.1-1

On the figure, draw a velocity vector for each position of the particle. Don't worry about the magnitudes, just attempt to identify the directions.

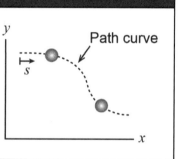

What is the difference between *speed* and *velocity*?

Conceptual Example 3.1-2

In the following situations, determine whether the *speed* or *velocity* is changing, both are changing, or neither is changing. A car's cruise control is set to a constant 50 mph and ...

1) the car is driving on a level straight-away.
2) the car rounds a curve.
3) the car drives over a hill.
4) the car is driving up a hill of constant grade.
5) the car is driving on a level straight-away when the driver applies the brakes.

Figure 3.1-7 shows a particle following a dashed path within the x-y plane. Vector $\mathbf{r}_1$ represents the position of the particle at one instant and the vector $\mathbf{r}_2$ represents the position of the particle an instant later. The displacement vector $\Delta\mathbf{r}$ drawn from the tip of $\mathbf{r}_1$ to the tip of $\mathbf{r}_2$ is the change in position of the particle, which is mathematically the difference $\Delta\mathbf{r} = \mathbf{r}_2 - \mathbf{r}_1$. Dividing this quantity by the change in time (Δt) gives the average velocity vector $\mathbf{v}_{ave}$ over that time interval (Equation 3.1-8). As the time interval in Equation 3.1-8 approaches zero, the average velocity vector approaches the instantaneous velocity vector $\mathbf{v}$ as shown in Equation 3.1-9. Note that taking the derivative of $\mathbf{r}$ to obtain $\mathbf{v}$ (Equation 3.1-9) involves taking the derivatives of the magnitudes in each direction and the derivatives of the unit direction vectors. Since the unit direction vectors $\mathbf{i}$, $\mathbf{j}$ and $\mathbf{k}$ are constant (i.e. the magnitude and direction do not change with time) their derivatives are zero.

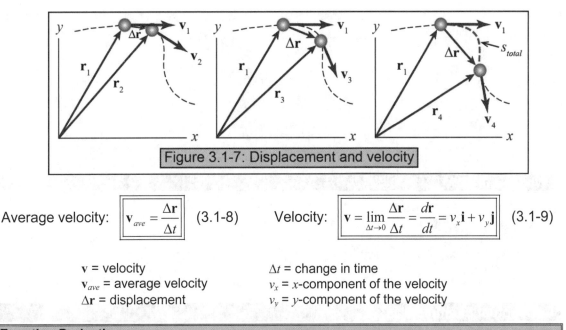

Figure 3.1-7: Displacement and velocity

Average velocity: $\mathbf{v}_{ave} = \dfrac{\Delta \mathbf{r}}{\Delta t}$ (3.1-8)   Velocity: $\mathbf{v} = \lim\limits_{\Delta t \to 0} \dfrac{\Delta \mathbf{r}}{\Delta t} = \dfrac{d\mathbf{r}}{dt} = v_x \mathbf{i} + v_y \mathbf{j}$ (3.1-9)

$\mathbf{v}$ = velocity Δt = change in time
$\mathbf{v}_{ave}$ = average velocity v_x = x-component of the velocity
$\Delta \mathbf{r}$ = displacement v_y = y-component of the velocity

Equation Derivation

　　　Velocity is the time rate of change of position as shown in Equation 3.1-9. This means that to get velocity ($\mathbf{v}$) we need to take the derivative of position ($\mathbf{r}$). We have already had practice taking derivatives of position to obtain velocity in the section on rectilinear motion. When dealing with one-dimensional motion, we assumed position was a scalar, now position is a vector. Does this change how we take the derivative? The answer is yes and no. It depends on what coordinate system you are using. For now, let's stick with Cartesian coordinates. Since the unit direction vectors $\mathbf{i}$, $\mathbf{j}$ and $\mathbf{k}$ are constant (i.e. magnitude and direction do not change with time) their derivatives are zero. For example, if we take the derivative of position by applying the product rule, we obtain the following expression for velocity.

$$\mathbf{v} = \frac{d\mathbf{r}}{dt} = \frac{d}{dt}(x\mathbf{i} + y\mathbf{j}) = \frac{d(x)}{dt}\mathbf{i} + x\,\frac{d(\mathbf{i})}{dt} + \frac{d(y)}{dt}\mathbf{j} + y\,\frac{d(\mathbf{j})}{dt}$$

$$\frac{d(\mathbf{i})}{dt} = 0$$

$$\frac{d(\mathbf{j})}{dt} = 0$$

　　　This leads us to conclude that the derivative of a vector, expressed in rectangular coordinates, can be calculated from the derivatives of each of its scalar components. This is demonstrated below, where the unit vectors can be thought of as any constant and hence can be factored out of the derivatives.

$$\mathbf{v} = \frac{d\mathbf{r}}{dt} = \frac{d(x\mathbf{i} + y\mathbf{j})}{dt} = \dot{x}\mathbf{i} + \dot{y}\mathbf{j}$$

　　　Integrals can be solved in a similar component-wise manner as shown below. This is true because an integral can be split across addition and constants can be factored out of the integral.

$$\mathbf{v} = \frac{d\mathbf{r}}{dt} \quad \Rightarrow \quad \int d\mathbf{r} = \int \mathbf{v}\, dt$$

$$\int (dx\mathbf{i} + dy\mathbf{j}) = \int (\dot{x}\mathbf{i} + \dot{y}\mathbf{j})\, dt$$

$$\left(\int_{x_o}^{x} dx \right)\mathbf{i} + \left(\int_{y_o}^{y} dy \right)\mathbf{j} = \left(\int_{t_o}^{t} (\dot{x})\, dt \right)\mathbf{i} + \left(\int_{t_o}^{t} (\dot{y})\, dt \right)\mathbf{j}$$

Why is $di/dt = 0$?

Conceptual Example 3.1-3

A fisherman sits in a motor boat at the shore of a river flowing downstream at 4 m/s. The fisherman wishes to reach the other side of the river. If the river is 60 meters wide and the motor boat has a maximum speed of 3 m/s with respect to the water, what is the shortest amount of time it could take the fisherman to perform the crossing? Note that the flow of the river is parallel to the shore on both sides of the river.

    a) 20 seconds
    b) 15 seconds
    c) 60/5 seconds
    d) 60/7 seconds

3.1.6) ACCELERATION

Acceleration is the rate at which velocity is changing and has units of length per time squared. The relationship between acceleration and velocity is similar to that between velocity and position. An illustration for $\Delta\mathbf{v}$, analogous to Figure 3.1-7, can be

> Units of Acceleration (length/time^2)
> Metric Units:
> • meters per second squared (m/s^2)
> US Customary Units:
> • feet per second squared (ft/s^2)

drawn using velocity vectors instead of position vectors. The instantaneous acceleration vector ($\mathbf{a}$) always points in the direction of $d\mathbf{v}$ and the average acceleration vector always points in the direction of $\Delta\mathbf{v}$. In other words, the acceleration vector is tangent to the path generated by the particle's velocity vector, called a *hodograph*. However, a particles acceleration is generally not tangent to the path of the particle's position. Acceleration is expressed as a vector because it has magnitude and direction, stemming from the fact that velocity can change both magnitude and direction. Mathematically, the average acceleration ($\mathbf{a}_{ave}$) is given by Equation 3.1-10 and the instantaneous acceleration ($\mathbf{a}$) is given by Equation 3.1-11.

Average acceleration: $$\mathbf{a}_{ave} = \frac{\Delta \mathbf{v}}{\Delta t}$$ (3.1-10)

Acceleration: $$\mathbf{a} = \frac{d\mathbf{v}}{dt} = \frac{d^2\mathbf{r}}{dt^2} = a_x\mathbf{i} + a_y\mathbf{j}$$ (3.1-11)

$\mathbf{a}$, $\mathbf{a}_{ave}$ = instantaneous and average
acceleration respectively
a_x = x-component of the acceleration
a_y = y-component of the acceleration
$\mathbf{v}$ = velocity

$\Delta\mathbf{v}$ = change in velocity
$\mathbf{r}$ = position
t = time
Δt = change in time

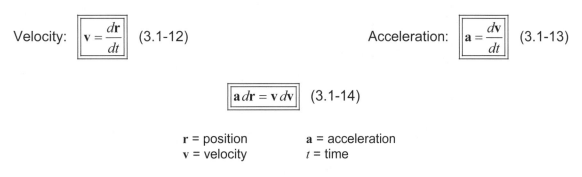

Conceptual Example 3.1-4

If a car rounds a turn at a constant speed, is its acceleration zero?

3.1.7) SOLVING PLANE CURVILINEAR PROBLEMS

Most plane curvilinear problems in the Cartesian coordinate system can be solved using Equations 3.1-12 through 3.1-14. Equations 3.1-12 and 3.1-13 come directly from the definitions of velocity and acceleration. Equation 3.1-14 is an algebraic combination of Equation 3.1-12 and 3.1-13.

Velocity: $$\mathbf{v} = \frac{d\mathbf{r}}{dt}$$ (3.1-12)   Acceleration: $$\mathbf{a} = \frac{d\mathbf{v}}{dt}$$ (3.1-13)

$$\mathbf{a}\,d\mathbf{r} = \mathbf{v}\,d\mathbf{v}$$ (3.1-14)

$\mathbf{r}$ = position $\mathbf{a}$ = acceleration
$\mathbf{v}$ = velocity t = time

The relationships given in Equations 3.1-12 through 3.1-14 can be used to determine displacement, velocity and acceleration of a particle given initial conditions through integration (see Equations 3.1-15, 3.1-16 and 3.1-17).

$$\left[\int_{t_o}^{t} \mathbf{v}\,dt = \int_{\mathbf{r}_o}^{\mathbf{r}} d\mathbf{r}\right] \quad (3.1\text{-}15) \qquad \left[\int_{t_o}^{t} \mathbf{a}\,dt = \int_{\mathbf{v}_o}^{\mathbf{v}} d\mathbf{v}\right] \quad (3.1\text{-}16) \qquad \left[\int_{\mathbf{r}_o}^{\mathbf{r}} \mathbf{a}\,d\mathbf{r} = \int_{\mathbf{v}_o}^{\mathbf{v}} \mathbf{v}\,d\mathbf{v}\right] \quad (3.1\text{-}17)$$

$\mathbf{r}$, $\mathbf{r}_o$ = position and initial position respectively
$\mathbf{v}$, $\mathbf{v}_o$ = velocity and initial velocity respectively

$\mathbf{a}$, $\mathbf{a}_o$ = acceleration and initial
acceleration respectively
t, t_o = time and initial time respectively

Equation Summary

Abbreviated variable definition list

x = x-direction position
y = y-direction position
$\mathbf{r}$ = position

v = velocity
a = acceleration
t = time

Position

$$\mathbf{r} = x\mathbf{i} + y\mathbf{j}$$

Displacement

$$\Delta\mathbf{r} = \mathbf{r}_{final} - \mathbf{r}_{initial}$$

Velocity

$$\mathbf{v} = \frac{d\mathbf{r}}{dt} \qquad \mathbf{v}_{ave} = \frac{\Delta\mathbf{r}}{\Delta t}$$

Acceleration

$$\mathbf{a} = \frac{d\mathbf{v}}{dt} \qquad \mathbf{a}_{ave} = \frac{\Delta\mathbf{v}}{\Delta t} \qquad \mathbf{a}\,d\mathbf{r} = \mathbf{v}\,d\mathbf{v}$$

Integral form of the basic equations

$$\int_{t_o}^{t} \mathbf{v}\,dt = \int_{\mathbf{r}_o}^{\mathbf{r}} d\mathbf{r} \qquad \int_{t_o}^{t} \mathbf{a}\,dt = \int_{\mathbf{v}_o}^{\mathbf{v}} d\mathbf{v} \qquad \int_{\mathbf{r}_o}^{\mathbf{r}} \mathbf{a}\,d\mathbf{r} = \int_{\mathbf{v}_o}^{\mathbf{v}} \mathbf{v}\,d\mathbf{v}$$

Example Problem 3.1-5

A fishing boat wishes to reach the other side of a briskly flowing river. The boat attempts to travel straight across the river, but the strength of the current pushes the boat downstream during the crossing. As a result of the river's current, the velocity of the boat has components in the y- as well as the x-direction. If the velocity of the boat relative to the shore is a constant $\mathbf{v}_b = (5\mathbf{i} - 2\mathbf{j})$ m/s and the river is 100 meters wide, what is the distance d that the boat is pushed downstream during its crossing?

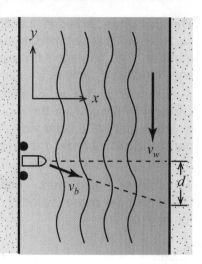

Given:

Find:

Solution:

Example Problem 3.1-6

The carriage, shown in the figure, has a vertical slot that is driven in an oscillatory manner. The carriage's horizontal position as a function of time is $x = 3\sin(2t)$, where x is in cm and t is in seconds.   This carriage in turn drives the motion of pin A which is further constrained to move in the parabolic slot whose shape is described by the function $y = -x^2$, where the y-coordinate also has units of cm. Determine the overall velocity of pin A as a function of time.

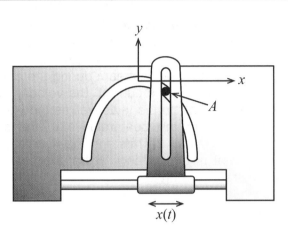

Given:

Find:

Solution:

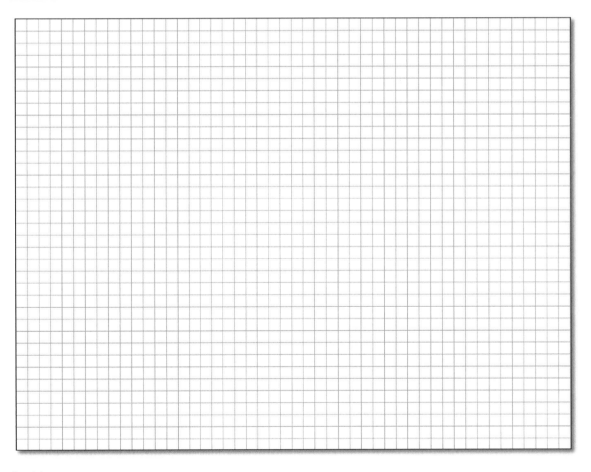

Example Problem 3.1-7

A girl operates a radio-controlled model car on a smooth flat cement playground area. The girl stands at the origin of the x-y coordinate system and the surface of the cement lies in the x-y plane.   The car's position is given by $\mathbf{r} = x\mathbf{i} + y\mathbf{j}$ and the x- and y-components of the velocity are given by $\dot{x} = 2 + 3t^2$ m/s and $\dot{y} = 6 + 0.5t^3$ m/s, where t is in seconds. At time $t = 0$, the car is located at $x = 2$ m and $y = 6$ m.

    a) Determine, as a function of time, the position, velocity and acceleration of the car.

    b) Plot the position of the car for the time period of $t = 0$ to $t = 2$ seconds.

    c) Determine the car's displacement $\Delta\mathbf{r}$ during the interval from $t = 0$ to $t = 2$ s.

    d) Determine the velocity and acceleration of the car at $t = 1$ second and superimpose these vectors on the position graph. Don't worry that the units don't match, just estimate the vector length and direction.

    e) Determine the average velocity and acceleration of the car during the interval from $t = 0$ to $t = 2$ s.

Given:

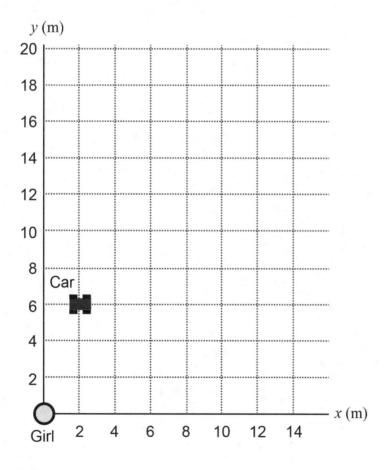

Find:

Solution:

Solved Problem 3.1-8

A particle moves on a plane with a velocity of $\mathbf{v}(t) = (5t\,\mathbf{i} - t^2\mathbf{j})$ cm/s. If the particle's initial position is $\mathbf{r}_o = (\mathbf{i} + \mathbf{j})$ cm at time t_o = 0, determine the position and acceleration of the particle as a function of time.   Also, calculate the displacement, average velocity, average acceleration and total distance traveled between 0 and 10 seconds.

<u>Given:</u> $\mathbf{v}(t) = 5t\,\mathbf{i} - t^2\mathbf{j}$ cm/s

$\mathbf{r}_o = \mathbf{i} + \mathbf{j}$ cm @ t_0 = 0

<u>Find:</u> $\mathbf{r}(t)$, $\mathbf{a}(t)$

$\Delta\mathbf{r}$, $\mathbf{v}_{\text{ave}}$, $\mathbf{a}_{\text{ave}}$, s_{total} between 0 and 10 s

<u>Solution:</u>

Acceleration

The easiest quantity to calculate is the acceleration, so let's start there. The acceleration is simply the time derivative of velocity.

$$\mathbf{a} = \frac{d\mathbf{v}}{dt} = \frac{d(5t\,\mathbf{i} - t^2\mathbf{j})}{dt} \qquad \boxed{\mathbf{a} = 5\mathbf{i} - 2t\,\mathbf{j}\ \frac{\text{cm}}{\text{s}^2}}$$

$$\mathbf{a}_{ave} = \frac{\Delta\mathbf{v}}{\Delta t} = \frac{\mathbf{v}_f - \mathbf{v}_o}{t_f - t_o} = \frac{(5(10)\mathbf{i} - 10^2\mathbf{j}) - (5(0)\mathbf{i} - 0^2\mathbf{j})}{10} \qquad \boxed{\mathbf{a}_{ave} = 5\mathbf{i} - 10\mathbf{j}\ \frac{\text{cm}}{\text{s}^2}}$$

Position

To find the position, we need to integrate the velocity.

$$\mathbf{v} = \frac{d\mathbf{r}}{dt} \qquad \rightarrow \qquad \int \mathbf{v}\,dt = \int d\mathbf{r}$$

When integrating, we need to take care to use the correct limits of integration. Even if you are not asked to generate a generalized equation (e.g. the position as a function of time), you should always use variables for the upper limit. You may need the resulting expression later in the problem.

$$\int_0^t (5t\,\mathbf{i} - t^2\mathbf{j})\,dt = \int_{\mathbf{r}_o}^{\mathbf{r}} d\mathbf{r} \qquad \left(\frac{5t^2}{2}\mathbf{i} - \frac{t^3}{3}\mathbf{j}\right)_0^t = \mathbf{r}\Big|_{\mathbf{r}_o}^{\mathbf{r}} \qquad \frac{5t^2}{2}\mathbf{i} - \frac{t^3}{3}\mathbf{j} = \mathbf{r} - \mathbf{r}_o$$

$$\mathbf{r} = \left(\frac{5t^2}{2}\right)\mathbf{i} - \left(\frac{t^3}{3}\right)\mathbf{j} + \mathbf{i} + \mathbf{j} = \left(\frac{5t^2 + 2}{2}\right)\mathbf{i} - \left(\frac{t^3 - 3}{3}\right)\mathbf{j} \qquad \boxed{\mathbf{r} = \left(\frac{5t^2 + 2}{2}\right)\mathbf{i} - \left(\frac{t^3 - 3}{3}\right)\mathbf{j}\ \text{cm}}$$

Displacement

Now that we have determined the position as a function of time, we can use it to determine the displacement between any two times.

$$\Delta \mathbf{r} = \mathbf{r}_{final} - \mathbf{r}_{initial} = \left(\frac{5t^2 + 2}{2} \mathbf{i} - \frac{t^3 - 3}{3} \mathbf{j} \right) - (\mathbf{i} + \mathbf{j}) = \left(\frac{5t^2}{2} \right) \mathbf{i} - \left(\frac{t^3}{3} \right) \mathbf{j}$$

$$\Delta \mathbf{r} = \mathbf{r}_f - \mathbf{r}_o = \left(\frac{5(10^2)}{2} \mathbf{i} - \frac{10^3}{3} \mathbf{j} \right) - \left(\frac{5(0^2)}{2} \mathbf{i} - \frac{0^3}{3} \mathbf{j} \right) \qquad \boxed{\Delta \mathbf{r} = (250\mathbf{i} - 333.33\mathbf{j}) \text{ cm}}$$

Average velocity

$$\mathbf{v}_{ave} = \frac{\Delta \mathbf{r}}{\Delta t} \qquad \boxed{\mathbf{v}_{ave} = (25\mathbf{i} - 33.33\mathbf{j}) \frac{\text{cm}}{\text{s}}}$$

Total distance traveled

The total distance traveled is different from the displacement as shown in the following graph.

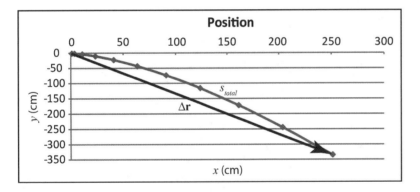

The speed of the particle is related to the distance traveled on the path by the following equation. The speed of the particle is simply the magnitude of the velocity.

$$v = \frac{ds}{dt} \qquad v = \sqrt{(5t)^2 + t^4} = t\sqrt{25 + t^2}$$

We can integrate the speed to determine the total distance traveled since the speed is always positive.

This integration is performed using the substitution technique.
$u = 25 + t^2$
$du = 2t\, dt$

$$v = \frac{ds}{dt} \quad \rightarrow \quad \int v\, dt = \int ds$$

$$\int_0^t \left(t\sqrt{25 + t^2} \right) dt = \int_0^s ds \qquad s = \frac{1}{3}(t^2 + 25)^{3/2}$$

We substitute the final time of 10 seconds to determine the total distance traveled.

$$s_{total} = \frac{1}{3}(10^2 + 25)^{3/2} \qquad \boxed{s_{total} = 465.8 \text{ cm}}$$

In comparison, we can calculate the magnitude of the displacement. Note that the magnitude of the displacement is less than the total distance traveled as shown in the graph.

$$\Delta r = \sqrt{250^2 + 333.33^2} = 416.7 \text{ cm}$$

3.2) PROJECTILE MOTION

3.2.1) PROJECTILE MOTION

A particular type of curvilinear motion problem that is especially well-suited to rectangular coordinates is projectile motion. A **projectile** is a particle that is given an initial velocity and is allowed to move through the air on its own as shown in Figure 3.2-1. Examples of projectiles include a cannonball in flight after it has left the cannon, or a football in flight after it has left the arm of the quarterback. The only forces acting on a projectile are its own weight and air resistance. However, it is often a reasonable approximation to neglect the forces due to air resistance. This is an example of how an engineer or scientist may abstract away an element of the physical world in order to make their model simpler and easier to analyze. Ultimately the engineer must determine whether or not this simplification is reasonable for the specific purpose of the model.

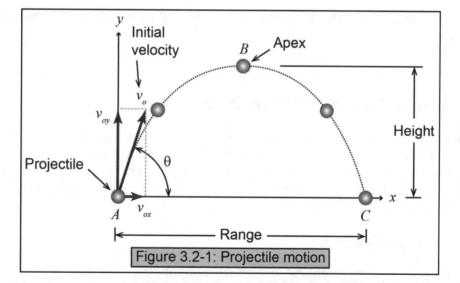

Figure 3.2-1: Projectile motion

What is a projectile?

3.2.2) FALLING

Once a projectile is aloft, it is not flying like an airplane that has lift forces, or how Buzz Lightyear flies in the words of Woody, "falling with style." A projectile just falls. So understanding how an object falls is important when analyzing projectile motion. All objects in freefall do so with a constant acceleration, which is the acceleration due to gravity (when air resistance is neglected). Mass has no influence on the speed with which the object falls and because we are neglecting air resistance, neither does the object's size. This concept will become clearer when we learn how to analyze the forces that act on a falling body. The basic idea, however, is that the increased weight force due to the increased mass is offset by the increased inertia of the object and hence its mass does not end up affecting the rate at which it falls.

If air resistance (i.e. aerodynamics drag) is considered, the acceleration at which an object falls is affected. Usually the aerodynamic drag depends on the density of air, the frontal area of the object, a drag coefficient and the square of the velocity. This means the bigger and rougher the object is and the faster the object is moving, the more air resistance it will experience.

Conceptual Example 3.2-1

Answer the following questions about falling objects. Neglect air resistance.

a) What is the acceleration that all objects fall with?

b) Does mass influence a falling object's velocity?

c) When would the size and shape of an object affect the speed at which an object falls?

3.2.3) SOLVING PROJECTILE PROBLEMS

Falling objects have certain special properties that help us solve projectile motion problems. Below is a list of these properties. Each item in the list will be explained in turn. Also, each item will be explored to show how it can be used to help us obtain information about the projectile's motion. Each of these properties assumes that air resistance is negligible.

Properties of projectile motion

1. The x- and y-components of motion are independent.
2. The x-direction velocity is constant.
3. The y-direction acceleration is constant.
4. The y-direction velocity at the apex of motion is zero.
5. If the starting height of the projectile is the same as the ending height, then the motion of the projectile is symmetric.

The first property states that the x- and y-components of motion are independent. This derives from the fact that the only force acting on the body is its weight. The weight acts entirely in the y-direction, therefore, the motion along each coordinate axis can be analyzed separately. It also allows us to apply the kinematic equations to each axis independently. This has one big advantage. It allows a curvilinear problem to be treated as two rectilinear problems. This property is similar to how we were able to take derivatives and integrals of vectors in x-y coordinates in a component-wise manner.

In order to address the two axes separately, it is necessary to take the initial velocity of the projectile and decompose it into its x- and y-components as shown in Equation 3.2-1 (see Figure 3.2-1). Furthermore, Equations 3.2-2 through 3.2-5 are the equations used to analyze projectile motion. Notice that these equations are the same kinematic relationships we are already familiar with. They are just written in terms of the x- and y-coordinates that we have defined.

$$\text{Initial velocity:} \quad \boxed{\mathbf{v}_o = v_{xo}\mathbf{i} + v_{yo}\mathbf{j} = v_o(\cos\theta\,\mathbf{i} + \sin\theta\,\mathbf{j})} \quad (3.2\text{-}1)$$

$$x\text{-direction velocity:} \quad \boxed{v_x = \frac{dx}{dt}} \quad (3.2\text{-}2) \qquad\qquad y\text{-direction velocity:} \quad \boxed{v_y = \frac{dy}{dt}} \quad (3.2\text{-}3)$$

$$y\text{-direction acceleration:} \quad \boxed{a_y = \frac{dv_y}{dt}} \quad (3.2\text{-}4) \qquad\qquad \boxed{a_y\,dy = v_y\,dv_y} \quad (3.2\text{-}5)$$

x, y = x-direction position and y-direction position respectively

v_x, v_y = x-direction velocity and y-direction velocity respectively

$\mathbf{v}_o$ = initial velocity

v_{xo}, v_{yo} = initial x-direction velocity and initial y-direction velocity respectively

θ = angle of $\mathbf{v}_o$ with respect to the x-axis

a_y = y-direction acceleration

t = time

The second property states that the x-direction velocity is constant. If we ignore air drag, there are no forces acting on the projectile in the horizontal or x-direction. Therefore, $\sum F_x = ma_x = 0$ and the acceleration in the x-direction is zero (Equation 3.2-6). It follows that the velocity in the x-direction is constant throughout the projectile's flight (Equation 3.2-7). This allows us to integrate Equation 3.2-7 to obtain an expression for the x-direction position of the projectile as given in Equation 3.2-8.

$$x\text{-direction acceleration:} \quad \boxed{a_x = 0} \quad (3.2\text{-}6)$$

$$x\text{-direction velocity:} \quad \boxed{v_x = \frac{dx}{dt} = \text{constant}} \quad (3.2\text{-}7)$$

x = x-direction position

a_x = x-direction acceleration

v_x = x-direction velocity

t = time

x-direction position: $\boxed{\boxed{x = x_o + v_{xo}t = x_o + (v_o \cos\theta)t}}$ (3.2-8)

x = x-direction position
x_o = initial x-direction position
v_o = initial speed

v_{xo} = initial x-direction velocity
θ = angle of $\mathbf{v}_o$ with respect to the x-axis
t = time

Equation Derivation

To obtain the x-position of a projectile, we will integrate the velocity equation as shown.

$$v_x = \frac{dx}{dt} \quad \Rightarrow \quad \int v_x dt = \int dx$$

The x-direction velocity of a projectile is constant if air resistance is neglected. Therefore, it will come out of the integral. Assuming that the initial time is zero and the initial x-position is x_o, we obtain the x-position of the projectile (Equation 3.2-8).

$$\int_0^t v_x dt = \int_{x_o}^x dx \quad \Rightarrow \quad v_x \int_0^t dt = \int_{x_o}^x dx$$

$$v_x t = x - x_o \quad \Rightarrow \quad x = v_x t + x_o$$

The third property states that the y-direction acceleration is constant. Based on the negligible air drag assumption, the only force acting on the particle in the y-direction is the object's weight due to gravity. Therefore, the acceleration in the y-direction is the acceleration due to gravity ($-g$) (Equation 3.2-9). In the y-direction only the acceleration is constant. The velocity changes with time (Equation 3.2-10). We can integrate Equation 3.2-10 to obtain an expression for the height of the projectile given in Equation 3.2-11.

y-direction acceleration: $\boxed{\boxed{a_y = \frac{dv_y}{dt} = -g}}$ (3.2-9)

y-direction velocity: $\boxed{\boxed{v_y = \frac{dy}{dt} = v_{yo} - gt = v_o \sin\theta - gt}}$ (3.2-10)

Height: $\boxed{\boxed{y = y_o + v_{yo}t - \frac{gt^2}{2} = y_o + (v_o \sin\theta)t - \frac{gt^2}{2}}}$ (3.2-11)

y, y_o = y-direction position and initial y-direction
    position respectively
v_y, v_{yo} = y-direction velocity and initial y-direction
    velocity respectively
v_o = initial speed

a_y = y-direction acceleration
g = acceleration due to gravity
θ = angle of $\mathbf{v}_o$ with respect to the x-axis
t = time

Equation Derivation

To obtain the y-direction velocity of a projectile, we will integrate the acceleration equation as shown.

$$a_y = \frac{dv_y}{dt} \quad \Rightarrow \quad \int a_y dt = \int dv_y$$

The y-direction acceleration of a projectile is constant and equal to $-g$ if air resistance is neglected. Therefore, it will come out of the integral. Assuming that the initial time is zero and the initial y-direction velocity is v_{yo}, we obtain the y-direction velocity of the projectile (Equation 3.2-10).

$$-g\int_0^t dt = \int_{v_{yo}}^{v_y} dv_y \quad \Rightarrow \quad -gt = v_y - v_{yo} \quad \Rightarrow \quad v_y = v_{yo} - gt$$

To obtain the y-position of a projectile, we will integrate the velocity equation as shown.

$$v_y = \frac{dy}{dt} \quad \Rightarrow \quad \int v_y dt = \int dy$$

The y-direction velocity of a projectile is given above and in Equation 3.2-10. Assuming that the initial time is zero and the initial y-position is y_o, we obtain the y-position of the projectile (Equation 3.2-11).

$$\int_0^t (v_{yo} - gt)\,dt = \int_{y_o}^y dy \quad \Rightarrow \quad v_{yo}t - \frac{gt^2}{2} = y - y_o \quad \Rightarrow \quad y = v_{yo}t - \frac{gt^2}{2} + y_o$$

The fourth property states that the y-direction velocity at the apex of motion equals zero (Equation 3.2-12). The apex is the absolute top of a projectile's travel and it is the point at which the y-component of the particle's velocity goes from being positive to being negative, hence, it must equal zero at this point. Figure 3.2-1 illustrates the path of the projectile throughout its flight. In the idealized case, the velocity in the x-direction will be constant and the velocity in the y-direction will decrease at a constant rate until it reaches zero at the apex. Using this property is pretty straightforward. It is just another known value that can be used to solve for unknowns.

y-direction apex velocity: $\boxed{v_{y(apex)} = 0}$ (3.2-12)

Further details can also be employed to help analyze projectile motion, namely that of flight symmetry which is stated in property five. If the projectile starts and ends at the same y-direction position, the projectile's flight will be completely symmetric unless an obstruction interferes with the projectile's path. Useful information comes out of this property. This means, for example, that the magnitude of the particle's velocity in the y-direction will be the same at a given altitude on the particle's way up as it will be on the particle's way down, just with opposite direction. The symmetry property also provides that the time it takes the particle to ascend between two heights is the same amount of time it takes to descend between the same two heights. If the projectile starts and ends at different altitudes, the symmetry property can also be applied but only at points where the particle has two points of motion along a constant altitude line.

Conceptual Example 3.2-2

Answer the following questions about projectile motion.

a) What is the x-direction *acceleration* of a projectile?

b) What is special about the x-direction *velocity* of a projectile?

c) What is the y-direction *velocity* of a projectile at the apex of its travel?

d) What is the y-direction *acceleration* of a projectile?

e) What is important not to forget about the y-direction *velocity* of a projectile?

Conceptual Example 3.2-3

Imagine that you are a passenger in one of those very tall monster trucks. The truck is driving safely at a constant speed down a long straight flat highway. Imagine that you have a ball. Now roll down the truck window and drop the ball outside of the truck. Neglecting air resistance, how does the ball behave before it hits the ground? How would your answer change if we took air resistance into account? How would your answer change if, at the moment the ball was dropped, the driver releases the gas pedal and the truck starts to coast (do not neglect air resistance)?

a) The truck moves ahead of the ball.
b) The ball and truck travel side-by-side and remain even.
c) The ball passes the truck.
d) The ball passes the truck briefly and then starts to lag behind.
e) Not enough information.

Conceptual Example 3.2-4

If you drop an object, we know that its acceleration is g downward. What happens if, with all your might, you throw an object up? What's the object's acceleration after you let go of it?

   a) Less than g.
   b) Equal to g.
   c) More than g.
   d) Less than g for a while and then g.
   e) More than g for a while and then g.

Conceptual Example 3.2-5

Imagine that you and a buddy are standing next to each other holding baseballs. Your buddy throws his baseball up with an initial velocity of 30 mph and you throw your baseball down with an initial velocity of 30 mph from the same initial height. Neglecting air resistance, which ball hits the ground with greater speed?

   a) Your ball.
   b) Your buddy's ball.
   c) Both your ball and your buddy's ball hit the ground with the same speed.

Equation Summary

Abbreviated variable definition list

x = x-direction position
y = y-direction position
$\mathbf{v}$ = velocity
θ = angle of $\mathbf{v}_o$ with respect to the x-axis

a_y = y-direction acceleration
t = time
g = acceleration due to gravity

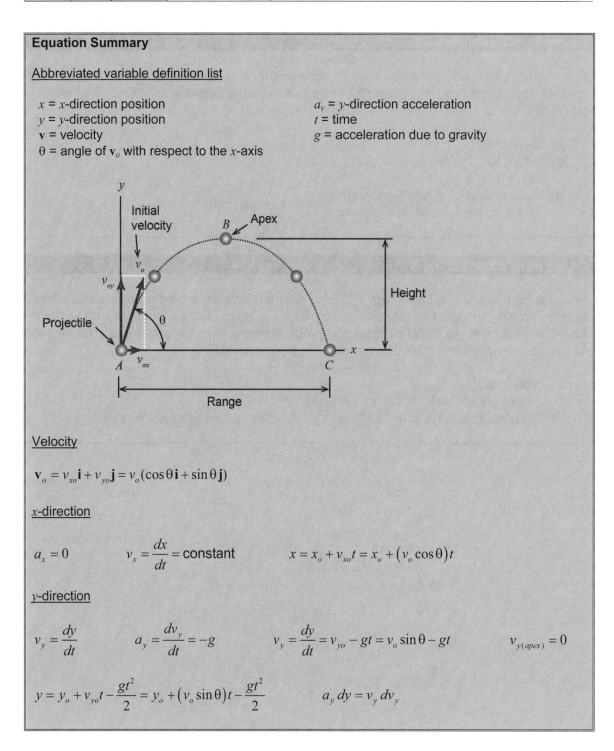

Velocity

$$\mathbf{v}_o = v_{xo}\mathbf{i} + v_{yo}\mathbf{j} = v_o(\cos\theta\,\mathbf{i} + \sin\theta\,\mathbf{j})$$

x-direction

$$a_x = 0 \qquad v_x = \frac{dx}{dt} = \text{constant} \qquad x = x_o + v_{xo}t = x_o + (v_o\cos\theta)t$$

y-direction

$$v_y = \frac{dy}{dt} \qquad a_y = \frac{dv_y}{dt} = -g \qquad v_y = \frac{dy}{dt} = v_{yo} - gt = v_o\sin\theta - gt \qquad v_{y(apex)} = 0$$

$$y = y_o + v_{yo}t - \frac{gt^2}{2} = y_o + (v_o\sin\theta)t - \frac{gt^2}{2} \qquad a_y\,dy = v_y\,dv_y$$

Example Problem 3.2-6

From a videotape, it was observed that a pro golfer hit a golf ball 126 yards for a hole in one. The duration of the golf ball's flight was measured to be 5.6 seconds. Determine the initial speed of the ball and the angle θ at which it was struck.

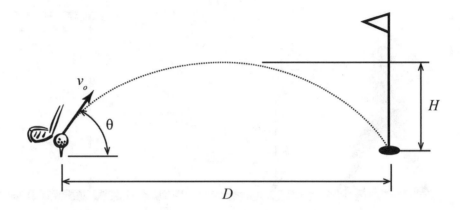

Given:

Find:

Solution:

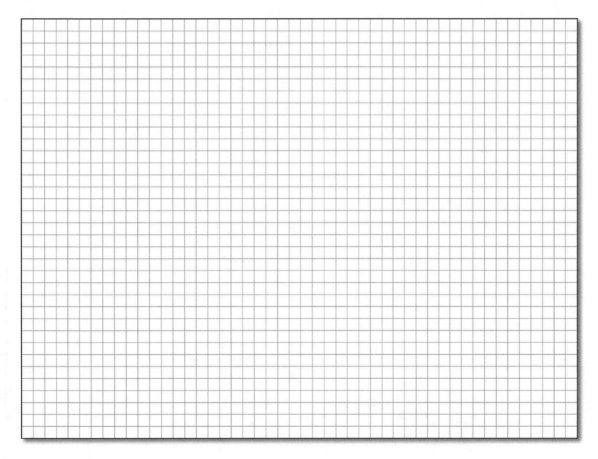

Example Problem 3.2-7

A pitcher throws a baseball horizontally with a speed of 90 mph from a height of 6 ft. If the batter is 60 ft away, determine the time needed for the ball to arrive at the batter and the height H at which it passes the batter.

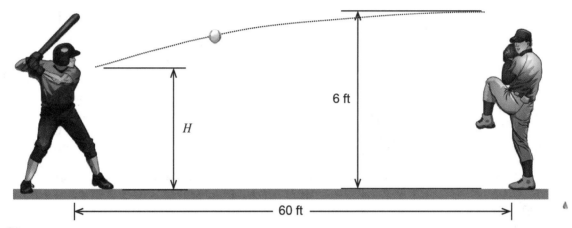

Given:

Find:

Solution:

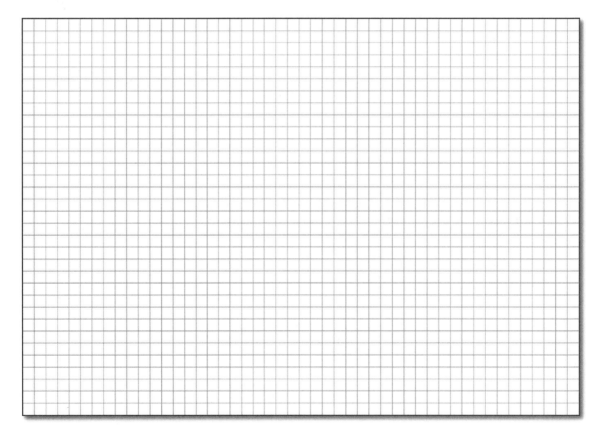

Example Problem 3.2-8

Consider the following projectile motion of the ball shown in the figure. The initial velocity of the ball is v_o = 30 m/s and it is thrown at an angle θ = 40° above the horizontal. Find the maximum height H, the time it takes the particle to reach point C and the total horizontal range L that the particle travels.

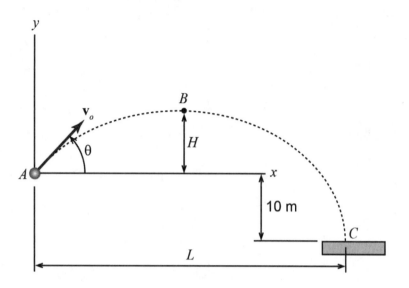

Given:

Find:

Solution:

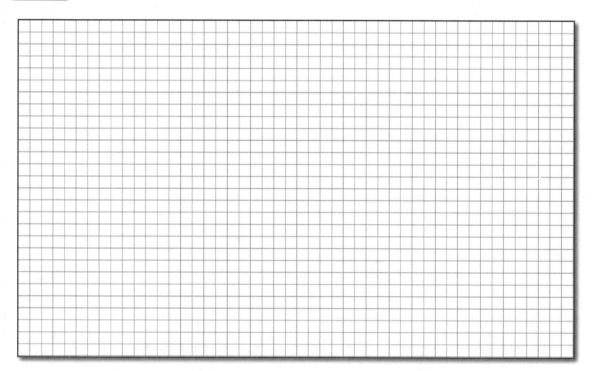

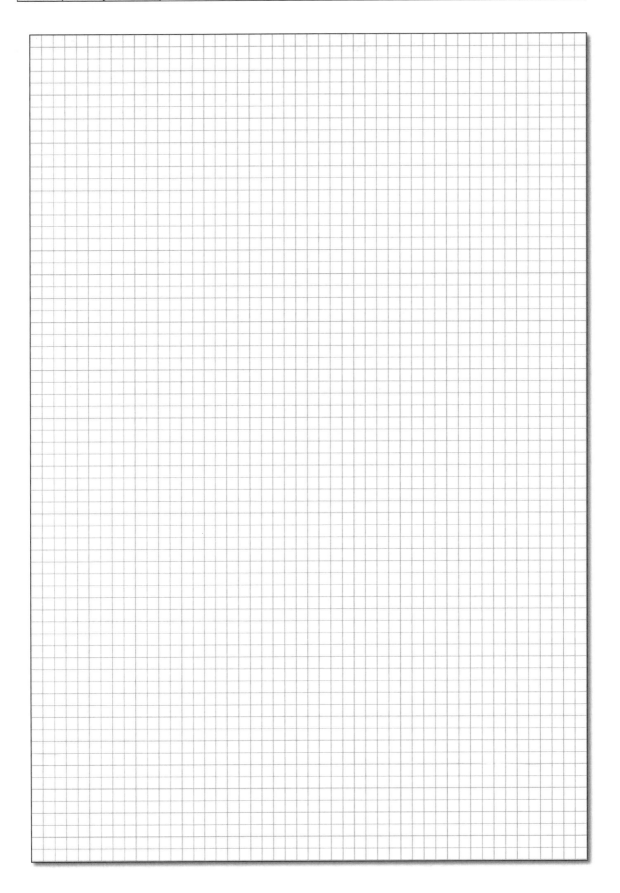

A projectile is launched with an initial velocity of 100 m/s at an angle of θ = 45 degrees with respect to an inclined hill. If the hill makes an angle of ϕ = 30 degrees with respect to the horizontal, how far down the hill D does the projectile land?

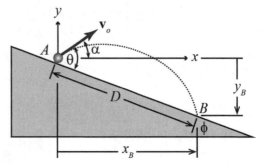

Given: v_o = 100 m/s $\theta = 45°$
          $\phi = 30°$

Find: D

Solution:

Initial velocity

One special property of a projectile problem is directional independence. This means that the x- and y-directions are independent and may be analyzed as two separate rectilinear problems. That being said, the first step when analyzing a projectile problem should be to decompose the initial velocity into its x- and y-components.

If $\theta = 45°$ and $\phi = 30°$, then $\alpha = 15°$.

$$\mathbf{v}_o = v_o(\cos\alpha\,\mathbf{i} + \sin\alpha\,\mathbf{j}) = 96.59\,\mathbf{i} + 25.88\,\mathbf{j}\ \frac{m}{s}$$

Range

We also know that if we neglect air resistance, the x-direction velocity is constant. We can use this fact to determine the range.

$$x = v_x t + x_o \qquad\qquad x_B = v_x t_B$$

The range x_B may be related to D through the geometry of the problem.

$$x_B = D\cos\phi = v_x t_B \qquad\qquad t_B = \frac{D\cos\phi}{v_x} \qquad\qquad (1)$$

Altitude

The acceleration in the y-direction is equal to $-g$.

$$y = -\frac{1}{2}gt^2 + v_{yo}t + y_o \qquad\qquad y_B = -\frac{1}{2}gt_B^2 + v_{yo}t_B$$

The altitude y_B may be related to D through the geometry of the problem.

$$y_B = -D\sin\phi = -\frac{1}{2}gt_B^2 + v_{yo}t_B \qquad\qquad (2)$$

Plugging equation (1) into equation (2) we get the following.

$$-D\sin\phi = -\frac{1}{2}gt_B^2 + v_{yo}t_B \qquad\qquad t_B = \frac{D\cos\phi}{v_x}$$

$$-D\sin\phi = -\frac{1}{2}g\left(\frac{D\cos\phi}{v_x}\right)^2 + v_{yo}\left(\frac{D\cos\phi}{v_x}\right)$$

Solving for D.

$$D = \frac{2}{g}\left(\frac{v_x}{\cos\phi}\right)^2\left[v_{yo}\left(\frac{\cos\phi}{v_x}\right) + \sin\phi\right] \qquad\qquad \boxed{D = 1857 \text{ m}}$$

3.3) PLANE CURVILINEAR MOTION: *n-t* COORDINATES

3.3.1) NORMAL - TANGENTIAL COORDINATE SYSTEM

When the path of a moving particle is known, it is often convenient to describe its motion in terms of its normal and tangential (n-t) coordinates. Consider a Ferris wheel, like that shown in Figure 3.3-1 and the following problem statement: "Find the acceleration of car A given that the Ferris wheel is rotating with constant angular speed ω." Solving this problem is a quick and easy task using the n-t coordinate system and fairly cumbersome using the x-y coordinate system. This is due to the fact that the x- and y-components of the car's acceleration are coupled. The x-y system is suited best for straight-line

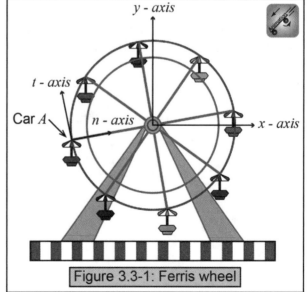

Figure 3.3-1: Ferris wheel

motion or for situations where the x- and y-components of the body's acceleration is independent and can be solved separately. In this problem, the path of the particle (car A) is known and curved making it well suited to n-t coordinates. Also, the origin of the coordinate system moves with the particle (car A) and the axes rotate to match the curvature of the particles path. This is illustrated in the accompanying animation.

Unlike the Cartesian coordinate system which is an inertial coordinate frame, the origin of the n-t coordinate system is attached to and moves with the particle as shown in Figure 3.3-2. This is known as a body-fixed coordinate frame. A body-fixed coordinate system may translate and rotate with the body, whereas, for our purposes, an inertial coordinate system is fixed.

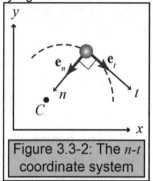

Figure 3.3-2: The n-t coordinate system

The n-t coordinate system has a tangential and normal axis. These axes are represented by the unit direction vectors $\mathbf{e}_t$ and $\mathbf{e}_n$ as shown in Figure 3.3-2. The tangential axis is tangent to the particle's path and positive in the direction of the particle's motion. The normal axis is perpendicular (i.e. normal) to the tangential axis and is positive in the direction of the path's center of curvature (C). Therefore, the direction of $\mathbf{e}_n$ can flip 180 degrees as the particle moves along its path.

Example 3.3-1

Draw the n-t axes for both particles. C denotes the center of curvature.

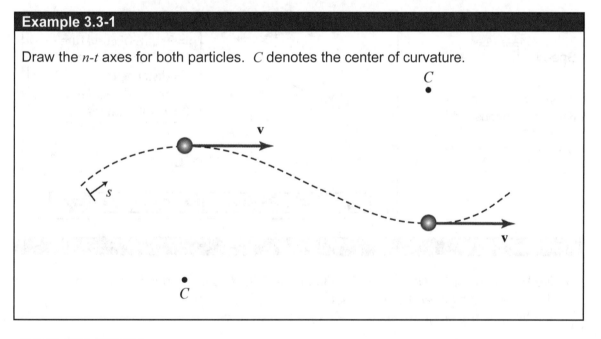

3.3.2) VELOCITY

The n-t coordinate system is used to describe a particle's velocity and acceleration. It does not make sense to define position in a body-fixed coordinate system. This is because a body-fixed coordinate system moves with the particle, therefore, the particle is always at the origin.

Units of Velocity (length/time)
Metric Units:
• meters per second (m/s)
• kilometers per hour (kph = km/h)
US Customary Units:
• feet per second (ft/s)
• miles per hour (mph = mi/h = 1.4667 ft/s)

We discovered earlier in this chapter that a particle's velocity is always tangent to its path of motion. Therefore, in terms of the n-t coordinate system, $\mathbf{v}$ is always in the tangential direction or the $\mathbf{e}_t$ direction as shown in Figure 3.3-2. We know from our previous study that velocity is the time rate of change of position. This means that the magnitude of the velocity is equal to the rate at which the particle's displacement along its path is changing. This gives us an understanding of the direction and magnitude of the velocity in the n-t coordinate system as shown in Equation 3.3-1.

Velocity: $$\mathbf{v} = \frac{ds}{dt}\mathbf{e}_t = v\mathbf{e}_t \quad (3.3\text{-}1)$$

$\mathbf{v}$ = velocity
v = magnitude of $\mathbf{v}$
s = position along the path
t = time

The magnitude of the velocity can also be written in terms of the displacement angle using the arc length equation (*arc length = radius x swept angle*) as shown in Equation 3.3-2. This equation is important when it comes time to derive the equation for acceleration in n-t coordinates. Equation 3.3-2 can be derived using the arc length equation and Figure 3.3-3 as a reference. Note that angular speed is also written as ω (i.e. $\dot{\theta} = \omega$).

Speed: $$\boxed{v = \rho\dot{\theta}} \quad (3.3\text{-}2)$$

v = speed
ρ = instantaneous radius of curvature
$\dot{\theta}$ = angular speed (rad/s)

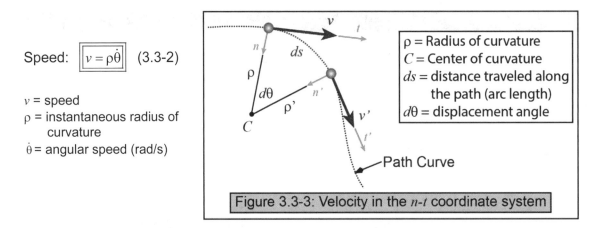

ρ = Radius of curvature
C = Center of curvature
ds = distance traveled along the path (arc length)
$d\theta$ = displacement angle

Figure 3.3-3: Velocity in the n-t coordinate system

Example 3.3-2

Derive Equation 3.3-2. Use Figure 3.3-3 and the following questions to guide you.
- What is the general calculus equation for the velocity?
- What is the length of ds?

3.3.3) ACCELERATION

Acceleration is defined as the time rate of change of velocity, therefore, in order to find the acceleration in n-t coordinates (i.e. path coordinates) we need to take the derivative of the velocity, $\mathbf{a} = d\mathbf{v}/dt$.   This process is different than in the case of rectangular

Units of Acceleration (length/time^2)
Metric Units:
• meters per second squared (m/s^2)
US Customary Units:
• feet per second squared (ft/s^2)

coordinates because the n-t unit direction vectors are not constant. The magnitudes of $\mathbf{e}_t$ and $\mathbf{e}_n$ don't change with time, but their directions do. Therefore, their derivatives are not zero.

Does $d\mathbf{e}_t/dt = 0$ like $d\mathbf{i}/dt = 0$?

The acceleration is the time derivative of velocity (i.e. $\mathbf{a} = d(v\mathbf{e}_t)/dt = \dot{v}\mathbf{e}_t + v\dot{\mathbf{e}}_t$). Substituting a more useful value for $\dot{\mathbf{e}}_t$, we get Equation 3.3-3. The radius of curvature (ρ) in Equation 3.3-3 can be determined from Equation 3.3-4 if the equation of the path curve is known in terms of the x-y coordinate system.

Acceleration: $$\mathbf{a} = a_t\mathbf{e}_t + a_n\mathbf{e}_n = \dot{v}\mathbf{e}_t + \left(\frac{v^2}{\rho}\right)\mathbf{e}_n$$ (3.3-3)

Radius of curvature: $$\rho = \frac{\left[1 + \left(\dfrac{dy}{dx}\right)^2\right]^{3/2}}{\left|\dfrac{d^2y}{dx^2}\right|}$$ (3.3-4)

$\mathbf{a}$ = total acceleration
a_t, a_n = tangential and normal components of the acceleration
    respectively
$\dot{v}$ = time rate of change of speed (i.e. tangential acceleration)

v = speed
ρ = radius of curvature
x = x-coordinate
y = y-coordinate

Equation Derivation

Acceleration in n-t coordinates is a very useful equation. However, it is not very useful when it has derivatives of unit direction vectors in it as shown below.

$$\mathbf{a} = \frac{d\mathbf{v}}{dt} = \frac{d(v\mathbf{e}_t)}{dt} = \dot{v}\mathbf{e}_t + v\dot{\mathbf{e}}_t$$

We need to take the above acceleration equation and transform it into a useful form. This requires that we determine the direction and magnitude of the vector $\dot{\mathbf{e}}_t$. The derivative of the unit direction vector $\dot{\mathbf{e}}_t$ can be understood based on the illustration shown in Figure 3.3-4. The vectors $\mathbf{e}_t$ and $\mathbf{e}_n$ represent the unit directions vectors of the n-t coordinate frame at one instant and the unit direction vectors $\mathbf{e}_t'$ and $\mathbf{e}_n'$ represent the tangential and normal directions an instant later. The difference between the two tangential unit direction vectors is $d\mathbf{e}_t$. If $d\mathbf{e}_t$ is small, a good approximation for its magnitude is the arc length equation. Recall that arc length is equal to the product of the radius and the angle swept through. Which means that for this situation, the magnitude of $d\mathbf{e}_t$ is given by the equation below, where 1 is the length of the unit vector and $d\theta$ is the angle made between the two sets of axes.

$$\left|d\mathbf{e}_t\right| = \left|\mathbf{e}_t\right|d\theta = 1 \cdot d\theta$$

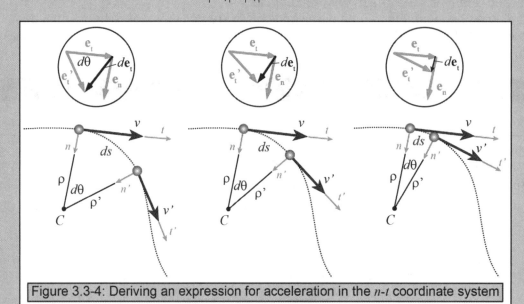

Figure 3.3-4: Deriving an expression for acceleration in the n-t coordinate system

Figure 3.3-4 also indicates that as $d\theta$ becomes small, $d\mathbf{e}_t$ points in the direction of $\mathbf{e}_n$.

$$d\mathbf{e}_t = d\theta\,\mathbf{e}_n\;.$$

Dividing both sides of the above equation by dt we then have the expression.

$$\dot{\mathbf{e}}_t = \dot{\theta}\,\mathbf{e}_n$$

Using above result and the speed equation $v = \rho\dot{\theta}$ (Equation 3.3-2), an expression for the particle's acceleration in n-t coordinates can be obtained and is given below.

$$\mathbf{a} = a_t\mathbf{e}_t + a_n\mathbf{e}_n = \dot{v}\,\mathbf{e}_t + \left(\frac{v^2}{\rho}\right)\mathbf{e}_n$$

It is important to remember that even if the speed is constant, it does not necessarily mean that there is no acceleration.   Velocity is a vector, therefore, the acceleration is due to a change in the velocity's magnitude as well as a change in the velocity's direction. If the speed is constant and the radius of curvature is finite, only the tangential component of the acceleration is zero (see Figure 3.3-5 and Equation 3.3-3). To give physical meaning to the expression for the acceleration given in Equation 3.3-3, imagine that you are driving a car. The tangential acceleration is the acceleration created when you press on the gas pedal and the normal acceleration is the acceleration created when you turn the steering wheel.

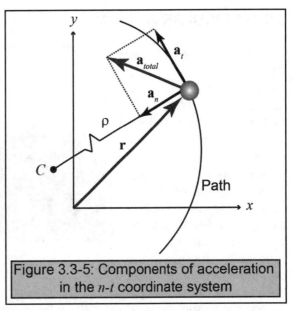

Figure 3.3-5: Components of acceleration in the n-t coordinate system

A particle's acceleration is always nonzero if it is moving on a curve, even if its speed is constant. Why?

What does $\dot{v}$ physically represent in the acceleration equation?
What does v^2/ρ physically represent in the acceleration equation?

$$\mathbf{a} = \dot{v}\,\mathbf{e}_t + \left(\frac{v^2}{\rho}\right)\mathbf{e}_n$$

3.3.4) CIRCULAR MOTION

Consider a car traveling around a circular test track of radius r. The car's location around the track can be defined by its angular position θ as shown in Figure 3.3-6. Many real-life situations, such as the car on a test track, can be described by the circular motion of a particle. Just as in the general curvilinear case, the velocity of the particle is tangent to the circle at

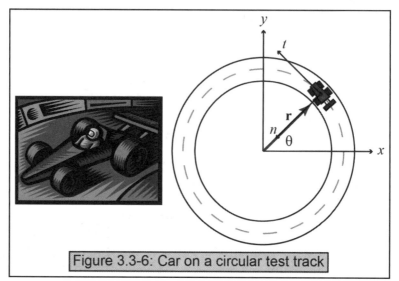

Figure 3.3-6: Car on a circular test track

any instant. Furthermore, the rate at which the particle's speed is changing as it travels around the circle is equal to the tangential component of the particle's acceleration a_t, while the normal component a_n is due to the fact that the direction of the particle's velocity is changing with time.

We will use the car analogy again to illustrate the components of the n-t coordinate acceleration. The tangential acceleration is what you feel when the driver steps on the gas and it seems like you are getting pressed back against the seat. The normal acceleration is the acceleration created when the driver turns the steering wheel and, as a passenger, it feels like you slam into the car door.* Therefore, even

*Technical Note: Actually, the seat pushes you forward and the car door pushes you around the curve.

if the particle was traveling with constant speed, it would still have non-zero acceleration because of the fact that it is traveling in a circle. Furthermore, the normal acceleration always points towards the center of the circle. Equations 3.3-5 through 3.3-8 summarize the kinematic equations for circular motion that you may remember from physics. Note how they can be related to the expressions given in Equations 3.3-2 and 3.3-3. Combing Equations 3.3-5 and 3.3-7, the following alternate representation for the normal component of the acceleration can also be derived.

$$\text{Speed:} \quad \boxed{v = r\dot{\theta}} \quad (3.3\text{-}5)$$

$$\text{Tangential acceleration:} \quad \boxed{a_t = r\ddot{\theta}} \quad (3.3\text{-}6) \qquad \text{Normal acceleration:} \quad \boxed{a_n = \frac{v^2}{r}} \quad (3.3\text{-}7)$$

$$\text{Normal acceleration:} \quad \boxed{a_n = r\dot{\theta}^2} \quad (3.3\text{-}8)$$

a_t, a_n = tangential acceleration and normal
acceleration respectively
v = speed

r = circle radius
$\dot{\theta}$ = angular speed (rad/sec)
$\ddot{\theta}$ = angular acceleration (rad/sec^2)

Conceptual Example 3.3-3

Consider the ski jumper shown in the figure. Neglect surface friction and air resistance. Determine the acceleration direction of the skier at positions A, B and C. Use the arrow diagram to describe the direction.

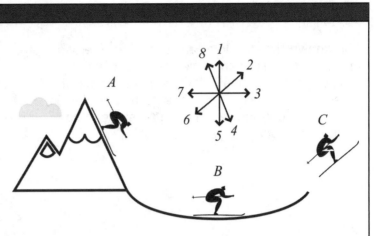

Conceptual Example 3.3-4

Consider a ball in the following situations represented in the figures. For the instant represented in each figure, rank the situations in order from largest to smallest magnitude of the total acceleration.

Largest _____ Next _____ Next _____ Next _____ Next _____ Smallest _____

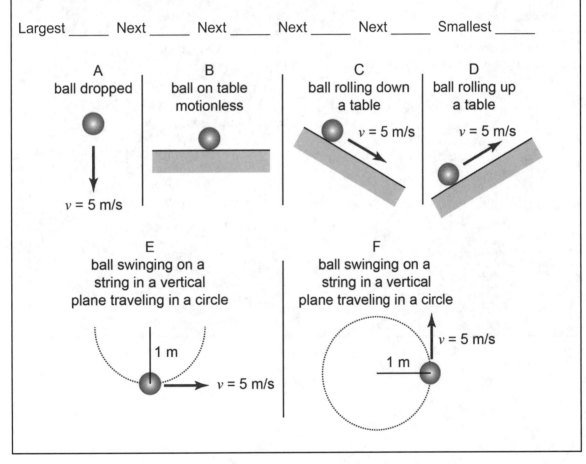

Conceptual Example 3.3-5

It is known that the total amount of acceleration that a racecar tire can achieve before skidding is limited. Which of the following paths should the racecar driver take through a turn in order to maximize the amount of speed that she can carry through the curve?

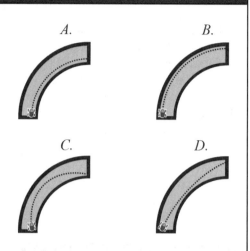

Equation Summary

Abbreviated variable definition list

x = x-coordinate
y = y-coordinate
s = position along the path
ρ = instantaneous radius of curvature
r = circle radius

t = time
v = velocity
a = total acceleration
$\dot{\theta}$ = angular speed (rad/s)
$\ddot{\theta}$ = angular acceleration (rad/sec^2)

Velocity

$$\mathbf{v} = v\,\mathbf{e}_t \qquad v = \frac{ds}{dt} = \rho\dot{\theta} = r\omega$$

Acceleration

$$\mathbf{a} = a_t\mathbf{e}_t + a_n\mathbf{e}_n = \dot{v}\mathbf{e}_t + \left(\frac{v^2}{\rho}\right)\mathbf{e}_n$$

Radius of curvature

$$\rho = \frac{\left[1 + \left(\frac{dy}{dx}\right)^2\right]^{3/2}}{\left|\frac{d^2y}{dx^2}\right|}$$

Circular motion

$$v = r\dot{\theta}$$

$$a_t = r\ddot{\theta} \qquad a_n = \frac{v^2}{r} = r\dot{\theta}^2$$

Example Problem 3.3-6

An automobile's speed increases at a rate of 8 m/s^2 around a curve of radius 300 m. The car's speed at this instant is 35 m/s. What is the car's total acceleration?

<u>Given:</u>

<u>Find:</u>

<u>Solution:</u>

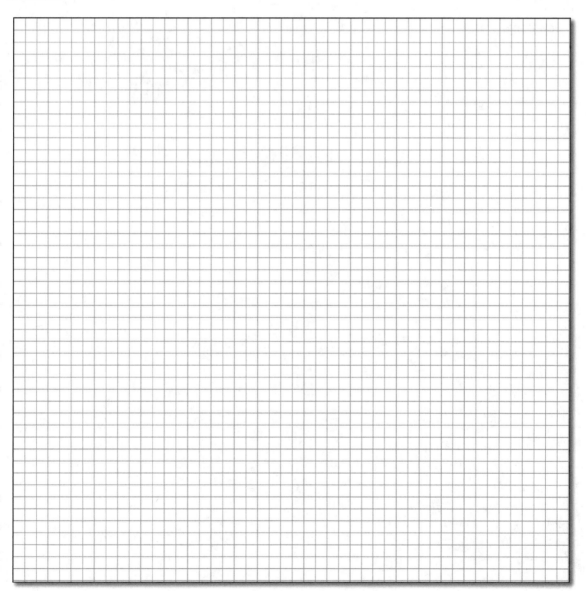

Example Problem 3.3-7

A motorcycle travels in a circular path having a radius of 200 ft at a speed of 45 mph. Starting from its initial speed at a location of $s = 0$, the motorcycle increases its speed according to $\dot{v} = 0.5s$ ft/s^2, where s is in feet. Determine its speed and the magnitude of its acceleration when it has reached $s = 30$ feet.

Given:

Find:

Solution:

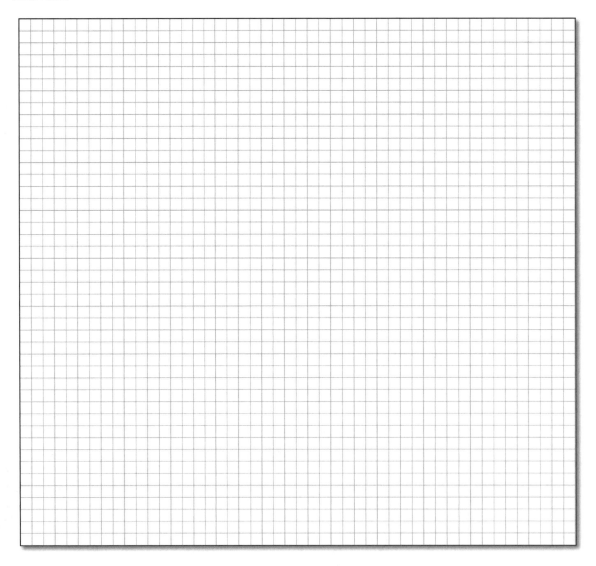

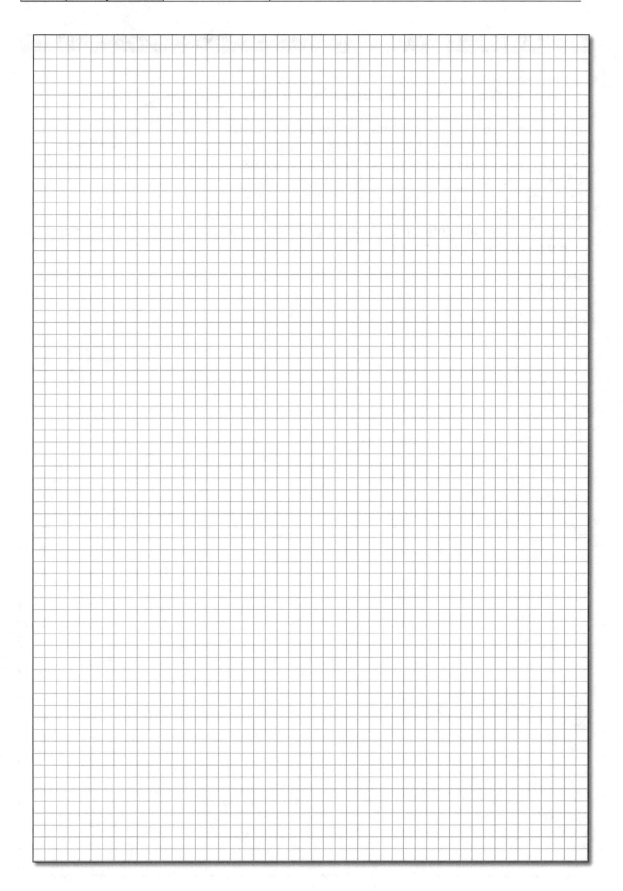

Example Problem 3.3-8

A racecar is initially travelling at 75 mph at point A as it enters the S-curve shown. In order to successfully traverse the curve, the racecar driver applies his brakes and decelerates uniformly between point A and B. Point B is located 750 ft down the track from point A. After point B, the racecar driver accelerates uniformly for 850 ft until he returns the vehicle's speed to 75 mph at point C. If the total acceleration of the

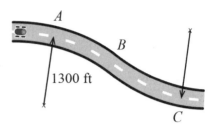

vehicle at point A and point C is 10 ft/s^2, determine (a) the tangential acceleration at point A, (b) the total acceleration and velocity of the car at point B and (c) the radius of curvature of the track at point C.

Given:

Find:

Solution:

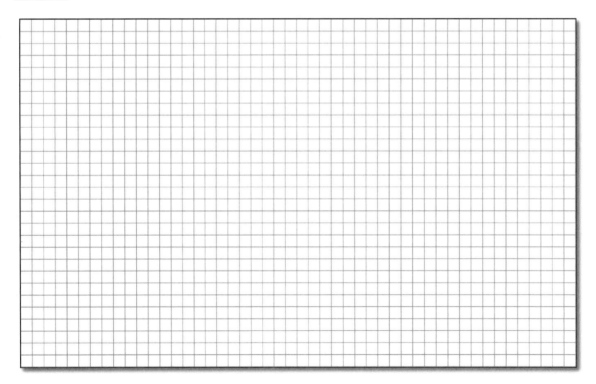

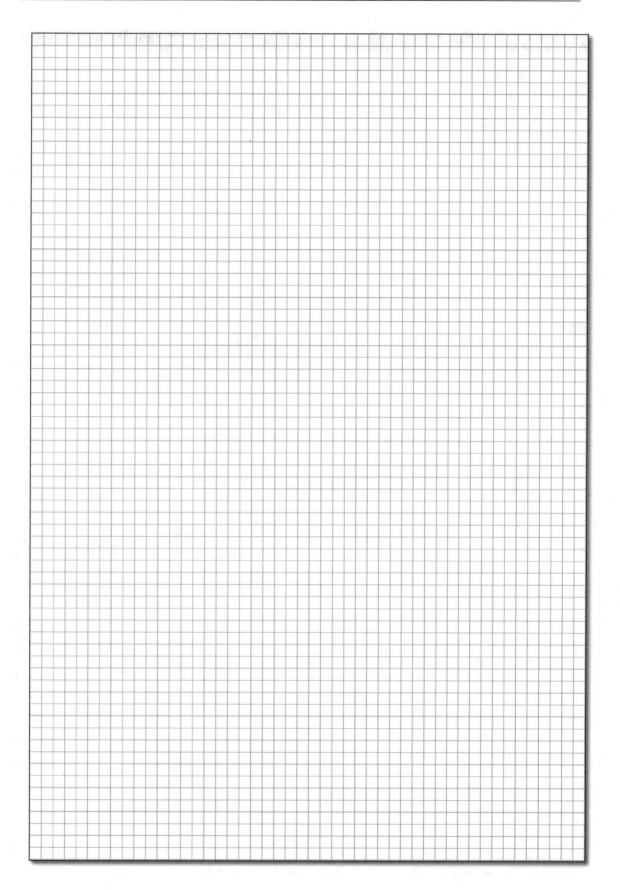

Example Problem 3.3-9

The angular speed of a 15-in diameter circular saw blade is 1500 rpm when the power to the saw is turned off. The speed decreases at a constant rate until the blade comes to rest 15 seconds later. Determine the acceleration of a saw tooth, on the circumference of the saw blade, 5 seconds after the power is turned off.

Given:

Find:

Solution:

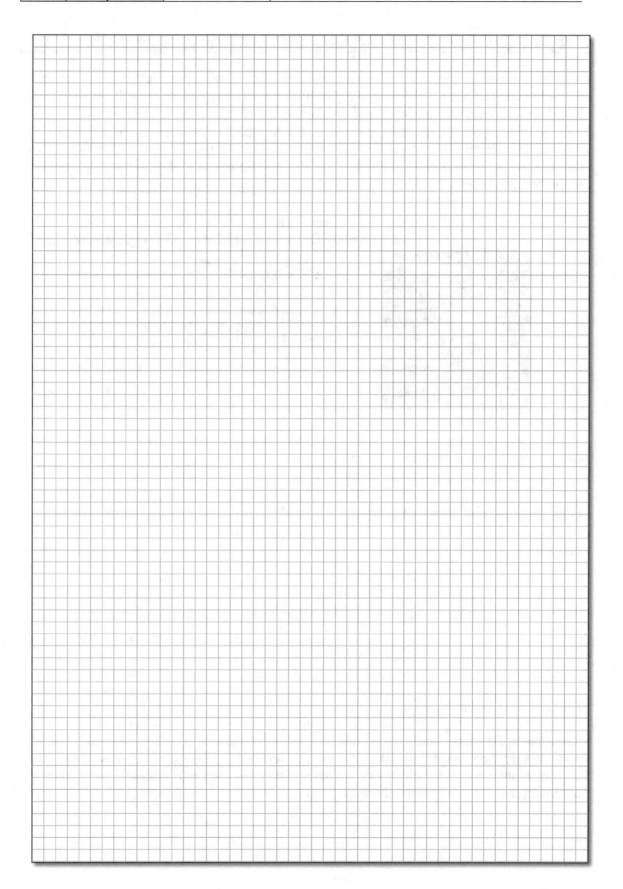

Solved Problem 3.3-10

A train is traveling through the mountains with a constant speed of 60 mph along a curved section of track. At time $t = 0$, the train passes a small town. After passing the town, the track can be described by the function $y = 10\ln(x/100)$ with the town representing a point on the track defined by $x = 0.00001$ ft. Determine the magnitude of the acceleration of the front of the train at the instant it reaches points along the track defined by $x = 500$ ft and $x = 1500$ ft. Explain the differences in the normal components of the two accelerations.

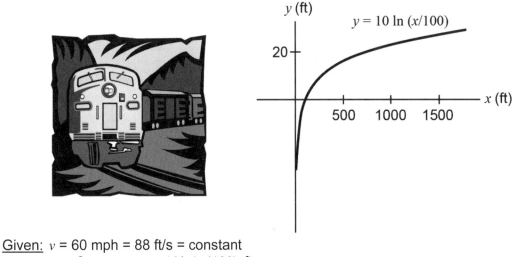

Given: $v = 60$ mph = 88 ft/s = constant
$\dot{v} = 0$ $\qquad$ $y = 10\ln(x/100)$ ft

Find: $\mathbf{a}$ at $x = 500$ ft and 1500 ft

Solution:

Acceleration

We need to find the total acceleration of the train given by the following equation.

$$\mathbf{a} = \dot{v}\,\mathbf{e}_t + \frac{v^2}{\rho}\,\mathbf{e}_n$$

Radius of curvature

Since the speed is constant, we know that $\dot{v}$ is equal to zero. The only unknown that we need to find, before we can calculate the acceleration, is the radius of curvature. The path of motion is given in terms of x and y. Therefore, we can calculate the radius of curvature using the following equation.

$$\rho = \frac{\left[1+\left(\dfrac{dy}{dx}\right)^2\right]^{3/2}}{\left|\dfrac{d^2y}{dx^2}\right|}$$

$$y = 10\ln(x/100) = 10\left(\ln(x) - \ln(100)\right) \qquad \frac{dy}{dx} = \frac{10}{x} = 10x^{-1} \qquad \frac{d^2y}{dx^2} = -10x^{-2} = -\frac{10}{x^2}$$

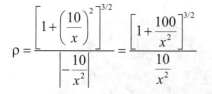

$$\rho = \frac{\left[1+\left(\dfrac{10}{x}\right)^2\right]^{3/2}}{\left|-\dfrac{10}{x^2}\right|} = \frac{\left[1+\dfrac{100}{x^2}\right]^{3/2}}{\dfrac{10}{x^2}}$$

$$\rho(x = 500) = 25015 \text{ ft} \qquad\qquad \rho(x = 1500) = 225015 \text{ ft}$$

Now we have all the information we need to calculate the total acceleration of the train.

$$\mathbf{a} = \cancel{\dot{v}}\,\mathbf{e}_t + \frac{v^2}{\rho}\mathbf{e}_n = \frac{v^2}{\rho}\mathbf{e}_n$$

$$\boxed{\mathbf{a}(x = 500) = 0.31\,\mathbf{e}_n\ \frac{\text{ft}}{\text{s}^2}} \qquad\qquad \boxed{\mathbf{a}(x = 1500) = 0.034\,\mathbf{e}_n\ \frac{\text{ft}}{\text{s}^2}}$$

Solved Problem 3.3-11

A ball, constrained in a cat toy, rolls in a circular path. At a position of $\theta = 20°$, the ball has an angular velocity of 2π rad/s and an angular acceleration of -1.4 rad/s^2. If the radius of the cat toy is 6.5 inches, determine the x- and y-components of the velocity and acceleration of the ball at $\theta = 20°$.

Given: $\theta = 20°$ $\qquad$ $\dot{\theta} = 2\pi$ rad/s

$\qquad$ $\ddot{\theta}$ = -1.4 rad/s^2 $\qquad$ $r = 6.5$ in

Find: $\mathbf{v}$, $\mathbf{a}$

Solution:

Get familiar with the system

Let's get familiar with the problem. We know that the velocity is tangent to the path curve. The ball is also slowing down based on the angular acceleration value. We also know that if a particle travels in a curved path, there will be a normal component of the acceleration. Let's draw the angular velocity, angular acceleration, velocity and acceleration directions on the figure. Note that the acceleration direction will just be an estimation at this point.

Velocity

The ball moves in a circular path, therefore, we can use the circular motion equations.

$$v = r\dot{\theta} = 40.84 \frac{\text{in}}{\text{s}}$$

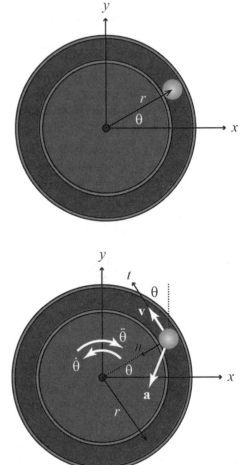

We know that the velocity is in the t-direction. Therefore, we can decompose $\mathbf{e}_t$ into the x- and y-directions.

$$\mathbf{v} = v\mathbf{e}_t = v(-\sin\theta\,\mathbf{i} + \cos\theta\,\mathbf{j}) = 40.84(-\sin 20\,\mathbf{i} + \cos 20\,\mathbf{j})$$

$$\boxed{\mathbf{v} = -13.97\,\mathbf{i} + 38.38\,\mathbf{j}\,\frac{\text{in}}{\text{s}}}$$

Acceleration

Similarly, we can decompose $\mathbf{e}_n$ in to the x- and y-directions to get the x-y acceleration.

$$\mathbf{a} = r\ddot{\theta}\mathbf{e}_t + r\dot{\theta}^2\mathbf{e}_n = r\ddot{\theta}(-\sin\theta\,\mathbf{i} + \cos\theta\,\mathbf{j}) + r\dot{\theta}^2(-\cos\theta\,\mathbf{i} - \sin\theta\,\mathbf{j})$$

$$= (-r\ddot{\theta}\sin\theta - r\dot{\theta}^2\cos\theta)\mathbf{i} + (r\ddot{\theta}\cos\theta - r\dot{\theta}^2\sin\theta)\mathbf{j}$$

$$= 6.5(-1.4\sin 20 - (2\pi)^2\cos 20)\mathbf{i} + 6.5(-1.4\cos 20 - (2\pi)^2\sin 20)\mathbf{j}$$

$$\boxed{\mathbf{a} = -238.02\,\mathbf{i} - 96.32\,\mathbf{j}\ \frac{\text{in}}{\text{s}}}$$

3.4) PLANE CURVILINEAR MOTION: POLAR COORDINATES

3.4.1) POLAR COORDINATE SYSTEM

A third coordinate system that can be useful for solving various dynamics problems is the *polar coordinate system*. Polar coordinates are useful for situations where the particle's motion is described with respect to an inertial reference frame. The **polar coordinate system** is a body-fixed coordinate frame that is attached to and moves with the particle. The particle is located using a radial coordinate (r) and a transverse coordinate (θ). These coordinates are defined with respect to a reference inertial coordinate system which is usually the x-y coordinate frame.

3.4.1.1) Radial coordinate & the r-axis

The radial coordinate (r) measures the straight-line distance between the origin of the inertial coordinate frame and the particle, as shown in Figure 3.4-1. You can think of the origin of the inertial coordinate frame as a central point and r as a radius. The path of the particle may or may not be moving in a circular path, but thinking of the radial coordinate as a radius originating at the origin of the inertial frame is an easy way to remember it.

If you draw a line from the origin of the inertial coordinate system to the particle, this is the positive direction of the r-axis. The r-axis starts at the particle and points away from the origin. As the particle moves, the direction of the r-axis will change because the line

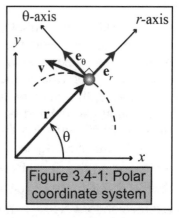

Figure 3.4-1: Polar coordinate system

connecting the origin to the particle changes. The unit direction vector $\mathbf{e}_r$ is used to define the direction of the r-axis.

3.4.1.2) Transverse coordinate & the θ-axis

The transverse coordinate (θ) measures the angle between a fixed axis (the positive x-axis in this case) and the positive r-axis as shown in Figure 3.4-1. The θ-axis is perpendicular to the r-axis and is positive in the direction of increasing θ. The unit direction vector $\mathbf{e}_\theta$ is used to define the positive direction of the θ-axis.

Example 3.4-1

Draw the r-θ axes and the radial and transverse coordinates for both positions of the particle shown.

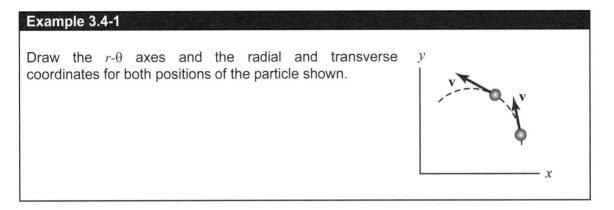

3.4.2) POSITION

The position (**r**) of a particle in the polar coordinate frame is represented by a vector that starts at the origin of the inertial reference frame and points to the particle as shown in Figure 3.4-1 and given in Equation 3.4-1. This also defines the direction of the r-axis.

Units of Position (length)
Metric Units:
• meter (m)
• centimeter (cm = 0.01 m)
• millimeter (mm = 0.001 m)
• kilometer (km = 1000 m)
US Customary Units:
• feet (ft = 12 in)
• inch (in)
• mile (mi = 5280 ft)

$$\text{Position:} \quad \boxed{\mathbf{r} = r\mathbf{e}_r} \quad (3.4\text{-}1)$$

r = position vector
r = radial coordinate (i.e. magnitude of the position vector)

3.4.3) VELOCITY

Taking the derivative of the position vector (Equation 3.4-1) will result in an expression for velocity $\mathbf{v} = d\mathbf{r}/dt$. As was the case with n-t coordinates, the unit direction vectors $\mathbf{e}_r$ and $\mathbf{e}_\theta$ move and their derivatives are, in general, not zero. Therefore, the

Units of Velocity (length/time)
Metric Units:
• meters per second (m/s)
• kilometers per hour (kph = km/h)
US Customary Units:
• feet per second (ft/s)
• miles per hour (mph = mi/h = 1.4667 ft/s)

product rule must again be applied, $\dot{\mathbf{r}} = \dot{r}\mathbf{e}_r + r\dot{\mathbf{e}}_r$. Using similar logic to that employed while finding the derivatives of the unit direction vectors in the n-t coordinate frame, we find $\dot{\mathbf{e}}_\theta = -\dot{\theta}\mathbf{e}_r$ and $\dot{\mathbf{e}}_r = \dot{\theta}\mathbf{e}_\theta$. Using these expressions for the unit vector derivatives, an expression for velocity can be obtained and is given in Equation 3.4-2.

$$\text{Velocity:} \quad \boxed{\mathbf{v} = \dot{r}\mathbf{e}_r + r\dot{\theta}\mathbf{e}_\theta} \quad (3.4\text{-}2)$$

v = velocity
r = radial coordinate
$\dot{r}$ = radial coordinate velocity
$\dot{\theta}$ = angular velocity (rad/s)

In Equation 3.4-2, the r-component of the velocity (v_r) is due to the changing radius while the θ-component (v_θ) is due to the changing angle. If the particle is traveling in a circle there will only be a component of velocity in the θ-direction. This is because the radius does not change magnitude and its derivative with respect to time is zero (i.e. $\dot{r} = 0$).

3.4.4) ACCELERATION

Taking the derivative of the velocity (Equation 3.4-2) leads to an expression for acceleration $(\mathbf{a} = \ddot{r}\mathbf{e}_r + \dot{r}\dot{\mathbf{e}}_r + \dot{r}\dot{\theta}\mathbf{e}_\theta + r\ddot{\theta}\mathbf{e}_\theta + r\dot{\theta}\dot{\mathbf{e}}_\theta)$. Employing the equations for the derivatives of the unit direction vectors ($\dot{\mathbf{e}}_\theta = -\dot{\theta}\mathbf{e}_r$, $\dot{\mathbf{e}}_r = \dot{\theta}\mathbf{e}_\theta$),

> **Units of Acceleration** (length/time^2)
> Metric Units:
> - meters per second squared (m/s^2)
>
> US Customary Units:
> - feet per second squared (ft/s^2)

an expression for acceleration in polar coordinates is obtained (Equation 3.4-3).

Acceleration: $$\boxed{\mathbf{a} = (\ddot{r} - r\dot{\theta}^2)\mathbf{e}_r + (r\ddot{\theta} + 2\dot{r}\dot{\theta})\mathbf{e}_\theta}$$ (3.4-3)

$\mathbf{a}$ = acceleration
r = radial coordinate
$\dot{r}$ = radial coordinate velocity

$\ddot{r}$ = radial coordinate acceleration
$\dot{\theta}$ = angular velocity (rad/s)
$\ddot{\theta}$ = angular acceleration (rad/s^2)

There is one term in Equation 3.4-3 that merits discussion. The term $2\dot{r}\dot{\theta}$ is sometimes referred to as the Coriolis acceleration and arises only when considering rotating reference frames, as we have here. If you have ever attempted to walk outward on a spinning platform such as a merry-go-round, you have perceived the Coriolis acceleration. As you walk outward on the platform, walking is difficult. You feel like you are tipping over and maybe feel a little nauseous.

If you consider the situation where a particle is traveling in a circle or on a path of constant radius, the equations for velocity and acceleration reduce to Equations 3.4-4 and 3.4-5. Note that these equations are derived by realizing that the radial coordinate is constant and $\dot{r} = \ddot{r} = 0$. Note that the polar and n-t coordinate axes are parallel in the case of circular motion, but not in general. The n-t versions of these equations will be discussed in more detail in the "Kinematics of Rigid Bodies" chapter.

Velocity for circular motion: $$\boxed{\mathbf{v} = r\dot{\theta}\mathbf{e}_\theta}$$ (3.4-4)

Acceleration for circular motion: $$\boxed{\mathbf{a} = -r\dot{\theta}^2\,\mathbf{e}_r + r\ddot{\theta}\mathbf{e}_\theta}$$ (3.4-5)

$\mathbf{v}$ = velocity for circular motion
$\mathbf{a}$ = acceleration for circular motion
r = radial coordinate

$\dot{\theta}$ = angular velocity (rad/s)
$\ddot{\theta}$ = angular acceleration (rad/s^2)

Conceptual Example 3.4-2

For the following situations, write an expression for the velocity and acceleration in polar coordinates.

a) A car travelling on a circular track.

b) A car on a Ferris wheel turning at a constant rate.

c) Body A held by the end of an articulating robotic arm.

Conceptual Example 3.4-3

Consider the following mechanism consisting of a slotted bar rotating about the fixed point O that is driving pin A which is constrained to move along the horizontal guide shown. Note that θ is the angle made by the slotted bar with a horizontal reference and r is the distance from the fixed point O to pin A. If the slotted bar is rotating counter-clockwise at a decreasing rate, what are the signs of the first two time derivatives of θ and r?

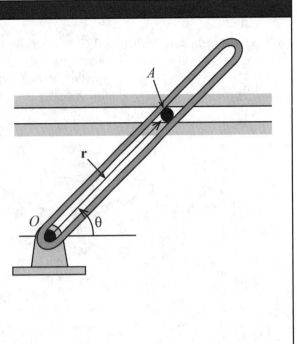

a) $\dot\theta, \ddot\theta$ positive and $\dot r, \ddot r$ negative

b) $\dot\theta, \dot r$ positive and $\ddot\theta, \ddot r$ negative

c) $\dot\theta, \ddot r$ positive and $\ddot\theta, \dot r$ negative

d) $\ddot\theta, \dot r$ positive and $\dot\theta, \ddot r$ negative

3.4.5) POLAR COORDINATES VERSUS N-T COORDINATES

It is important to note that the r-θ and n-t coordinate systems do not necessarily coincide. One important distinction is that in the n-t coordinate system, the velocity has only a t-component. In the polar coordinate system, the velocity generally has both an r- and θ-component.

Example 3.4-4

On the figure, draw the r-θ and n-t axes. Also, include the radial position vector, the velocity vector and the transverse coordinate.

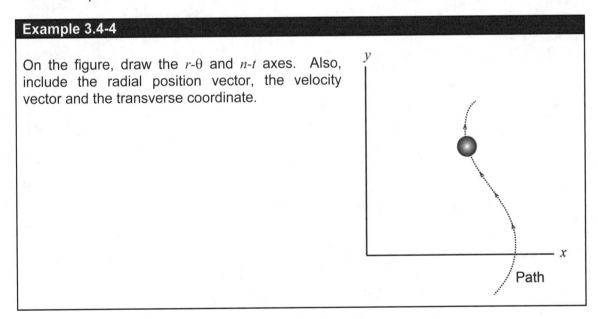

Equation Summary

<u>Abbreviated variable definition list</u>

$\mathbf{r}$ = position vector
r = radial coordinate
$\dot{r}$ = radial coordinate velocity
$\ddot{r}$ = radial coordinate acceleration
θ = transverse coordinate

$\dot{\theta}$ = angular velocity (rad/s)
$\ddot{\theta}$ = angular acceleration (rad/s^2)
$\mathbf{v}$ = velocity
$\mathbf{a}$ = acceleration

<u>Position</u>

$$\mathbf{r} = r\,\mathbf{e}_r$$

<u>Velocity</u>

$$\mathbf{v} = \dot{r}\,\mathbf{e}_r + r\dot{\theta}\,\mathbf{e}_\theta$$

<u>Acceleration</u>

$$\mathbf{a} = (\ddot{r} - r\dot{\theta}^2)\,\mathbf{e}_r + (r\ddot{\theta} + 2\dot{r}\dot{\theta})\,\mathbf{e}_\theta$$

<u>Circular motion</u>

$$\mathbf{v} = r\dot{\theta}\,\mathbf{e}_\theta$$

$$\mathbf{a} = -r\dot{\theta}^2\,\mathbf{e}_r + r\ddot{\theta}\,\mathbf{e}_\theta$$

Example Problem 3.4-5

A robotic arm extends along a path $r = (1 + 0.5\cos\theta)$ m.   At $\theta = \pi/4$, $\dot{\theta} = 0.5$ rad/s and $\ddot{\theta} = 0.7$ rad/s^2, find the velocity and acceleration of point A.

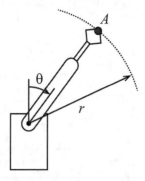

Given:

Find:

Solution:

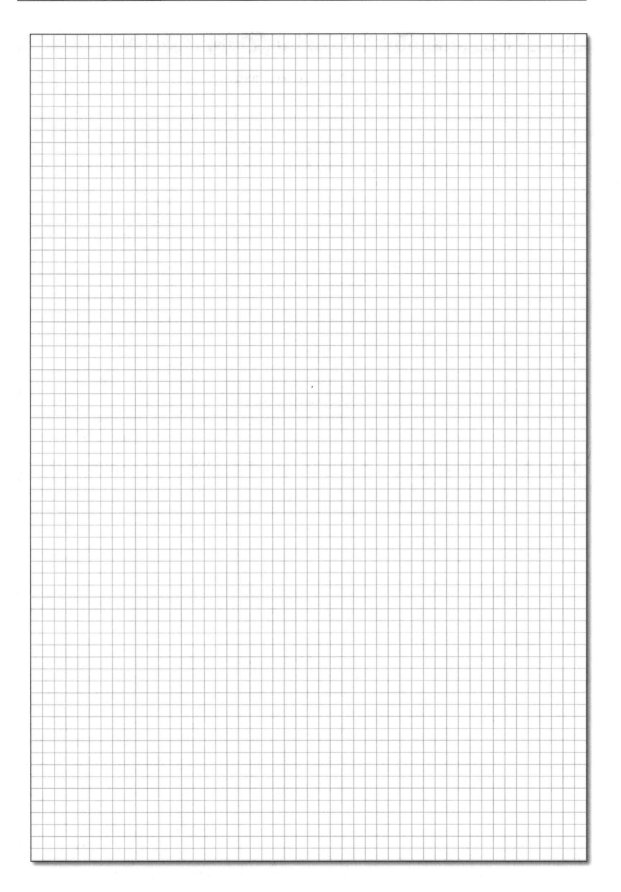

Example Problem 3.4-6

A slotted bar pinned at point O rotates with an angle described by the function $\theta(t) = 5t$ where θ is in radians and t is in seconds. As the bar rotates, it forces pin A to travel along the vertical guide shown.   At the point when $t = 0.2$ seconds determine (a) the magnitude of the velocity of pin A, and (b) the magnitude of the acceleration of pin A.

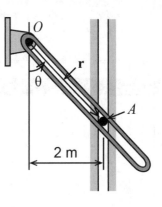

Given:

Find:

Solution:

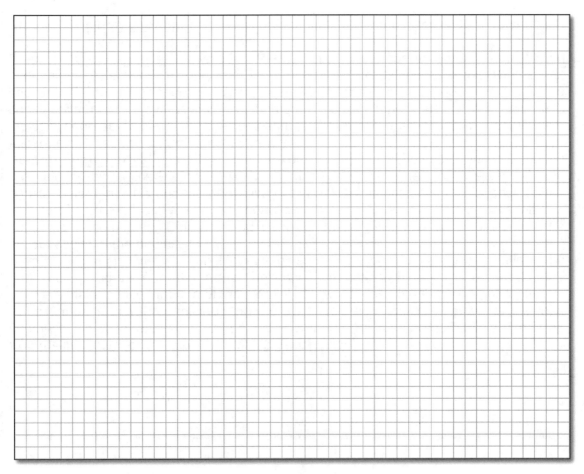

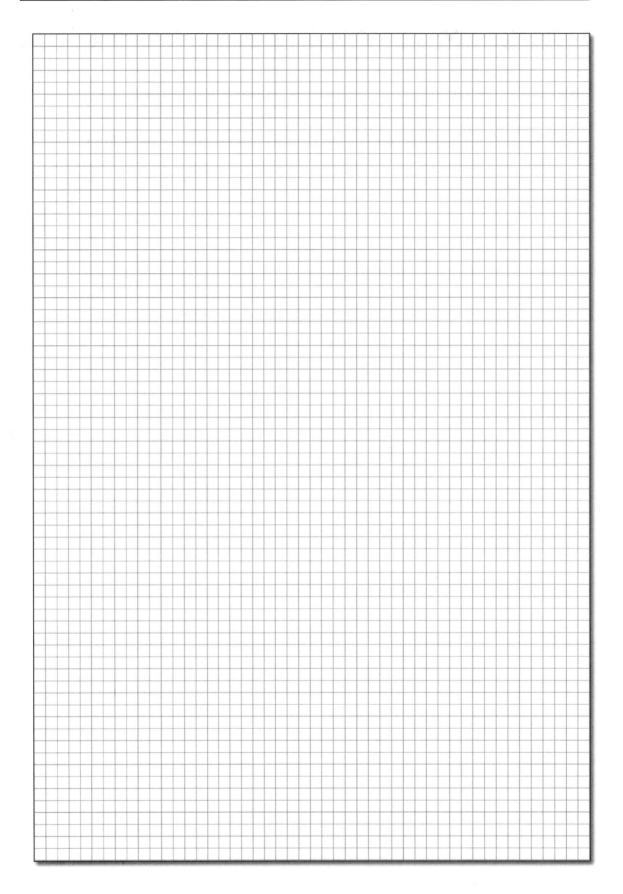

As a train travels on a track around a curve, the radial position of the train is described by $r = (900 / \theta)$ m, where θ is in radians. The radial position is relative to a fixed reference frame. If the train maintains a constant speed v = 15 m/s, determine the train's velocity when θ = 60°.

Given: $r = \dfrac{900}{\theta}$ m v = 15 m/s = constant

θ = 60°

Find: $\mathbf{v}$

Solution:

Getting familiar with the problem

The problem statement indicates that r is dependent on θ and, therefore, many of the other motion parameters will be dependent on θ. Also, r is relative to a fixed reference frame. This is a good indication that polar coordinates would be well suited to use in this problem.

Velocity

We need to find the velocity of the train. Our first step should be to write down the polar coordinate velocity equation. This will give us a good idea of what we need to find before the velocity can be determined.

$$\mathbf{v} = \dot{r}\mathbf{e}_r + r\dot{\theta}\mathbf{e}_\theta$$

The parameters that we need to calculate are $\dot{r}$ and $\dot{\theta}$. We can determine $\dot{r}$ by taking the derivative of r with respect to time, however, we do not have an equation for θ. We will need to find another way to determine $\dot{\theta}$.

Radial position

$$r = \frac{900}{\theta} = 900\,\theta^{-1}$$

When calculating the radial position, we need to make sure that θ is in radians.

θ = 60° = $\pi/3$ rad

$$r = \frac{900}{\pi / 3} = 859.437 \text{ m}$$

Since θ depends on time, we need to employ the chain rule to perform the time derivative on r.

$$\dot{r} = 900(-1)\theta^{-2}\dot{\theta} = \frac{900}{\theta^2}\dot{\theta} = -\frac{900}{(\pi/3)^2}\dot{\theta} = -820.7\,\dot{\theta}$$

Angular velocity

We will find the $\dot{\theta}$ using the information we have about the speed of the train.

$$\mathbf{v} = \dot{r}\mathbf{e}_r + r\dot{\theta}\mathbf{e}_\theta$$

$$v = \sqrt{(\dot{r})^2 + (r\dot{\theta})^2} \qquad\qquad 15 = \sqrt{(-820.7\,\dot{\theta})^2 + (859.437\,\dot{\theta})^2}$$

$$\dot{\theta} = 0.013\,\frac{\text{rad}}{\text{s}}$$

Velocity

$$\mathbf{v} = \dot{r}\mathbf{e}_r + r\dot{\theta}\mathbf{e}_\theta \qquad\qquad \boxed{\mathbf{v} = -10.7\mathbf{e}_r + 11.2\mathbf{e}_\theta\ \frac{\text{m}}{\text{s}}}$$

3.5) CHOOSING A COORDINATE SYSTEM

In this chapter we have introduced three different types of coordinate frames that may be employed for modeling and answering questions about dynamic systems. The choice of which coordinate frame to use can greatly affect the ease with which you are able to arrive at a solution. Equations 3.5-1 through 3.5-9 summarize the expressions for position, velocity and acceleration in each of the three coordinate frames. The paragraph preceding each set of equations gives some guidelines to help you choose the best reference frame for your particular situation.

3.5.1) RECTANGULAR COORDINATES

Rectangular coordinates are useful for straight-line motion. When using rectangular coordinates, it can be useful to tilt the x-y coordinate frame so that one of the axes is oriented in the direction of motion. Rectangular coordinates are also useful when the x- and y-components of a body's acceleration are independent (e.g. projectile motion).

Position: $\boxed{\mathbf{r} = x\mathbf{i} + y\mathbf{j}}$ (3.5-1)   Velocity: $\boxed{\mathbf{v} = \dot{x}\mathbf{i} + \dot{y}\mathbf{j}}$ (3.5-2)

Acceleration: $\boxed{\mathbf{a} = \ddot{x}\mathbf{i} + \ddot{y}\mathbf{j}}$ (3.5-3)

$\mathbf{r}$ = position
$\mathbf{v}$ = velocity
$\mathbf{a}$ = acceleration
x, y = x-coordinate position and y-coordinate position respectively

$\dot{x}$, $\dot{y}$ = x-coordinate velocity and y-coordinate velocity respectively
$\ddot{x}$, $\ddot{y}$ = x-coordinate acceleration and y-coordinate acceleration respectively

3.5.2) NORMAL AND TANGENTIAL COORDINATES

Normal and tangential coordinates are helpful when a particle's path is curved and the particle's motion is described along its path. Information, such as the path's radius of curvature or the particle's speed and acceleration along the path, may indicate that the problem is best analyzed with n-t coordinates. If a particle's path is given as a function $y = f(x)$, then its radius of curvature can be calculated by Equation 3.5-6.

Velocity: $\boxed{\mathbf{v} = v\,\mathbf{e}_t}$ (3.5-4)   Acceleration: $\boxed{\mathbf{a} = \dot{v}\mathbf{e}_t + \left(\dfrac{v^2}{\rho}\right)\mathbf{e}_n}$ (3.5-5)

Radius of curvature: $\boxed{\rho = \dfrac{\left[1 + \left(\dfrac{dy}{dx}\right)^2\right]^{3/2}}{\left|\dfrac{d^2 y}{dx^2}\right|}}$ (3.5-6)

$\mathbf{v}$ = velocity
v = speed
$\mathbf{a}$ = total acceleration

$\dot{v}$ = tangential acceleration
ρ = radius of curvature

3.5.1) POLAR COORDINATES

As with n-t coordinates, the polar coordinate system is useful when a particle's path is known. The difference is that polar coordinates are most useful when the particle's motion is described with respect to an inertial reference frame. This means that some or all of the information about the particle's distance from the inertial reference frame (r) as well as the rotation of the polar coordinate frame with respect to the inertial reference frame (θ) are given. One common example is circular motion. In this case, the motion is described with respect to the circle's center. It is also useful when angular velocities ($\dot{\theta}$) and accelerations ($\ddot{\theta}$) are given.

Position: $\boxed{\mathbf{r} = r\mathbf{e}_r}$ (3.5-7)   Velocity: $\boxed{\mathbf{v} = \dot{r}\mathbf{e}_r + r\dot{\theta}\mathbf{e}_\theta}$ (3.5-8)

Acceleration: $\boxed{\mathbf{a} = \left(\ddot{r} - r\dot{\theta}^2\right)\mathbf{e}_r + \left(r\ddot{\theta} + 2\dot{r}\dot{\theta}\right)\mathbf{e}_\theta}$ (3.5-9)

$\mathbf{a}$ = total acceleration	$\dot{r}$ = radial velocity
$\mathbf{v}$ = total velocity	$\ddot{r}$ = radial acceleration
$\mathbf{r}$ = position	$\dot{\theta}$ = angular velocity
r = radial position along the r-axis	$\ddot{\theta}$ = angular acceleration

Conceptual Example 3.5-1

The rocket shown in the figure below travels vertically upward with velocity $\mathbf{v}$. The radar is located a known distance D from the rocket's launch site and tracks the rocket's flight. If the angular velocity $\dot{\theta}$ and angular acceleration $\ddot{\theta}$ of the radar dish are known, which coordinate system would you employ if you wished to find the rocket's velocity $\mathbf{v}$ when the rocket has reached an altitude of h?

    a) Rectangular coordinates
    b) Normal-tangential coordinates
    c) Polar coordinates

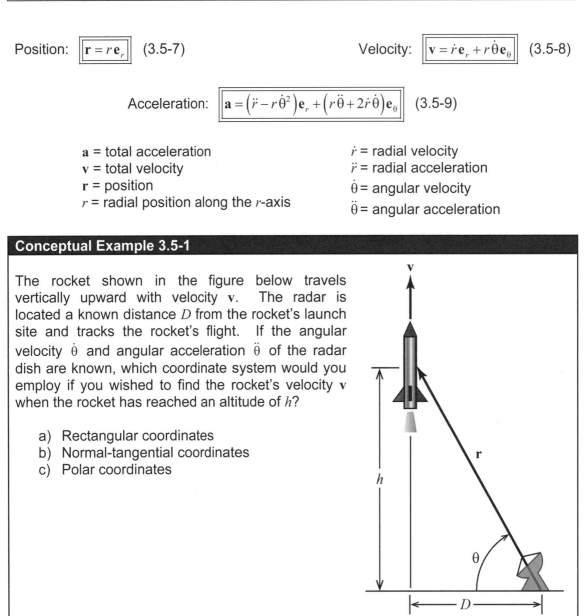

Conceptual Example 3.5-2

The tank shown in the figure fires a shell with initial velocity $\mathbf{v}$ at an angle θ with the horizontal from a height h. If the horizontal distance D traveled by the shell is known, which coordinate system would you employ if you wished to find the shell's initial velocity $\mathbf{v}$ and firing angle θ?

    a) Rectangular coordinates
    b) Normal-tangential coordinates
    c) Polar coordinates

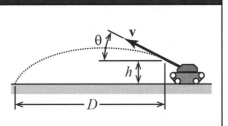

Conceptual Example 3.5-3

Which coordinate system would be best suited in solving the following problems?

- Rectangular coordinates
- Normal-tangential coordinates
- Polar coordinates

A car is traveling around a circular track and ...

a) we are given the rate of increasing speed, the radius of the track and the car's speed. We want to find the total acceleration.

b) we are given the rate of increasing angular velocity, the radius of the track and the car's speed. We want to find the total acceleration.

c) we are given the radius of the track and the car's speed as a function of time. We want to find the total acceleration.

3.6) RELATIVE MOTION

Imagine that you are standing on the side of a freeway watching cars go by. You see a very cool sports car passing a small hybrid car as shown in Figure 3.6-1. The sports car is going around 100 mph and the hybrid is going around 60 mph. The 100 and 60 mph speeds stated are considered absolute speeds because they are relative to some fixed non-moving object. In this case, you standing on the side of the freeway. But, what does the speed of the sports car appear to be relative to a passenger in the hybrid? The hybrid already has some speed, so the sports car appears to be going slower to a passenger in the hybrid than it does to you because you are not moving. To a passenger in the hybrid the sports car appears to be going (100-60) mph or 40 mph. In other words, the sports car is going 40 mph relative to the moving hybrid.

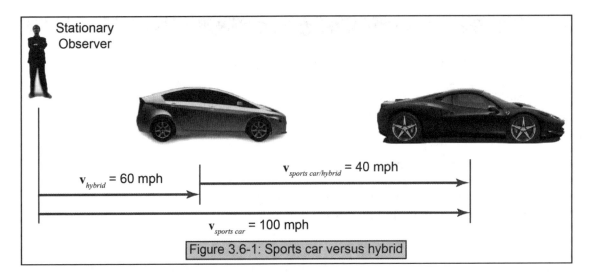

Figure 3.6-1: Sports car versus hybrid

3.6.1) REFERENCE FRAME DEFINITIONS

So far, we have introduced reference frames that are either inertially fixed or that are attached to and move with the particle of interest. When analyzing relative motion, we will be using both inertial and body-fixed coordinate systems simultaneously. Therefore, this is a good place to state a more rigorous definition of inertial and body-fixed coordinate systems.

An **inertial coordinate system** is a reference frame in which Newton's laws hold true. We have not learned about Newton's laws yet, therefore, you can think of an inertial coordinate system as one where the origin is not accelerating and the axes are not rotating. The orientation of the axes is arbitrary, apart from the fact that they must be orthogonal and not accelerating or rotating. In this book, for the most part, the inertial frames

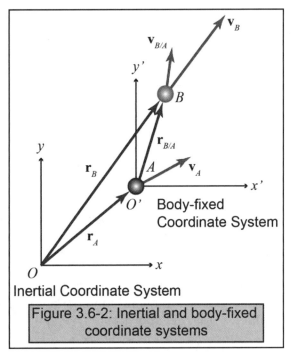

Figure 3.6-2: Inertial and body-fixed coordinate systems

that we will be dealing with will be non-moving, though this is not a requirement.

A **non-inertial coordinate system** is one that is accelerating and/or rotating. Newton's laws do not hold true in this type of coordinate system unless terms are added to the equations. For example, imagine that you and a friend are in a car that is sharply rounding a curve. If you do not have your seatbelt on but your friend does, you will appear to our friend to mysteriously slide away from him. In reality, you are actually sitting still and it is the car that is moving underneath you. The fictitious force in the car reference frame is called the centrifugal force. Your acceleration in the car reference frame is the normal acceleration described in the section on n-t coordinates (i.e. v^2/ρ) or the radial acceleration described in the section on r-θ coordinates (i.e. $\dot{r}\dot{\theta}^2$).

A **body-fixed coordinate system** has its origin fixed to a body that may or may not be moving. We are mostly interested in situations where the coordinate system is fixed to a moving body. The motion of a body-fixed coordinate system and the body that it is fixed to, can be described with respect to an inertial coordinate system. The motion of any other body in the space may be described in terms of the inertial coordinate system or the body-fixed coordinate system (see Figure 3.6-2).

3.6.2) ABSOLUTE AND RELATIVE MOTION

Position, velocity or acceleration are termed **absolute** if they are relative to a fixed coordinate system. They are termed **relative** if they refer back to something that is usually not stationary, such as, another object or a body-fixed coordinate system.

Consider two moving particles A and B as shown in Figure 3.6-3. The absolute position of particle A relative to the inertial coordinate frame x-y is described by the vector $\mathbf{r}_A$ and the absolute position of particle B is described by the vector $\mathbf{r}_B$. What if we wanted to measure the position of particle A relative to particle B? Particle B would be considered a moving origin. We would write this position vector as $\mathbf{r}_{A/B}$. This indicates the position of particle A relative to the translating reference frame whose origin corresponds to particle B. Mathematically, the absolute and relative positions can be related through vector addition as given in Equation 3.6-1.

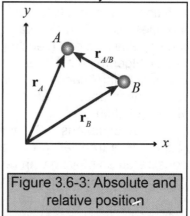

Figure 3.6-3: Absolute and relative position

What is the difference between absolute position and relative position?

Taking the first and second derivative of Equation 3.6-1 we can obtain expressions for the relationships between the absolute and relative velocities and accelerations as given in Equations 3.6-2 and 3.6-3.

Relative Position: $\boxed{\mathbf{r}_A = \mathbf{r}_B + \mathbf{r}_{A/B}}$ (3.6-1)   Relative Velocity: $\boxed{\mathbf{v}_A = \mathbf{v}_B + \mathbf{v}_{A/B}}$ (3.6-2)

Relative Acceleration: $\boxed{\mathbf{a}_A = \mathbf{a}_B + \mathbf{a}_{A/B}}$ (3.6-3)

$\mathbf{r}_A$ = absolute position of particle A
$\mathbf{r}_B$ = absolute position of particle B
$\mathbf{r}_{A/B}$ = position of particle A relative to B
$\mathbf{v}_A$ = absolute velocity of particle A
$\mathbf{v}_B$ = absolute velocity of particle B

$\mathbf{v}_{A/B}$ = velocity of A relative to B
$\mathbf{a}_A$ = absolute acceleration of particle A
$\mathbf{a}_B$ = absolute acceleration of particle B
$\mathbf{a}_{A/B}$ = acceleration of A relative to B

Each of the above equations is essentially two equations, one for each coordinate direction. These can be solved algebraically by expressing the vectors in terms of their components. Furthermore, any reference frame can be chosen (e.g rectangular, *n-t*, polar) but be careful that you are consistent and only add or subtract components that are in the same direction. For example, we are all well aware that $4\mathbf{i} + 7\mathbf{j} \neq 11$.

The equation which relates the absolute and relative positions of two particles may be solved through the use of geometry. Equation 3.6-3 may be obtained through vector addition (see Figure 3.6-3) where vector $\mathbf{r}_{A/B}$ is equal to the sum of vectors $\mathbf{r}_A$ and $\mathbf{r}_B$. Drawing a vector triangle is very helpful when relating the vectors through vector addition. When employing any geometrical relations, be aware that the resulting triangles will not, in general, be right triangles. The law of sines and the law of cosines are especially useful (see Appendix B). Even if you choose to solve the equations using vector algebra rather than geometry, it is useful to draw a rough approximation of the corresponding triangles in order to double check your calculations. The velocity and acceleration equations (Equation 3.6-2 and 3.6-3) equations may also be derived using vector addition.

In Equations 3.6-1 through 3.6-3, particle A was considered to be moving with respect to a translating reference frame with its origin at B. Similarly, point B could be considered as moving with respect to a translating reference frame with its origin at A. In this case, the equations can be re-written as follows.

$$\mathbf{r}_B = \mathbf{r}_A + \mathbf{r}_{B/A}$$

$$\mathbf{v}_B = \mathbf{v}_A + \mathbf{v}_{B/A}$$

$$\mathbf{a}_B = \mathbf{a}_A + \mathbf{a}_{B/A}$$

Comparing the two different sets of equations it becomes apparent that $\mathbf{r}_{A/B} = -\mathbf{r}_{B/A}$, $\mathbf{v}_{A/B} = -\mathbf{v}_{B/A}$ and $\mathbf{a}_{A/B} = -\mathbf{a}_{B/A}$.

Equation Summary

Abbreviated variable definition list

$\mathbf{r}_A$ = absolute position of particle A
$\mathbf{r}_B$ = absolute position of particle B
$\mathbf{r}_{A/B}$ = position of particle A relative to B
$\mathbf{v}_A$ = absolute velocity of particle A
$\mathbf{v}_B$ = absolute velocity of particle B

$\mathbf{v}_{A/B}$ = velocity of A relative to B
$\mathbf{a}_A$ = absolute acceleration of particle A
$\mathbf{a}_B$ = absolute acceleration of particle B
$\mathbf{a}_{A/B}$ = acceleration of A relative to B

Position

$$\mathbf{r}_A = \mathbf{r}_B + \mathbf{r}_{A/B}$$

Acceleration

$$\mathbf{a}_A = \mathbf{a}_B + \mathbf{a}_{A/B}$$

Velocity

$$\mathbf{v}_A = \mathbf{v}_B + \mathbf{v}_{A/B}$$

Conceptual Example 3.6-1

Consider a train moving along a track. The speed of the train is given in the figures. A person is walking inside the train in the direction shown with the speed shown relative to the train. A stationary observer watches the walking person. In which situation does the person walking inside the train appear to the stationary observer to be walking the fastest? In which situation does the walker appear to be slowest?

Fastest _____
Next _____
Next _____
Slowest _____

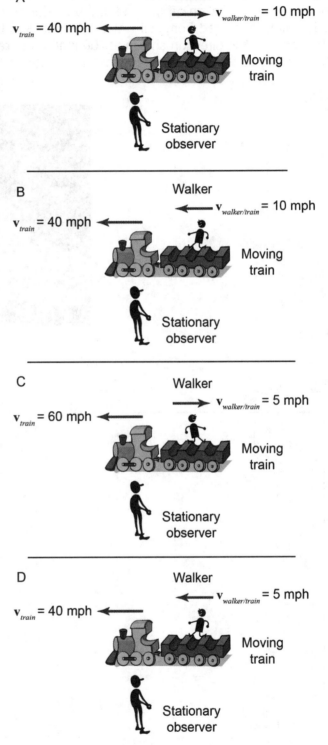

Example Problem 3.6-2

An aircraft carrier is traveling forward with a velocity of 60 km/h. At the instant shown, plane A has just taken off the end of the carrier and has a forward horizontal velocity of 250 km/h, measured from still water. If plane B is traveling along a different runway of the carrier in the direction shown at 200 km/h, with respect to the carrier deck, determine the velocity of plane A as seen by the pilot of plane B.

Given:

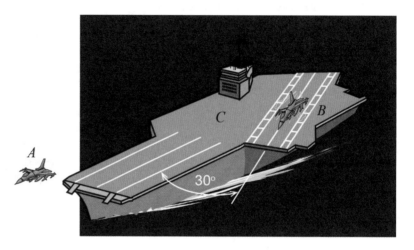

Find:

Solution:

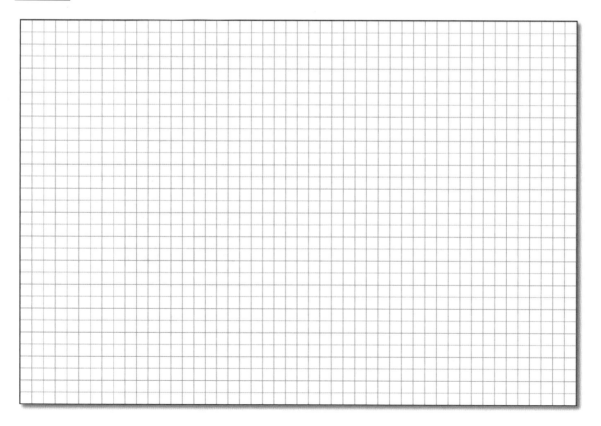

Example Problem 3.6-3

The passenger plane A is flying east with a velocity of 620 km/h when a smaller aircraft B passes underneath heading south-east in the direction shown. To passengers in plane A, however, plane B appears to be flying sideways and moving due south. Determine the actual velocity of plane B and the velocity which plane B appears to have relative to passengers in plane A.

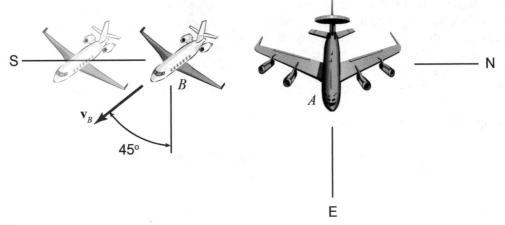

Given:

Find:

Solution:

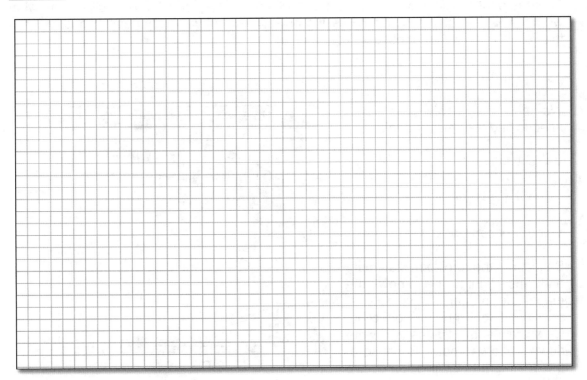

Solved Problem 3.6-4

Car B merges onto a freeway beside car A as shown in the figure. At state 1, car A is traveling at a constant 70 mph and car B is traveling at 40 mph where $\phi = 45°$ as shown in the figure. Car B increases its speed at a constant rate and travels 500 ft in 6 seconds at which point he is traveling parallel to car A at position 2. The radius of curvature of the on ramp is 1000 ft and it has a 30% grade at state 1 and a 0% grade at position 2. Determine the velocity that car B appears to be traveling to a passenger in car A in both states.

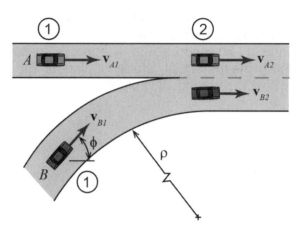

Given: $v_A = 70$ mph = constant
    $\Delta s = 500$ ft
    $\rho = 1000$ ft
    $\theta_1 = 4\%$ grade

$v_{B1} = 40$ mph = 58.67 ft/s
$\Delta t = 6$ s
$\phi = 45°$
$\theta_2 = 0\%$ grade

Find: $\mathbf{v}_{B/A}$ at state 1 and 2

Solution:

Getting familiar with the problem

There is a lot of information given in this problem. It may be difficult to determine what we need to do first. Let's start by asking ourselves what we really need to find which is the velocity of car B with respect to car A ($\mathbf{v}_{B/A}$).

$$\mathbf{v}_{B/A} = \mathbf{v}_B - \mathbf{v}_A$$

We know the velocity of car A but need to find the velocity of car B.

Velocity of car A

$$\mathbf{v}_A = v_A\mathbf{i} = 70\mathbf{i} \text{ mph}$$

Velocity of car B at state 1

Based on the problem statement and the figure, we know the magnitude of car B's velocity at position 1 but not the direction. Let's draw some figures and write down what we know. Because of the road grade at position 1, we

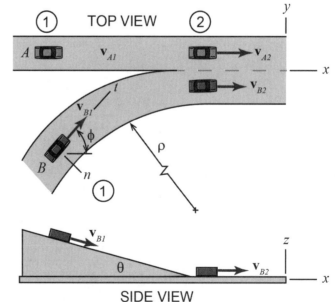

need to project car B's velocity onto the x-y plane in order to determine the x- and y-components of the velocity.

$$v_{B1,x-y} = v_{B1} \cos \theta$$

The grade of the onramp is converted into degrees as follows.

$$\theta = \tan^{-1} 0.3 = 16.7^{\circ}$$

Now we can decompose car B's velocity onto all the axes.

$$\begin{aligned}
\mathbf{v}_{B1} &= v_{B1} \cos \theta (\cos \phi \mathbf{i} + \sin \phi \mathbf{j}) - v_{B1} \sin \theta \mathbf{k} \\
&= 40 \cos 16.7 (\cos 45 \mathbf{i} + \sin 45 \mathbf{j}) - 40 \sin 16.7 \mathbf{k} \\
&= 27.1 \mathbf{i} + 27.1 \mathbf{j} - 11.5 \mathbf{k} \text{ mph}
\end{aligned}$$

Velocity of car B at position 2

We know the direction of car B's velocity at position 2 but not the magnitude.

$$\mathbf{v}_{B2} = v_{B2} \mathbf{i}$$

The problem statement says that car B accelerates uniformly, therefore, we can apply the constant acceleration equations in the t-direction.

$$s = \frac{1}{2} a_o (t - t_o)^2 + v_o (t - t_o) + s_o$$

$$\Delta s = \frac{1}{2} a \Delta t^2 + v_o \Delta t \qquad 500 = \frac{1}{2} a(6)^2 + 58.67(6) \qquad a = 8.22 \frac{\text{ft}}{\text{s}^2}$$

$$v = a_o (t - t_o) + v_o \qquad v_{B2} = 8.22(6) + 58.67 = 108 \frac{\text{ft}}{\text{s}} = 73.63 \text{ mph}$$

Relative velocity

Now that we know all the absolute velocities, we can calculate the relative velocity.

$$\mathbf{v}_{B/A} = \mathbf{v}_B - \mathbf{v}_A$$

$$\mathbf{v}_A = 70 \mathbf{i} \text{ mph}$$

$$\mathbf{v}_{B2} = 73.63 \mathbf{i} \text{ mph} \qquad \mathbf{v}_{B1} = 27.1 \mathbf{i} + 27.1 \mathbf{j} - 11.5 \mathbf{k} \text{ mph}$$

$$\mathbf{v}_{B/A,1} = 27.1 \mathbf{i} + 27.1 \mathbf{j} - 11.5 \mathbf{k} - 70 \mathbf{i} \qquad \boxed{\mathbf{v}_{B/A,1} = -42.9 \mathbf{i} + 27.1 \mathbf{j} - 11.5 \mathbf{k} \text{ mph}}$$

$$\mathbf{v}_{B/A,2} = 73.63 \mathbf{i} - 70 \mathbf{i} \qquad \boxed{\mathbf{v}_{B/A,2} = 3.63 \mathbf{i}}$$

3.7) CONSTRAINED AND DEPENDENT MOTION

3.7.1) CONSTRAINED AND DEPENDENT MOTION

What is constrained and dependent motion? Let's use the example of a train driving down a track. The train is *constrained* to move along the track. It must move in the direction of the track and cannot make a turn unless the track makes a turn. This is constrained motion. Now imagine that a train car is attached to the engine by a flexible coupling. Both the engine and the train car are constrained to move in the direction of the track, but in addition to this, the train car is *dependent* on the motion of the engine. This means that if the engine moves, the train car will move. They may not move with the exact same velocity because of the flexible coupling, but the motion of the train car is related to the motion of the engine.

Constrained motion is when a particle is constrained to move in a particular direction. Constraints provide information about the geometry of the motion. **Dependent motion**, on the other hand, is when points on two separate bodies are related to one another in some way. Many mechanisms that you will see in real life have a combination of constrained and dependent motion. Knowledge about the constraints/dependencies of the motion of one of the interconnected bodies can be used to help find unknown information.

Complex systems can often be modeled as simple mechanisms or connecting elements. Examples of some types of mechanisms that constrain or connect bodies are ropes and pulleys, linear bearings, joints and surface contacts. We will give examples of how to analyze systems that include these types of elements in this section.

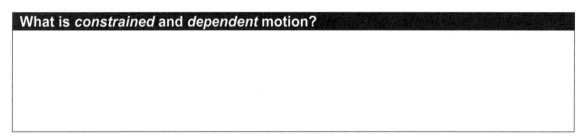

What is *constrained* and *dependent* motion?

3.7.2) ROPES AND PULLEYS

One of the most commonly analyzed dependent-motion situations involves ropes and pulleys. An example of a rope and pulley system is shown in Figure 3.7-1. This seemingly simple problem teaches many important concepts about analyzing dependent motion. Figure 3.7-1 depicts two bodies connected through a system of pulleys and inextensible ropes. From simple observation it is apparent that the velocities and accelerations of the bodies will be related. If A moves, B must move, but the more interesting question is, "If A move 1 m/s, how fast will B move?"

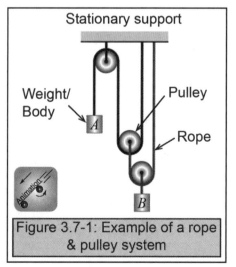

Figure 3.7-1: Example of a rope & pulley system

In Figure 3.7-1, *A* moves. Will *B* move? Why?

In the Figure 3.7-1 animation, did *B* move faster or slower than *A*? Why?

3.7.3) SOLVING ROPE AND PULLEY PROBLEMS

In the animation associated with Figure 3.7-1, we saw that body *A* moved faster than body *B*, but we don't know by how much. A systematic approach can be used to determine the motion relationships between the connected bodies. The approach laid out in this section is tailored to rope and pulley problems, but many of the concepts can be translated to other systems.

Step 1) Choose datums: Choose a datum line which is a fixed reference usually attached to something that does not move. It will be used as an origin from which distances are measured. Identify one datum for every direction of motion.

Step 2) Position coordinates: Assign coordinates to the distances from each datum to the respective moving particles.

Step 3) Rope lengths: Write down the length of each rope in terms of the position coordinates. Sections of rope that are not included in the position coordinates (i.e. sections of the rope that do not change lengths), can be lumped together into a single constant term. For systems where there are multiple ropes, the systems will have multiple degrees of freedom, one for each rope. In these situations, there will be a separate expression for each length of rope.

Step 4) Time derivatives: Take the time derivative of each length equation to obtain the velocity and acceleration equations. Since the overall length of each cable is constant, its derivative and the derivative of the unknown constants with respect to time are zero. Differentiating a length equation once gives a relationship between the velocities of the bodies. Differentiating it a second time gives a relationship between the accelerations of the bodies.

Step 5) Solve and verify: Substitute the given information into the motion relationships from Step 4 and solve for the desired unknown information. Finally, make sure that the answers make sense in terms of the signs and magnitudes. In terms of sign, a positive velocity means that the particle is moving away from its respective datum (or the rope segment is getting longer), while a negative velocity means that the particle is moving towards its datum (or the rope segment is getting shorter).

Example 3.7-1

Use the steps described above to set up the following dependent motion problem.

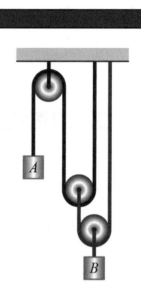

Step 1) Choose your datum(s).
Step 2) Draw your position coordinates.
Step 3) Write the length equation(s) / position coordinate equation(s).

Step 4) Differentiate the position coordinate equation with respect to time to determine the velocity and acceleration equations.

Step 5) Verify to see if the solution makes sense.

Example 3.7-2

Use the steps described to set up the following dependent-motion problem.

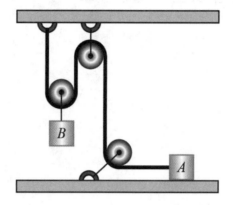

Step 1) Choose your datum(s).
Step 2) Draw your position coordinates.
Step 3) Write the length equation(s) / position coordinate equation(s).

Step 4) Differentiate the position coordinate equation with respect to time to determine the velocity and acceleration equations.

Step 5) Verify to see if the solution makes sense.

Example Problem 3.7-3

If the end of rope A is pulled down with a speed of 1 m/s, determine the speed at which block B rises.

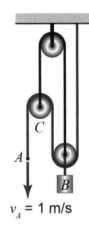

v_A = 1 m/s

Given:

Find:

Solution:

3.7.3) LINEAR BEARINGS / COLLARS

Linear bearings and collars are sleeves that fit around or on shafts as shown in Figure 3.7-3.   The bearing or collar is constrained to move along the direction of the shaft, therefore, so is the end of any link or other pinned mechanism attached to the collar or bearing.

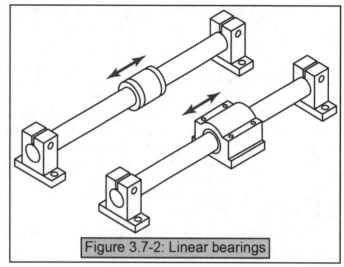

Figure 3.7-2: Linear bearings

3.7.4) SLOTS

A slot constrains the motion of a pin to move in the direction of the slot. The velocity of the pin is always tangent to the path of the slot as shown in Figure 3.7-3.   An example of constrained slot motion is "slot cars." The cars run in a slot around the car track (see Figure 3.7-4). The cars are constrained to move along the direction of the track.

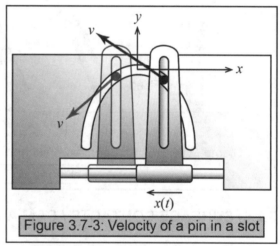

Figure 3.7-3: Velocity of a pin in a slot

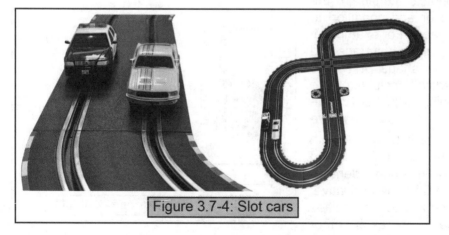

Figure 3.7-4: Slot cars

3.7.5) JOINTS

Joints can restrict the motion of mechanisms that they are attached to from 1 to 6 degrees of freedom. In total, a rigid body has 6 degrees of freedom or 6 ways in which it can move. A particle has 3 degrees of freedom or 3 ways in which it can move. A particle can translate, for example, in the x-, y- and z-directions. Along with the translational degrees of freedom, a rigid body can also rotate about each of these three axes adding 3 rotational degrees of freedom.

There are four main types of joints: a clamped or welded joint, a pinned or hinged joint, a ball and socket joint and a roller joint. A clamped joint restricts all 3 directions of translation and all 3 directions of rotation. This means that the end condition attached to a clamped joint does not move. A pinned joint restricts all 3 directions of translation at the joint and allows for free rotation in one direction about the joint. A ball and socket restricts all 3 directions of translation and allows for free rotation in all three directions about the joint. The last type of joint is a roller joint. This joint may allow up to 3 directions of translational motion and all directions of rotation. All types of joints are illustrated in Figure 3.7-5.

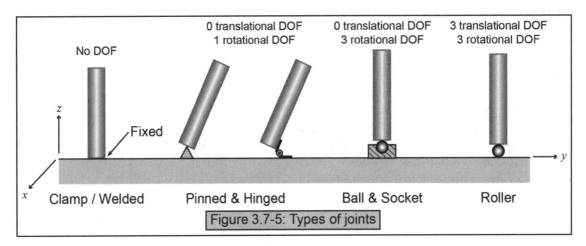

Figure 3.7-5: Types of joints

3.7.6) SURFACE CONTACTS / CAM AND FOLLOWER

Surface contacts constrain the motion of a particle or end condition to the profile of the surface. For example, if a roller joint is rolling along a surface and the surface is in the shape of a sine wave, then the vertical position of the end of the mechanism attached to the roller joint would be $y = A\sin(nx)$, where $n = 1/\lambda$ is given in radians per length (see Figure 3.7-6). The x- and y-directions of motion are dependent on each other and if we take the derivative, we will find that the x- and y-velocities are also dependent on each other (i.e. $\dot{y} = \dot{x}An\cos(nx)$).

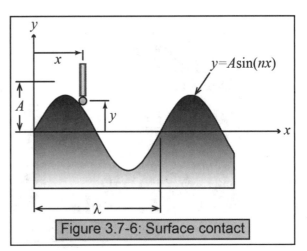

Figure 3.7-6: Surface contact

Cams and followers work in a similar way. The follower rides on the cam surface and is constrained to move according to the profile of the cam as shown in Figure 3.7-7.

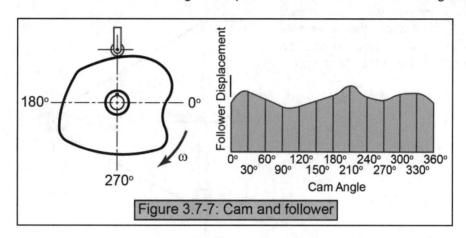

Figure 3.7-7: Cam and follower

Example Problem 3.7-4

If collar A moves with a constant speed v_A, determine the speed of collar B as a function of the position and velocity of A and the distance h.

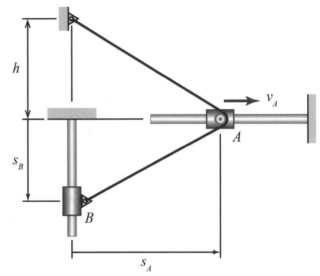

Given:

Find:

Solution:

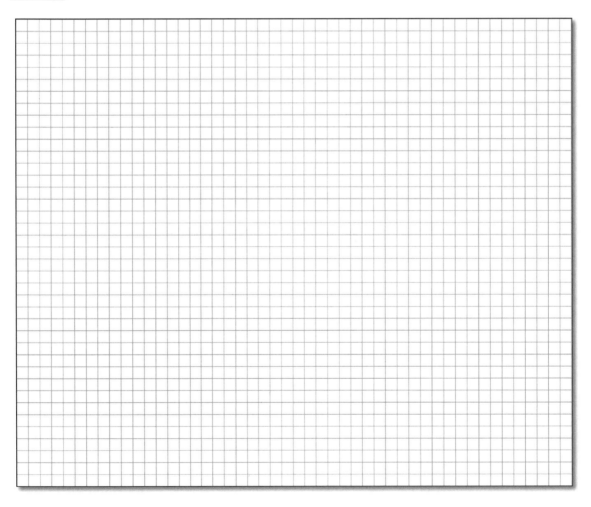

Collars A and B slide along a vertical rod while connected via a cord that passes over a fixed pulley that has a radius of 5 cm. If collar B is sliding downward at 4 cm/s at the instant shown, determine the corresponding velocity of collar A. The distance between the pulley and the rod (D) is 30 cm.

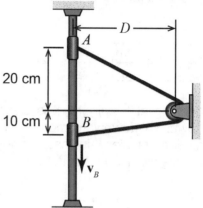

Given: v_B = 4 cm/s s_B = 10 cm
$\qquad$ s_A = 20 cm D = 30 cm
$\qquad$ r = 5 cm

Find: s_A

Solution:

Step 1) Identify a datum

The datum should be attached to a non-moving entity. In this case, the fixed pulley is a good choice.

Step 2) Add position coordinates

The position coordinates (s_A and s_B) measure the distance, from the datum, to every moving particle. In this case, the two collars.

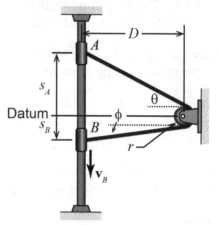

Step 3) Rope length

We need to determine the length of the rope in terms of the position coordinates.

$$L = \sqrt{(s_A - r)^2 + D^2} + \sqrt{(s_B - r)^2 + D^2} + \text{constant}$$

The constant term is the length of the rope that wraps around the pulley. This never changes.

Step 4) Time derivatives

To determine the velocity of collar A, we need to take the derivative of the position coordinate equation (i.e. the rope length equation). The length of the rope and value of the constant do not change, therefore, their time derivatives are zero.

$$\frac{dL}{dt} = 0 = ((s_A - r)^2 + D^2)^{-1/2} 2(s_A - r)\dot{s}_A + ((s_B - r)^2 + D^2)^{-1/2} 2(s_B - r)\dot{s}_B$$

Step 5) Solve and verify

To solve for the velocity of collar A, we need to rearrange the position coordinate equation and plug in the values for the instant shown.

$$\dot{s}_A = -\frac{((s_A - r)^2 + D^2)^{1/2}(s_B - r)\dot{s}_B}{((s_B - r)^2 + D^2)^{1/2}(s_A - r)}$$

$$\boxed{\dot{s}_A = -1.47 \ \frac{\text{cm}}{\text{s}}}$$

Looking at the figure, it is clear that if collar B is pulled down collar A will also move down. A downward motion for collar B indicates a positive velocity and a downward motion of collar A indicates a negative velocity. This is because a velocity is positive if it moves away from the datum.

CHAPTER 3 REVIEW PROBLEMS

RP3-1) Consider a projectile that is only acted on by gravity (i.e. no air resistance). The x-direction velocity is constant. (True, False)

RP3-2) Consider a projectile that is only acted on by gravity (i.e. no air resistance). The y-direction velocity is constant. (True, False)

RP3-3) Consider a projectile that is only acted on by gravity (i.e. no air resistance). The initial value of the y-direction velocity is always the same as the final value of the y-direction velocity. (True, False)

RP3-4) Consider a projectile that is only acted on by gravity (i.e. no air resistance). The y-direction velocity at the peak is zero. (True, False)

RP3-5) Consider a projectile that is only acted on by gravity (i.e. no air resistance). The x-direction acceleration is zero. (True, False)

RP3-6) Consider a projectile that is only acted on by gravity (i.e. no air resistance). The y-direction acceleration is constant. (True, False)

RP3-7) An astronaut performs an experiment on the surface of the moon to measure the acceleration of a falling object. The astronaut obtains an estimated acceleration of 3.1 m/s^2 from her experiment. It is known that the acceleration due to gravity on the moon's surface is approximately 1.6 m/s^2. What is the most likely explanation of the difference in values?

    a) air resistance
    b) local variations in gravity
    c) experimental error
    d) effect of wind blowing the object sideways as it fell

RP3-8) Ball A and ball B are dropped from the same height and are photographed at constant intervals of time as they fall. Which of the following answers most likely explains the difference in the trajectory of ball A as compared to the trajectory of ball B?

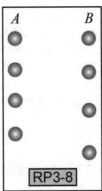

a) Ball A has more mass.
b) Ball A has a smaller initial velocity.
c) Ball A experiences more air drag.
d) Ball A is dropped after ball B.

RP3-9) The five balls shown are thrown with the same initial velocity at the same angle with respect to the horizontal and from the same initial height. Each ball weighs a different amount. Neglecting air resistance, which ball will travel the farthest?

Farthest _____   Next _____   Next _____   Next _____   Shortest _____

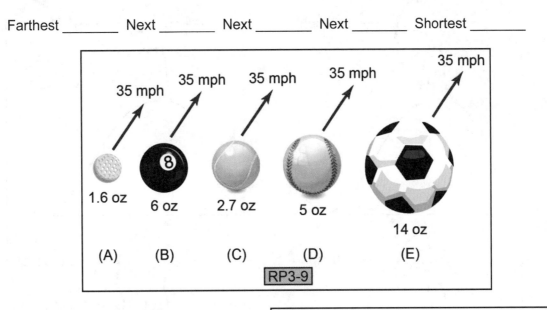

RP3-10)   Consider   a   bi-athlete shooting a target as shown in the figure. The target is far away from the athlete, but still within the gun's range. He aims directly at the center of the target and shoots the instant the target support snaps and the target begins to fall to the ground. Neglecting air resistance, where does the bullet hit the target?

a) Above the center of the target.
b) The center of the target.
c) Below the center of the target.
d) Does not hit the target at all because by the time the bullet reaches the target, the target will have already hit the ground.

RP3-11) Shown are six figures of archers shooting arrows from the tops of hills. The arrows are all the same and all are shot horizontally. Rank each arrow in order from longest to shortest time to reach the ground. Then rank each arrow in order from furthest horizontal distance traveled to shortest horizontal distance traveled.

Time to reach ground:

Longest _____   Next _____   Next _____   Next _____   Next _____   Shortest _____

Horizontal distance traveled:

Furthest _____   Next _____   Next _____   Next _____   Next _____   Shortest _____

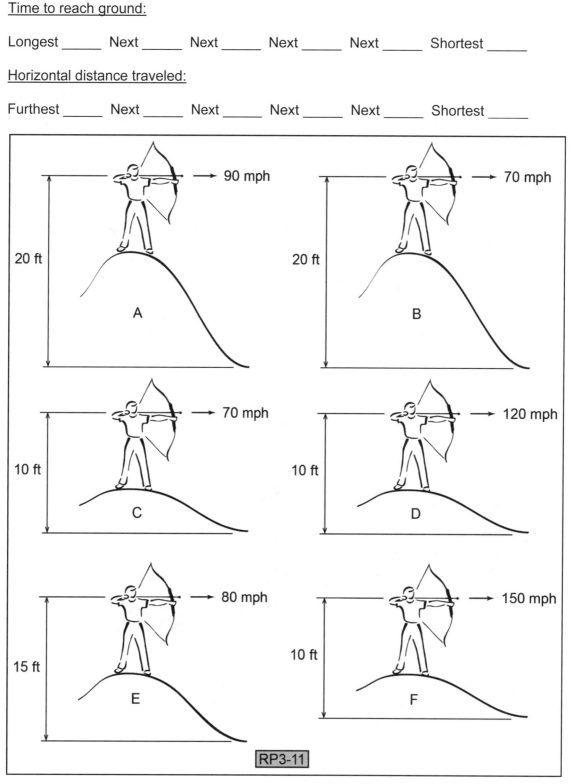

RP3-12) A car is rounding a curve at constant speed. Is the velocity also constant?

RP3-13) A car is rounding a curve at constant speed. Is the acceleration zero? If the acceleration is non-zero, what is the acceleration direction?

RP3-14) Consider a particle moving in 2-D space on a known path. What direction is the velocity?

RP3-15) Consider the acceleration $\mathbf{a} = \dot{v}\, \mathbf{e}_t + \dfrac{v^2}{\rho} \mathbf{e}_n$ of a particle moving in 2-D space. Explain the physical meanings of the two acceleration terms.

RP3-16) The position of a particle is given by $\mathbf{r} = b\sin(ct)\mathbf{i} + d\cos(ft)\mathbf{j}$, where $b = 5$ m/rad, $c = 3$ rad/s, $d = 1$ m/rad, and $f = 5$ rad/s. What is the magnitude of the particle's velocity and acceleration at $t = 4$ s?

Given: $\mathbf{r} = b\sin(ct)\mathbf{i} + d\cos(ft)\mathbf{j}$ $b = 5$ m/rad

         $c = 3$ rad/s $d = 1$ m/rad

         $f = 5$ rad/s $t = 4$ s

Find: v

         a

Solution:

Find the particle's velocity. | Find the particle's acceleration.

$\mathbf{v} = $ _____ | $\mathbf{a} = $ _____

Find the magnitude of the velocity. | Find the magnitude of the acceleration.

$v = $ _____ | $a = $ _____

RP3-17) A track for motorcycle racing was designed so that riders jump off a slope at 20° from a height of 5 meters. During a race it was observed that the rider and bike left the ramp at 150 km/s and remained in the air for 5 seconds. Determine the horizontal distance the rider traveled before striking the ground and the maximum height he attained.

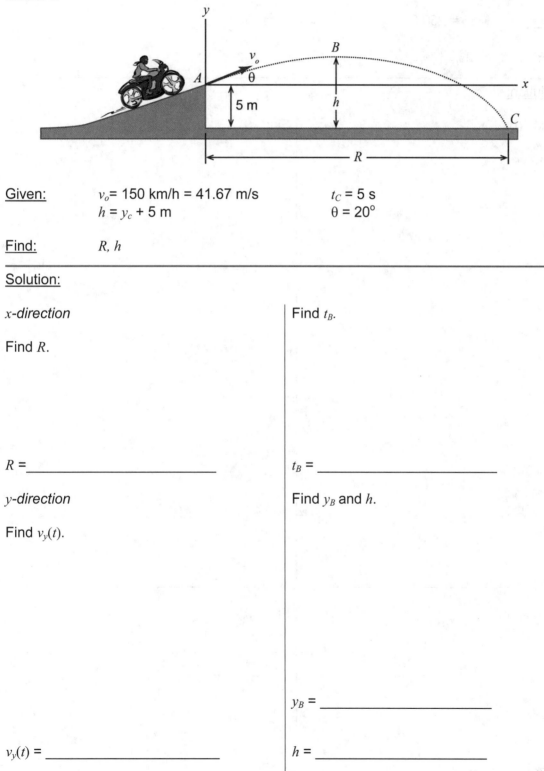

Given: v_o= 150 km/h = 41.67 m/s t_C = 5 s
              $h = y_c + 5$ m $\theta = 20°$

Find: R, h

Solution:

x-direction | Find t_B.

Find R. |

$R =$ _____ | $t_B =$ _____

y-direction | Find y_B and h.

Find $v_y(t)$. |

                                         | $y_B =$ _____

$v_y(t) =$ _____ | $h =$ _____

RP3-18)   A car drives on a circular track of radius 250 ft.   The car's speed is $v = 3(t + t^2)$ ft/s for the time period $0 \le t \le 2$ s where t is in seconds. Determine the magnitude of the car's acceleration when $t = 2$ s. How far has the car traveled during this period of time?

<u>Given:</u> $\rho = 250$ ft $v = 3(t + t^2)$ ft/s

<u>Find:</u> $|\mathbf{a}_{t=2}|$, $s_{t=2}$

<u>Solution:</u>

Find $\mathbf{a}(t)$ | Find $s(t)$

$\mathbf{a}(t) =$ _____ | $s(t) =$ _____

Find $\mathbf{a}_{t=2}$ | Find $s_{t=2}$

$|\mathbf{a}_{t=2}| =$ _____ | $s_{t=2} =$ _____

RP3-19) Consider the pendulum shown. On the upswing of its motion and at the instant when $\theta = 20°$, $\dot{\theta} = 3$ rad/s, and $\ddot{\theta} = 5$ rad/sec^2 in the directions shown, determine the acceleration of the pendulum's head in both r-θ and x-y coordinates. The length of the pendulum bar is $L = 0.5$ m.

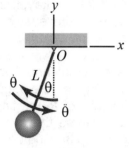

Given: $\theta = 20°$ $\dot{\theta} = 3$ rad/s
              $\ddot{\theta} = -5$ rad/sec^2 $L = 0.5$ m

Find: $\mathbf{a}_{\theta = 20}$ in both r-θ and x-y coordinates

---

Solution:

Draw the r- and θ-coordinate axes on the figure.

Find the acceleration of the pendulum in the r-θ coordinate system.

Project the r-θ unit direction vectors onto the x-y coordinate axes.

$\mathbf{e}_r = $ _____

$\mathbf{e}_\theta = $ _____

Find the acceleration of the pendulum in the x-y coordinate system.

a = _____(r-θ)   **a** = _____(x-y)

RP3-20) A kayak travels from the west bank to the east bank of a river with a constant velocity relative to the river current of $\mathbf{v}_{B/R} = v_{B/R}\,\mathbf{i}$. If the river current is constant and flowing in the direction shown in the figure, determine the velocity of B, the time it takes the kayak to traverse the river and the distance it ends up landing downstream (d) in terms of $v_{B/R}$, v_R and D.

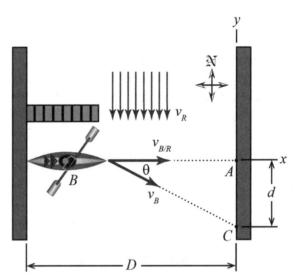

Given: $\mathbf{v}_{B/R} = v_{B/R}\,\mathbf{i} = $ constant
           $\mathbf{v}_R = $ constant

Find: $\mathbf{v}_B$, t, d as function of ($v_{B/R}$, v_R and D)

Solution:

Determine the velocity of the boat.

$\mathbf{v}_B = $ _____

Determine the time it takes the boat to cross the river.

$t = $ _____

Determine d.

$d = $ _____

RP3-21) A crate C is being pulled up an incline using motor M and the rope and pulley system shown. Determine the speed of the crate if the motor is pulling the rope in at a constant speed of 6 m/s.

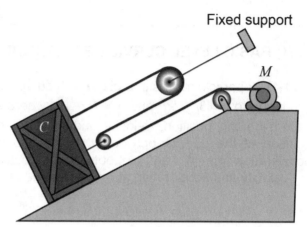

Fixed support

M

C

Given: v_M = 6 m/s = constant

Find: v_C

_____

Solution:

Identify the datum and draw the position coordinates on the figure.

Write the length of the rope in terms of the position coordinates.

$L =$ _____

Determine the velocity of the crate.

$v_C = \dot{s}_C =$ _____

CHAPTER 3 PROBLEMS

P3.1) BASIC LEVEL CURVILINEAR PROBLEMS

P3.1-1) A particle moves in a plane with curvilinear motion. Its position varies with time according to $x = 3t^3 - 5t^2$ and $y = 2t^2 - 10$, where t is in seconds and x and y are in feet. Determine the position, velocity and acceleration as functions of time and plot the position of the particle between $t = 0$ and 5 seconds. Also, find the velocity and acceleration at $t = 4$ s and plot both on the position graph. What are the magnitudes of the velocity and acceleration at $t = 4$ s and the angle that they make with the x-axis?

Ans: $\mathbf{v} = 104\mathbf{i} + 16\mathbf{j}$ ft/s, $v = 105.2$ ft/s, $\theta = 8.75°$, $\mathbf{a} = 62\mathbf{i} + 4\mathbf{j}$ ft/s^s, $a = 62.1$ ft/s^2, $\theta = 3.7°$

P3.1-2)^{fe} The position of a particle is given by $x = t^4 - 1$ and $y = t^3 + t + 3$, where t is in seconds and, x and y are in meters. What is the particle's velocity?

a) $\mathbf{v} = 3t^3\mathbf{i} + (3t^2 + 1)\mathbf{j}$ m/s

b) $\mathbf{v} = 4t^3\mathbf{i} + (3t^2 + 1)\mathbf{j}$ m/s

c) $\mathbf{v} = (4t^5 - t)\mathbf{i} + (3t^4 + t^2 + t)\mathbf{j}$ m/s

d) $\mathbf{v} = 4t^4\mathbf{i} + (3t^3 + 1)\mathbf{j}$ m/s

P3.1-3) A car is driving up a 25% grade hill. The car drives with a constant acceleration of 0.5 m/s^2. At time $t = 0$, the car is going 10 kph. What is the car's final velocity?

Ans: $\mathbf{v}_f = 114.5\mathbf{i} + 28.5\mathbf{j}$ kph

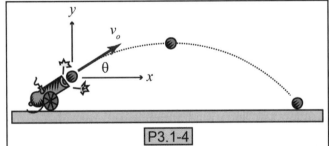

P3.1-3

P3.1-4)^{fe} A cannon fires a cannon ball as shown in the figure. The position of the cannon ball is given by $x = v_{xo}t$ and $y = v_{yo}t - (1/2)gt^2$, where v_{xo} and v_{yo} are the initial velocities in the x- and y-directions respectively. Also, g is the acceleration due to gravity and t is given in seconds. If the initial velocity of the cannon ball is 90 mph and $\theta = 40°$, determine the cannon ball's velocity and acceleration at $t = 5$ s.

P3.1-4

a) $\mathbf{v} = 101\mathbf{i} - 76.2\mathbf{j}$ ft/s, $\mathbf{a} = -32.2\mathbf{j}$ ft/s^2

b) $\mathbf{v} = 68.9\mathbf{i} - 103.1\mathbf{j}$ ft/s, $\mathbf{a} = -32.2\mathbf{j}$ ft/s^2

c) $\mathbf{v} = 101\mathbf{i} + 35.8\mathbf{j}$ ft/s, $\mathbf{a} = -9.81\mathbf{j}$ ft/s^2

d) $\mathbf{v} = 101\mathbf{i} + 35.8\mathbf{j}$ ft/s, $\mathbf{a} = -32.2\mathbf{j}$ ft/s^2

P3.1-5) A particle moves along a circular path where its position is given by the equation $\mathbf{r} = 5\cos(\pi t/4)\mathbf{i} + 5\sin(\pi t/4)\mathbf{j}$ where $\mathbf{r}$ is in meters and t is in seconds. Show that the magnitudes of the particle's velocity and acceleration are constant. Also determine the magnitude and direction of the particle's position, velocity and acceleration when $t = 2$ s.

Ans: $\mathbf{r} = 5\mathbf{j}$ m, $\mathbf{v} = -5\pi/4\ \mathbf{i}$ m/s, $\mathbf{a} = -5\pi^2/16\ \mathbf{j}$ m/s^2

P3.2) INTERMEDIATE LEVEL CURVILINEAR PROBLEMS

P3.2-1) The velocity of a particle is given by $\mathbf{v} = 4t^2\mathbf{i} + 3t\mathbf{j}$ m/s, where t is in seconds. At $t = 0$ seconds the particle's position is $\mathbf{r} = 2\mathbf{i} - \mathbf{j}$ m. Determine the particle's position at $t = 2$ seconds. Also, determine the magnitude and direction of the particle's acceleration at $t = 2$ seconds.

Ans: $\mathbf{r} = 12.7\mathbf{i} + 5\mathbf{j}$ m, $a = 16.3$ m/s^2, $\theta = 10.6°$

P3.2-2) A particle moves counterclockwise in a circle that has its center located at the origin of the x-y coordinate system. The radius of the circle is $r = 0.8$ ft and the particle goes completely around the circle once every second. Determine the velocity of the particle when $\theta = 30°$ (i.e. the angle between the x-axis and the position vector $\mathbf{r}$).

Ans: $\mathbf{v} = -2.51\mathbf{i} + 4.35\mathbf{j}$ ft/s

P3.2-3) A particle having a constant velocity starts at the point within the Cartesian coordinate system defined by (1 in, 2 in) and then moves to point (5 in, 4 in). The particle starts at the first point at $t = 0$ s and it takes 2 seconds to reach the second point. Where is the particle located when $t = 3$ s?

Ans: $\mathbf{r} = 7\mathbf{i} + 5\mathbf{j}$ in

P3.2-4) A spacecraft is drifting in orbit in the direction shown with a constant velocity of 7000 m/s. When turned on, the thrusters of the spacecraft accelerate the vehicle according to the equation $a = 20 + 30t$ in a direction perpendicular to the craft's current direction of motion where a is in m/s^2 and t is in seconds. Determine the spacecraft's velocity and its displacement 10 seconds after the thrusters are first fired.

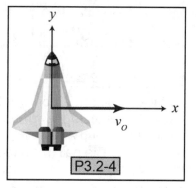

P3.2-4

Ans: $\mathbf{v} = 7000\mathbf{i} + 1700\mathbf{j}$ m/s, $\Delta\mathbf{r} = 70,000\mathbf{i} + 6000\mathbf{j}$ m

P3.2-5) A boy throws a baseball straight up in the air with an initial velocity of 10 m/s. As the ball is released, there is a wind gust that imparts the ball with a constant horizontal acceleration of 0.3 m/s^2 for the duration of the ball's flight. Assuming the boy catches the ball at the same height from which it was thrown and that air resistance is negligible, determine the time the ball is in the air and how far the boy must run from his initial position to catch the ball.

Ans: $t = 2.04$ s, $\Delta x = 0.62$ m

P3.3) ADVANCED LEVEL CURVILINEAR PROBLEMS

P3.3-1) A truck travels at a constant speed of 60 mph up a hill whose elevation is given as a function of the horizontal distance traveled, $y = 0.001x^3$. Determine the truck's velocity and acceleration when the truck passes the horizontal position, $x = 1000$ ft.

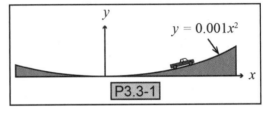

Ans: $\mathbf{v} = 39.35\mathbf{i} + 78.70\mathbf{j}$ ft/s, $\mathbf{a} = -1.24\mathbf{i} + 0.62\mathbf{j}$ ft/s^2

P3.3-2) The velocity and acceleration of a particle are given by $\mathbf{v} = 2\mathbf{i} - \mathbf{j} + 3\mathbf{k}$ m/s and $\mathbf{a} = 5\mathbf{i} + \mathbf{j} - 4\mathbf{k}$ m/s^2. Determine the rate of change in speed $\dot{v}$ (i.e. the acceleration that is tangent to the path curve).

Ans: $\dot{v} = -0.8$ m/s^2

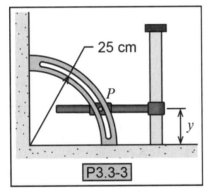

P3.3-3) Pin P shown is constrained to move in a circular slot with radius 25 cm. If the collar constrained to move vertically has a constant upward speed of 5 cm/s, determine the magnitude of the velocity and acceleration of pin P when $y = 10$ cm.

Ans: $v = 5.5$ cm/s, $a = 1.31$ cm/s^2

P3.3-4) A ball rolls without slip on a horizontal surface. Point A, a point on the outer surface of the ball, contacts the ground at the origin and then rolls up and away. The position of point A is given by $\mathbf{r}$. The ball has a radius of R and rolls to the right with an angular velocity and acceleration of $\dot{\theta}$ and $\ddot{\theta}$ respectively. Determine the position, velocity and acceleration of point A when $\theta = 60°$, $\dot{\theta} = 6$ rad/s, $\ddot{\theta} = -1$ rad/s^2 and $R = 10$ cm.

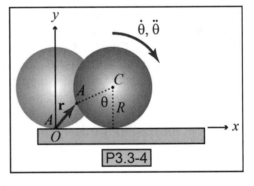

Ans: $\mathbf{r} = 0.18\mathbf{i} + 0.05\mathbf{j}$ m, $\mathbf{v} = 0.3\mathbf{i} + 0.52\mathbf{j}$ m/s, $\mathbf{a} = 3.07\mathbf{i} + 1.71\mathbf{j}$ m/s^2

P3.4) BASIC LEVEL PROJECTILE PROBLEMS

P3.4-1) A man standing on the edge of a cliff shoots a bullet from a gun. The bullet leaves the gun with an initial horizontal velocity of 750 mph. If the cliff is h = 1000 ft high, what is the bullet's range (R)?

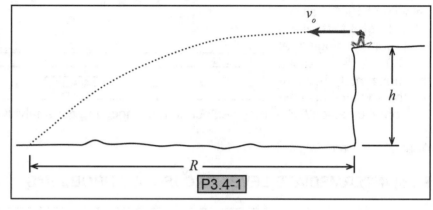

Ans: R = 1.64 mi

P3.4-2) A projectile is launched on level terrain with an initial velocity of 10 mph at an angle of $25°$ with the horizontal. Determine the time of flight, the range and the maximum attained height of the projectile.

Ans: h = 0.597 ft, t = 0.385 s, R = 5.12 ft

P3.4-3) A quarterback throws a football with an initial velocity of 40 mph at an angle of $30°$ relative to the horizontal. If the height of the football relative to the ground as it leaves the quarterback's hand is

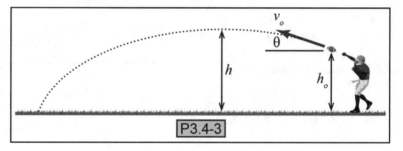

h_o = 5 ft, determine the maximum attained height (h) of the football.

Ans: h = 18.4 ft

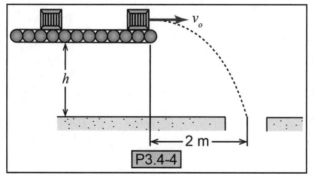

P3.4-4) The conveyor belt shown delivers packages such that they fall into a chute located a horizontal distance of 2 meters away from the edge of the belt. If the conveyor is located a height h = 1.5 m above the ground, what speed does the conveyor need to operate with so that the packages fall into the chute?

Ans: v_o = 3.62 m/s

P3.4-5) A tank fires a shell with an initial velocity of 1000 ft/s from a height of 8 feet. If the tank gun is angled at $\theta = 30$ degrees above the horizontal, how far will the shell travel in the

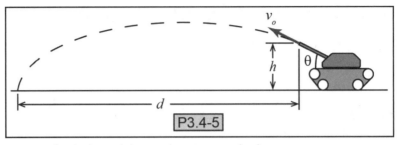

horizontal direction? You may neglect air resistance in your analysis.

Ans: $d = 5.1$ mi

P3.5) INTERMEDIATE LEVEL PROJECTILE PROBLEMS

P3.5-1) A student team designed a catapult to launch a ball at a target of stacked cans. If the distance to the target ($R = 8$ ft) and the initial launch angle ($\theta = 45°$) are known, determine the range of initial launch speeds (v_o) that will enable the ball to hit the target if the height of the stacked cans is $h = 2$ ft.

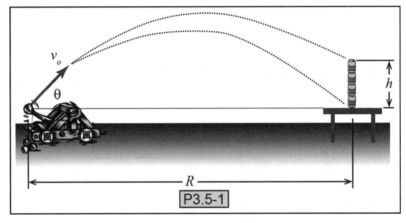

Ans: $v_o = 16.05 - 18.53$ ft/s

P3.5-2) If an arrow is aimed at point A as shown, compute the distance h below the intended target where the arrow will hit at point B. The initial velocity v_o of the arrow is 44 m/s, the angle that the velocity makes with the horizontal is $\theta = 15°$ and the distance between the arrow and the target is $L = 50$ m.

Ans: $h = 6.79$ m

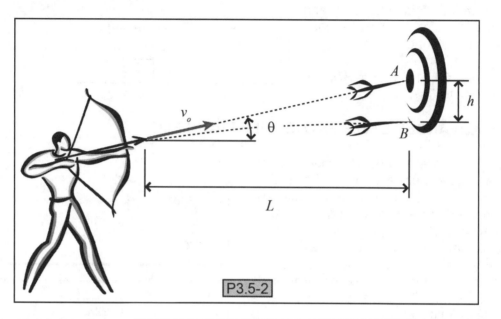

P3.5-2

P3.5-3) Two men are having a long jump contest as illustrated in the figure.   The first man starts his jump with an initial speed of $v_o = 12$ ft/s and reaches a maximum height of $h = 1$ ft.   The second man starts his jump with an initial speed of 14 ft/s and reaches a height of h = 1.5 ft. Which man will win the contest and their jump distances?

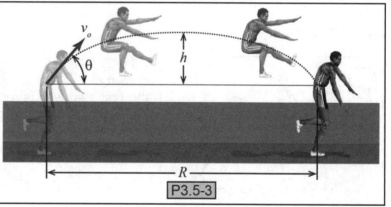

P3.5-3

Ans: $R_1 = 4.45$ ft, $R_2 = 6.09$ ft

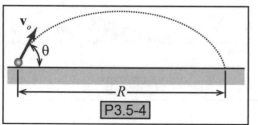

P3.5-4

P3.5-4) Calculate the minimum speed (v_o) a projectile must have in order to hit a target $R = 500$ feet away. The target is located on the same horizontal plane as the starting location of the projectile.

Ans: $v_o = 126.9$ ft

P3.5-5) The tank shown fires a shell with an initial velocity of 1500 ft/s at an angle of 30 degrees. Neglecting the size of the tank and air resistance, determine the range D, maximum attained height H and the total time for the shell to reach its target at point C.

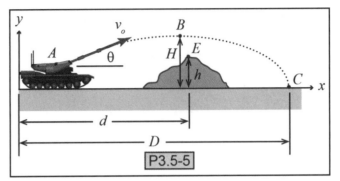

P3.5-5

Ans: D = 60,514 ft, H = 8741 ft, t_C = 46.6 s

P3.6) ADVANCED LEVEL PROJECTILE PROBLEMS

P3.6-1) A skeet shooter aims at a clay pigeon. The muzzle velocity of the rifle is 700 mph and the man holds the rifle such that the end of the gun is 6 ft off the ground. Assuming that the speed of the skeet is insignificant relative to the speed of the bullet, determine the angle at which the man needs to hold the rifle in order to hit a clay pigeon that is 50 yards away and at an altitude of 30 ft.

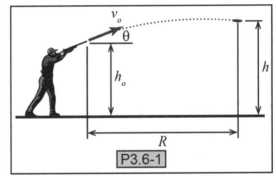

P3.6-1

Ans: θ = 9.22°

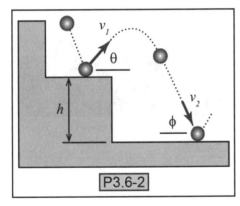

P3.6-2

P3.6-2) A ball bounces down some stairs as shown in the figure. The velocity the ball leaves the first step with is related to the velocity the ball hits the second step with by the equation $v_{y2} = 2v_{y1}$. Prove that the height of the step is equal to $h = (3v_2^2 \sin^2 \phi)/(8g)$.

P3.6-3) A projectile is launched from the base of a hill at an angle of $\theta = 70°$ with an initial velocity of 20 m/s. If the shape of the hill is described by the function $y = 0.02x^2$ where x and y are both in meters, determine the height h and range d of travel of the projectile where it impacts the hill.

Ans: d = 22.0 m, h = 9.68 m

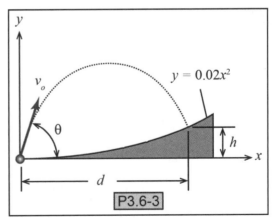

P3.6-3

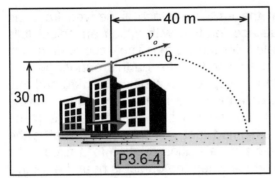

P3.6-4

P3.6-4) You and some friends are launching water balloons off the roof of your dormitory. You determine that the stiffness of the launcher and the effect of gravity causes each balloon to accelerate in the launcher according to the equation $a = -25s + 9.81\sin\theta$ in m/s^2, where s is the distance the launcher is stretched in meters and θ is the angle with which the launcher is aimed.   If your dormitory is 30 meters tall and you wish to hit one of your friends on the ground a horizontal distance of 40 meters away from your dormitory, determine how far back you must pull your water balloon launcher if you launch the balloon at an angle of $\theta = 45°$. You may neglect air resistance.

Ans: $s = 3.29$ m

P3.7) BASIC LEVEL N-T COORDINATE PROBLEMS

P3.7-1) A car rounds a curve of radius 100 ft at a speed of 25 mph. If the magnitude of its total acceleration is 15 ft/s^2, determine the rate at which its speed is changing.

Ans: $\dot{v} = 6.65$ ft/s^2

P3.7-2) A truck travels at a constant speed of 60 mph up a hill whose elevation is given as a function of the horizontal distance traveled, $y = 0.001x^2$. Determine the truck's velocity and acceleration when the truck passes the horizontal position, $x = 1000$ ft.

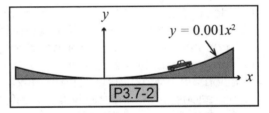

P3.7-2

Ans: $\mathbf{v} = 88\mathbf{e}_t$ ft/s, $\mathbf{a} = 1.385\mathbf{e}_n$ ft/s^2

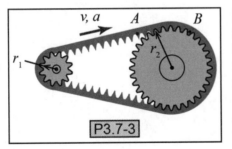

P3.7-3

P3.7-3) In the given figure, a small gear ($r_1 = 50$ mm) drives a larger gear ($r_2 = 150$ mm) via a belt. If the belt speed is initially 1 m/s and it is accelerating uniformly at 0.5 m/s^2, determine the acceleration magnitudes of point A and point B after 2 seconds.

Ans: $a_A = 0.5$ m/s^2, $a_B = 26.67$ m/s^2

P3.7-4) The centrifuge, shown in the figure, is used to test human g tolerance. The arm length (ρ) of the centrifuge is 30 ft. If the experimenters wanted to test a tolerance of $5g$, what velocity would the pod attain?

Ans: $v = 47.5$ mph

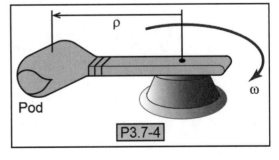

P3.7-4

P3.7-5) The acceleration due to gravity g at the earth's surface is derived from the Universal Law of Gravitation. This is the true acceleration with which an object falls relative to a fixed set of axes that do not rotate with the earth. The acceleration with which an object falls relative to a point that is attached to the earth's surface will be slightly less than g. If you assume that the earth is a sphere of radius 6371 km and that the earth makes 1 revolution every 24 hours, determine the normal component of the acceleration of a point on the earth's surface on the celestial equator (on a plane perpendicular to the axis of rotation). If it is assumed that $g = 9.81$ m/s^2, what is the percentage difference between the falling object's acceleration relative to a fixed point and relative to a point attached to the earth's surface when the object is near the earth's surface?

Ans: $a_n = 0.0337$ m/s^2, %difference = 0.343%

P3.7-6)^{fe} A particle travels in a circle of radius 3 m with a constant velocity of 10 m/s. Determine the magnitude of the particle's total acceleration.

a) 0.33 m/s^2 b) 3.33 m/s^2 c) 33.3 m/s^2 d) 333.3 m/s^2

P3.7-7)^{fe} The autonomous ground robot shown is equipped with a digital compass that measures the robot's heading θ as well as two accelerometers, one aligned with the robot's longitudinal axis (a_t) and one aligned with the lateral axis (a_n). It is desired to use these sensors to locate the robot in the inertial x-y reference frame. Write equations for the robot's x- and y-components of acceleration in terms of the outputs of the three available sensors (θ, a_t, a_n). Now express these equations in matrix form. The matrix (**A**) that converts the accelerations from the body-fixed reference frame to the inertial frame is called the **direction cosine matrix**.

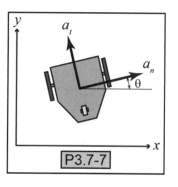

P3.7-7

$$\begin{bmatrix} a_x \\ a_y \end{bmatrix} = \mathbf{A} \begin{bmatrix} a_t \\ a_n \end{bmatrix}$$

a) $\mathbf{A} = \begin{bmatrix} \cos\theta & -\sin\theta \\ \sin\theta & \cos\theta \end{bmatrix}$ b) $\mathbf{A} = \begin{bmatrix} -\sin\theta & \cos\theta \\ \cos\theta & \sin\theta \end{bmatrix}$

c) $\mathbf{A} = \begin{bmatrix} -\cos\theta & \sin\theta \\ \sin\theta & \cos\theta \end{bmatrix}$ d) $\mathbf{A} = \begin{bmatrix} \cos\theta & -\sin\theta \\ -\sin\theta & -\cos\theta \end{bmatrix}$

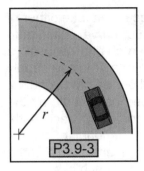

P3.9-3

P3.7-8)[fe] A vehicle is traversing a constant-radius curve. The magnitude of the car's acceleration is limited by the 'friction circle' of the vehicle's tires, that is $\sqrt{a_t^2 + a_n^2} \le a_{max}$, where a_t is the longitudinal acceleration of the vehicle (for braking and accelerating) and a_n is the lateral acceleration of the vehicle (for cornering). If the radius of the curve is $r = 80$ m and the maximum acceleration the tires are able to provide is $a_{max} = 8$ m/s^2, determine the maximum speed the vehicle can achieve on the curve before skidding.

a) $v_{max} = 640$ m/s   b) $v_{max} = 11.5$ m/s   c) $v_{max} = 25.3$ m/s   d) $v_{max} = 32.4$ m/s

P3.8) INTERMEDIATE LEVEL N-T COORDINATE PROBLEMS

P3.8-1) A man drives a car on a 100-m radius circular horizontal test track. The driver starts the car from rest and increases its speed at a uniform rate for 20 seconds until the car reaches a speed of 36 m/s. Determine the magnitude of the total acceleration of the car 10 seconds after the driver starts accelerating the car.

Ans: $a = 3.71$ m/s^2

P3.8-2) A truck starts from rest on a road with a constant radius of curvature ($\rho = 50$ m) The truck increases its speed at a constant rate until it reaches 10 m/s after 5 seconds. Determine the total acceleration of the truck 4 seconds after it starts increasing its speed.

Ans: $\mathbf{a} = 2\mathbf{e}_t + 1.28\mathbf{e}_n$ m/s^2

P3.8-3) Consider the racing oval shown. A car begins from rest at point A and increases its speed at a rate of $\dot{v} = 0.2t$ m/s^2, where t is in seconds. Determine (a) the time it takes the car to reach the apex of the curve at point B and (b) the racecar's velocity and total acceleration as it passes point B.

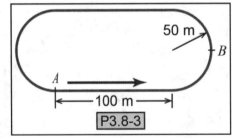

P3.8-3

Ans: $t = 17.5$ s, $\mathbf{v} = 30.6\mathbf{e}_t$ m/s,

P3.8-4) A projectile is launched at an initial velocity of 40 m/s at an angle of 35°. What is the radius of curvature of the projectile's path at the apex of travel?

Ans: $\rho = 109.4$ m

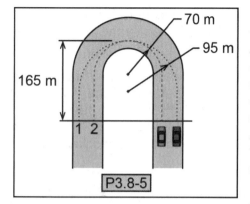

P3.8-5

P3.8-5) Consider that a racecar can take either of the paths shown through a 180° curve. Each path consists of straight lines into and out of a constant-radius curve. If the car's tires can supply only 1.2g of total acceleration before skidding, determine the maximum speed achievable by the car for the constant radius portion of each path. Also, determine the minimum amount of time it takes the car to traverse path 1 and path 2 if the car travels with constant speed in each case.

Ans: v_{max1} = 33.4 m/s, v_{max2} = 28.7 m/s, t_{min1} = 13.1 s, t_{min2} = 14.3 s

P3.9) ADVANCED LEVEL N-T COORDINATE PROBLEMS

P3.9-1) A stretch of rollercoaster track has a dip as shown in the figure. The bottom of the dip has a radius of r. If the cars have negligible velocity at the top of the hill (point A), the velocity of the cars when they hit the bottom (point B) is given by the relationship $(1/2)v_B^2 = gh$, where g is the acceleration due to gravity. If the designers limited the acceleration experienced by the riders to 3g, what is the minimum radius of the dip?

Ans: $r = 2h/3$

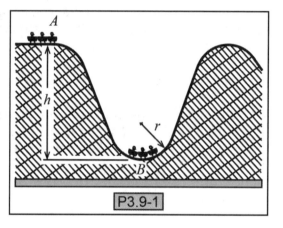

P3.9-1

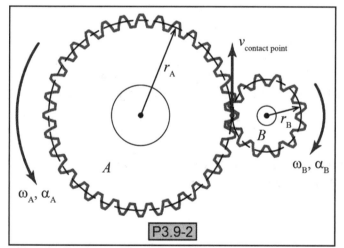

P3.9-2

P3.9-2) Drive gear A rotates gear B, as shown in the figure. The rate of change of speed of the contact point is given by $\dot{v} = r\alpha$, where r is the gear radius and α is the angular velocity given in rad/s. The angular acceleration of gear A is α_A = 13 rad/s^2 and the angular acceleration of gear B is α_B = 21.67 rad/s^2. The velocity of the contact points is 22.361 in/s and the normal acceleration of the contact point on gear A is 50 in/s^2. Note that the gears mesh without slipping. Therefore, the contact points between the two gears have the same velocity and the same tangential acceleration. Determine the radii of the gears and total acceleration of the gear teeth.

Ans: r_A = 10 in, r_B = 6 in, $\mathbf{a}_A = 130\mathbf{e}_t + 50\mathbf{e}_n$ in/s^2, $\mathbf{a}_B = 130\mathbf{e}_t + 83.3\mathbf{e}_n$ in/s^2

P3.10) BASIC LEVEL POLAR COORDINATE PROBLEMS

P3.10-1) A particle's motion is defined by the equation $cr = \theta^2$ as shown in the graph, where r is in meters and $c = 40$ 1/m. If the particle is traveling with constant angular velocity ($\dot{\theta} = 2$ rad/s), determine the particle's velocity and acceleration as a function of θ.

Ans: $\mathbf{v} = (\theta/20)(2\mathbf{e}_r + \theta\mathbf{e}_\theta)$ m/s, $\mathbf{a} = (-1/10)((2-\theta^2)\mathbf{e}_r + 4\theta\mathbf{e}_\theta)$ m/s^2

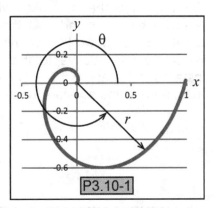

P3.10-1

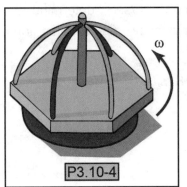

P3.10-2

P3.10-2)[fe] A little girl is playing with a skipball as shown in the figure. If the girl manages to rotate the skip ball around her leg at a constant rate ($\dot{\theta}$ = constant), what is the ball's velocity and acceleration?

a) $\mathbf{v} = \dot{r}\mathbf{e}_r$, $\mathbf{a} = -r\dot{\theta}^2\mathbf{e}_r$

b) $\mathbf{v} = r\dot{\theta}\mathbf{e}_\theta$, $\mathbf{a} = -r\dot{\theta}^2\mathbf{e}_\theta$

c) $\mathbf{v} = -r\dot{\theta}\mathbf{e}_\theta$, $\mathbf{a} = r\ddot{\theta}\mathbf{e}_\theta$

d) $\mathbf{v} = r\dot{\theta}\mathbf{e}_\theta$, $\mathbf{a} = -r\dot{\theta}^2\mathbf{e}_r$

P3.10-3) The shape of the stationary cam shown in the given mechanism is described by the equation, $r = 1.5 + 0.5\sin(3\theta)$ inches. If the slotted bar rotates with constant angular velocity $\dot{\theta} = 5$ rad/sec, determine the velocity of the cam follower A when $\theta = 60°$ (a) in terms of polar coordinates and (b) in terms of Cartesian coordinates.

Ans: $\mathbf{v} = -7.5\mathbf{e}_r + 7.5\mathbf{e}_\theta = -10.25\mathbf{i} - 2.75\mathbf{j}$ in/s

P3.10-3

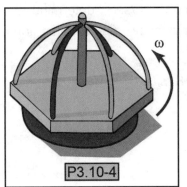

P3.10-4

P3.10-4)[fe] You are standing at the center of a merry-go-round while your friend pushes the platform such that it turns at a constant rate of 0.4 rad/sec. You then begin walking out toward the edge at a constant rate of 0.6 m/s. Determine the acceleration you experience when you reach a distance of 0.5 m from the center of the platform.

a) $\mathbf{a} = 0.08\mathbf{e}_r + 0.48\mathbf{e}_\theta$ m/s^2

b) $\mathbf{a} = -0.08\mathbf{e}_r + 0.24\mathbf{e}_\theta$ m/s^2

c) $\mathbf{a} = -0.08\mathbf{e}_r + 0.48\mathbf{e}_\theta$ m/s^2

d) $\mathbf{a} = -0.06\mathbf{e}_r - 0.24\mathbf{e}_\theta$ m/s^2

P3.10-5) Rod R rotates about pinned joint A. The angle that the rod makes with the x-axis is given by $\theta = t^2$ rad. At the same time, collar C slides down the rod with a speed of $v_{C/R} = 2\theta$ m/s relative to the rod. At the instant $t = 1$ s and $r = 0.8$ m, determine the magnitude of the acceleration of the collar.

Ans: $a = 8.5$ m/s^2

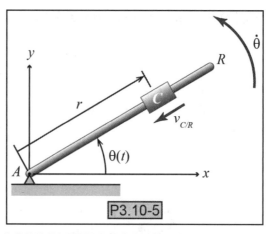

P3.10-5

P3.11) INTERMEDIATE LEVEL POLAR COORDINATE PROBLEMS

P3.11-1) The bar shown in the figure rotates counterclockwise with an angular velocity of $\dot{\theta} = 1.3$ rad/s and an angular acceleration of $\ddot{\theta} = -0.5$ rad/s^2. Find the magnitude of the velocity and acceleration for the ball if $L = 2$ m when $\theta = 30°$.

Ans: $\mathbf{v} = 1.733\mathbf{e}_r + 3\mathbf{e}_\theta$ m/s,
$\mathbf{a} = 1.94\mathbf{e}_r + 3.35\mathbf{e}_\theta$ m/s^2

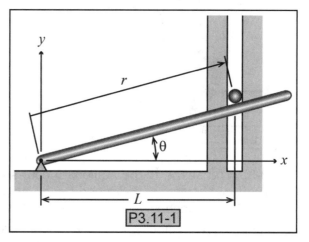

P3.11-1

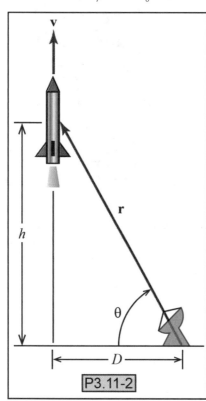

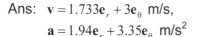

P3.11-2

P3.11-2) The rocket shown in the figure below travels vertically upward with velocity v. At this instant, the rocket is located a distance $D = 2$ miles from the radar station which is tracking the rocket's flight. When the rocket has reached an altitude of $h = 1$ mile, $\dot{\theta} = 0.2$ rad/sec, $\ddot{\theta} = 0.03$ rad/sec^2 determine the magnitude of the rocket's velocity $\mathbf{v}$ and acceleration $\mathbf{a}$ at this instant.

Ans: $v = 0.5$ mi/s, $a = 0.18$ mi/s^2

P3.11-3) The end of a cam follower rides along the surface of a cam with a constant angular velocity of $\dot\theta$ = 1.5 rad/s. The radial coordinate of the follower is given by , where t is given in seconds and r is in inches. Plot out the cam profile on the x-y coordinate axes. Also, determine the velocity and acceleration of the cam follower.

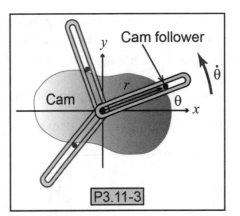

Ans: $\mathbf{v}$ = -6πsin(πt)$\mathbf{e}_r$ + 9(3 + cos(πt))$\mathbf{e}_\theta$ in/s,
$\mathbf{a}$ = (-40.5 - (6π^2 + 13.5)cos(πt))$\mathbf{e}_r$ - 18πsin(πt)$\mathbf{e}_\theta$ in/s^2

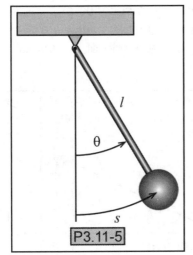

P3.11-4) Car A is traveling along the straight road shown below at a speed of 100 ft/s when the driver spots a police car O on the shoulder. Recognizing that he is speeding, the driver then begins to decelerate at a rate of 10 ft/s^2. The radar gun being used by the policeman at O in essence measures the rate at which the distance r is changing. Determine the speed registered by the radar gun for the instant shown where r = 100 ft and θ = 60°. Also determine whether the registered speed is increasing or decreasing.

Ans: $\dot{r}$ = 52.2 mph, $\ddot{r}$ = 49 ft/s^2, decreasing

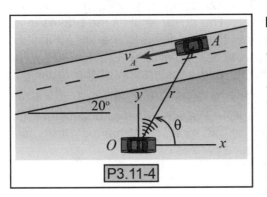

P3.11-5) The tangential acceleration of the end of the pendulum shown depends on its position s and is equal to -$gsin\theta$. Determine the pendulum's angular acceleration as a function of θ, g and l. When the pendulum is at its lowest position, its position is s_o = 0 and the velocity of the pendulum end is v_o. Determine, as a function of the given parameters, the total distance traveled by the pendulum bob between its lowest position and the point where the pendulum momentarily comes to rest.

Ans: $\ddot\theta = \dfrac{-g\sin\theta}{l}$, $s_{max} = l\cos^{-1}\left[1 - \dfrac{v_o^2}{2gl}\right]$

P3.11-6) An airplane flying horizontally at an altitude of $h = 3$ miles is being tracked by a radar station on the ground as shown. The radar's tracking data shows that $\dot{\theta} = 0.01$ rad/sec and $\ddot{\theta} = -0.05$ rad/sec² when θ equals $60°$. Determine the airplane's velocity and acceleration at this instant.

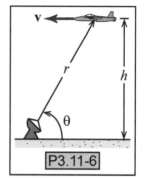

Ans: v = 144 mph $\leftarrow$, a = 1058.4 ft/s² $\rightarrow$

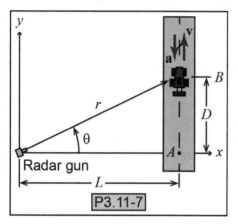

P3.11-7) A policeman is tracking a speeder with a radar gun at a distance of L = 15 yards away from the road. The car spots the policeman when it passes point A. The driver applies the brakes in an attempt to slow the car down to the speed limit by the time he reaches point B where $\theta = 30°$. The angular velocity and acceleration of the radar gun (which is perfectly tracking the car) is $\dot{\theta}$ = 2 rad/s and $\ddot{\theta}$ = -3 rad/s². Was the driver successful if the speed limit is 70 mph? What is the car's acceleration as it passes point B?

Ans: v = 82 mph (not successful), a = 163.3 ft/s²

P3.12) ADVANCED LEVEL POLAR COORDINATE PROBLEMS

P3.12-1) A particle moves in two-dimensional space. Its distance away from the origin of the x-y coordinate system is defined by $r = t - e^{-t/5}$, where t is in seconds and r is in meters. The angle that r make with the x-axis is given by $\theta = 3 + 2e^{-t/5}$, where θ is in radians. Plot the radial and transverse coordinates and the magnitude of the velocity and acceleration versus time from t = 0 to 10 s.

P3.12-2) A radar gun is used to track a race car's speed. The race car has constant acceleration and when it passes position A, it is traveling at 50 mph. As it passes position B, it is traveling at 60 mph. If L is 15 yards, determine the velocity and acceleration of the car as a function of θ in the r-θ coordinate system. Also, if the speed of the microwave radiation emitted from the gun is 186,000 mi/s, determine the time it takes for the radiation to travel from the gun to the car and back again when the car is at position B.

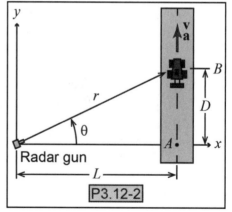

P3.13) BASIC LEVEL RELATIVE MOTION PROBLEMS

P3.13-1) Kayak A approaches the dock with a speed of v_A relative to still water in the direction shown. Kayak B leaves the dock with a speed of v_B relative to still water in the direction shown. Find the speed of kayak A relative to kayak B.

Ans: $v_{A/B} = \sqrt{v_A^2 + 2v_A v_B \cos\theta + v_B^2}$

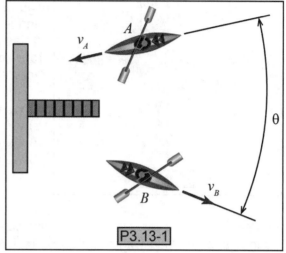

P3.13-1

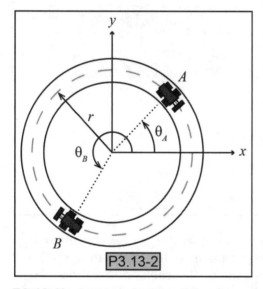

P3.13-2

P3.13-2) Two race cars are running on a circular track of radius r as shown in the figure. The speed of car A and car B are equal to v and both are constant. Determine the acceleration of car A with respect to car B if the cars are in the position shown in the figure.

Ans:

$$\mathbf{a}_{A/B} = \frac{-v^2}{r}((\cos\theta_A - \cos\theta_B)\mathbf{i} + (\sin\theta_A - \sin\theta_B)\mathbf{j})$$

P3.13-3) A paper boy is delivering papers. He is biking due North at 10 mph on the public sidewalk. If the boy throws a newspaper from his bike due West at 30 mph relative to himself, what is the distance prior to the path leading to the porch, d, that he needs to release the paper for the paper to actually reach the porch? The porch is 25 ft from the sidewalk.

Ans: $d = 8.3$ ft

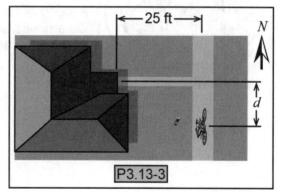

P3.13-3

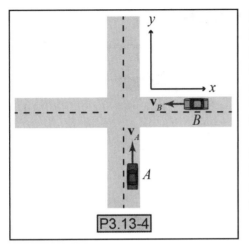

P3.13-4

P3.13-4)[fe] Two cars are approaching an intersection as shown in the accompanying figure. Car A is traveling with a speed of 15 m/s and is slowing down at a rate of 3 m/s^2, while car B has a speed of 20 m/s which is decreasing at a rate of 5 m/s^2. Determine the velocity and acceleration of car B relative to car A.

a) $v_{B/A} = -20i - 15j$ m/s, $a_{B/A} = 5i + 3j$ m/s^2
b) $v_{B/A} = 20i - 15j$ m/s, $a_{B/A} = 5i - 3j$ m/s^2
c) $v_{B/A} = -20i + 15j$ m/s, $a_{B/A} = -5i + 3j$ m/s^2
d) $v_{B/A} = 20i + 15j$ m/s, $a_{B/A} = -5i - 3j$ m/s^2

P3.13-5)[fe] The driver of car A is traveling along a straight road with a speed of 15 m/s while accelerating at 1 m/s^2 when he notices car B directly to his right as shown. If at this instant car B is traveling around a 90-m curve with a speed of 5 m/s while slowing down at a rate 1 m/s^2, what acceleration does car B appear to have to the driver of car A?

a) $a_{B/A} = -2i$ m/s^2
b) $a_{B/A} = -2i - 0.278j$ m/s^2
c) $a_{B/A} = - 0.278j$ m/s^2
d) $a_{B/A} = 2i + 0.278j$ m/s^2

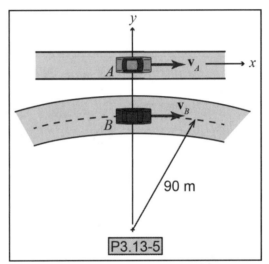

P3.13-5

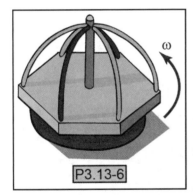

P3.13-6

P3.13-6)[fe] You are standing at the center of a merry-go-round while your friend pushes the platform such that it turns at a constant rate of 0.4 rad/sec. You then begin walking out toward the edge at a constant rate of 0.6 m/s. Determine your speed when you reach a distance of 0.5 m from the center of the platform.

a) $v = 0.63$ m/s   b) $v = 0.60$ m/s

c) $v = 0.20$ m/s   d) $v = 0.52$ m/s

P3.14) INTERMEDIATE LEVEL RELATIVE MOTION PROBLEMS

P3.14-1) A balloon is ascending at a constant rate of v_b = 30 ft/s when a ballast bag is dropped from the basket of the balloon. Five seconds later the balloon is still ascending at the same rate when a gust of wind with a velocity of v_w = 50 ft/s begins to blow the balloon horizontally (without appreciably affecting the ballast bag). At this instant, what is the velocity of the ballast bag as seen by an observer in the balloon?

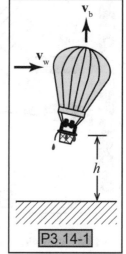

Ans: $\mathbf{v}_{bag/balloon} = -50\mathbf{i} - 161\mathbf{j}$ ft/s

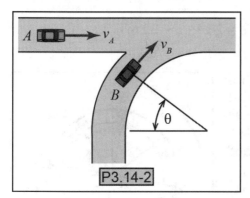

P3.14-2) Car A is traveling along a straight road with a speed of 20 m/s increasing at a rate of 8 m/s^2 when it sees car B attempting to merge. Car B is traveling with constant speed 10 m/s around a curve with radius 20 m. What velocity and acceleration does car B appear to have to the driver of car A at the instant when θ = 40 degrees?

Ans: $\mathbf{v}_{B/A}$ = -13.57$\mathbf{i}$ + 7.66$\mathbf{j}$ m/s, $\mathbf{a}_{B/A}$ = -4.17$\mathbf{i}$ - 3.21$\mathbf{j}$ m/s^2

P3.14-3) A boat takes t_1 seconds to travel a distance of d upriver parallel to the flow of the river. The boat then turns around and travels downriver for t_2 seconds before it returns to its initial starting position. If the velocity of the river and the speed of the boat relative to the river are constant, prove that $v_{boat/river} = d(t_1 + t_2)/(2t_1t_2)$

P3.14-4) Kayak A approaches the dock with a speed of 3 knots relative to the flow of the river in the direction shown ($\theta = 60°$). Kayak B leaves the dock with a speed of 5 knots relative to the flow of the river in the direction shown ($\phi = 40°$). The flow of the river may be assumed to be straight and in the direction shown. Calculate the speed of kayak A relative to kayak B and prove that this value does not depend on the river speed (v_R). Also, calculate the absolute speed of both kayaks.

Ans: $v_{A/B}$ = 7 knots, v_A = 4.13 knots, v_B = 6.46 knots.

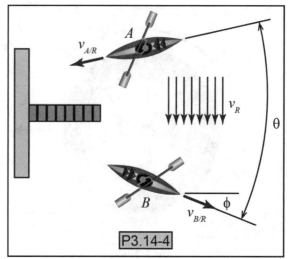

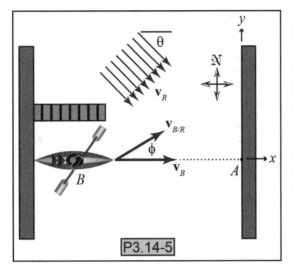

P3.14-5

P3.14-5) A kayak starts on the west river bank and would like to travel straight across to the east bank to point A. The river's current is traveling at a constant 2 knots in the south-east direction (i.e. $\theta = 45°$).   If the man paddling the kayak is able to maintain a constant 3 knots relative to the river's current and the river is 0.25 miles wide, what direction must he paddle in and how long will it take to cross the river? The length of the kayak may be neglected.

Ans: $\phi = 28.1°$, $t = 3.2$ min

3.14-6) Consider the Ferris wheel as shown with a radius of 10 meters. At this instant, a little girl in Car B drops her toy out of her car and, due to air resistance, it falls straight down with an acceleration of 5 m/s^2. If θ = 45°, $\dot\theta$ = -0.5 rad/sec and $\ddot\theta$ = 0.05 rad/sec^2, what acceleration does the toy appear to have to the passengers in Car A? You may assume that Car A is a particle concentrated at the point where it is attached to the Ferris wheel

Ans: $\mathbf{a}_{toy/A} = 2.12\mathbf{i} - 3.59\mathbf{j}$ m/s^2

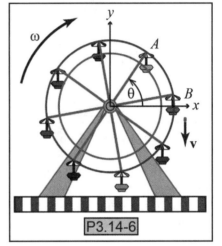

P3.14-6

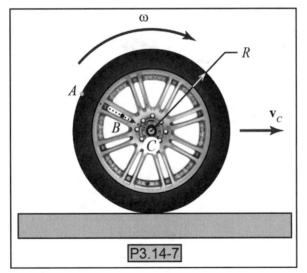

P3.14-7

P3.14-7) A tire rolls with an angular velocity of ω. The velocity of the wheel's center, C, is constant and its magnitude is equal to $v_C = R\omega$, in the direction shown, where R is the radius of the tire. The speed of any point, P, on the tire relative to the tire center is equal to $v_{P/C} = r_{P/C}\omega$, where $r_{P/C}$ is the distance of point P relative to the tire center.   The direction of $v_{P/C}$ is perpendicular to the line connecting C and P in the direction of tire rotation. Point A is on the tire surface that contacts the ground and point B is located on the radial line connecting points C and A. If the distance between points A and B is d, derive the speed of point A relative to point B as a function of d, v_C and R.

Ans: $v_{A/B} = v_C d / R$

P3.15) ADVANCED LEVEL RELATIVE MOTION PROBLEMS

P3.15-1) A horizontal surface and pin P move back and forth in the x-direction as described by the function $x = A\sin(\pi t)$, where t is time and A is the amplitude. A pinned slotted bar is constrained to move with pin P as shown in the figure. The distance from the origin O of the x-y coordinate system to pin P is r_P and the distance from O to a point Q on the bar is r. Determine the velocity of pin P with respect to point Q in terms of θ, $\dot{\theta}$, r, t and A.

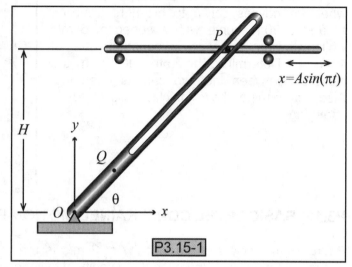

Ans: $\mathbf{v}_{P/Q} = (A\pi\cos(\pi t) + r\dot{\theta}\sin\theta)\mathbf{i} - r\dot{\theta}\cos\theta\mathbf{j}$

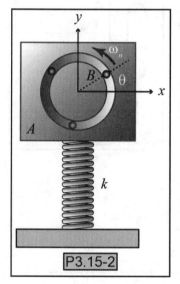

P3.15-2) Block A, having a mass of $m = 15$ kg, rests on a spring of spring constant $k = 200$ N/m. A groove of radius 10 cm is cut into block A allowing ball B (of negligible mass) to move in a circle with a frequency that is equal to the natural frequency of the block A/spring system which is $\omega_n = \sqrt{k/m}$ rad/s. This causes block A to move as $y_A(t) = A_1\sin(\omega_n t) + A_2\cos(\omega_n t)$, where y_A is in meters, t is in seconds and constants A_1 and A_2 are determined by block A's initial conditions. If $v_{B/A} = r\omega_n$ and the initial conditions are $y_A(0) = 0.03$ m and $\dot{y}_A(0) = 0$, what is the velocity of ball B when $\theta = 20°$ (i.e. the angle between the ball and the x-axis) and $t = 20$ s?

P3.15-3) The driver of car A has a speed of 15 m/s and is accelerating at a constant 1 m/s^2 when he notices car B to his right traveling around a 90-m curve with a speed of 5 m/s while slowing down at a constant rate of 1 m/s^2. Determine the rate at which the distance between the two cars is changing. Also, determine how quickly this rate is changing.

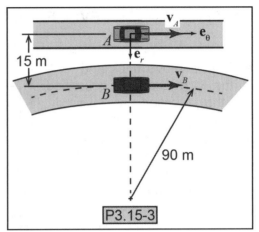

P3.15-3

P3.16) BASIC LEVEL CONSTRAINED MOTION PROBLEMS

P3.16-1) If block B has a velocity of 2 m/s in the direction shown on the figure, determine the velocity of cylinder A.

Ans: $v_A = (2/3) \downarrow$ m/s

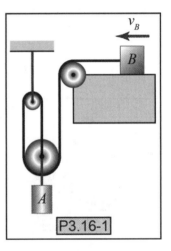

P3.16-1

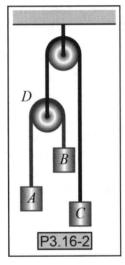

P3.16-2

P3.16-2) Block C is moving downward with a velocity of 5 m/s, which is increasing at a rate of 1 m/s^2, while block A is moving upward with a velocity of 7 m/s, which is decreasing at a rate of 2 m/s^2. Determine the velocity and acceleration of block B at this instant.

Ans: $v_B = 3 \uparrow$ m/s, $a_B = 4 \uparrow$ m/s^2

P3.16-3) In the figure shown, the mass B is falling at a constant rate of 5 m/s. Determine the velocity of mass A as a function of the position of mass B.

Ans: $v_A = -10 s_B / \sqrt{25 + s_B^2} \uparrow$ m/s

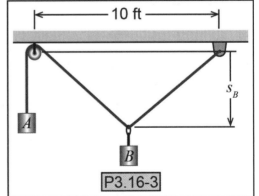

P3.16-3

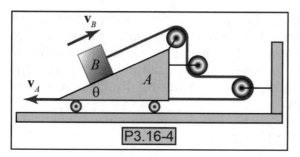

P3.16-4)

P3.16-4) Cart A moves to the left causing block B to slide up on the cart, as shown in the figure. Determine the velocity of B in terms of v_A and θ.

Ans: $v_B = v_A(1 - 2\cos\theta) \uparrow$

P3.17) INTERMEDIATE LEVEL CONSTRAINED MOTION PROBLEMS

P3.17-1)   If the end of the rope at A has an upward velocity of $v_A = 12$ m/s, determine the velocity of block B.

Ans: $v_B = 1.71 \uparrow$ m/s

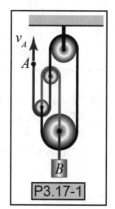

P3.17-1

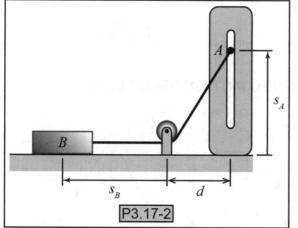

P3.17-2

P3.17-2) Pin A is moving upward with a velocity of $v_A = 10$ m/s and a downward acceleration of $a_A = 3$ m/s^2.   Determine the velocity and acceleration of Block B at the instant $s_A = 0.5$ m and $s_B = 0.7$ m ($d = 0.25$ m).

Ans: $v_B = 8.94$ m/s $\rightarrow$,
         $a_B = 33.1$ m/s$^2 \rightarrow$

P3.17-3)   Consider the 6 m long ladder shown. If point A of the ladder contacting the wall is sliding down with a constant velocity of 6 m/s, then determine the acceleration of point B of the ladder contacting the floor at the instant when y is equal to 2 m.

Ans: $\ddot{x} = -7.16$ m/s^2

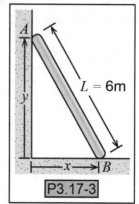

P3.17-3

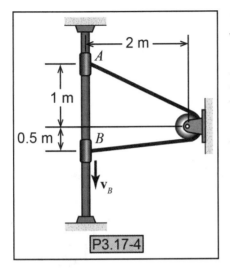

P3.17-4)

P3.17-4) Collars A and B slide along a vertical rod while connected via a cord that passes over the fixed pulley shown. If collar B is sliding downward at 20 cm/s at the instant shown, determine the corresponding velocity of collar A. Neglect the size of the pulley.

Ans: v_A = 0.11 m/s $\downarrow$

P3.17-5) Load B is moving down with velocity v_B. Determine the velocity and acceleration of crate A in terms of the motion of load B and the parameters shown.

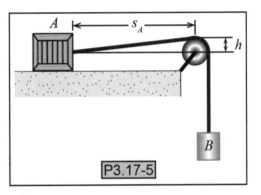

P3.17-5

Ans: $v_A = \dfrac{v_B\sqrt{s_A^2 + h^2}}{s_A} \rightarrow$,

$a_A = a_B\sqrt{1 + \left(\dfrac{h}{s_A}\right)^2} + \dfrac{v_B^2 h^2}{s_A^3} \rightarrow$

P3.18) ADVANCED LEVEL CONSTRAINED MOTION PROBLEMS

P3.18-1) A man uses a pulley system to lift a heavy collar A as shown in the figure. The man pulls down on the rope maintaining a constant angle. At the instant represented in the figure, the velocity of collar A is 1 m/s upward, its acceleration is 2 m/s² downward, height h = 5 m and distance D = 4 m. Determine the velocity and acceleration with which the man is pulling the rope.

Ans: v_B = 0.78 m/s $\downarrow$, a_B = 1.62 m/s² $\uparrow$

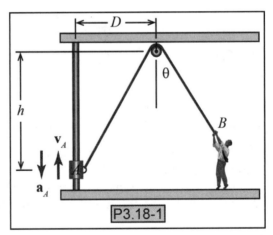

P3.18-1

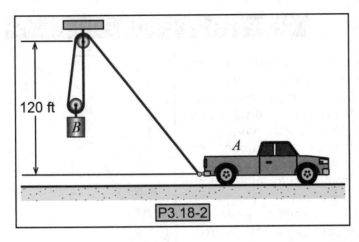

P3.18-2 0.112 ft/s

P3.18-2) A pickup truck is initially at rest below the pulley system shown when it begins accelerating at a constant rate of 0.5 ft/s^2. Determine the velocity of load B when $t = 6$ seconds.

Ans: $v_B = 0.112$ ft/s ↑

CHAPTER 3 COMPUTER PROBLEMS

C3-1) A golf ball is struck with an initial speed of 112.6 ft/s at an angle of 53.2°, resulting in a hole in one. Find the total horizontal distance traveled by the ball in the air and the total time the ball is in the air. Usually we negligible air resistance. This assumption is helpful since it allows analysis of the x- and y-directions of motion separately. For this problem, assume that in addition

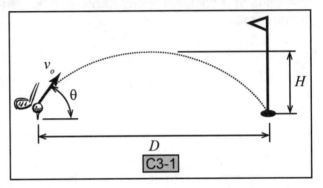

C3-1

to the acceleration due to gravity, there is acceleration due to air resistance that has a magnitude equal to $a = 0.002v^2$ ft/s^2 and is directed opposite the ball's direction of motion. Since the x- and y-directions are now coupled, use a numerical approach to solve this problem. If we had assumed no air resistance, would this have been a good assumption?

C3-2) A vehicle is traversing a constant-radius curve. The magnitude of the car's acceleration is limited by the 'friction circle' of the vehicle's tires, that is $\sqrt{a_t^2 + a_n^2} \leq a_{max}$, where a_t is the longitudinal acceleration of the vehicle (for braking and acceleration) and a_n is the lateral acceleration of the vehicle (for cornering). If the radius of the curve is $r = 80$ m and the maximum acceleration the tires are able to provide is $a_{max} = 8$ m/s^2, determine the minimum possible stopping distance (measured along the vehicle's curved trajectory) for the vehicle beginning

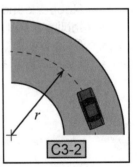

C3-2

from an initial speed of 20 m/s. Note: you must take into account that the friction-limited braking deceleration increases as the vehicle speed is reduced. Use a numerical approach for determining the answer.

CHAPTER 3 DESIGN PROBLEMS

D3-1) Consider that you are designing a freeway off-ramp that transitions cars from traveling on a straight section of roadway to a circular curve of radius 500 ft. If there is no transition between these road segments, the car passengers will experience an abrupt and uncomfortable change in acceleration even if the speed of the car remains constant. This change will be due to the normal component of the car's acceleration which will jump from zero to a large non-zero value between the two segments. Design an intermediate curve of the form $x=(y/a)^3$ (commonly employed in practice) that will transition the car's acceleration from some nominally small value at $x = 0$ ft to the acceleration experienced by the car on the circular segment of the off-ramp (radius = 500 ft) at $x = 1000$ ft. Assume the vehicle maintains at a constant speed of 50 mph as it travels along the off-ramp.

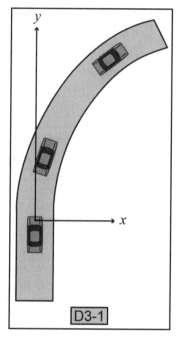

D3-1

CHAPTER 3 ACTIVITIES

A3-1) This activity gives the student experience measuring the range of a projectile and then using that range to determine its initial velocity.

Group size: Minimum of 3

Supplies

- Small metal ball or a marble
- Rubber band
- Inclined surface
- Stop watch
- Measuring tape

Experimental setup

Wrap the rubber band around the end of the incline so that the ball can be shot up and off the incline using the rubber band like a sling shot. Place evenly space marks on the top of the incline.

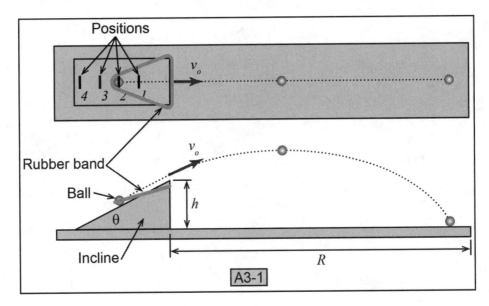

Procedure

- Measure the angle (θ) and the height (h) of the incline.
- Place the ball in the rubber band and stretch it to position 1.
- While student 1 releases the ball, student 2 measures the time it takes for the ball to hit the ground and student 3 measures the ball's range.
- Repeat for the other 3 positions.

Calculations and discussion

- Calculate the initial velocity of the projectile for each experimental run.
- Plot the initial velocity versus range.
- List all of the assumptions that were applied in your calculations.
- Discuss what experimental errors may have influenced your results.
- Will these error result in a lower or higher value of the initial velocity than you calculated?

NOTES

PART II: KINEMATICS

CHAPTER 4: KINEMATICS OF RIGID BODIES

CHAPTER OUTLINE

CHAPTER SUMMARY

In this chapter we begin to analyze the motion of rigid bodies. Recall that rigid bodies have size as well as mass. Therefore, the rotation of a rigid body must be considered in addition to its translation. We will begin to look, not only at a body's linear motion, but also at its angular motion characteristics (e.g. angular velocity and angular acceleration). We will start simple with pure rotation and then move on to general planar motion.

4.1) RIGID-BODY MOTION

4.1.1) WHAT IS A RIGID BODY?

There are certain circumstances where a particle representation of a body in motion is not sufficiently accurate. A clear indication that a particle is not suited for the job is if rotation is an important part of the body's motion. In this case and in other circumstances, we will need to represent the object under consideration as a rigid body. A **rigid body** is different from a particle because it has both mass and size, whereas, a particle only has mass. Therefore, with a rigid-body model, the body's rotation must be considered in addition to its translation.

4.1.2) HOW DOES A RIGID BODY MOVE?

A rigid body can move in pure translation, pure rotation, or general planar motion. **Pure translation** occurs when the body goes from one location to another without changing its orientation. All points on the rigid body undergoing pure translation have the same velocity and acceleration (i.e. $\mathbf{v}_A = \mathbf{v}_B$, $\mathbf{a}_A = \mathbf{a}_B$). This is because the paths taken by each of the points are parallel to one another. **Pure rotation** occurs when the body is rotating about some fixed axis and only the orientation of the body is changing. In this case, all parts of the body move in circular paths about the fixed axis. All points on the body rotate through the same angular displacement at the same time. If the body is undergoing **general planar motion**, it is simultaneously translating and rotating.

Our prior study of translational kinematics involved the investigation of the relationship between the displacement, velocity, and acceleration of a body. Completely analogous relationships exist for the rotational kinematics of a rigid body and are shown below. Since a rigid body can both translate and rotate, both sets of equations apply, given a certain set of circumstances. We will look at these sets of equations in detail in the upcoming sections. Specifically, we will investigate the relationship between a body's angular displacement θ, its angular velocity ω, and its angular acceleration α.

Rotational Equations	**Translational Equations**
$$\omega = \frac{d\theta}{dt}$$	$$v = \frac{ds}{dt}$$
$$\alpha = \frac{d\omega}{dt} = \frac{d^2\theta}{dt^2}$$	$$a = \frac{dv}{dt} = \frac{d^2s}{dt^2}$$
$$\alpha\, d\theta = \omega\, d\omega$$	$$a\, ds = v\, dv$$

What is the difference between the motion of a particle and the motion of a rigid body?

Conceptual Example 4.1-1

Draw possible paths for the motion of bar AB as it moves from point 1 to point 2 in pure translation.

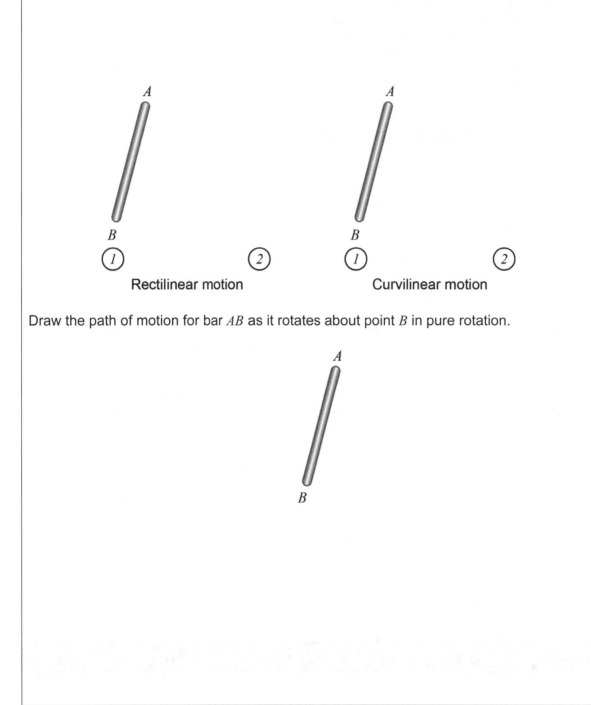

Rectilinear motion   Curvilinear motion

Draw the path of motion for bar AB as it rotates about point B in pure rotation.

4.2) PURE ROTATION

4.2.1) ROTATION ABOUT A FIXED AXIS

We will first consider the case of a rigid body rotating about a fixed axis as shown in Figure 4.2-1. **Pure rotation** occurs when a body rotates about a fixed non-moving axis. Under the rigid-body assumption, the distance between any two points on the body remains constant. Therefore, each point on the body can be considered to be moving in concentric circles about the fixed axis. Fixed-point O, shown in Figure 4.2-1, is a point the coincides with the fixed axis, in this case, the z-axis.

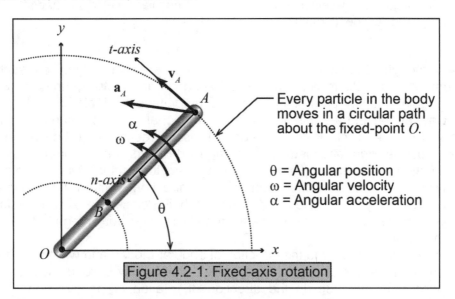

Every particle in the body moves in a circular path about the fixed-point O.

θ = Angular position
ω = Angular velocity
α = Angular acceleration

Figure 4.2-1: Fixed-axis rotation

4.2.2) ANGULAR KINEMATIC RELATIONSHIPS

In previous chapters, we have studied linear position (s), linear velocity (v), and linear acceleration (a). Each of these motion parameters are related to each other through kinematic relationships (i.e., $dv = ds/dt$, $a = dv/dt$, $ads = vdv$). In this section, we will learn about the parameters that describe the rotational characteristics of a body's motion. These motion parameters are angular position (θ), angular velocity (ω) and angular acceleration (α). As one would expect, angular velocity is the time derivative of angular

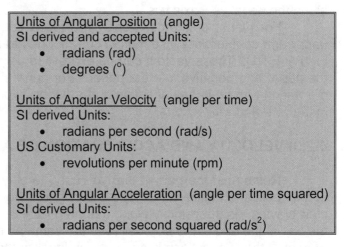

Units of Angular Position (angle)
SI derived and accepted Units:
- radians (rad)
- degrees ($^\circ$)

Units of Angular Velocity (angle per time)
SI derived Units:
- radians per second (rad/s)
US Customary Units:
- revolutions per minute (rpm)

Units of Angular Acceleration (angle per time squared)
SI derived Units:
- radians per second squared (rad/s^2)

displacement (i.e. $\omega = \dot{\theta}$), and angular acceleration is the time derivative of angular velocity (i.e. $\alpha = \dot{\omega}$) as shown in Equations 4.2-1 through 4.2-3. In general, units of radians, rad/sec and rad/sec^2 are employed for angular displacement, angular velocity and angular acceleration, respectively. However, degrees and revolutions per minute (rpm) are also common units for angular position and angular velocity, respectively. It is always a good idea to convert non-radian units into radians before performing calculations to avoid errors due to unit mismatch.

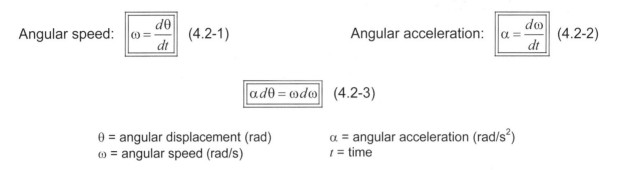

Angular speed: $\boxed{\omega = \dfrac{d\theta}{dt}}$ (4.2-1)   Angular acceleration: $\boxed{\alpha = \dfrac{d\omega}{dt}}$ (4.2-2)

$$\boxed{\alpha\, d\theta = \omega\, d\omega}$$ (4.2-3)

θ = angular displacement (rad) α = angular acceleration (rad/s^2)
ω = angular speed (rad/s) t = time

Even though Equations 4.2-1 through 4.2-3 are not written as such, angular displacement, angular velocity and angular acceleration are vector quantities. Specifically, all of the angular motion parameters have a magnitude and direction of rotation. A simplified way of looking at this is that the rotational direction can be clockwise or counterclockwise. This clockwise/counterclockwise direction can be expressed as a negative or positive sign attached to the magnitude. For more complex systems, the rotational direction is expressed with unit direction vectors, usually **k** for a planar motion (i.e. 2-D) system, or as a combination of unit direction vectors for motion in three dimensions. Whether you indicate the direction as being clockwise or counterclockwise, or with unit direction vectors such as **k**, these quantities have both magnitude and direction making them vectors.

An easy way to determine the direction of angular motion and whether it is positive or negative is through the use of the right-hand rule. Here's how it works: take the fingers on your right hand and curl them in the direction of rotation. The direction of the vector representing the angular motion is indicated by the direction of your thumb. For example, if your thumb is pointing along the positive y-axis, the direction is **j**. If your thumb is pointing along the negative x-axis it is -**i**.

For planar problems, clockwise motion corresponds to a vector pointing in to the paper and counterclockwise motion corresponds to a vector pointing out of the paper (as in Figure 4.2-1). These vectors can be expressed using the unit vector **k** that is positive out of the paper and negative into the paper. For general three-dimensional motion, the direction of rotation can be in any direction and must be expressed as a combination of components in the **i**-, **j**- and **k**-directions.

4.2.3) VELOCITY AND ACCELERATION OF A POINT

Remember the games that you used to play outside as a kid, such as jumping rope or skip ball, see Figure 4.2-2. Think about how the jump rope and skip ball move when you start playing. Is the velocity along the jump rope the same at every point? Is the velocity of the ball at the end of the skip ball rope the same as the ring around your ankle? The answer to both of these questions is no. We can idealize these games and assume that the velocity at the hands (in the case of jumping rope) and the ankle (in the case of skip ball) is very near zero. This makes these situations similar to fixed-axis rotation. As we move out from the hands or ankle, the velocity will increase until we reach a maximum velocity at the apex of the jump rope or at the ball. The key point that these two examples illustrate is that the velocity of every point on a rigid body undergoing fixed-axis rotation may be different. The velocity of a particular point depends on the distance it is away from the fixed axis. The magnitude of the velocity increases the further away you get from the fixed axis of rotation.

Figure 4.2-2: Jump rope and skip ball

The velocity of each point on a rigid body can be derived by considering the bar, shown in Figure 4.2-3, that moves from position *1* to *2*. The arc length scribed by point *A* is given by the arc length equation $ds = rd\theta$. Taking the time derivative of the arc length equation gives us an equation for the velocity of point *A*, which is $v_A = r_A\dot{\theta}$. This equation shows the relationship between velocity and distance from the fixed axis as previously mentioned. This equation may be extended to any point on the body. We also know that the velocity is always in the tangential direction. The velocity of any point on a rigid body, in pure rotation, is given by Equation 4.2-4.

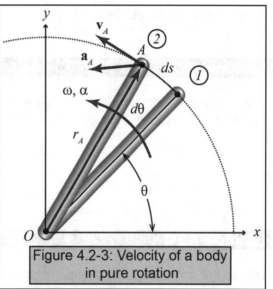

Figure 4.2-3: Velocity of a body in pure rotation

The acceleration of each point on a rigid body can be obtained by using the normal and tangential coordinate acceleration equation presented in the chapter on *Kinematics of Particles - Curvilinear Motion* and making the substitution for the velocity with the one given in Equation 4.2-4. The normal and tangential acceleration equation is $\mathbf{a} = \dot{v}\mathbf{e}_t + (v^2/\rho)\mathbf{e}_n$. Substituting $v = r\omega$ and $\dot{v} = r\dot{\omega} = r\alpha$, we get the Equation 4.2-5. This equation may be used to calculate the acceleration of any point on a rigid body undergoing pure rotation. This equation can also be derived from the expression for acceleration in the polar coordinate frame when r is constant.

Velocity: $\boxed{\mathbf{v}_A = r_A\omega\mathbf{e}_t}$ (4.2-4)   Acceleration: $\boxed{\mathbf{a}_A = r_A\alpha\mathbf{e}_t + r_A\omega^2\mathbf{e}_n}$ (4.2-5)

$\mathbf{v}_A$ = linear velocity of point *A* ω = angular speed
$\mathbf{a}_A$ = linear acceleration of point *A* α = angular acceleration
r_A = distance of point *A* from the fixed axis

Note that while velocity and acceleration on the rigid body are dependent on radial location (i.e. distance away from the fixed axis of rotation), the body as a whole has only a single angular velocity ω and angular acceleration α. Consider for example the body shown in Figure 4.2-1. A line drawn from point O to point A will sweep out the same angle as a line drawn from point O to point B for the same time interval. The distance travelled by point A, however, will be significantly larger than the distance travelled by point B during that same time interval. This is because point A is located at a radius that is farther from point O than it is from point B.

Conceptual Example 4.2-1

What is the velocity direction of any point on a rigid body undergoing planar pure rotation? Choose all that are true.

    a) Tangent to its path of motion.
    b) Perpendicular to the line drawn from the fixed axis to the point in question.
    c) $\mathbf{e}_t$
    d) $\mathbf{e}_\theta$
    e) None of the above.

Example 4.2-2

Use the arc length equation to derive the velocity of point A on the rigid body shown in Figure 4.2-3 which is undergoing pure rotation.

Example 4.2-3

Derive the acceleration of any point on a rigid body undergoing pure rotation in terms of the radial distance (r), angular speed (ω), and angular acceleration (α) starting with the two equations given below.

$$\mathbf{a} = \dot{v}\mathbf{e}_t + \frac{v^2}{\rho}\mathbf{e}_n \qquad\qquad \mathbf{v} = \dot{r}\mathbf{e}_r + r\dot{\theta}\mathbf{e}_\theta$$

Equation 4.2-4 and 4.2-5 gives the velocity and acceleration for an arbitrary point on a rigid body. When you use these equations, you are forced to work in the normal and tangential or the polar coordinate systems. If you wished to express the velocity and acceleration of a particle as a vector in rectangular coordinates, geometry could be employed to decompose given vectors into their rectangular components, but this can be very tedious. Another approach for finding the vector expressions for velocity and acceleration in any non-rotating coordinate system is to use the following relationships involving the cross product. For instructions on how to perform a cross product, refer to Appendix B.

Velocity: $\boxed{\mathbf{v}_A = \boldsymbol{\omega} \times \mathbf{r}_{A/O}}$ (4.2-6)

Acceleration for 2-D applications: $\boxed{\mathbf{a}_A = \boldsymbol{\alpha} \times \mathbf{r}_{A/O} - \omega^2 \mathbf{r}_{A/O}}$ (4.2-7)

Acceleration for 3-D applications: $\boxed{\mathbf{a}_A = \boldsymbol{\alpha} \times \mathbf{r}_{A/O} + \boldsymbol{\omega} \times (\boldsymbol{\omega} \times \mathbf{r}_{A/O})}$ (4.2-8)

$\mathbf{v}_A$ = linear velocity of point A
$\mathbf{a}_A$ = linear acceleration of point A
$\boldsymbol{\omega}$ = angular velocity

$\boldsymbol{\alpha}$ = angular acceleration
$\mathbf{r}_{A/O}$ = position of point A relative to the fixed axis through O

Conceptual Example 4.2-4

Consider the figure of the wind turbine shown.

1) What is the velocity of the center O?

2) The angular velocity of point A (a point midway between the center O and the tip of the blade) is _____ that of point B (a point at the tip of the blade).

  a) half
  b) the same as
  c) twice
  d) None of the above.

3) The velocity of point A is _____ that of point B.

  a) half
  b) the same as
  c) twice
  d) None of the above.

Conceptual Example 4.2-5

A cat sits without slipping on the edge of a spinning disk. The disk is turning in the direction shown and is slowing down.

1) At the instant shown, what is the direction of the radial component of the cat's acceleration?
2) At the instant shown, what is the direction of the tangential component of the cat's acceleration?
3) At the instant shown, what is the direction of the angular velocity of the cat?

a) $+x$ b) $-x$ c) $+y$ d) $-y$ e) $+z$ f) $-z$

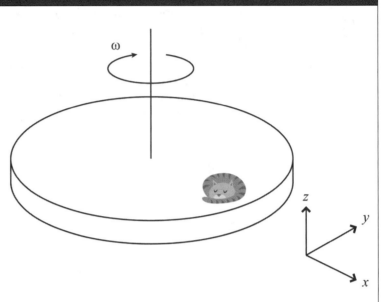

Equation Summary

Abbreviated variable definition list

θ = angular displacement (rad)
ω = angular speed (rad/s)

α = angular acceleration (rad/s^2)
t = time

Rotation

$$\omega = \frac{d\theta}{dt} \qquad \alpha = \frac{d\omega}{dt} \qquad \alpha\, d\theta = \omega\, d\omega$$

Constant angular acceleration

$$\alpha = \alpha_o \qquad \omega = \alpha_o(t - t_o) + \omega_o \qquad \theta = \frac{1}{2}\alpha_o(t - t_o)^2 + \omega_o(t - t_o) + \theta_o$$

$$\omega^2 = \omega_o^2 + 2\alpha_o(\theta - \theta_o)$$

Non-constant angular acceleration

$$\int_{t_o}^{t} \omega(t)\,dt = \int_{\theta_o}^{\theta} d\theta \qquad \int_{t_o}^{t} \alpha(t)\,dt = \int_{\omega_o}^{\omega} d\omega \qquad \int_{\theta_o}^{\theta} \alpha(\theta)\,d\theta = \int_{\omega_o}^{\omega} \omega\,d\omega$$

Example Problem 4.2-6

The blades on the wind turbine are turning with an angular speed of ω_o = 30 rpm in the clockwise direction. If the blade length is L = 120 ft and they are given a constant angular acceleration of α = 0.2 rad/s^2 in the clockwise direction, determine the angular velocity and the magnitude of the acceleration of point A on the tip of the blade when t = 1 minute.

Given:

Find:

Solution:

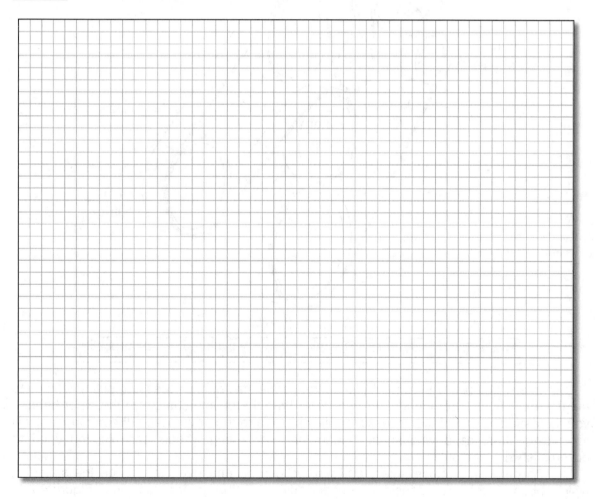

4.2.4) GEARS AND BELTS

Shafts are a common rotational motion component in machines. Often it is necessary to transfer the rotation of one shaft to another. Transferring the motion of one shaft to another on a one-to-one basis is not always useful. This means that if the first shaft is rotating with an angular speed of ω_1, the second shaft will rotate with the same angular speed. Many times we want to transfer motion and change the rate at which the shaft is rotating. Take, for example, the accessory components in your car (e.g. compressor, water pump, alternator). These accessory components are driven by the crankshaft, which in turn, is driven by the piston motion. The crankshaft rotates at one speed and the other components that run off of it need to run at different speeds. Therefore, the crankshaft rotation needs to be transferred and the speed needs to be modified. This is accomplished by using a belt (i.e. the serpentine belt) and a series of pulleys (see Figure 4.2-4). It is easy to see how the belt transfers the rotational motion, but how does the speed change if the belt is continuously running at a constant speed? This is accomplished by using different sized pulleys.

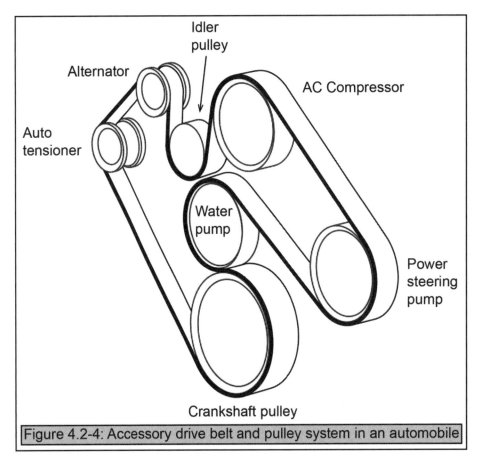

Figure 4.2-4: Accessory drive belt and pulley system in an automobile

In any belt-pulley system, if you decrease the size of the accessory pulley relative to the drive pulley, that component will run faster than the driving shaft. On the other hand, if you increase the size of the accessory pulley relative to the drive pulley, the component will run slower than the driving shaft. Gears and contacting wheels can also be used to accomplish similar results.

As discussed above, gears and pulleys can be employed to change the speed at which a body is rotating. Consider one gear driving another gear as shown in Figure 4.2-5.

The driven gear is given an angular speed of ω_1 and will turn the other gear with an angular speed of ω_2. The two angular speeds are related through the ratio of their radii. This follows from the fact that the linear velocity of the contact points for each of the gears are equal; this is assuming that the gears do not slip with respect to one another.

$$v_{1,\text{point of contact}} = v_{2,\text{point of contact}} \qquad \Rightarrow \qquad r_1\omega_1 = r_2\omega_2 \qquad \Rightarrow \qquad \omega_2 = \frac{r_1}{r_2}\omega_1$$

The above relationship works for gears, pulleys with belts, and non-slipping wheels in contact. The logic for a pulley-belt system follows from the fact that the belt has the same speed at every point. Which means that every point on the rim of a pulley in contact with the belt will have the same speed. The speed shared by the belt and pulley is proportional to the radius of the pulley (i.e. $v = r\omega$). Gears have an additional relationship shown below. Their angular speeds can be related through the ratio of their number of teeth (N). If the two

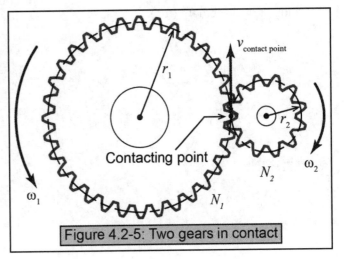

Figure 4.2-5: Two gears in contact

gears where designed to mesh together, the ratio of the gear radii is equal to the gear teeth ratio.

$$\omega_2 = \frac{N_1}{N_2}\omega_1$$

A similar relationship for the angular acceleration may be derived. The tangential acceleration ($a_t = r\alpha$) of the gear's contacting point or the belt/pulley interface must be shared and equal. Therefore, the following relationship for the angular acceleration between two gears or a belt and two pulleys are as follows.

$$a_{t1,\text{point of contact}} = a_{t2,\text{point of contact}} \qquad \Rightarrow \qquad r_1\alpha_1 = r_2\alpha_2 \qquad \Rightarrow \qquad \alpha_2 = \frac{r_1}{r_2}\alpha_1$$

Equation Summary

Abbreviated variable definition list

s = arc length

$\mathbf{v}$ = velocity

$\mathbf{a}$ = acceleration

$\mathbf{r}$ = position

θ = angular displacement (rad)

ω = angular velocity (rad/s)

α = angular acceleration (rad/s^2)

r = gear/pulley 1 radius

N = number of teeth

Pure rotation

$$s_A = r_A \theta \qquad \mathbf{v}_A = r_A \omega \mathbf{e}_t \qquad \mathbf{v}_A = \boldsymbol{\omega} \times \mathbf{r}_{A/O}$$

$$\mathbf{a}_A = r_A \alpha \mathbf{e}_t + r_A \omega^2 \mathbf{e}_n \qquad \mathbf{a}_A = \boldsymbol{\alpha} \times \mathbf{r}_{A/O} - \omega^2 \mathbf{r}_{A/O} \qquad \mathbf{a}_A = \boldsymbol{\alpha} \times \mathbf{r}_{A/O} + \boldsymbol{\omega} \times (\boldsymbol{\omega} \times \mathbf{r}_{A/O})$$

Gears and belts

$$v_{common} = r_1 \omega_1 = r_2 \omega_2 \qquad N_1 \omega_1 = N_2 \omega_2 \qquad a_t = r_1 \alpha_1 = r_2 \alpha_2$$

Example Problem 4.2-7

In the following figure, mass C is attached to a rope that is wrapped around the outer edge of drum B. As mass C falls, the rope around the drum unrolls without slipping, causing the drum to rotate. This, in turn, causes pulley A to rotate due to the belt connecting the drum's inner radius and the pulley. Consider that mass C is released from rest and reaches a velocity of 1 m/s after it has fallen 30 cm. If the mass accelerates at a constant rate, find for the instant shown
   a) the magnitude of the acceleration of point P on the outer surface of the drum,
   b) the angular velocity and angular acceleration of drum B, and
   c) the angular velocity and angular acceleration of pulley A.

Given:

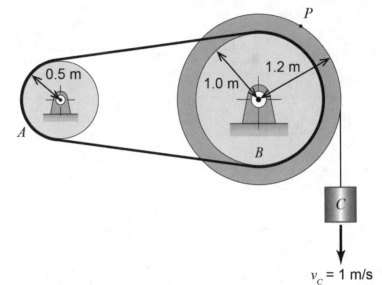

Find:

$v_C = 1$ m/s

Solution:

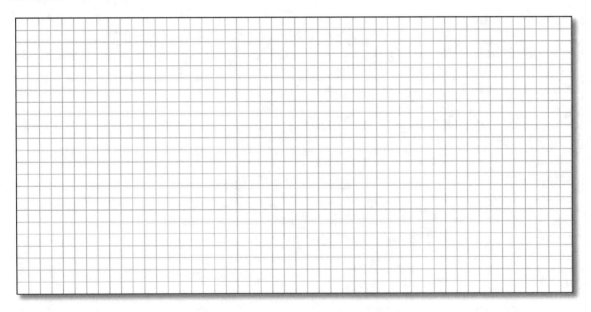

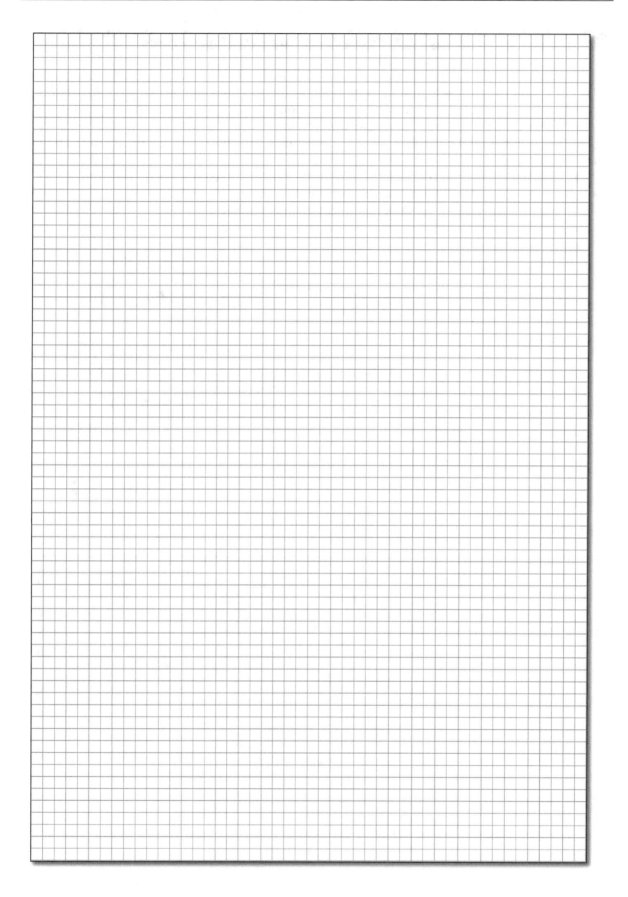

Example Problem 4.2-8

The square plate rotates about the fixed pivot O. At the instant represented, the direction of the angular velocity, ω, and angular acceleration, α, of the plate are shown in the figure.   Determine the velocity and acceleration of points A and B in the x-y coordinate frame.

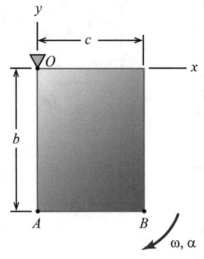

Given:

Find:

Solution:

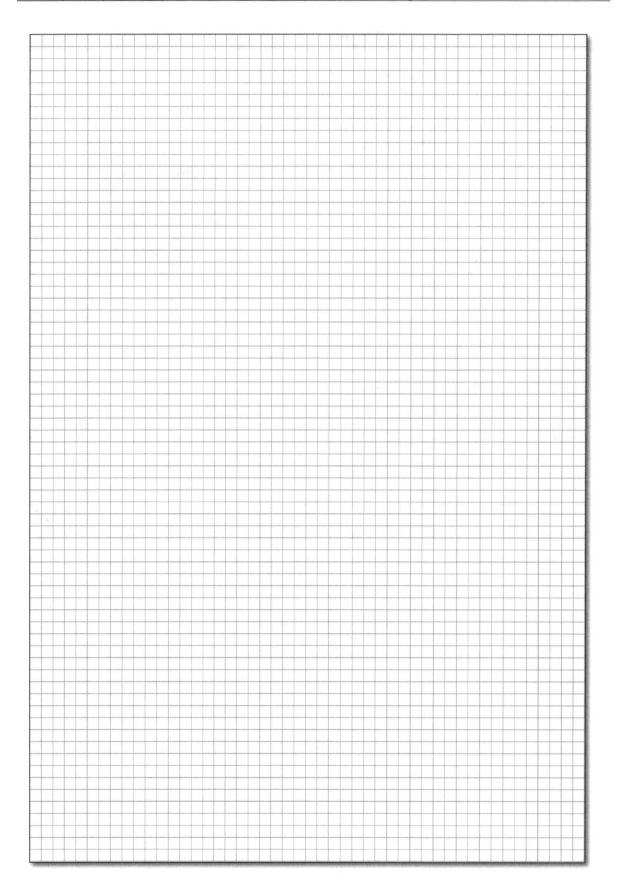

Solved Problem 4.2-9

Disk B rotates with an angular velocity of $\omega_B = 3t^2$ about a fixed point O_B in the direction shown, where t is in seconds and ω is in revolutions per minute. Disk B rotates disk A through a friction interface. Disk A also rotates about a fixed point O_A. Determine the angular displacement, angular velocity and angular acceleration of disk B after 10 seconds. At the start of motion, point b (on the circumference of disk B) is located at $\theta_B = 0°$. After 10 second, calculate the velocity and acceleration of point b in both n-t and x-y coordinates. Also, after 10 seconds, determine the angular velocity and angular acceleration of disk A assuming there is no slip between the disks. The radii of disk A and B are 4 inches and 8 inches respectively.

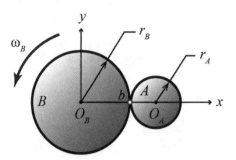

Given: $\omega_B = 3t^2$ rpm $r_B = 8$ in
    $r_A = 4$ in $\theta_{bo} = 0°$
    $t_f = 10$ s

Find: $\Delta\theta_B,\ \omega_{Bf},\ \alpha_{Bf}$
    $v_{bf},\ a_{bf}$ in n-t and x-y coordinates
    $\omega_{Af},\ \alpha_{Af}$

Solution:

We are given the angular velocity ω_B of disk B and need to find both the angular displacement $\Delta\theta_B$ and angular acceleration α_B after 10 seconds. We will solve for the unknowns by using the basic relationships shown below. Since ω_B is a function of time, α_B is not constant and we cannot use the constant acceleration relationships.

$$\omega = \frac{d\theta}{dt} \qquad \alpha = \frac{d\omega}{dt}$$

We will start by finding the final angular velocity. We first need to convert its units to radians per second.

$$\omega_B = 3t^2 \frac{\text{rev}}{\text{min}}\left(\frac{2\pi\,\text{rad}}{\text{rev}}\right)\left(\frac{\text{min}}{60\,\text{s}}\right) = 0.1\pi t^2 \frac{\text{rad}}{\text{s}} \qquad \omega_{Bf} = 0.1\pi t_f^2 \frac{\text{rad}}{\text{s}}$$

$$\boxed{\omega_{Bf} = 31.4 \frac{\text{rad}}{\text{s}}}$$

Now that we have ω_B in radians per second, we can calculate the final angular acceleration and angular displacement. The angular acceleration is obtained by differentiating the angular velocity and the angular displacement is obtained by integrating the angular velocity.

$$\alpha_B = \frac{d\omega_B}{dt} = 0.2\pi t \,\frac{rad}{s^2}$$

$$\alpha_{Bf} = 0.2\pi t_f \,\frac{rad}{s^2}$$

$$\boxed{\alpha_{Bf} = 6.28 \,\frac{rad}{s^2}}$$

$$\omega_B = \frac{d\theta_B}{dt} \quad \rightarrow \quad \int_0^{t_f} \omega_B \, dt = \int_0^{\theta_{Bf}} d\theta_B$$

$$\int_0^{t_f} 0.1\pi t^2 \, dt = \int_0^{\theta_{Bf}} d\theta_B$$

$$\left.\frac{0.1\pi t^3}{3}\right|_0^{t_f} = \left.\theta_B\right|_0^{\theta_{Bf}}$$

$$\frac{0.1\pi t_f^3}{3} = \Delta\theta_B$$

$$\Delta\theta_B = 104.72 \,\text{rad}\left(\frac{180^o}{\pi \,\text{rad}}\right)$$

$$\boxed{\Delta\theta_B = 6000^o}$$

Now that we know the angular displacement, we can determine the final position of point b by subtracting out the sixteen full rotations that disk B made.

$$\theta_{bf} = 240^o$$

Now that we know the final position of point b, we can calculate its velocity and acceleration. Let's start with the n-t coordinate system.

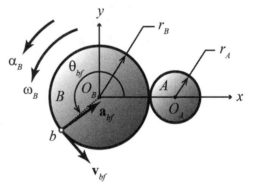

$$\mathbf{v}_{bf} = r_B \omega_{Bf} \, \mathbf{e}_t$$

$$\boxed{\mathbf{v}_{bf} = 251.3 \, \mathbf{e}_t \,\frac{in}{s}}$$

$$\mathbf{a}_{bf} = r_B \alpha_{Bf} \, \mathbf{e}_t + r_B \omega_{Bf}^2 \, \mathbf{e}_n$$

$$\boxed{\mathbf{a}_{bf} = 50.3 \, \mathbf{e}_t + 7895.7 \, \mathbf{e}_n \,\frac{in}{s^2}}$$

Now let's calculate the velocity and acceleration of point b in the x-y coordinate system.

$$\mathbf{v}_b = \boldsymbol{\omega}_{Bf} \times \mathbf{r}_{b/O_B} = \omega_{Bf}\mathbf{k} \times r_B(\cos\theta_{bf}\mathbf{i} + \sin\theta_{bf}\mathbf{j})$$
$$= \omega_{Bf} r_B(\cos\theta_{bf}\mathbf{j} - \sin\theta_{bf}\mathbf{i})$$

$$\boxed{\mathbf{v}_{bf} = 217.7\mathbf{i} - 125.7\mathbf{j}\,\frac{in}{s}}$$

$$\mathbf{a}_{bf} = \boldsymbol{\alpha}_{Bf} \times \mathbf{r}_{b/O} - \omega_{Bf}^2 \mathbf{r}_{b/O} = \alpha_{Bf}\mathbf{k} \times r_B(\cos\theta_{bf}\mathbf{i} + \sin\theta_{bf}\mathbf{j}) - \omega_{Bf}^2 r_B(\cos\theta_{bf}\mathbf{i} + \sin\theta_{bf}\mathbf{j})$$
$$= \alpha_{Bf} r_B(\cos\theta_{bf}\mathbf{j} - \sin\theta_{bf}\mathbf{i}) - \omega_{Bf}^2 r_B(\cos\theta_{bf}\mathbf{i} + \sin\theta_{bf}\mathbf{j})$$

$$\boxed{\mathbf{a}_{bf} = 3987.3\mathbf{i} + 6805.8\mathbf{j}\,\frac{in}{s^2}}$$

Disk B transfers motion to disk A through a friction contact. For every revolution of disk B, disk A needs to complete two revolutions (based on their radii). Therefore, disk A will have an angular velocity and angular acceleration that is twice that of disk B. But, let's confirm this with some calculations.

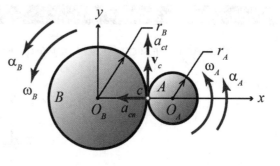

The point of contact between the two disks must have the same velocity.

$$v_{c,B} = v_{c,A} \qquad r_B \omega_B = r_A \omega_A \qquad \omega_A = \frac{r_B}{r_A}\omega_B \qquad \boxed{\omega_{Af} = 62.8 \ \frac{\text{rad}}{\text{s}}}$$

The contact point between the two disks must have the same tangential acceleration.

$$a_{c,Bt} = a_{c,At} \qquad r_B \alpha_B = r_A \alpha_A \qquad \alpha_A = \frac{r_B}{r_A}\alpha_B \qquad \boxed{\alpha_{Af} = 12.6 \ \frac{\text{rad}}{\text{s}^2}}$$

4.3) GENERAL PLANAR MOTION

4.3.1) RELATIVE VELOCITY AND ACCELERATION

Unrestricted, a rigid body can both rotate and translate. In the last section, we looked at the simplified case of pure rotation. We essentially said that the body was restricted to rotate about a fixed axis and, therefore, could only rotate and not translate. In this section, we will look at general planar motion. **General planar motion** allows for simultaneous rotational and translational motion, however, the body is restricted to move within a 2-D plane. This may sound complex, but it can be shown that general planar motion can be considered as a simple superposition of translation plus rotation about a point on the body as shown in Figure 4.3-1.

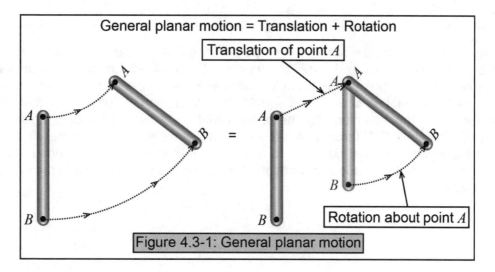

Figure 4.3-1: General planar motion

Conceptual Example 4.3-1

Watch the animation on *Rigid-body Motion* and answer the following questions.

What kind of motion is this?

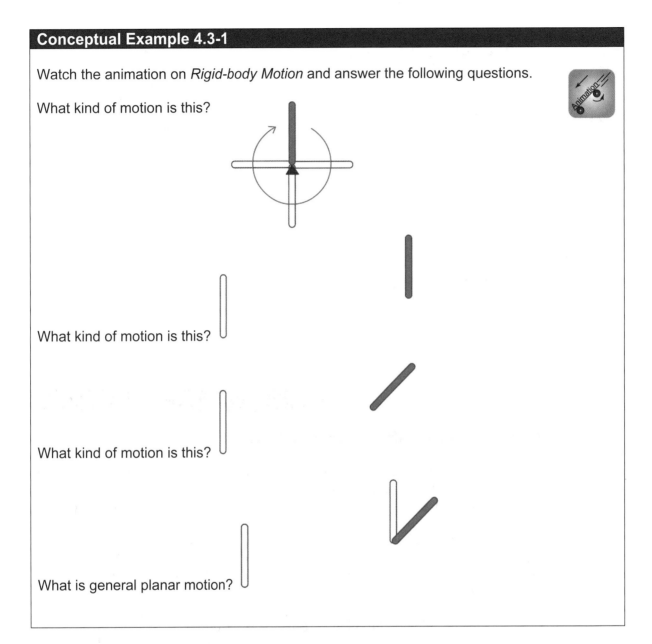

What kind of motion is this?

What kind of motion is this?

What is general planar motion?

As we have seen, the motion of a rigid body can be considered as a combination of translation and rotation. Consider the bar shown in Figure 4.3-2. In general planar motion, the movement of some point A on the rigid body can be separated into a translational and a rotational part. The motion of point A is equal to the translation of point B plus the rotation of point A about B. The equations for the relative motion of two points on a rigid body are given by Equations 4.3-1 through 4.3-3 when the vectors are given in a non-rotating reference frame. It is always important to remember that when we speak about the motion of a rigid body, the velocity and acceleration of each point on the rigid body may differ if the body is rotating.

Relative velocity: $\boxed{\mathbf{v}_A = \mathbf{v}_B + \mathbf{v}_{A/B} = \mathbf{v}_B + \boldsymbol{\omega} \times \mathbf{r}_{A/B}}$ (4.3-1)

Relative acceleration 2D: $\boxed{\mathbf{a}_A = \mathbf{a}_B + \mathbf{a}_{A/B} = \mathbf{a}_B + \boldsymbol{\alpha} \times \mathbf{r}_{A/B} - \omega^2 \mathbf{r}_{A/B}}$ (4.3-2)

Relative acceleration 3D: $\boxed{\mathbf{a}_A = \mathbf{a}_B + \mathbf{a}_{A/B} = \mathbf{a}_B + \boldsymbol{\alpha} \times \mathbf{r}_{A/B} + \boldsymbol{\omega} \times (\boldsymbol{\omega} \times \mathbf{r}_{A/B})}$ (4.3-3)

$\mathbf{r}_{A/B}$ = position of point A relative to point B
$\mathbf{v}_A$, $\mathbf{v}_B$ = absolute velocity of points A and B respectively
$\mathbf{v}_{A/B}$ = velocity of point A relative to point B

$\mathbf{a}_A$, $\mathbf{a}_B$ = absolute acceleration of points A and B respectively
$\mathbf{a}_{A/B}$ = acceleration of point A relative to point B
ω = angular velocity of the rigid body (rad/s)
α = angular acceleration of the rigid body (rad/s^2)

Let's also look at this in another way. To describe the motion of a point on a rigid body that is undergoing general planar motion, we will use the concept of a body-fixed coordinate system. A body-fixed coordinate system is a coordinate system whose origin is attached to and moves with some point on the rigid body. Consider the body shown in Figure 4.3-2. A body-fixed coordinate system (x'-y') is attached to and moves with point B. Point B moves relative to the fixed/inertial coordinate system (x-y) with velocity $\mathbf{v}_B$ and acceleration $\mathbf{a}_B$. Then, consider point B as being on a fixed axis that point A rotates about. In this way, one can envision general planar motion as a superposition of translation ($\mathbf{v}_B$) and rotation ($\mathbf{v}_{A/B}$). Since the distance between points A

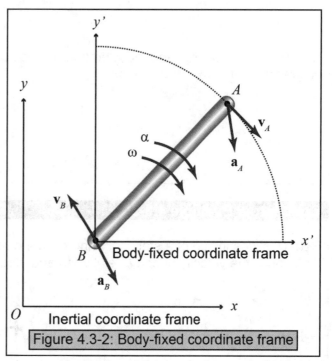

Figure 4.3-2: Body-fixed coordinate frame

and B are always constant in a rigid body, the motion of point A relative to point B is circular (i.e. $\boldsymbol{\omega} \times \mathbf{r}_{A/B}$) and the relative velocity is always perpendicular to the line connecting A and B. Note, the general planar motion of the body could be considered as the translation and rotation about any point on the body, not just point B.

Example 4.3-2

Referring to Figure 4.3-2, derive the equation for the velocity and acceleration of point A using the following equation for fixed-axis rotation and relative motion.

Fixed-axis rotation equations:
$$\mathbf{v}_A = \boldsymbol{\omega} \times \mathbf{r}_{A/O}$$
$$\mathbf{a}_A = \boldsymbol{\alpha} \times \mathbf{r}_{A/O} - \omega^2 \mathbf{r}_{A/O}$$

Relative motion equations:
$$\mathbf{v}_A = \mathbf{v}_B + \mathbf{v}_{A/B}$$
$$\mathbf{a}_A = \mathbf{a}_B + \mathbf{a}_{A/B}$$

What is the velocity and acceleration of point A if point B is fixed and A is rotating about B?

Knowing the velocity and acceleration of point B, what is the absolute velocity and acceleration of A?

Conceptual Example 4.3-3

Rod AB slides inside of a stationary ring C. If the rod never loses contact with the ring, what direction is the velocity of end B relative to end A at the instant shown?

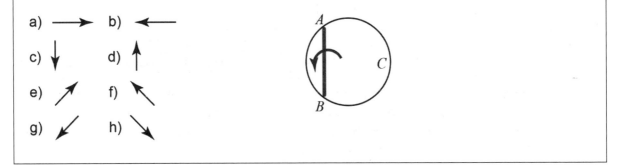

a) ⟶   b) ⟵

c) ↓   d) ↑

e) ↗   f) ↖

g) ↙   h) ↘

4.3.2) MECHANISMS

Many mechanical systems involve rigid bodies connected together in various ways to convert motion into different forms. For example, the linear motion of a piston in an automobile engine is transferred into rotary motion as shown in Figure 4.3-3. Another example is the bicycle mechanism that converts your rotary motion into straight line motion as shown in Figure 4.3-4. Other mechanisms, such as gears, are used to change the speed and direction of rotation (see Figure 4.3-5).

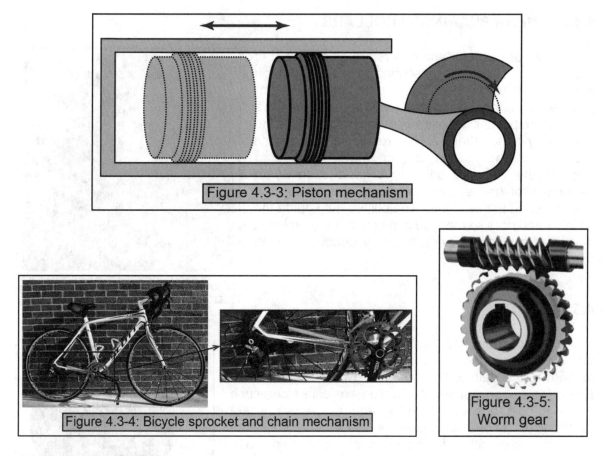

Figure 4.3-3: Piston mechanism

Figure 4.3-4: Bicycle sprocket and chain mechanism

Figure 4.3-5:
Worm gear

4.3.3) CONSTRAINTS

We examined constrained motion in the chapter on *kinematics of particles - curvilinear motion* and saw how various constraints affected the motion of particles. In this section, we will extend this knowledge and look at some examples of the constrained motion of rigid bodies.

Many mechanisms that you will see in real life have constrained motion. For example, a piston is constrained to move in the direction of the cylinder. A car is constrained to move along the road profile on which it is driving (that is, unless you are Steve McQueen in the film Bullet.) If we can identify and characterize each constraint in a mechanism, it will aid in its analysis.

Many times complex systems can be modeled as simple mechanisms or connecting elements. Examples of some types of mechanisms that constrain or connect bodies are linear bearings, joints, and surface contacts.

4.3.4) LINEAR BEARINGS / COLLARS

Linear bearings and collars are sleeves that fit around or on shafts as described in the chapter on *kinematics of particles - curvilinear motion*. If a rod or linkage is connected to the bearing or collar, the connection point is constrained to move along the direction of the shaft. The rotational constraints imposed on the rod depend on the joint used to connect the rod and collar. A good example of bearing/collar use is a weight machine, such as, the one shown in Figure 4.3-6. The collars help guide the weights and bars so that you can concentrate on lifting the weights without having to worry about balancing the weights.

Figure 4.3-6: Excercise machine

4.3.5) SLOTS

A slot constrains the motion of a pin to move in the direction of the slot. The velocity of the pin is always tangent to the path of the slot as shown in Figure 2.8-3. Examples of constrained slot motion are coin slots, such as in a coin operated laundry, and a Geneva wheel mechanism (shown in Figure 4.3-7). The coin slot constrains the coins to move along a certain path, however; the coins are free to rotate. The Geneva wheel transfers rotational motion between two shafts at prescribed intervals. The interval frequency depends on the wheel's design and the angular velocity of the driving shaft.

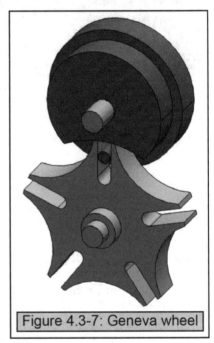

Figure 4.3-7: Geneva wheel

4.3.6) JOINTS

Joints can restrict the motion of mechanisms that they are attached to from 1 to 6 degrees of freedom. In total, a rigid body has 6 degrees of freedom or 6 ways in which it can move. A completely unconstrained body can translate in the x-, y- and z-directions and can rotate about the x-, y- and z-directions for a total of 6 degrees of freedom.

There are four main types of joints: a clamped or welded joint, a pinned or hinged joint, a ball and socket joint and a roller joint. A clamped joint restricts all 3 directions of translation and all 3 directions of rotation. This means that the end condition attached to a clamped joint does not move. A pinned joint restricts all 3 directions of translation at the joint and allows for free rotation in one direction about the joint. A ball and socket restricts all 3 directions of translation and allows for free rotation in all three directions about the joint. The last type of joint is a roller joint. This joint may allow up to 3 directions of translational motion and all directions of rotation. All types of joints are illustrated in Figure 4.3-8.

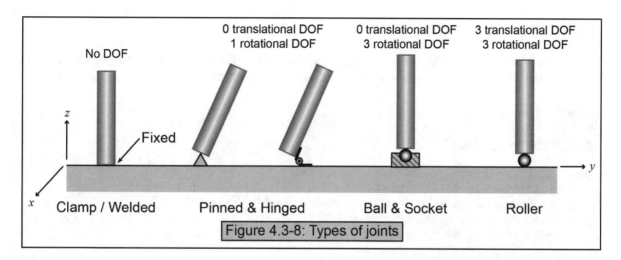

Figure 4.3-8: Types of joints

4.3.7) SURFACE CONTACTS

Surface contacts constrain the motion of some point on the rigid body to the profile of the surface. For example, if a car is riding along a road, the tires are forced to move up and down with the undulations of the road (see Figure 4.3-9).

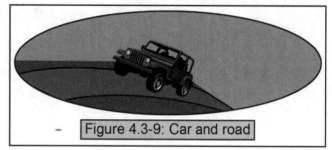

Figure 4.3-9: Car and road

4.3.8) ANALYZING MECHANISMS

We can analyze the planar motion of rigid-body mechanisms using the relative velocity and acceleration equations given in Equations 4.3-1 and 4.3-2. We will first present an overall strategy for analyzing these systems before examining some more commonly encountered examples.

Step 1: Get familiar with the system

- Determine if there are any fixed axes/points of rotation in the system or any points of known velocity.
- Determine if there are any geometric constraints. For example, a point sliding in a channel or along a surface, or a rolling contact.
- Assign angular velocity directions. Your assignment may be incorrect initially, but if you are consistent in your analysis, this error will be discovered in your results.

Step 2: Analyze the velocities

- Start at a position that is fixed or that has known velocity and work outward.
- Sometimes two different relationships can be found for the motion of a single point by starting from different points and meeting in the middle.
- Apply known information and geometric constraints to eliminate unknowns.

Step 3: Analyze the accelerations

- Repeat the process presented for the velocity. Typically the expressions used to calculate the accelerations of the system depend on the velocities of the system.

Equation Summary

Abbreviated variable definition list

$\mathbf{v}_A$, $\mathbf{v}_B$ = absolute velocity of points A and B respectively
$\mathbf{v}_{A/B}$ = velocity of point A relative to point B
$\mathbf{a}_A$, $\mathbf{a}_B$ = absolute acceleration of points A and B respectively
$\mathbf{a}_{A/B}$ = acceleration of point A relative to point B
$\mathbf{r}_{A/B}$ = position of point A relative to point B
ω = angular velocity of the rigid body (rad/s)
α = angular acceleration of the rigid body (rad/s^2)

General planar motion

$$\mathbf{v}_A = \mathbf{v}_B + \omega \times \mathbf{r}_{A/B}$$

$$\mathbf{a}_A = \mathbf{a}_B + \alpha \times \mathbf{r}_{A/B} - \omega^2 \mathbf{r}_{A/B} \qquad \mathbf{a}_A = \mathbf{a}_B + \alpha \times \mathbf{r}_{A/B} + \omega \times (\omega \times \mathbf{r}_{A/B})$$

Example Problem 4.3-4

Consider the following diagram of a piston assembly in an internal combustion engine. The crankshaft OA is rotating at a constant rate of 2000 rpm counter-clockwise. If the piston is at top dead center position ($\theta = 0°$), determine (a) the velocity and acceleration of piston head B, (b) the angular velocity of connecting rod AB, and (c) the angular acceleration of connecting rod AB.

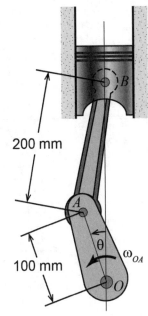

Given:

200 mm

100 mm

Find:

Solution:

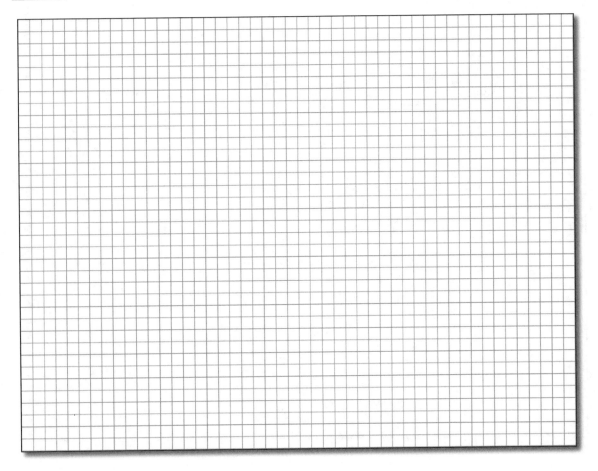

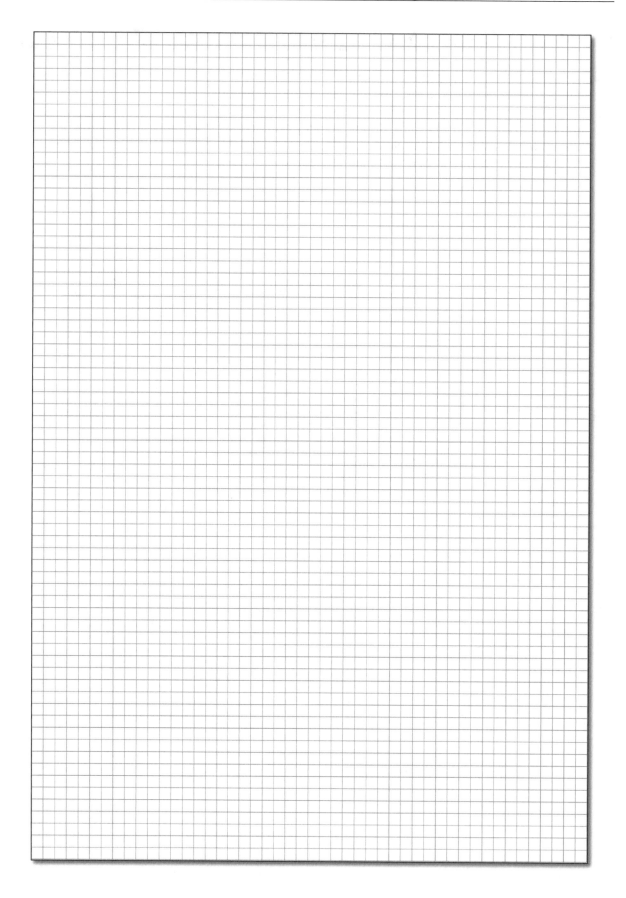

Example Problem 4.3-5

Consider the four-bar linkage shown. If link CD is being driven with an angular velocity of 4 rad/s, determine the velocity of point P on link BC and the angular velocity of link AB at the instant shown.

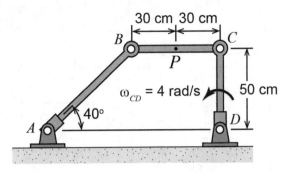

Given:

Find:

Solution:

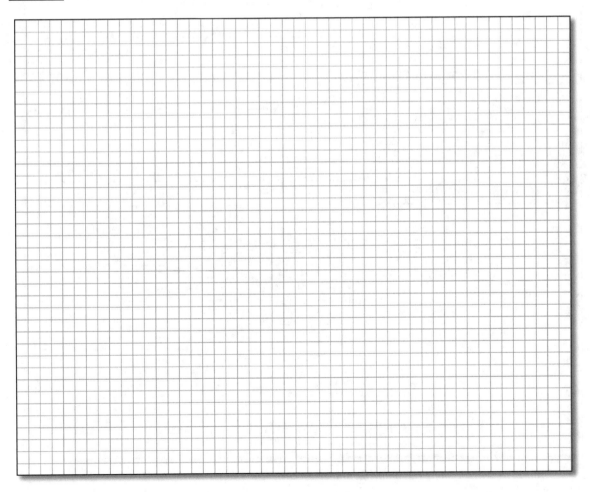

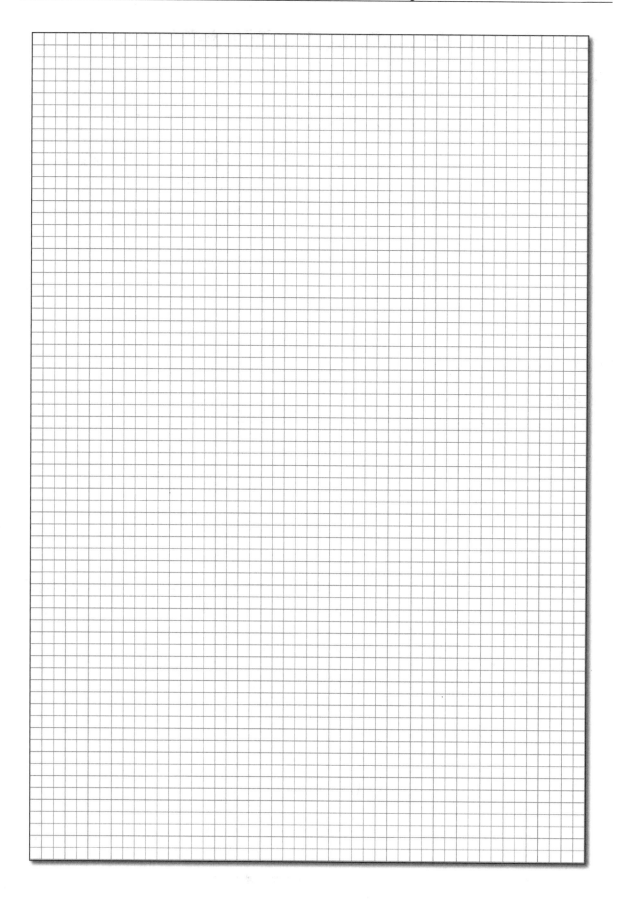

The three-bar mechanism shown is put into motion by an actuator. The actuator applies a known velocity (v_C) and acceleration (a_C) to the end of link C. Link C is constrained to move horizontally by a collar. Links AB and BC are the same length and end A is fixed by a pin joint. Determine the angular velocity and angular acceleration of links AB and BC when the links make an angle of θ with the horizontal, as shown.

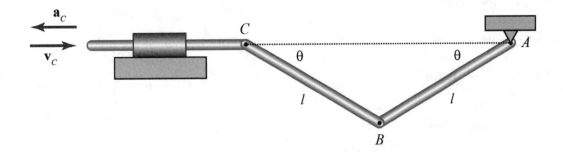

Given: v_C, a_C, The length of link AB and BC, θ

Find: ω_{AB}, ω_{BC}, α_{AB}, α_{BC}

Solution:

Step 1: Getting familiar with the system

Assign all linear velocity, angular velocity and angular acceleration directions. If you are uncertain of the direction of the angular velocity or angular acceleration, that's okay, take a guess. If you are incorrect, the angular velocity or angular acceleration result will be negative letting you know that you guessed wrong. Also, you will need to add a coordinate system if one is not specified for you.

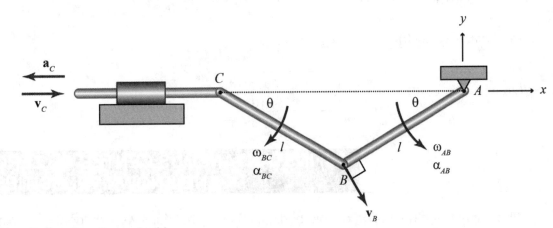

Step 2: Analyze the velocities

Our strategy for solving this problem is to work from a known velocity (joint A) to joint C. We can then use the constraint applied to link C to determine the unknowns.

Let's look at link AB.

$$\mathbf{v}_B = \mathbf{v}_A + \boldsymbol{\omega}_{AB} \times \mathbf{r}_{B/A} = 0 + \omega_{AB}\mathbf{k} \times l(-\cos\theta\mathbf{i} - \sin\theta\mathbf{j}) = \omega_{AB}l(-\cos\theta\mathbf{j} + \sin\theta\mathbf{i})$$

Now let's look at link BC and get an expression for the velocity of C.

$$\mathbf{v}_C = \mathbf{v}_B + \boldsymbol{\omega}_{BC} \times \mathbf{r}_{C/B} = \mathbf{v}_B - \omega_{BC}\mathbf{k} \times l(-\cos\theta\mathbf{i} + \sin\theta\mathbf{j}) = \mathbf{v}_B + \omega_{BC}l(\cos\theta\mathbf{j} + \sin\theta\mathbf{i})$$

$$= l\left[(\omega_{AB} + \omega_{BC})\sin\theta\mathbf{i} + (-\omega_{AB} + \omega_{BC})\cos\theta\mathbf{j}\right]$$

We will use the fact that $\mathbf{v}_C$ does not have a $\mathbf{j}$ component to solve for the unknowns.

$$\mathbf{j}: 0 = l(-\omega_{AB} + \omega_{BC})\cos\theta \qquad\qquad \omega_{AB} = \omega_{BC} = \omega$$

$$\mathbf{i}: v_C = l(\omega_{AB} + \omega_{BC})\sin\theta = 2\omega l\sin\theta \qquad \boxed{\omega = \dfrac{v_C}{2l\sin\theta}}$$

Step 3: Analyze the accelerations

We will use a similar procedure to solve for the angular accelerations.

$$\mathbf{a}_B = \mathbf{a}_A + \boldsymbol{\alpha}_{AB} \times \mathbf{r}_{B/A} - \omega^2\mathbf{r}_{B/A} = 0 + \alpha_{AB}\mathbf{k} \times l(-\cos\theta\mathbf{i} - \sin\theta\mathbf{j}) - \omega^2(-\cos\theta\mathbf{i} - \sin\theta\mathbf{j})$$

$$= \alpha_{AB}l(-\cos\theta\mathbf{j} + \sin\theta\mathbf{i}) - \omega^2(-\cos\theta\mathbf{i} - \sin\theta\mathbf{j}) = l\left[(\alpha_{AB}\sin\theta + \omega^2\cos\theta)\mathbf{i} + (-\alpha_{AB}\cos\theta + \omega^2\sin\theta)\mathbf{j}\right]$$

$$\mathbf{a}_C = \mathbf{a}_B + \boldsymbol{\alpha}_{BC} \times \mathbf{r}_{C/B} - \omega^2\mathbf{r}_{C/B} = \mathbf{a}_B - \alpha_{BC}\mathbf{k} \times l(-\cos\theta\mathbf{i} + \sin\theta\mathbf{j}) - \omega^2(-\cos\theta\mathbf{i} + \sin\theta\mathbf{j})$$

$$= \mathbf{a}_B + \alpha_{BC}l(\cos\theta\mathbf{j} + \sin\theta\mathbf{i}) - \omega^2(-\cos\theta\mathbf{i} + \sin\theta\mathbf{j})$$

$$= l\left[(\alpha_{AB}\sin\theta + \omega^2\cos\theta + \alpha_{BC}\sin\theta + \omega^2\cos\theta)\mathbf{i} + (-\alpha_{AB}\cos\theta + \omega^2\sin\theta + \alpha_{BC}\cos\theta - \omega^2\sin\theta)\mathbf{j}\right]$$

$$= l\left[(\alpha_{AB}\sin\theta + 2\omega^2\cos\theta + \alpha_{BC}\sin\theta)\mathbf{i} + (-\alpha_{AB}\cos\theta + \alpha_{BC}\cos\theta)\mathbf{j}\right]$$

$$\mathbf{j}: 0 = l(-\alpha_{AB}\cos\theta + \alpha_{BC}\cos\theta) \qquad\qquad \alpha_{AB} = \alpha_{BC} = \alpha$$

$$\mathbf{i}: a_C = l(2\alpha\sin\theta + 2\omega^2\cos\theta) \qquad \boxed{\alpha = \dfrac{a_C - 2l\omega^2\cos\theta}{2l\sin\theta}}$$

4.4) INSTANTANEOUS CENTER OF ZERO AND KNOWN VELOCITIES

When analyzing the motion of rigid bodies, it is very helpful if we can identify a point of known velocity. This velocity gives us a starting point and helps us solve for relevant unknown information. We have already discussed some instances where points have known velocities. A fixed axis/point of rotation has a known velocity that is equal to zero. If the end of a link is constrained to slide or move along another component of the mechanism, then we may not know the velocity. But, we know something about its direction. In this section, we will focus on how to identify other points of known velocity.

The first type of point we will look at is a specialized case called the *instantaneous center of zero velocity* or *IC.* Then, we will look at special cases, such as rolling, where points of known velocities can be determined by looking at the motion of other components in the system.

4.4.1) INSTANTANEOUS CENTER OF ZERO VELOCITY

The **instantaneous center of zero velocity** (IC), for a rigid body, is a point on or off the rigid body that has zero velocity for an instant. The IC may be treated as a fixed axis of rotation for that instant. Even though the IC has zero velocity at a particular instant, it usually has non-zero acceleration. This means that the location of the IC may change with time. Utilizing the IC of a rotating body often saves calculation effort. If the location of the IC is known, solving for the velocity of any other point on the rigid body is simplified as shown in Equation 4.4-1.

Velocity relative to the IC: $\boxed{\mathbf{v}_P = \mathbf{v}_{IC} + \mathbf{v}_{P/IC} = \boldsymbol{\omega} \times \mathbf{r}_{P/IC}}$ (4.4-1)

$\mathbf{v}_P$ = velocity of point P $\boldsymbol{\omega}$ = angular velocity
$\mathbf{v}_{IC}$ = velocity of the IC $\mathbf{r}_{P/IC}$ = position of point P relative to the IC

Not only can the IC simplify the necessary calculations, but it can also help in visualizing the magnitude and direction of the velocity for any point on the rigid body. The direction of the velocity of any point on the body is perpendicular to the line connecting the IC and the point in question. The magnitude of the velocity is equal to the body's angular velocity multiplied by this length. For example, Figure 4.4-1 shows a disk rolling without slip on a stationary ground. The IC is located at the point where the disk contacts the ground. This is

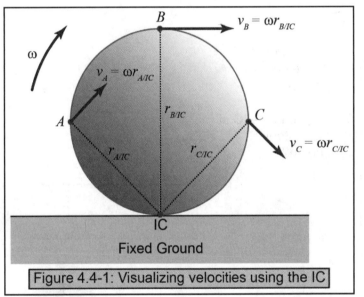

Figure 4.4-1: Visualizing velocities using the IC

true because there cannot be a discontinuity between the velocity of the ground and the velocity of the rolling body if there is no slip. Which means that, if the ground has zero velocity, the contacting point will also have zero velocity. This can be extended to say that whatever velocity the ground has (zero or non-zero) the contacting point will also have if there is no slip. Referring to Figure 4.4-1, the velocity of point A is perpendicular to the line connecting the IC and point A and its speed is equal to $\omega r_{A/IC}$. Also, the speed of point B is greater than the speed of point A because it is farther away from the IC. When trying to identify an IC for a given situation, keep the following two things in mind.
1. The velocity at any point on an object is always perpendicular to its position vector as measured from the IC.
2. The velocities are proportional to the distance that the point is from the IC.

Example 4.4-1

Watch the *IC* animation and answer the following questions.

1)   What direction is the velocity of point A?

2)   What is the velocity of B?

3)   Is the velocity of point D smaller or larger than the velocity of point A?

4)   What direction is the velocity of point A?

5)   Where is the IC at the instant shown?

6)   In which case does point A have the larger velocity?

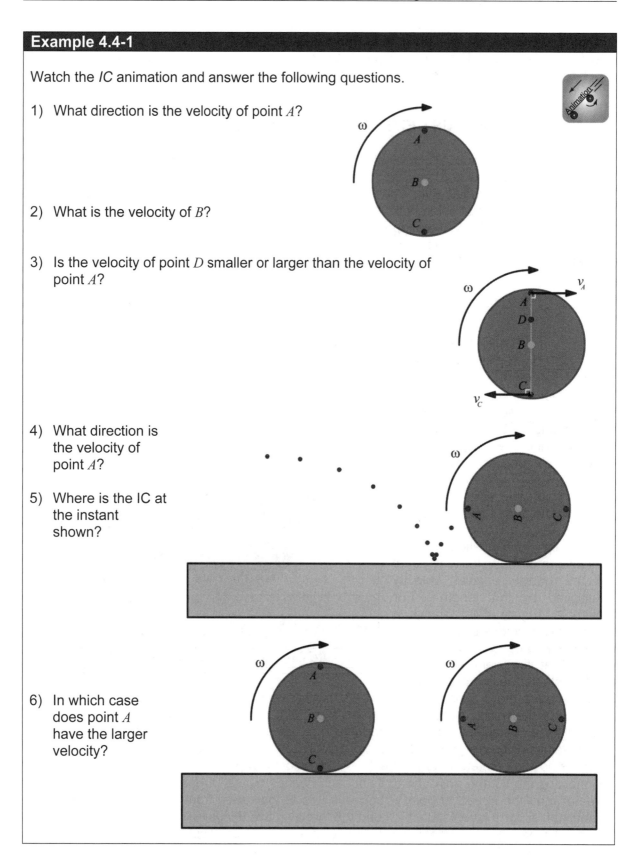

Example 4.4-2

Draw the direction of the velocity of point A and point B. Which velocity is larger?

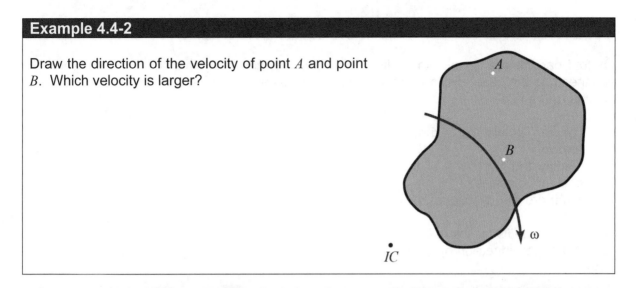

Example 4.4-3

Given the velocities of point A and B, locate the IC.

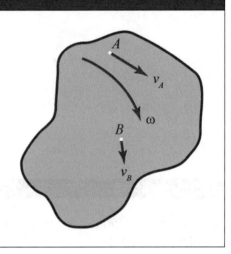

Example 4.4-4

A four-bar linkage is set into motion as shown. It has pin joints at A, C and where it connects to the base. Following the steps below, sketch the directions of the velocities for points A, B and C.

1. Locate points of known velocity.
2. Identify the directions of the angular velocities?
3. What are the directions of $\mathbf{v}_A$ and $\mathbf{v}_C$?
4. Identify the IC for link AC.
5. What is the direction of $\mathbf{v}_B$?

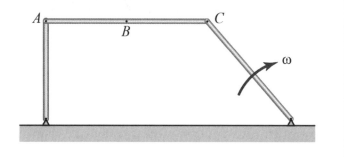

Example 4.4-5

A disk rolls, without slip, on a stationary ground. What is the velocity direction of point B relative to point A?

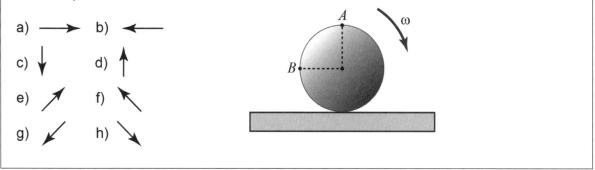

Equation Summary

Abbreviated variable definition list

$\mathbf{v}_A$, $\mathbf{v}_{IC}$ = absolute velocity of points A and IC respectively
$\mathbf{v}_{A/IC}$ = velocity of point A relative to point IC
$\mathbf{a}_A$, $\mathbf{a}_{IC}$ = absolute acceleration of points A and IC respectively
$\mathbf{a}_{A/IC}$ = acceleration of point A relative to point IC
$\mathbf{r}_{A/IC}$ = position of point A relative to point IC
ω = angular velocity of the rigid body (rad/s)
α = angular acceleration of the rigid body (rad/s^2)

General planar motion

$$\mathbf{v}_{IC} = 0 \qquad \mathbf{v}_A = \mathbf{v}_{IC} + \omega \times \mathbf{r}_{A/IC} = \omega \times \mathbf{r}_{A/IC}$$

$$\mathbf{a}_A = \mathbf{a}_{IC} + \alpha \times \mathbf{r}_{A/IC} - \omega^2 \mathbf{r}_{A/IC} \qquad \mathbf{a}_A = \mathbf{a}_{IC} + \alpha \times \mathbf{r}_{A/IC} + \omega \times (\omega \times \mathbf{r}_{A/IC})$$

Example Problem 4.4-6

The rectangular plate is confined by rollers within slots at A and B. When $\theta = 20^\circ$, point B is moving at $v_B = 10$ m/s. Determine the velocity of points C and D at this instant. The dimensions of the plate are $a = 100$ mm and $b = 50$ mm.

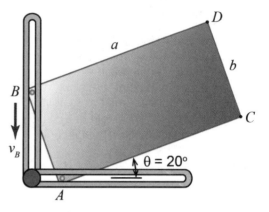

Given:

Find:

Solution:

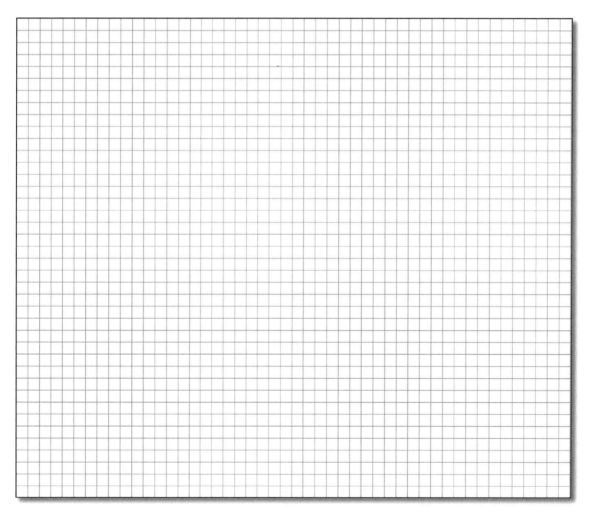

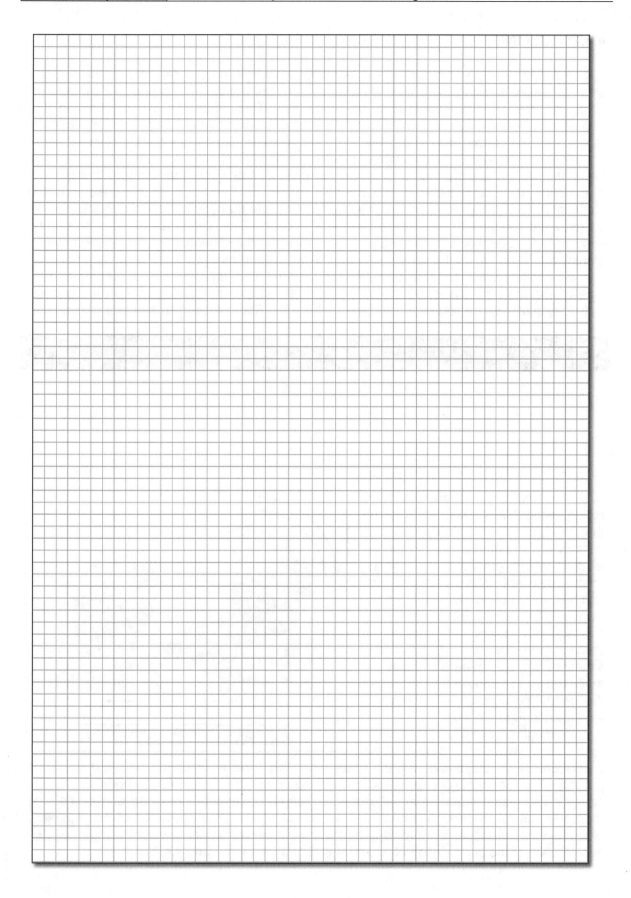

4.4.2) ROLLING

A specific type of motion that exemplifies how rigid bodies can rotate and translate simultaneously is rolling. This type of motion is relevant to many types of real life problems, such as, tires on cars, motorcycles and bicycles. This relates to any mechanism that transfers motion through a rolling contact to create linear motion. There are two cases of motion that may occur for rolling contacts. Each case requires a completely different set of equations for analysis. First, a wheel or other rolling object can roll without slip. In this situation, we can identify the IC as being the point of the wheel that touches the ground. The friction that occurs between the ground and the rolling contact, in this case, is called static friction. Second, the wheel rolls with slip. Imagine that you are driving your car on an icy road and you slam on your brakes and lose traction. This is what happens in the second case. There is relative motion between the wheel and the ground. This means we cannot identify the IC as the contact point between the wheel and the ground. The friction between the two bodies, in this case, is called kinetic friction. It is important to understand and be able to analyze both cases, but in this section we will only be looking at the no slip case. Rolling with slip will be addressed in detail in the *Rigid Body Newtonian Mechanics* chapter.

What is the difference between a wheel that rolls without slip and one that rolls with slip?

In the case where we have rolling without slip, you can imagine that the point of the body that has contact with the ground does not move relative to the ground. The contact point does not slide, rather it just touches the ground before it rotates away. The next point on the body subsequently takes its position in contact with the ground. For this situation, we can generate equations that express the translational motion of the geometric center of the body (O) as shown in Figure 4.4-2.

Imagine, for example, that a rope is wrapped around the

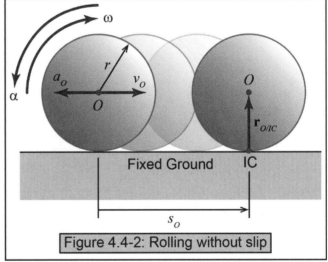

Figure 4.4-2: Rolling without slip

circumference of the body. As the body rolls, this rope is unwound and laid upon the ground. The length of rope laid down represents the displacement of the body's center. The displacement is equal to the arc length corresponding to the angle the body has rolled through (Equation 4.4-2). Taking the time derivative of this expression for displacement provides equations for the translational velocity and acceleration of the body's geometric center O (Equations 4.4-3 and 4.4-4).

Distanced traveled by body center O: $\boxed{s_O = r\theta}$ (4.4-2)

Speed of body center O: $\boxed{v_O = r\omega}$ (4.4-3)

Acceleration magnitude of body center O: $\boxed{a_O = r\alpha}$ (4.4-4)

r = radius
s_O = distance traveled by the center of the rolling body
v_O = magnitude of the velocity of the rolling body's center
a_O = magnitude of the acceleration of the rolling body's center

θ = angle the body has rolled through (rad)
ω = angular velocity (rad/s)
α = angular acceleration (rad/s^2)

The velocity and acceleration of a rolling body's center may also be expressed in terms of the cross product as shown in Equations 4.4-5 and 4.4-6.

Velocity of body center O: $\boxed{\mathbf{v}_O = \boldsymbol{\omega} \times \mathbf{r}_{O/IC}}$ (4.4-5)

Acceleration of body center O: $\boxed{\mathbf{a}_O = \boldsymbol{\alpha} \times \mathbf{r}_{O/IC}}$ (4.4-6)

$\mathbf{r}_{O/IC}$ = position of the body center O relative to the IC
$\mathbf{v}_O$ = velocity of the body center O
$\mathbf{a}_O$ = acceleration of the body center O

α = angular acceleration (rad/s^2)
ω = angular velocity (rad/s)

Equations 4.4-2 through 4.4-6 are only valid for a body that is rolling without slip. If the body is slipping, the contact point would not be an IC. The assumption of $v_{IC} = 0$ is essential in the derivation of the equations as shown below. Without this assumption, the above equations do not hold.

Once the motion of the center of a rolling body is known, then the translational motion of other points can be determined. This is possible through the use of the relative velocity and acceleration equations. For example, if it were desired to determine the velocity and acceleration of point A on a rolling body as shown in Figure 4.4-3, we could use the following equations.

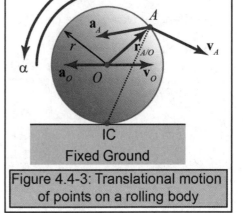

$$\mathbf{v}_A = \mathbf{v}_O + \mathbf{v}_{A/O} = \mathbf{v}_O + \boldsymbol{\omega} \times \mathbf{r}_{A/O}$$

$$\mathbf{a}_A = \mathbf{a}_O + \mathbf{a}_{A/O} = \mathbf{a}_O + \boldsymbol{\alpha} \times \mathbf{r}_{A/O} - \omega^2 \mathbf{r}_{A/O}$$

Figure 4.4-3: Translational motion of points on a rolling body

Conversely, if information about the body's rotation is unknown, we can use the velocity and acceleration of the body center to determine the angular velocity and acceleration as shown below.

$$\omega = \frac{v_O}{r} \qquad\qquad \alpha = \frac{a_O}{r}$$

Equation derivation

We can also derive the velocity and acceleration for the center of a body rolling without slip in a slightly different way. Consider the relative velocity equation that we use to analyze rigid-body motion ($\mathbf{v}_B = \mathbf{v}_A + \boldsymbol{\omega} \times \mathbf{r}_{B/A}$). If we apply this equation between the body center O and the IC, we get the following equation (see Figure 4.4-2).

$$\mathbf{v}_O = \mathbf{v}_{IC} + \boldsymbol{\omega} \times \mathbf{r}_{O/IC}$$

The velocity of the IC is zero, therefore, we end up with the velocity of the body center given below.

$$\mathbf{v}_O = \boldsymbol{\omega} \times \mathbf{r}_{O/IC}$$

If we take the derivative of the above velocity equation, we obtain an equation for the acceleration of the body center.

$$\mathbf{a}_O = \frac{d\mathbf{v}_O}{dt} = \frac{d(\boldsymbol{\omega} \times \mathbf{r}_{O/IC})}{dt} = \frac{d\boldsymbol{\omega}}{dt} \times \mathbf{r}_{O/IC} + \boldsymbol{\omega} \times \frac{d\mathbf{r}_{O/IC}}{dt}$$

Looking at the above derivative, it is easy to see that the derivative, with respect to time of the angular velocity, is the angular acceleration (i.e. $d\boldsymbol{\omega}/dt = \boldsymbol{\alpha}$). But, what is the derivative of the position vector of O with respect to the IC ($d\mathbf{r}_{O/IC}/dt = ?$)? If you think about it, the size of $\mathbf{r}_{O/IC}$ never changes because the distance between O and IC never changes. It is also true that the direction of $\mathbf{r}_{O/IC}$ never changes because the IC is always the contact point and the body center O is always directly above the contact point. Therefore, the derivative of $\mathbf{r}_{O/IC}$ with respect to time is zero ($d\mathbf{r}_{O/IC}/dt = 0$). The acceleration of the body center is then given by the following equation.

$$\mathbf{a}_O = \boldsymbol{\alpha} \times \mathbf{r}_{O/IC}$$

Equation Summary

Abbreviated variable definition list

s_O = distance traveled by the center of the rolling body

$\mathbf{v}_O$ = velocity of the body center O

$\mathbf{a}_O$ = acceleration of the body center O

$\mathbf{r}_{O/IC}$ = position of the body center O relative to the IC

r = radius

θ = angle the body has rolled through (rad)

ω = angular velocity (rad/s)

α = angular acceleration (rad/s^2)

Rolling without slip

$$s_O = r\theta \qquad v_O = r\omega \qquad a_O = r\alpha$$

$$\mathbf{v}_O = \boldsymbol{\omega} \times \mathbf{r}_{O/IC} \qquad \mathbf{a}_O = \boldsymbol{\alpha} \times \mathbf{r}_{O/IC}$$

Example Problem 4.4-7

Center O of the disk has the velocity and acceleration shown in the figure. If the disk rolls without slipping on the horizontal surface, determine the velocity and acceleration of points A, B and C for the instant represented.

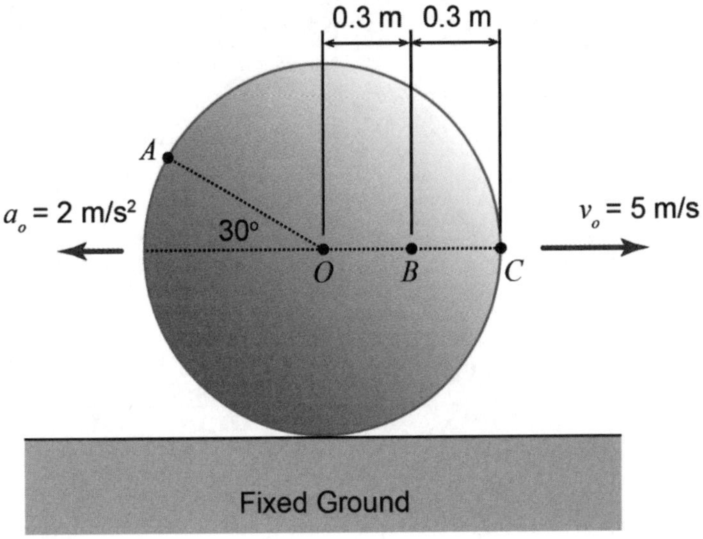

Given:

Find:

Solution:

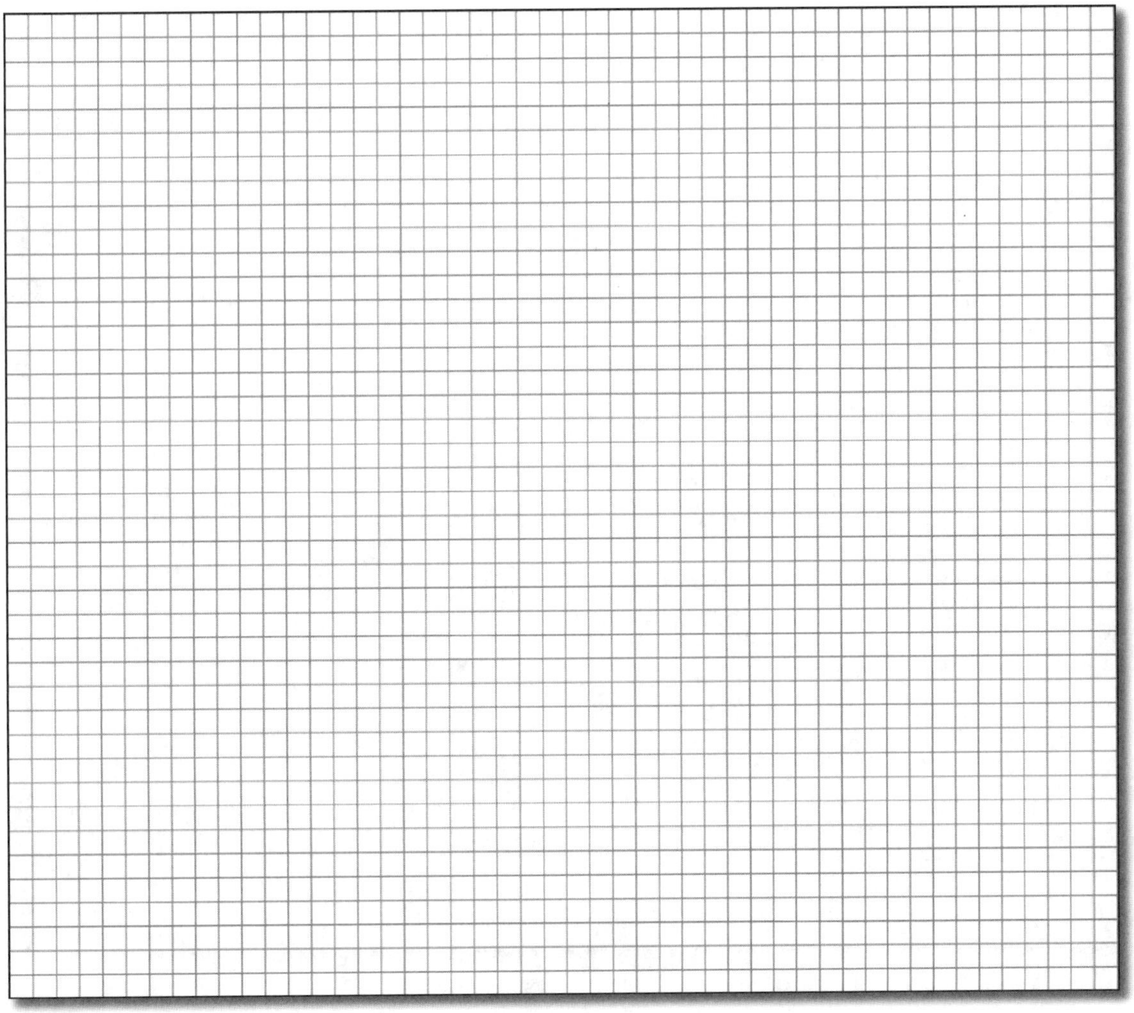

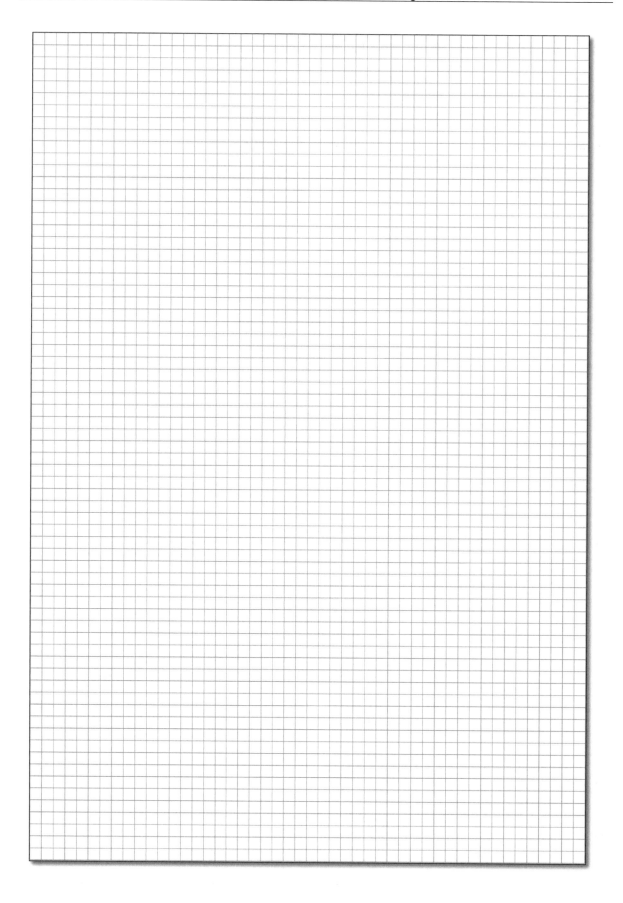

Example Problem 4.4-8

Shown below, the autonomous ground robot is geometrically symmetric about the y axis. The two large wheels of the robot are driven by their own electric motors that cause the wheels to turn, propelling the vehicle forward. The vehicle is able to turn by commanding each of the two wheels to rotate at different speeds. If the left wheel is commanded to rotate at 4 rad/sec and the right wheel is commanded to rotate at 6 rad/sec, determine for the instant shown (a) the translational velocity of the center of the left wheel (point A) and the translational velocity of the center of the right wheel (point B), (b) the translational velocity of the vehicle's mass center G, and (c) the vehicle's overall angular velocity ω. Assume that each of the two wheels roll without slip.

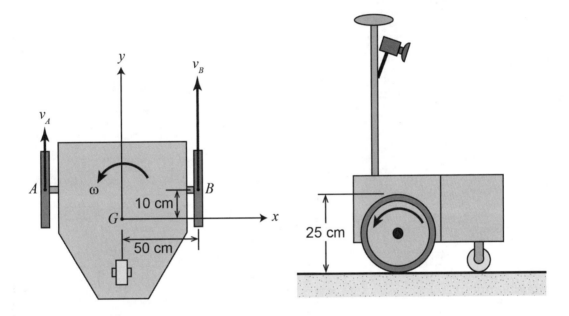

Given:

Find:

Solution:

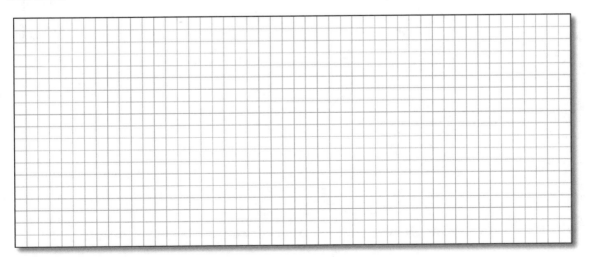

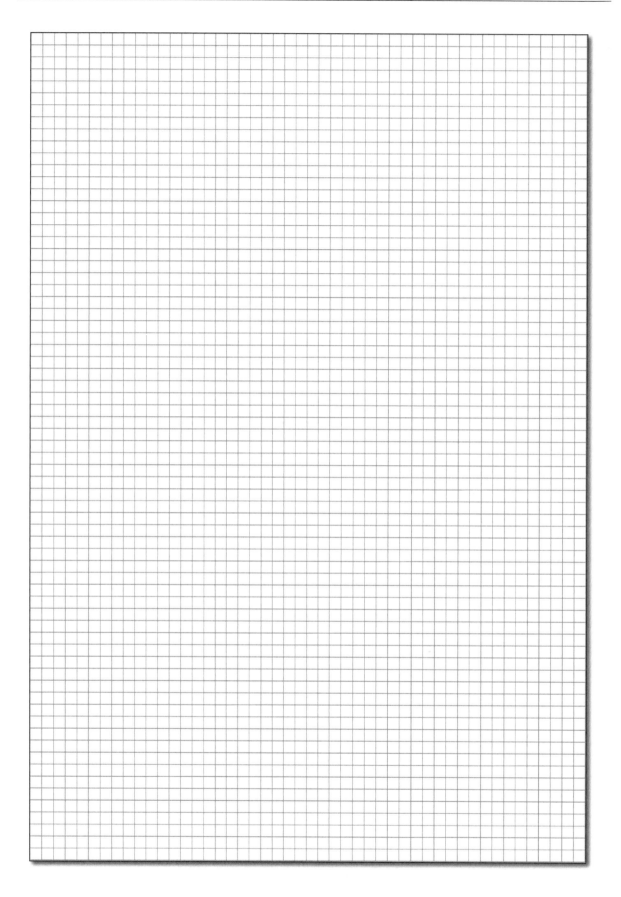

Example Problem 4.4-9

Disk D, of radius $r = 0.5$ m, rolls without slipping on a horizontal floor with a constant clockwise angular velocity of $\omega = 3$ rad/s. Rod R, is hinged to D at A, and the end of the rod B, slides along the floor. Determine the angular velocity of R when the line OA joining the center of the disk to the hinge at A is horizontal as shown.

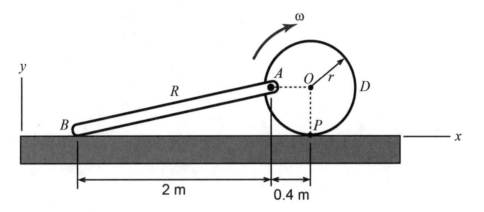

Given:

Find:

Solution:

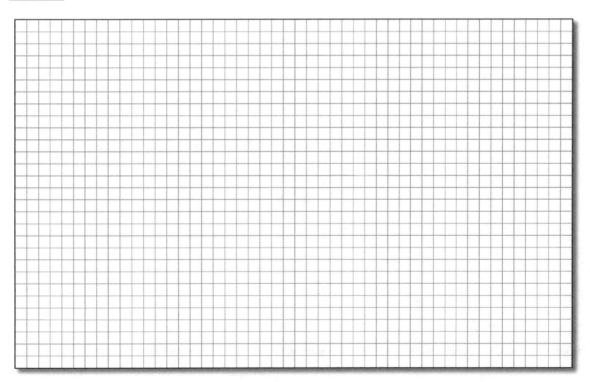

Solved Problem 4.4-10

Wheels of radius 0.17 m are attached to the end of a bar of length 1 m. Each wheel runs on a 30-degree incline without slip. The velocity of the wheel attached to end A is a constant 0.5 m/s, in the direction shown. When the bar is horizontal, determine the angular velocities

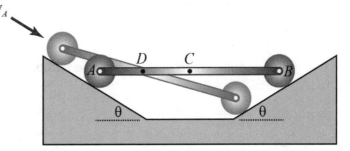

of the bar and wheels. Also, determine the velocity of points C and D. Point C is located at the center of the bar and point D is located halfway between point C and point A.

Given: $L = 1$ m, $r = 0.17$ m
       $\theta = 30°$, $v_A = 0.5$ m/s
       No slip

Find: ω_{bar}, ω_{wheel}, v_C, v_D

Solution:

Getting familiar with the problem

We will solve this problem through the use of the bar's instantaneous center of zero velocity (IC). Let's draw the location of the bar's IC and use this to determine the directions of the velocities of various points of interest. If we know the direction of two velocities, we can locate the IC. We know the directions of v_A and v_B since they are parallel to their respective inclines.

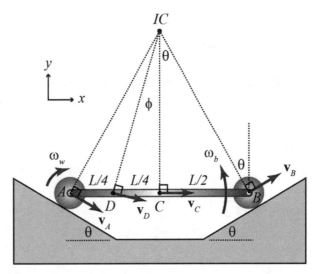

Angular velocity of the bar

Since the IC gives us the directions of the unknown velocities, we can use the scalar version of the velocity equation to calculate their sizes (i.e. $v_P = r_{P/IC}\, \omega$). Before we can calculate the unknown velocities, we need to determine the bar's angular velocity.

$$v_A = r_{A/IC}\omega_{bar} = \frac{L}{2\sin\theta}\omega_{bar}$$

Note: The triangle A-B-IC is an equilateral triangle. This would make $r_{A/IC}$ equal to L. However, the calculations will be carried out regardless of this fact.

$$\omega_{bar} = \frac{2v_A \sin\theta}{L} \qquad \boxed{\boldsymbol{\omega}_{bar} = 0.5\mathbf{k}\,\frac{\text{rad}}{\text{s}}}$$

Velocities

Knowing the angular velocity of the bar, we can determine the linear velocities.

$$v_C = r_{C/IC}\omega_{bar} = \frac{L}{2\tan\theta}\omega_{bar} \qquad\qquad \boxed{\mathbf{v}_C = 0.433\mathbf{i}\,\frac{\text{m}}{\text{s}}}$$

$$v_D = r_{D/IC}\omega_{bar} \qquad\qquad \mathbf{v}_D = v_D(\cos(\phi)\mathbf{i} - \sin(\phi)\mathbf{j})$$

We need to determine $r_{D/IC}$ and the angle ϕ before we can calculate the velocity of point D.

$$r_{D/IC}^2 = r_{C/IC}^2 + (L/4)^2 = \left(\frac{L}{2\tan\theta}\right)^2 + (L/4)^2 = \frac{L^2}{4\tan^2\theta} + \frac{L^2}{16} = \frac{L^2(4+\tan^2\theta)}{16\tan^2\theta}$$

$$r_{D/IC} = 0.9014\text{m} \qquad\qquad v_D = 0.45\,\frac{\text{m}}{\text{s}}$$

$$\tan\phi = \frac{(L/4)}{r_{C/IC}} = \frac{\tan\theta}{2} \qquad\qquad \phi = 16.1° \qquad\qquad \boxed{\mathbf{v}_D = 0.432\mathbf{i} - 0.125\mathbf{j}\,\frac{\text{m}}{\text{s}}}$$

Angular velocity of the wheels

The ICs for the wheels are located where each wheel touches the ground, provided that the wheels do not slip. We can use this fact to determine the angular velocity of the wheels.

$$v_A = r\omega_{wheel} \qquad\qquad \omega_{wheel} = \frac{v_A}{r} \qquad\qquad \boxed{\omega_{wheel} = 2.94\,\frac{\text{rad}}{\text{s}}}$$

4.5) RELATIVE SLIDING IN MECHANISMS

If a mechanism, such as that shown in Figure 4.5-1, has a component on one link that slides relative to another, the analysis takes on one more layer of complexity. In some mechanisms, it is necessary to express the motion of a single point in two different ways, that is, with respect to two different points whose motion is known. The mechanism shown in Figure 4.5-1 may be analyzed in a similar manner. However, care must be taken because the two bodies that make up the mechanism are allowed to slide relative to one another.

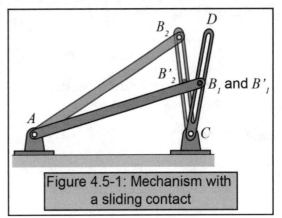

Figure 4.5-1: Mechanism with a sliding contact

The mechanism shown in Figure 4.5-1 has a sliding contact at end B. In order to analyze the motion of this mechanism, we need to express the motion of point B in two different ways. Specifically, we need to describe the motion of point B relative to the fixed point A and then relative to fixed point C. Relating the velocity and acceleration relative to fixed point A is relatively straightforward as shown in the following relations.

$$\mathbf{v}_B = \boldsymbol{\omega}_{AB} \times \mathbf{r}_{B/A}$$
$$\mathbf{a}_B = \boldsymbol{\alpha}_{AB} \times \mathbf{r}_{B/A} - \omega_{AB}^2 \mathbf{r}_{B/A}$$

How to express the motion of point B relative to the fixed point C may not be as clear. To help understand this situation better, imagine that there is a point B' that is attached to bar CD and is coincident with point B at a given instant. The motion of point B' can be expressed relative to point C in a similar manner to the relationships given above. However, point B (fixed to bar AB and sliding relative to bar CD), and B' (fixed to bar CD) do not necessarily have the same velocity or acceleration. This becomes clear by examining Figure 4.5-1 and realizing that even though points B and B' coincide at state 1, they do not at state 2, which means that they must have different velocities. Expressions for the motion of point B with respect to point C must include a term for the motion of point B relative to the motion of point B' as shown in Equation 4.5-1.

Velocity with relative sliding: $\boxed{\mathbf{v}_B = \mathbf{v}_C + \boldsymbol{\omega}_{CD} \times \mathbf{r}_{B/C} + \mathbf{v}_{B/B'}}$ (4.5-1)

$\mathbf{v}_B$ = linear velocity of point B
$\mathbf{v}_C$ = linear velocity of point C
$\boldsymbol{\omega}_{CD}$ = angular velocity of bar CD

$\mathbf{r}_{B/C}$ = position of point B (and B') relative to point C
$\mathbf{v}_{B/B'}$ = velocity of point B relative to point B'

The magnitude of the velocity of point B relative to point B' is unknown, but its direction may be determined from the geometry of the problem. It is known that vector $\mathbf{v}_{B/B'}$ is directed along the length of the arm CD. The equations for acceleration are not as straightforward since bar CD is rotating, therefore, the acceleration of point B relative to point B' may have a component that is not directed along bar CD. The motion of point B relative to point B' can more generally be understood as the motion of a particle relative to a rotating reference frame. This situation is described in detail in the section that follows.

Equation Summary

Abbreviated variable definition list

$\mathbf{v}_B$ = linear velocity of point B
$\mathbf{v}_{B/B'}$ = velocity of point B relative to point B'
$\mathbf{v}_C$ = linear velocity of point C
$\mathbf{r}_{B/C}$ = position of point B (and B') relative to point C
ω_{CD} = angular velocity of bar CD

Relative sliding

$$\mathbf{v}_B = \mathbf{v}_C + \boldsymbol{\omega}_{CD} \times \mathbf{r}_{B/C} + \mathbf{v}_{B/B'}$$

Example Problem 4.5-1

Consider the mechanism shown. Arm CD is rotating at 4 rad/sec in the counterclockwise direction and block C slides in the slot of bar AB. Determine the angular velocity of bar AB for the instant shown.

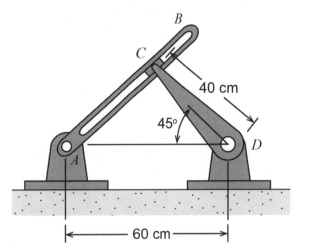

Given:

Find:

Solution:

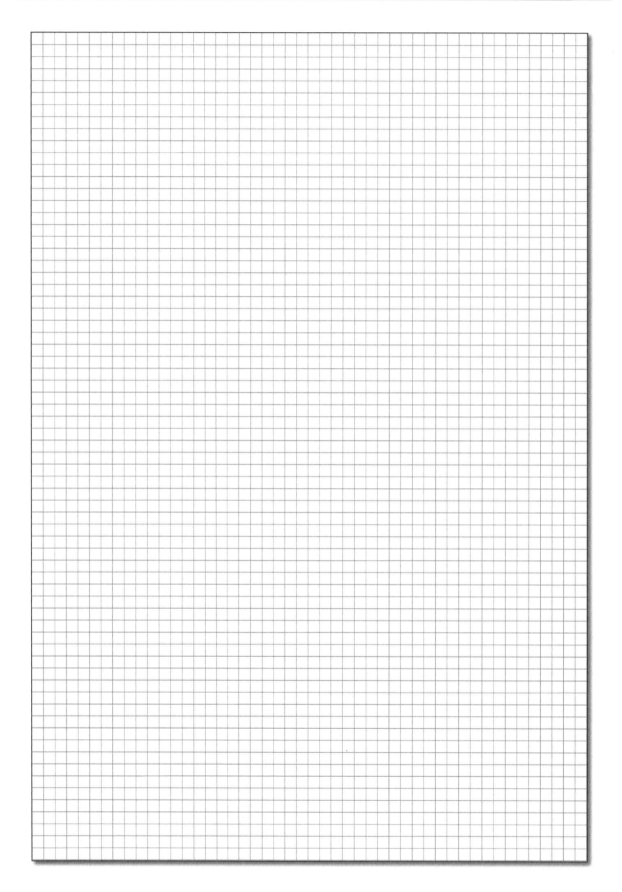

Solved Problem 4.5-2

A collar is pinned to the end of rod AB. The collar pivots about point B and is allowed to slide freely along rod CD. If rod CD is given a counterclockwise angular velocity of 3 rad/sec, determine the velocity of point B ($\mathbf{v}_B$) for the instant shown in the given figure.

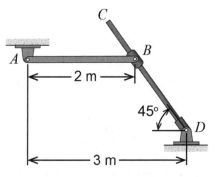

<u>Given:</u> $\omega_{CD} = 3$ rad/sec (ccw)

<u>Find:</u> $\mathbf{v}_B$

<u>Solution:</u>

Step 1: Get familiar with the system.

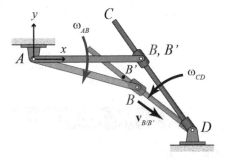

- We will assume a direction for ω_{AB} (cw) and will create a point B' on bar CD that is coincident with point B for the instant shown. From inspection, we can see that the velocity of point B relative to B' will be directed along rod CD.
- Also note that points A and D are fixed and their velocities are zero.

Step 2: Analyze the velocities.

Our approach for analyzing the velocities will be to find two different expressions for the velocity of point B. We will then set them equal and solve for the unknowns. Each of the two expressions will start from points of known velocity, that is, point A and point D.

Starting from point A:

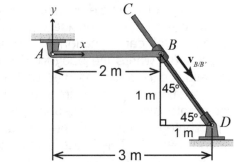

$$\mathbf{v}_B = \mathbf{v}_A + \mathbf{v}_{B/A} = \boldsymbol{\omega}_{AB} \times \mathbf{r}_{B/A}$$
$$= -\omega_{AB}\,\mathbf{k} \times 2\mathbf{i}$$
$$= -2\omega_{AB}\mathbf{j} \qquad\qquad (1)$$

Similarly, starting from point D:

$$\mathbf{v}_B = \mathbf{v}_D + \mathbf{v}_{B'/D} + \mathbf{v}_{B/B'} = \boldsymbol{\omega}_{CD} \times \mathbf{r}_{B'/D} + \mathbf{v}_{B/B'} = 3\mathbf{k} \times (-1\mathbf{i} + 1\mathbf{j}) + v_{B/B'}(\cos 45\mathbf{i} - \sin 45\mathbf{j})$$

$$= (-3 + \frac{v_{B/B'} \cdot \sqrt{2}}{2})\mathbf{i} + (-3 - \frac{v_{B/B'} \cdot \sqrt{2}}{2})\mathbf{j} \qquad\qquad (2)$$

If we set Equation (1) and Equation (2) equal, we have two unknowns (ω_{AB} and $v_{B/B'}$) which can then be solved for between the x- and y-direction equations.

$$\mathbf{i}: 0 = -3 + \frac{v_{B/B'}\sqrt{2}}{2} \qquad\qquad \mathbf{j}: -2\omega_{AB} = -3 - \frac{v_{B/B'}\sqrt{2}}{2}$$

In this instance, only the x-direction equation is needed to solve for $v_{B/B'}$.

$$\Rightarrow v_{B/B'} = \frac{6}{\sqrt{2}} = 3\sqrt{2}\frac{m}{s}$$

Substituting back into Equation (2) gives us the solution for v_B.

$$\mathbf{v}_B = (-3 + \frac{(3\sqrt{2})\sqrt{2}}{2})\mathbf{i} + (-3 - \frac{(3\sqrt{2})\sqrt{2}}{2})\mathbf{j} \qquad \boxed{\mathbf{v}_B = -6\mathbf{j}\frac{m}{s}}$$

For good measure, we can also find that ω_{AB} equals 3 rad/sec.

4.6) MOTION RELATIVE TO ROTATING REFERENCE FRAMES

4.6.1) POSITION

So far in this chapter we have described the general planar motion of rigid bodies with respect to reference frames that *do not rotate*. It may also be desirable to consider a body's motion in a frame that is rotating, for example, in terms of a polar coordinate frame. Consider the case of general planar motion shown in Figure 4.6-1 where the body-fixed x-y axes are allowed to rotate as well as translate. The position of point A in the body-fixed coordinate system attached to point B is represented in terms of the unit direction vectors $\mathbf{i}$ and $\mathbf{j}$. This is the same equation that was seen when the coordinate system did not rotate.

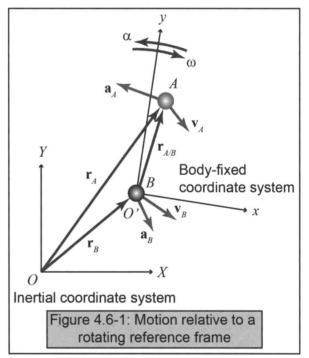

Figure 4.6-1: Motion relative to a rotating reference frame

$$\mathbf{r}_A = \mathbf{r}_B + \mathbf{r}_{A/B} = \mathbf{r}_B + (x\mathbf{i} + y\mathbf{j})$$

4.6.2) VELOCITY

To determine the absolute velocity of point A we can differentiate the position equation to generate the following expression. Note that since the body-fixed coordinate frame is rotating, the derivatives of the unit-direction vectors may be non-zero.

$$\mathbf{v}_A = \frac{d}{dt}\left(\mathbf{r}_B + (x\mathbf{i} + y\mathbf{j})\right) = \mathbf{v}_B + (\dot{x}\mathbf{i} + \dot{y}\mathbf{j}) + (x\dot{\mathbf{i}} + y\dot{\mathbf{j}})$$

If the x-y frame is non-rotating, the above reduces to $\mathbf{v}_A = \mathbf{v}_B + \mathbf{v}_{A/B}$. This is the equation that we introduced earlier. In this case, however, the x-y frame is rotating and hence the unit vectors may not have derivatives equal to zero. This was the situation we encountered when considering the n-t and polar coordinate frames which also are allowed to rotate. Employing the same logic used in the sections on n-t and polar coordinates, the derivative of $\mathbf{i}$ can be shown to equal $\omega\mathbf{j}$, where ω is the angular velocity with which the x-y coordinate frame is rotating. Likewise, the derivative of $\mathbf{j}$ can be shown to equal $-\omega\mathbf{i}$. These relations expressed employing the cross product are shown below.

$$\dot{\mathbf{i}} = \omega \times \mathbf{i} \quad \text{and} \quad \dot{\mathbf{j}} = \omega \times \mathbf{j}$$

Substituting the above expressions into the previous velocity equation, we get the following relationship for the absolute velocity of point A in terms of the rotating reference frame.

$$\begin{aligned}
\mathbf{v}_A &= \mathbf{v}_B + (\dot{x}\mathbf{i} + \dot{y}\mathbf{j}) + (x\dot{\mathbf{i}} + y\dot{\mathbf{j}}) \\
&= \mathbf{v}_B + (\dot{x}\mathbf{i} + \dot{y}\mathbf{j}) + x(\omega \times \mathbf{i}) + y(\omega \times \mathbf{j}) \\
&= \mathbf{v}_B + (\dot{x}\mathbf{i} + \dot{y}\mathbf{j}) + \omega \times (x\mathbf{i} + y\mathbf{j})
\end{aligned}$$

Rewriting the preceding equation one more time, we get Equation 4.6-1 for the absolute velocity of point A.

Velocity in a rotating reference frame: $\boxed{\mathbf{v}_A = \mathbf{v}_B + \omega \times \mathbf{r}_{A/B} + \mathbf{v}_{A,rel}}$ (4.6-1)

$\mathbf{v}_A$, $\mathbf{v}_B$ = absolute velocity of point A and B, respectively

$\mathbf{v}_{A,rel}$ = velocity of point A relative to the rotating reference frame

ω = angular velocity of the rotating reference frame (rad/s)

$\mathbf{r}_{A/B}$ = position of point A relative to point B

In Equation 4.6-1, the term $\omega \times \mathbf{r}_{A/B} + \mathbf{v}_{A,rel}$ represents the velocity of point A relative to point B, that is, $\mathbf{v}_{A/B}$. Intuitively, $\mathbf{v}_{A,rel}$ is the velocity of point A as seen by someone rotating with the body-fixed x-y frame, and $\omega \times \mathbf{r}_{A/B}$ is due to the rotation of this coordinate frame. The $\mathbf{v}_{A,rel}$ term can also be thought of as the velocity of point A relative to a point A' which is attached to the rotating reference frame but is coincident with point A at the given instant. This was the situation we explored in the previous section when we considered mechanisms with a relative sliding contact. The above derivation assumed all terms were expressed in the rotating reference frame (x-y). All of the terms do not need to be expressed in the x-y frame, but if you wish to combine terms, then they need to be converted to a consistent frame.

In order to better understand the general expression given in Equation 4.6-1, it may be helpful to consider a specific case of a rotating reference frame we have already encountered.   In the case of polar coordinates, $\mathbf{v}_{A,rel}$ is equal to $\dot{r}\mathbf{e}_r$ which is the instantaneous speed of the particle as seen by someone rotating with the polar axes.   The term $\omega \times \mathbf{r}_{A/B}$ is equal to $r\dot{\theta}\mathbf{e}_\theta$ and is due to the rotation of the polar axes.

4.6.3) ACCELERATION

To determine the absolute acceleration of point A we can differentiate Equation 4.6-1 to generate the following expression.

$$\mathbf{a}_A = \frac{d}{dt}\left(\mathbf{v}_B + \mathbf{v}_{A,rel} + \omega \times \mathbf{r}_{A/B}\right)$$

$$= \mathbf{a}_B + \dot{\mathbf{v}}_{A,rel} + \dot{\omega} \times \mathbf{r}_{A/B} + \omega \times \dot{\mathbf{r}}_{A/B}$$

Recognizing that $\mathbf{v}_{A,rel}$ and $\mathbf{r}_{A/B}$ are expressed in the rotating x-y reference frame, their time derivatives will include derivatives of the unit vectors $\mathbf{i}$ and $\mathbf{j}$ as described earlier. Therefore, the above equation is equal to the following.

$$\mathbf{a}_A = \mathbf{a}_B + (\omega \times \mathbf{v}_{A,rel} + \mathbf{a}_{A,rel}) + \dot{\omega} \times \mathbf{r}_{A/B} + \omega \times (\omega \times \mathbf{r}_{A/B} + \mathbf{v}_{A,rel})$$

Rearranging and combining like terms, we get Equation 4.6-2 for the absolute acceleration of point A.   Again referring back to the case of polar coordinates, $\mathbf{a}_{A,rel}$ is equal to $\ddot{r}\mathbf{e}_r$, $2\omega \times \mathbf{v}_{A,rel}$ is equal to $2\dot{r}\dot{\theta}\mathbf{e}_\theta$, $\alpha \times \mathbf{r}_{A/B}$ is equal to $r\alpha\mathbf{e}_\theta$ and $\omega \times (\omega \times \mathbf{r}_{A/B})$ is equal to $-r\dot{\theta}^2\mathbf{e}_r$. If, however, particle A is traveling a curved path in the rotating reference frame, then $\mathbf{a}_{A,rel}$ won't be equal to $\ddot{r}\mathbf{e}_r$, it will have a normal and tangential component.

Acceleration in a rotating reference frame:

$$\boxed{\mathbf{a}_A = \mathbf{a}_B + \mathbf{a}_{A/B} = \mathbf{a}_B + 2\omega \times \mathbf{v}_{A,rel} + \alpha \times \mathbf{r}_{A/B} + \omega \times (\omega \times \mathbf{r}_{A/B}) + \mathbf{a}_{A,rel}} \quad (4.6\text{-}2)$$

or

$$\boxed{\mathbf{a}_A = \mathbf{a}_B + \mathbf{a}_{A/B} = \mathbf{a}_B + 2\omega \times \mathbf{v}_{A,rel} + \alpha \times \mathbf{r}_{A/B} - \omega^2 \mathbf{r}_{A/B} + \mathbf{a}_{A,rel}}$$

$\mathbf{a}_A$, $\mathbf{a}_B$ = absolute acceleration of point A and B respectively

$\mathbf{v}_{A,rel}$ = velocity of point A relative to the rotating reference frame

$\mathbf{a}_{A,rel}$ = acceleration of point A relative to the rotating reference frame

ω = angular velocity of the rotating reference frame (rad/s)

α = angular acceleration of the rotating reference frame (rad/s^2)

$\mathbf{r}_{A/B}$ = position of point A relative to point B

Another type of body-fixed reference frame that we have employed is the n-t coordinate system.   Whereas the relative velocity and acceleration equations in polar coordinates have the exact form of Equations 4.6-1 and 4.6-2, the relative motion equations in n-t coordinates do not.   This follows from the fact that the polar coordinate frame is defined in terms of coordinates r and θ that are defined with respect to another reference frame. The n-t coordinate frame is not defined with respect to another reference frame. It is defined in terms of the particle's path (i.e. the particle's speed v along the path of travel), and the path's radius of curvature ρ.

Equation Summary

Abbreviated variable definition list

$\mathbf{v}_A$, $\mathbf{v}_B$ = absolute velocity of points A and B respectively
$\mathbf{v}_{A,rel}$ = velocity of point A relative to the rotating reference frame
$\mathbf{a}_A$, $\mathbf{a}_B$ = absolute acceleration of points A and B respectively
$\mathbf{a}_{A,rel}$ = acceleration of point A relative to the rotating reference frame
ω = angular velocity of the rotating reference frame (rad/s)
α = angular acceleration of the rotating reference frame (rad/s^2)
$\mathbf{r}_{A/B}$ = position of point A relative to point B

Relative sliding

$$\mathbf{v}_A = \mathbf{v}_B + \omega \times \mathbf{r}_{A/B} + \mathbf{v}_{A,rel}$$

$$\mathbf{a}_A = \mathbf{a}_B + \mathbf{a}_{A/B} = \mathbf{a}_B + 2\omega \times \mathbf{v}_{A,rel} + \alpha \times \mathbf{r}_{A/B} - \omega^2 \mathbf{r}_{A/B} + \mathbf{a}_{A,rel}$$

$$\mathbf{a}_A = \mathbf{a}_B + \mathbf{a}_{A/B} = \mathbf{a}_B + 2\omega \times \mathbf{v}_{A,rel} + \alpha \times \mathbf{r}_{A/B} + \omega \times (\omega \times \mathbf{r}_{A/B}) + \mathbf{a}_{A,rel}$$

A collar is pinned to the end of rod AB. The collar pivots about point B and is allowed to slide freely along rod CD. If rod CD is given a counterclockwise angular velocity of 3 rad/sec that is decreasing at a rate of 1 rad/sec^2, determine the velocity $\mathbf{v}_B$ and acceleration $\mathbf{a}_B$ of point B for the instant shown in the given figure.

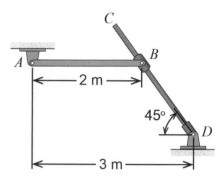

Given: ω_{CD} = 3 rad/sec (ccw)
          α_{CD} = 1 rad/sec^2 (cw)

Find: $\mathbf{v}_B$, $\mathbf{a}_B$

Solution:

This system is the same one we considered in Solved Problem 4.5-2. We will revisit the problem, but will solve it in terms of a rotating reference x-y frame attached to bar CD. We will also determine the acceleration of point B in this case.

Step 1: Get familiar with the system

We will again assume a direction for ω_{AB} (cw) and α_{AB} (ccw). From inspection, we can determine that the direction of $\mathbf{v}_{B,rel}$ and $\mathbf{a}_{B,rel}$ (the velocity and acceleration of point B relative to the rotating reference frame x-y) is along the bar CD. Note: If the rod CD were curved, then $\mathbf{a}_{B,rel}$ would also have a component normal to the rod.

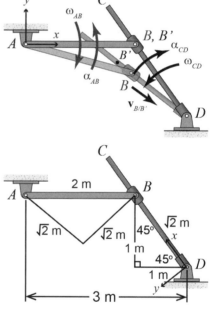

Step 2: Analyze the velocities

We will use the same basic approach from Solved Problem 4.5-2 where we started from two known velocities (for point A and point D) and met in the middle at point B.

Starting from point A:

$$\mathbf{v}_B = \cancel{\mathbf{v}_A} + \mathbf{v}_{B/A} = \boldsymbol{\omega}_{AB} \times \mathbf{r}_{B/A} = -\omega_{AB}\,\mathbf{k} \times (-2\cos 45\mathbf{i} - 2\sin 45\mathbf{j})$$

$$= -\omega_{AB}\sqrt{2}\mathbf{i} + \omega_{AB}\sqrt{2}\mathbf{j} \qquad (1)$$

Similarly, starting from point D:

$$\mathbf{v}_B = \cancel{\mathbf{v}_D} + \mathbf{v}_{B/D} = \boldsymbol{\omega}_{CD} \times \mathbf{r}_{B/D} + \mathbf{v}_{B,rel} = (3\mathbf{k} \times \sqrt{2}\mathbf{i}) - v_{B,rel}\mathbf{i}$$

$$= -v_{B,rel}\mathbf{i} + 3\sqrt{2}\mathbf{j} \qquad (2)$$

If we set Equation (1) and Equation (2) equal we have two unknowns (ω_{AB} and $v_{B,rel}$) which can then be solved for between the x- and y-direction equations.

$$\mathbf{i}: -\omega_{AB}\sqrt{2} = -v_{B,rel} \qquad\qquad \mathbf{j}: \omega_{AB}\sqrt{2} = 3\sqrt{2}$$

Simultaneously solving the two above equations provides that ω_{AB} = 3 rad/sec and $v_{B,rel}$ = 3√2 m/s. Then substituting back into either Equation (1) or Equation (2) provides the expression for $\mathbf{v}_B$. Notice that each of these values is consistent with the results from Solved Problem 4.5-2. The only difference is that $\mathbf{v}_B$ is represented in a different coordinate frame.

$$\boxed{\mathbf{v}_B = -3\sqrt{2}\mathbf{i} + 3\sqrt{2}\mathbf{j}\ \frac{m}{s}}$$

Step 3: Analyze the accelerations

Here we will employ an approach that is similar to the one we applied in analyzing the velocities.

Starting from point A:

$$\mathbf{a}_B = \mathbf{a}_A + \mathbf{a}_{B/A} = \boldsymbol{\alpha}_{AB} \times \mathbf{r}_{B/A} + \boldsymbol{\omega}_{AB} \times (\boldsymbol{\omega}_{AB} \times \mathbf{r}_{B/A})$$
$$= \alpha_{AB}\ \mathbf{k} \times (-2\cos 45\mathbf{i} - 2\sin 45\mathbf{j}) + -3\mathbf{k} \times (-3\mathbf{k} \times (-2\cos 45\mathbf{i} - 2\sin 45\mathbf{j}))$$
$$= (\alpha_{AB}\sqrt{2} + 9\sqrt{2})\mathbf{i} + (-\alpha_{AB}\sqrt{2} + 9\sqrt{2})\mathbf{j} \tag{3}$$

Similarly, starting from point D:

$$\mathbf{a}_B = \mathbf{a}_D + \mathbf{a}_{B/D} = 2\boldsymbol{\omega}_{CD} \times \mathbf{v}_{B,rel} + \boldsymbol{\alpha}_{cd} \times \mathbf{r}_{B/D} + \boldsymbol{\omega}_{CD} \times (\boldsymbol{\omega}_{CD} \times \mathbf{r}_{B/D}) + \mathbf{a}_{B,rel}$$
$$= (2(3\mathbf{k}) \times -3\sqrt{2}\mathbf{i}) + (-1\mathbf{k} \times \sqrt{2}\mathbf{i}) + (3\mathbf{k} \times (3\mathbf{k} \times \sqrt{2}\mathbf{i})) - a_{B,rel}\mathbf{i}$$
$$= (-9\sqrt{2} - a_{B,rel})\mathbf{i} - 19\sqrt{2}\mathbf{j} \tag{4}$$

If we set Equation (3) and Equation (4) equal we have two unknowns (α_{AB} and $a_{B,rel}$) which can then be solved for between the x- and y-direction equations.

$$\mathbf{i}: \alpha_{AB}\sqrt{2} + 9\sqrt{2} = -9\sqrt{2} - a_{B,rel} \qquad\qquad \mathbf{j}: -\alpha_{AB}\sqrt{2} + 9\sqrt{2} = -19\sqrt{2}$$

Simultaneously solving the two above equations provides that α_{AB} = 28 rad/sec^2 and $a_{B,rel} = -46\sqrt{2}$ m/s^2. The negative sign on $a_{B,rel}$ means that it is actually oriented opposite the direction we originally assumed. Substituting this value back into either Equation (3) or Equation (4) provides the expression for $\mathbf{a}_B$.

$$\boxed{\mathbf{a}_B = 37\sqrt{2}\mathbf{i} - 19\sqrt{2}\mathbf{j}\ \frac{m}{s^2}}$$

CHAPTER 4 REVIEW PROBLEMS

RP4-1) Explain the difference between relative and absolute velocities.

RP4-2) Consider a rigid body undergoing pure rotation. Every point on that body moves in a _____ path around the fixed axis.

RP4-3) Points on a body undergoing pure rotation that are at different distances from the fixed axis have different velocities. (True, False)

RP4-4) Points on a body undergoing pure rotation that are at different distances from the fixed axis have different angular velocities. (True, False)

RP4-5) Rigid-body motion has two components. What are they?

RP4-6) The velocity of any point on a body experiencing fixed-axis rotation is $v_A = r\omega$. What direction is this velocity?

RP4-7) Consider the velocity equation for point B on a rigid body $(\mathbf{v}_B = \mathbf{v}_A + \mathbf{v}_{B/A} = \mathbf{v}_A + \boldsymbol{\omega} \times \mathbf{r}_{B/A})$. Which component accounts for the translation of the body and which accounts for the rotation of the body about A?

RP4-8) How can the *instantaneous center of zero velocity* make our calculations easier?

RP4-9) The velocity of a point on a rigid body is always perpendicular to the line that connects the point to the _____.

RP4-10) Consider two points (A, B) on a rigid body and the instantaneous center of zero velocity (IC). If point A is twice as far away from the IC as B is, write an equation relating v_A to v_B.

RP4-11) A wheel of radius r is rolling while slipping at the point of contact between the wheel and ground. The speed of the wheel's center is $v = r\omega$. (True, False)

RP4-12) Consider the equation for the acceleration of point B on a rigid body $(\mathbf{a}_B = \mathbf{a}_A + \boldsymbol{\alpha} \times \mathbf{r}_{B/A} - \omega^2 \mathbf{r}_{B/A})$. Which component accounts for the translation of the body, which component accounts for the tangential acceleration and which component accounts for the normal acceleration?

RP4-13) The acceleration of an IC is generally zero. (True, False)

RP4-14) A cord is tightly wrapped around a cylinder of radius 0.2 meters, as shown in the figure. Starting from rest, the cord is pulled downward with a constant acceleration of 10 m/s^2. Determine the angular acceleration and angular velocity of the disk after it has completed 10 revolutions.

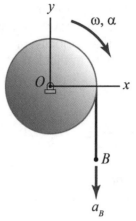

Given: $a_B = 10$ m/s^2 = constant, $\omega_o = 0$
                $\theta_f = 10$ revolutions, $r = 0.2$ m

Find: α_f, ω_f

Solution: | Find the final angular velocity.

Find the angular acceleration.

$\alpha_f =$ _____ $\omega_f =$ _____

RP4-15) When the slider block C is in the position shown, the link AB has a clockwise angular velocity of 2 rad/s. Determine the velocity of block C at this instant. The length of link AB is 1 meter and the length of link BC is 0.7 meters.

Given: ω_{AB} = 2 rad/s
l_{AB} = 1 m, l_{BC} = 0.7 m

Find: $\mathbf{v}_C$

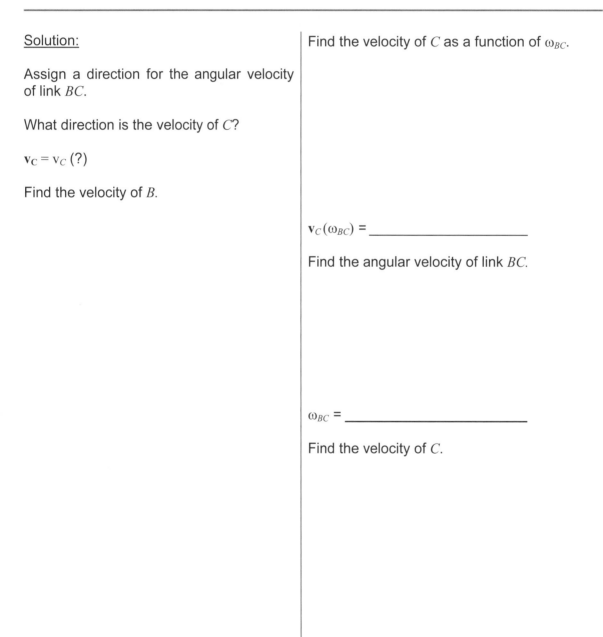

Solution:

Assign a direction for the angular velocity of link BC.

What direction is the velocity of C?

$\mathbf{v}_C = v_C \, (?)$

Find the velocity of B.

Find the velocity of C as a function of ω_{BC}.

$\mathbf{v}_C(\omega_{BC}) = \underline{\hspace{3cm}}$

Find the angular velocity of link BC.

$\omega_{BC} = \underline{\hspace{3cm}}$

Find the velocity of C.

$\mathbf{v}_B = \underline{\hspace{3cm}}$

$\mathbf{v}_C = \underline{\hspace{3cm}}$

RP4-16) A cord is tightly wrapped around the inner hub of a wooden spool. The end of the cord is attached to a fixed horizontal support. The spool is released and allowed to fall under the influence of gravity as the cord unwinds. Assuming that the spool does not sway back and forth, determine the velocity of point A for the instant shown where the speed of the center of the spool (point O) has reached 5 ft/s.

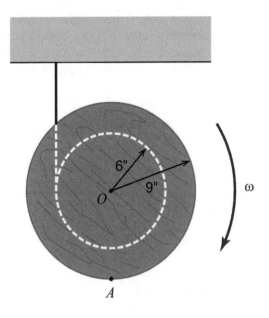

Given: $v_O = 5$ ft/s
          $r_i = 6$ in, $r_o = 9$ in

Find: $\mathbf{v}_A$

Solution:

Find the angular velocity.

Find the velocity of point A.

$\omega =$ _____

$\mathbf{v}_A$ _____

RP4-17) The ball shown rolls without slipping and has an angular velocity and angular acceleration of 6 rad/s and 4 rad/s^2, respectively, with the directions shown in the figure. Determine the acceleration of point B at this instant.

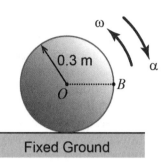

Given: ω = 6 rad/s
             α = 4 rad/s^2
             r = 0.3 m

Find: $\mathbf{a}_B$

Solution:

Locate the IC.

$v_{IC} = 0$ True   False
$a_{IC} = 0$ True   False

Find the acceleration of O.

Find the acceleration of B.

$\mathbf{a}_B =$ _____

Explain the differences between $\mathbf{a}_O$ and $\mathbf{a}_B$.

i-component:

j-component:

$\mathbf{a}_O =$ _____

P4.1) BASIC FIXED-AXIS ROTATION PROBLEMS

P4.1-1) A circular saw blade is rotating at 1500 rpm when power to the saw is turned off. The angular speed of the blade decelerates at a rate of $\alpha = 2t$ rad/s^2, where t is in seconds. Determine the number of revolutions it takes for the saw blade to come to rest.

Ans: $\theta = 209$ rev

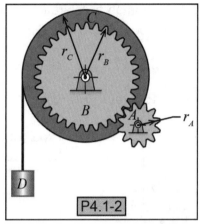

P4.1-2

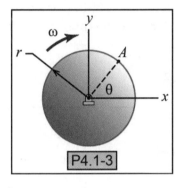

P4.1-3

P4.1-2)^{fe} It is desired that the shown hoisting mechanism be operated such that the load D is lifted at a constant rate of 2 ft/s. If the drum C, which is rigidly attached to gear B, has a radius of $r_C = 4$ ft and the gear ratio between B and A is 3:1, determine the angular velocity with which a motor must drive pinion gear A.

a) $\omega = 1.0$ rad/s ccw   b) $\omega = 1.5$ rad/s ccw

c) $\omega = 2.0$ rad/s ccw   d) $\omega = 2.5$ rad/s ccw

P4.1-3)^{fe} A 20-in diameter disk rotates about its center point with a constant angular velocity of 100 rpm in the clockwise direction. What is the velocity of point A on its rim if $\theta = 25°$? Solve for the velocity in the x-y coordinate systems.

a) $\mathbf{v}_A = 44.2\mathbf{i} - 94.9\mathbf{j}$ in/s   b) $\mathbf{v}_A = 34.1\mathbf{i} - 87.4\mathbf{j}$ in/s

c) $\mathbf{v}_A = 28.9\mathbf{i} - 76.9\mathbf{j}$ in/s   d) $\mathbf{v}_A = 56.2\mathbf{i} - 104.1\mathbf{j}$ in/s

P4.1-4)^{fe} A toy train travels around a circular track. Its angular position is given by $\theta(t) = t^3 - 10t + 2$, where θ is given in radians and t is in seconds. Determine the train's angular acceleration when $t = 10$ seconds.

a) $\alpha = 100$ rad/s^2 b) $\alpha = 580$ rad/s^2

c) $\alpha = 20$ rad/s^2 d) $\alpha = 40$ rad/s^2

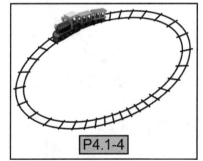

P4.1-4

P4.2) INTERMEDIATE FIXED-AXIS ROTATION PROBLEMS

P4.2-1)[fe] A circular disk of radius 0.5 m rotates about its center O in the direction shown with a constant angular acceleration. Point P on the rim has an acceleration of $\mathbf{a}_P = 2\mathbf{e}_t + 5\mathbf{e}_n$ m/s^2 at t = 0. Determine the number of revolutions made by the disk in 5 seconds.

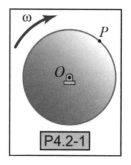

P4.2-1

a) θ = 5.5 rev

b) θ = 10.5 rev

c) θ = 15.5 rev

d) θ = 20.5 rev

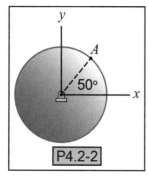

P4.2-2

P4.2-2) A circular disk of radius 0.2 m rotates about its center. The acceleration of point A on the rim of the disk is $\mathbf{a} = 5\mathbf{i} - 7\mathbf{j}$ m/s^2. Determine the angular velocity and angular acceleration of the disk at this instant.

Ans: ω = 3.28 $\mathbf{k}$ rad/s, α = 41.6 $\mathbf{k}$ rad/s^2

P4.2-3) The angled pendulum is swinging freely about the fixed axis perpendicular to the vertical plane that passes through O. On its upswing, and at the moment when its long arm makes an angle of 30 degrees with the vertical, the pendulum's angular velocity is 3 rad/s clockwise and its angular acceleration is 4 rad/s^2 counterclockwise. Determine the velocity and acceleration of the tip of the pendulum (point A) at this instant.

Ans: $\mathbf{v}_A = -2.48\mathbf{i} - 0.3\mathbf{j}$ ft/s, $\mathbf{a}_A = 2.41\mathbf{i} + 7.84\mathbf{j}$ ft/s^2

P4.2-3

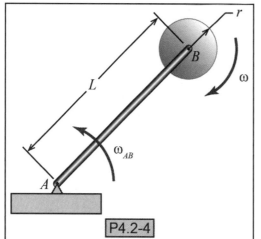

P4.2-4

P4.2-4) A bar rotates with an angular velocity of ω_{AB} = 4 rad/s in the direction shown. A disk, attached to the bar's end, rotates independently with an angular velocity of ω = 10 rad/s in the direction shown. If the radius of the disk is 10 cm and the length of the bar is 80 cm, determine the location on the body with the greatest speed and determine the value of that speed.

Ans: v = 420 cm/s

P4.2-5) The disk shown rotates about its center. The bearing supporting the disk at the center is old and applies a resistive friction moment. This friction causes the disk to decelerate with an angular acceleration of $\alpha = b\omega$ rad/s^2, where b = 2 per second and ω is the angular velocity of the disk in radians per second. If the disk has an initial angular velocity of 500 rpm, determine its angular velocity as a function of time.

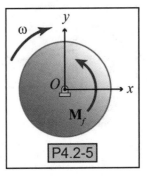

Ans: $\omega = 500e^{-2t}$ rpm

P4.3) BASIC GENERAL PLANAR MOTION PROBLEMS

P4.3-1) At the instant shown in the figure, link BC has a clockwise angular velocity of ω_{BC} = -2 $\mathbf{k}$ rad/s. Determine the angular velocity of links AB and CD.

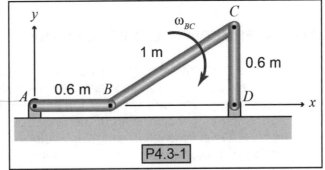

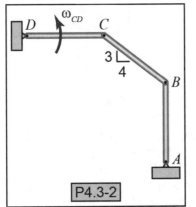

P4.3-2)^{fe} The bar linkage shown is set in motion by applying a counterclockwise angular velocity of 10 rad/s to bar CD. All links have the same length. Determine the angular velocity of bar AB at the instant represented in the figure.

a) ω = 12.5 rad/s   b) ω = 7.5 rad/s

c) ω = 5.5 rad/s   d) ω = 2.5 rad/s

P4.3-3) Consider the articulated robot arm shown. The motor at joint A (driving arm AB) has a constant counterclockwise angular velocity of 2 rad/s and the motor at joint B (driving arm BC) has a clockwise angular velocity of 3 rad/s that is decreasing at a rate of 1 rad/s^2. If arm AB has length 1.7 m and arm BC has length 2.5 m, determine the acceleration of joint C at the instant pictured.

Ans: $\mathbf{a}_C$ = 13.43$\mathbf{i}$ - 18.22$\mathbf{j}$ m/s^2

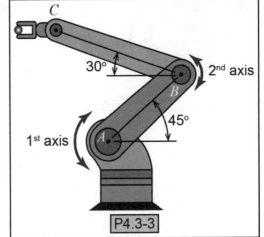

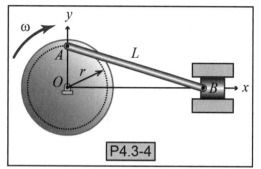

P4.3-4

P4.3-4)[fe] The disk in the reciprocating piston mechanism rotates at 1000 rpm in the clockwise direction. The connecting rod AB converts the rotational motion of the disk to translation of the piston. The connecting rod attaches to the disk 1 ft from its center and the length of rod AB is 3 ft. Determine the velocity of the piston. Back up your answer with mathematical formulations.

a) v_B = 52.3 ft/s   b) v_B = 1000 ft/s

c) v_B = 207 ft/s   d) v_B = 104.7 ft/s

P4.4) INTERMEDIATE GENERAL PLANAR MOTION PROBLEMS

P4.4-1)   The belt-driven pulley system shown has a smaller pulley (r_A = 3 in) rigidly attached to a larger pulley (r_B = 12 in). The belt speed of the smaller pulley is 4 ft/s and the belt acceleration of the larger pulley is 0.5 ft/s^2.   Determine the angular velocity and acceleration of the pulley system. Also, determine the belt speed of the larger pulley and the acceleration of point c located on the rim of the larger pulley.

Ans: ω = 16 rad/s, α = 0.5 rad/s^2,
$v_b = 16$ ft/s, $\mathbf{a}_c = 0.5\mathbf{e}_t + 256\mathbf{e}_n$ ft/s^2

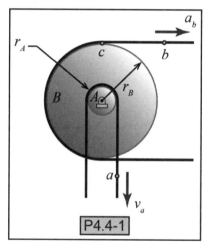

P4.4-1

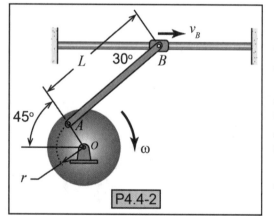

P4.4-2

P4.4-2) The disk shown is pinned at O and is rotating clockwise with an angular velocity of ω = 5 rad/sec. The disk is connected via rod AB (L = 4 m) to collar B. The A end of rod AB is attached to the disk at a distance of $r = 0.8$ m from point O.   Collar B is constrained to move along a horizontal guide. At the instant shown, determine the velocity of point B.

Ans: v_B = 4.46 m/s

P4.4-3) The bar linkage shown is set in to motion by applying a counter clockwise angular velocity of 10 rad/s to bar CD.   All links have the same length.   It is additionally known that bar CD has a clockwise angular acceleration of 2 rad/s^2.   Determine the angular acceleration of bar AB.

Ans: $\alpha_{AB} = 336$ rad/s^2 ccw

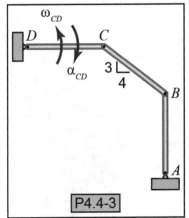

P4.4-3

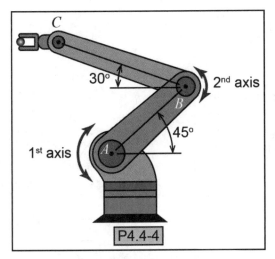

P4.4-4

P4.4-4) Consider the articulated robot arm shown. It is desired that the robot be controlled to move joint C vertically upward with a speed of 0.5 m/s at the instant shown. Determine the angular velocities that the motors at joint A (driving arm AB) and joint B (driving arm BC) must have to achieve this motion. Let arm AB have length 1.7 m and arm BC have length 2.5 m.

Ans: $\omega_{BC} = 0.146$ rad/s, $\omega_{AB} = 0.152$ rad/s

P4.5) ADVANCED GENERAL PLANAR MOTION PROBLEMS

P4.5-1) The drive link in the mechanism shown runs in a linear bearing with a velocity v_C directed to the left.   Determine the angular velocities of link AB and BC as a function of v_C, x and L.

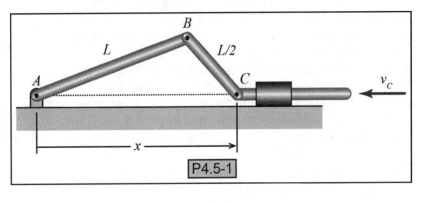

P4.5-1

Ans: $\omega_{AB} = \dfrac{2v_C}{L}\left(\dfrac{2x_2}{L}\right)$

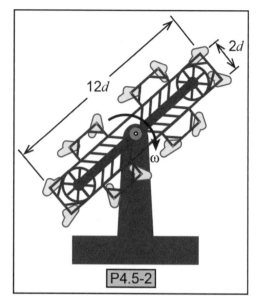

P4.5-2

P4.5-2) You are designing the carnival ride shown and wish to prevent riders from experiencing accelerations greater than three times the acceleration due to gravity g. The cars attached to the main body rotate only due to gravity and may be assumed to experience a maximum angular acceleration of 10 rad/sec² and a maximum angular velocity of 4 rad/sec. Determine the limitations on the dimension d of the ride and on the constant angular velocity ω with which the main body of the ride can be driven with while still maintaining the $3g$ limit

Ans: $\omega^2 d < 6.57$ ft/s²

P4.5-3) Rod AB contacts a wall and floor as shown in the figure. If these contacts are not broken and the velocity of end A is constant upward, prove that the acceleration of point P is always perpendicular to the line drawn from O to A.

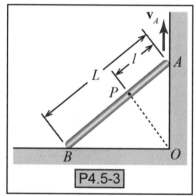

P4.5-3

P4.6) BASIC IC & ROLLING PROBLEMS

P4.6-1) The wheel with a rigidly attached bar shown, rolls without slip. Determine the velocity and acceleration of the wheel center (Point O) and the end of the bar (Point A) when the bar is parallel to the rolling surface. The radius of the wheel is 0.25 meters and the total length of the bar from end to end is 1 meter. At the instant the bar is horizontal, the angular velocity and angular acceleration directions are clockwise and have values of 4 rad/s and 5 rad/s², respectively.

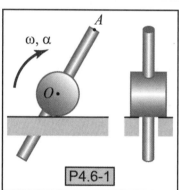

P4.6-1

Ans: $\mathbf{v}_O = \mathbf{i}$ m/s, $\mathbf{a}_O = 1.25\mathbf{i}$ m/s², $\mathbf{v}_A = \mathbf{i} - 2\mathbf{j}$ m/s,
   $\mathbf{a}_A = -6.75\mathbf{i} - 2.5\mathbf{j}$ m/s²

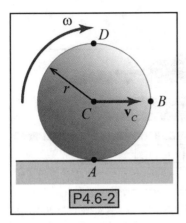

P4.6-2

P4.6-2)[fe] A disk with radius r rolls without slipping on a horizontal surface. If the velocity of the disk's center is v_C, in the direction shown, determine the velocity of point B and D whose locations are identified on the figure.

a) $\mathbf{v}_B = r\omega(\mathbf{i} - \mathbf{j})$, $\mathbf{v}_D = 2\omega r\mathbf{i}$ b) $\mathbf{v}_B = r\omega(\mathbf{i} - \mathbf{j})$, $\mathbf{v}_D = 2\omega r\mathbf{i}$

c) $\mathbf{v}_B = r\omega(\mathbf{i} - \mathbf{j})$, $\mathbf{v}_D = 2\omega r\mathbf{i}$ d) $\mathbf{v}_B = r\omega(\mathbf{i} - \mathbf{j})$, $\mathbf{v}_D = 2\omega r\mathbf{i}$

P4.6-3)   Given the piston assembly shown, determine the location of the connecting rod's instantaneous center of zero velocity for the instant $\theta = 35°$. The length of the connecting rod is $l_{BP} = 10$ in and the length of the crankshaft rod is $l_{AB} = 4$ in.

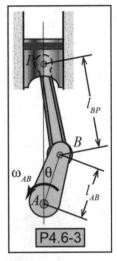

P4.6-3

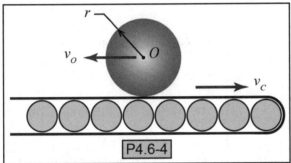

P4.6-4

P4.6-4)[fe] A ball, of radius 2 meters, rides on a conveyer belt as shown in the figure. If the velocity of the ball's center O is 1.5 m/s, directed to the left, and the velocity of the top of the conveyer belt is 3 m/s, directed to the right, determine the angular velocity of the ball. Assume that the ball rolls on the conveyor belt without slipping.

a) $\omega = 2.25$ rad/s ccw   b) $\omega = 3.25$ rad/s ccw

c) $\omega = 4.25$ rad/s ccw   d) $\omega = 5.25$ rad/s ccw

P4.6-5)[fe] A disk rolls down a hill without slipping. The disk is released from rest and accelerates down the hill with a constant angular acceleration of 8 rad/s². Determine the distanced traveled by the center of the disk when the speed of the center reaches 2 m/s.

a) $s = 0.125$ m   b) $s = 0.25$ m

c) $s = 0.50$ m   d) $s = 0.75$ m

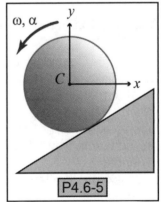

P4.6-5

P4.7) INTERMEDIATE IC & ROLLING PROBLEMS

P4.7-1) A car is driving on a road at 50 mph (22.4 m/s). The driver sees a squirrel and applies his brakes causing the car to decelerate at a constant rate of 4.5 m/s². The pavement is dry and the majority of the braking is done with the rear wheels such that the front wheels may

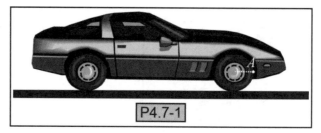

P4.7-1

be assumed to be rolling without slip. What is the velocity and acceleration of point A, on the car tire, 2 seconds after the driver applies the brakes? Point A is on the outer edge of the tire at the forward position that is level with the wheel's center at the end of this 2 seconds. The radius of the tire is 13 inches (0.33 m).

Ans: $\mathbf{v}_A = 13.4\mathbf{i} - 13.4\mathbf{j}$ m/s, $\mathbf{a}_A = -548.7\mathbf{i} - 4.5\mathbf{j}$ m/s²

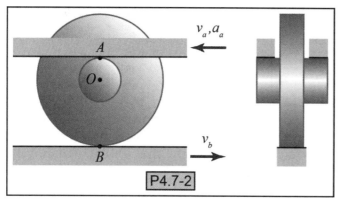

P4.7-2

P4.7-2) A wheel is trapped and rolls without slip between two moving surfaces. The wheel has an inner hub and an outer hub. The radius of the inner hub is 6 inches and the radius of the outer hub is 18 inches. The velocity and acceleration of the surface touching the inner hub is 0.5 ft/s and 0.2 ft/s² respectively, in the direction shown, and the velocity of the surface touching the outer hub is a constant 2 ft/s, in the direction shown. Find the angular velocity and acceleration of the wheel and the velocity and acceleration of the wheel's center at this instant.

Ans: ω = 1.25 **k** rad/s, α = 0.1 **k** rad/s², $\mathbf{v}_O = 0.125\mathbf{i}$ ft/s, $\mathbf{a}_O = -0.15\mathbf{i}$ ft/s²

P4.7-3) Consider the two disks of radius r = 10 cm shown that are rolling without slip on level ground. The disks are connected by bar AB of length L = 35 cm. If the disk on the left has an angular velocity of ω = 3 rad/s in the ccw direction, determine the angular velocity of the disk on the right for the instant shown.

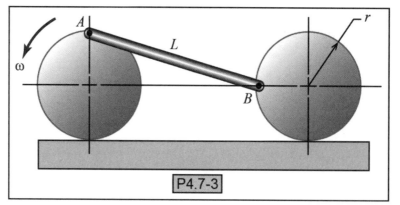

P4.7-3

Ans: ω_{right} = 9.28 rad/s cw

P4.8) ADVANCED IC & ROLLING PROBLEMS

P4.8-1) In hybrid electric vehicles a planetary gear set can be employed to couple the vehicle's multiple power plants. The planetary gear set is specifically referred to as a speed coupling device because the speeds of a power plant attached to the Sun gear (ω_s) and a power plant attached to the Ring gear (ω_r) sum to produce the motion of the Carrier (ω_c) according to the relation, $\omega_c = k_1\omega_s + k_2\omega_r$. Express the constants k_1 and k_2 in terms of the radii R_s, R_r and R_c as defined in the given figure. Assume that the Sun gear, the Ring gear and the Planet gears (attached to the Carrier) mesh without slipping.

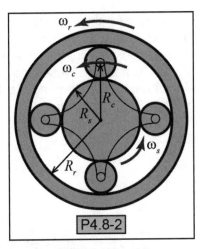

P4.8-2

Ans: $k_1 = R_s/2R_C$, $k_2 = R_r/2R_C$

P4.9 BASIC RELATIVE SLIDING AND ROTATING REFERENCE FRAME PROBLEMS

P4.9-1) At the instant shown, bar AB has an angular velocity of $\omega_{AB} = 2$ rad/sec in the counterclockwise direction. Also, pin B slides in the slot of bar CD. Determine the angular velocity of bar CD at this instant.

Ans: $\omega_{CD} = 2$ rad/s ccw

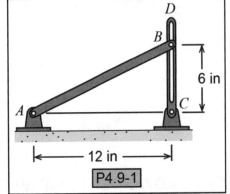

P4.9-1

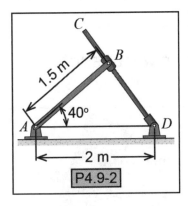

P4.9-2

P4.9-2) A collar is pinned to the end of rod AB. The collar pivots about point B and is allowed to slide freely along rod CD. If rod AB is given a clockwise angular velocity of 3 rad/sec, determine the angular velocity ω_{CD} of rod CD for the instant shown in the given figure.

Ans: $\omega_{CD} = 0.703$ rad/s ccw

P4.10) INTERMEDIATE RELATIVE SLIDING AND ROTATING REFERENCE FRAME PROBLEMS

P4.10-1)   At the instant shown, bar AB has an angular velocity of ω_{AB} = 2 rad/sec in the counterclockwise direction that is decreasing at a rate of 3 rad/sec^2. Also, the pin at B slides in the slot of bar CD. Determine the angular acceleration of bar CD at this instant.

Ans: α_{CD} = 5 rad/s ccw

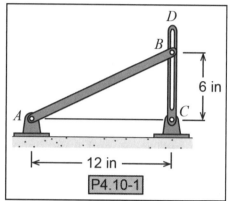

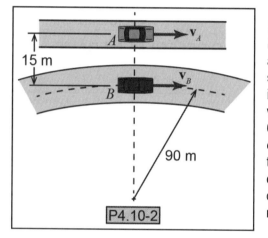

P4.10-2) Car B is traveling with a speed of 40 km/h that is decreasing at a rate of 3 m/s^2 around a curve with radius 90 m when its driver sees car A immediately to his left. At this instant, car A is traveling along a straight road with a speed of 60 km/h increasing at a rate of 6 m/s^2. What velocity and acceleration does car A appear to have to the driver of car B at the instant shown in the figure, presuming the driver of car B is rotating with his car (he doesn't turn his head)? Use the concept of rotating reference frames to solve this problem.

Ans: $\mathbf{v}_{A,rel}$ = 3.70$\mathbf{i}$ m/s, $\mathbf{a}_{A,rel}$ = 9.5$\mathbf{i}$ + 2.5$\mathbf{j}$ m/s^2

P4.11) ADVANCED RELATIVE SLIDING AND ROTATING REFERENCE FRAME PROBLEMS

P4.11-1)   Plate D is pinned at C and is rotating with an angular velocity of ω_D = 2 rad/sec and an angular acceleration of 1 rad/sec^2, both in the clockwise direction. Pin P is attached to bar AB and is constrained to move along a curved slot in plate D. Pin P is located a distance of 45 cm from point A and the slot has a radius of curvature of 35 cm. Determine the angular acceleration of bar AB at the instant shown. Note that at this instant the position vector of point P relative to point C ($\mathbf{r}_{P/C}$) is tangent to the slot.

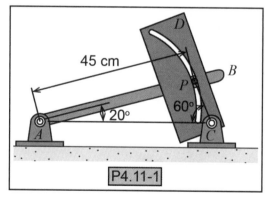

CHAPTER 4 COMPUTER PROBLEMS

C4-1) Perform a computer simulation of the motion of a particle attached to the rim of a disk that is rolling without slip. Determine the position, velocity and acceleration of the particle. Assume that the disk starts from rest and the angular acceleration is constant. Construct a scatter plot of the *x*- and *y*-components of the position, velocity and acceleration for various angular accelerations and disk radii.

CHAPTER 4 DESIGN PROBLEMS

D4-1) Consider the "tea-cup" amusement park ride shown. Each cup has an occupancy limit of three riders. The cups are rotated with respect to the platform by the riders through a wheel at the center of the teacup. The platform itself is turned by a large electric motor. Use your knowledge of dynamics to address the following issues: a) it is desired that the sustained acceleration of a rider should be less than $3g$ and b) the peak acceleration experienced by a rider should be less than $5g$.

- Recommend the maximum constant angular speed with which the platform should be allowed to rotate.
- Recommend the maximum angular acceleration with which the platform should be allowed rotate.

Note: There may be information that you need to complete your design which you do not have. Explain how you would obtain estimates of any missing information that you need.

A4-1) Ride on a merry-go-round and discuss why you "feel" different at different locations on the ride.

<u>Group size:</u> 2 students

<u>Supplies</u>

- Stop watch
- Measuring tape
- Masking tape

<u>Procedure</u>

1. Find a merry-go-round. This may be a ride in an amusement park or one at a park that you need to spin.
2. Mark and measure three different radial locations on the ride.
3. Throughout the experiment try to rotate the merry-go-round at a constant angular speed. One student should apply a moment to the merry-go-round as necessary to maintain the constant angular speed. Use the stop watch to determine the merry-go-round's angular speed.
4. Have the other student ride the merry-go-round at the center of the ride for at least 30 seconds. Make notes about how you "feel".
5. Walk to the next radial location. Note what you experience walking between the locations. Stay at this location for at least 30 seconds. Note how you "feel".
6. Repeat for each successive radial locations.

<u>Calculations and Discussion</u>

- Calculate the angular speed of the merry-go-round.
- Calculate your linear velocity and linear acceleration.
- List all of the assumptions you made in your calculations.
- Discuss how you felt during the experiment.
- Discuss why you think you experienced the feelings that you did.
- List possible sources of experimental error.

PART III: KINETICS - NEWTONIAN MECHANICS

CHAPTER 5: PARTICLE NEWTONIAN MECHANICS

CHAPTER OUTLINE

CHAPTER SUMMARY

Dynamics is the study of motion and is broken up into two major categories: *Kinematics* and *Kinetics*. In general, kinematics relates the position, velocity and acceleration of a body through the use of defining relations and calculus. Kinetics relates the motion of a body to the forces that are applied to it, that is, the cause of the motion. Kinetic analysis can be performed in three distinctly different ways: through *Newtonian mechanics*, *energy methods* or using *impulse and momentum* principles. The approach covered in this chapter is Newtonian mechanics as applied to a particle. This method uses Newton's laws to analyze the translational motion of an object that results from the applied forces.

5.1) NEWTONIAN MECHANICS

5.1.1) KINETICS

Kinematics is an analysis technique that relates the position, velocity and acceleration of a body through the use of defining relations and calculus. A common question posed in kinematic analysis is, "Given the position of a particle as a function of time, find the acceleration." **Kinetics**, on the other hand, is the study of how forces generate motion. A common question posed in kinetic analysis is, "A force **F** is applied to a particle. What is the resulting acceleration of the particle?" In this chapter, we will answer this type of question using Newtonian mechanics. In subsequent chapters we will learn other approaches for answering the same question.

5.1.2) NEWTONIAN MECHANICS

If you look around, you will see things in motion. In 1687, Newton published three laws of motion that are so powerful that they are still taught in every college engineering program nearly 300 years after they were conceived. One of Newton's assumptions was that space and time are absolute. If you are a student of Einstein, you know that this is not true. Newton's laws are very good at predicting how an object will move in most practical instances. However, they do break down at the atomic level or at velocities approaching the speed of light.

It should be further noted that Newton's laws only hold true when the particle's acceleration is described with respect to an inertial reference frame, that is, a reference frame that is not accelerating or rotating. If the acceleration of the reference frame is small compared to the acceleration of the body being examined, the reference frame can, for all intents and purposes, be considered inertial. For example, in many cases a reference frame attached to the earth's surface is used. This type of reference frame may be treated as inertial without contributing significant error, even though the earth itself is accelerating.

Newton's laws describe the relationship between the forces acting on a body and the resulting motion. This means that we can use Newton's laws to find the motion of a body if we know the forces acting on the body. Or, conversely, we can find the resultant force acting on the body if we know how the body is moving. Solving problems using Newtonian mechanics relies heavily on the concepts of mass, force and acceleration. Even though we have seen these concepts before, this is a good place to revisit them because of their importance to this particular analysis method.

5.1.3) MASS VERSUS WEIGHT

In the coming sections, we will describe all of Newton's laws of motion in detail. The particular law that ends up coming to the forefront is Newton's second law which states mathematically that the sum of all forces acting on a body equals the mass of a body times its acceleration.

$$\sum \mathbf{F} = m\mathbf{a}$$

As you can see, the above equation uses mass not weight. If you use weight in this equation, your results will be incorrect. Therefore, it is important to understand not only the difference between mass and weight, but also how to identify mass and weight units.

It is easy to confuse *mass* and *weight*, but they are different. **Mass** is the quantity of matter in an object and **weight** is the effect gravity has on that matter. Mass is the measure of how much a body resists acceleration (i.e. how much inertia it has). It determines an object's acceleration in the presence of an applied force. A body's mass stays the same no matter where it is located, here on earth or on the moon. A body's weight depends on its

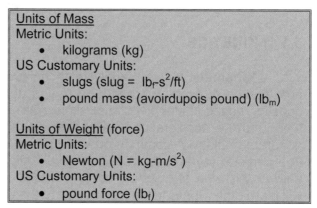

location. A body will weigh differently here on earth than it does on the moon. This is because weight is the force of gravity and gravity is different in each location. The relationship between mass and weight is given in Equation 5.2-1.

Mass has no direction, therefore, it is a scalar. Weight is determined by multiplying mass by the acceleration due to gravity. Gravity acts towards the center of the earth (or whatever body is being considered) making it a vector. Therefore, weight ($\mathbf{W}$) is a vector.

Weight / Mass relationship: $\boxed{\mathbf{W} = m\mathbf{g}}$ (5.2-1)

$\mathbf{W}$ = weight
m = mass
$\mathbf{g}$ = acceleration due to gravity = 9.807 m/s^2 = 32.1740 ft/s^2

What is the difference between *mass* and *weight*?

Example 5.1-1

Given below are several objects. Identify whether or not the number given next to each object is specifying mass or weight.

182 g

11 kN

400 lb

64 oz

7.347x10^{22} kg

0.3 slugs

Conceptual Example 5.1-2

Astronauts on the Moon can jump higher than on the Earth because

   a) they weigh less on the Moon.
   b) they have less mass on the Moon.
   c) there is no atmosphere on the Moon.

Conceptual Example 5.1-3

An astronaut floating weightlessly in space shakes the wrench in his hand back and forth. Which of the following is true.

   a) The shaking requires no effort because the wrench has no inertia to resist the motion.
   b) The shaking requires some effort, but less than on earth because the wrench weighs less in space.
   c) The shaking requires the same amount of effort as on earth because the inertial mass of the wrench is the same no matter the wrench's location.

5.1.4) FORCE

Newtonian mechanics, as it relates to dynamics, is all about relating force and acceleration. Understanding the different types of forces and how they act on a body is important in the application of Newton's laws of motion. Later in the chapter, we will discuss some common types of forces. For now, let's concentrate on what a force is and how it is applied to a body.

A **force** is the action of one body on another. This force can come from an actual physical contact between bodies or through a body force due to a gravitational or magnetic field. In this text, we will, in general, idealize forces to act in a well-defined direction and at a single location.

Units of Force
Metric Units:
• Newton (N = kg-m/s^2)
US Customary Units:
• pound force (lb$_f$)

5.1.5) ACCELERATION

As mentioned before, Newton's second law states, mathematically, that the sum of all forces acting on a body equals the mass of a body times its acceleration. Thus far, we have spent a lot of time learning about acceleration. Now we will learn how to relate it to force.

$$\sum \mathbf{F} = m\mathbf{a}$$

We have already learned that a body's **acceleration** is the rate at which its velocity changes and that acceleration is, mathematically, defined by the kinematic relationship $\mathbf{a} = d\mathbf{v}/dt$. Let's try thinking

Units of Acceleration (length/time^2)
Metric Units:
• meters per second squared (m/s^2)
US Customary Units:
• feet per second squared (ft/s^2)

about acceleration in a slightly different way. Imagine that you are a passenger in a car traveling with constant velocity. Close your eyes. If the road that you are traveling on is perfectly smooth, would you know that you are moving? No. You feel no forces acting on your body except the force of gravity pulling you down into your seat. When do you really feel that you are moving? You really feel that you are moving when the car accelerates in some way. If you step on the gas pedal to pass a car, you feel the car seat pushing you forward causing you to accelerate. If you take a turn, you have to apply a force to keep yourself upright. These are simple examples that show the relationship between acceleration and force.

How do you know that you are accelerating?

5.2) NEWTON'S LAWS

There are three laws first identified by Sir Isaac Newton that will help us in our analysis and understanding of kinetics problems.

5.2.1) NEWTON'S FIRST LAW

The first of Newton's laws pertains to objects in *equilibrium*. Mathematically, this law is expressed in Equation 5.3-1. This law is commonly applied to objects that are in static equilibrium, that is, to objects that are at rest and that are acted on by a balance of forces. A body that is in **equilibrium** implies that the resultant force or the sum of all forces acting on the body is zero. However, this law also applies to objects that are in motion and are not accelerating. In this case, the resultant force is again zero, but the body is moving in a straight line with a constant speed. There are many translated versions of the original text of Newton's laws. The following two paragraphs are two common translations.

Newton's First Law:
- An object at rest tends to stay at rest and an object in motion tends to stay in motion at constant speed in a straight line.
- In the absence of a net external force, a body either is at rest or moves in a straight line with constant speed.

Newton's first law: $\boxed{\sum \mathbf{F} = 0}$ (5.3-1) $\mathbf{F}$ = externally applied forces acting on the body

The part of Newton's first law that states a body may be in equilibrium if it is moving with constant velocity may be less intuitive than realizing a stationary body is in equilibrium. This is because in the real world there is always some force acting on a body, however small it may be. For example, even a hockey puck sliding on ice experiences some friction and air drag.

Newton's first law is sometimes referred to as the *Law of inertia*. **Inertia** is a measure of an object's resistance to changes in motion. The measure of a body's translational inertia is its mass.

What is *equilibrium*?

What is *inertia*?

Conceptual Example 5.2-1

Which of the following particles are in equilibrium?

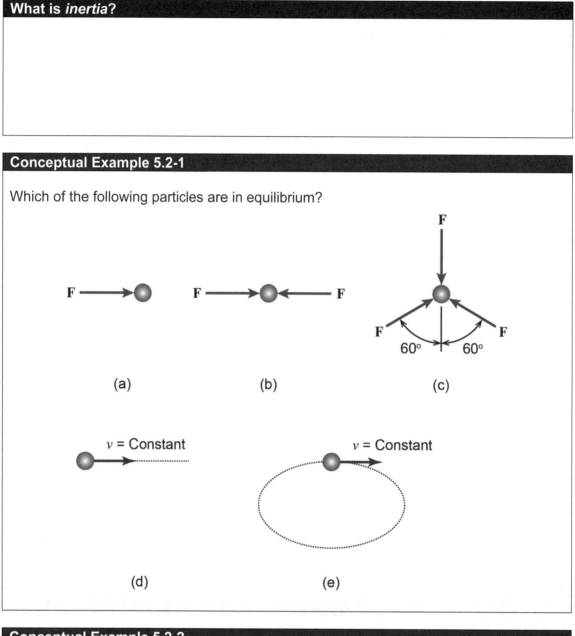

(a) (b) (c)

(d) (e)

Conceptual Example 5.2-2

Consider that you are swinging a weight on the end of a string around your head. If you let go, which of the following paths best approximate the direction the weight will travel?

5.2.2) NEWTON'S SECOND LAW

Newton's second law states that when a body's mass is constant, the product of its mass and acceleration is equal to the resultant force acting on the body. Mathematically, this is expressed in Equation 5.2-2.

Newton's second law: $\boxed{\sum \mathbf{F} = m\mathbf{a}}$ (5.2-2)

$\mathbf{F}$ = externally applied forces acting on the body
m = mass
$\mathbf{a}$ = acceleration

The equation that results from applying Newton's second law is usually called the body's **equation of motion**. Newton's second law, in paragraph form, is given as follows.

> **Newton's Second Law:** A body acted on by an imbalance of forces will accelerate. The amount of acceleration is inversely proportional to its mass.

According to Newton's second law, when an imbalance of forces act on a particle, it will experience acceleration. Since both the forces and acceleration are vector quantities, the body's acceleration is always in the direction of the resultant force. When there is more than one force acting on a particle, the resultant force can be determined by taking a summation of all the forces applied as shown in Figure 5.2-1.

Equation 5.2-2 (Newton's second law) is a more general version of Equation 5.2-1 (Newton's first law) in that it also

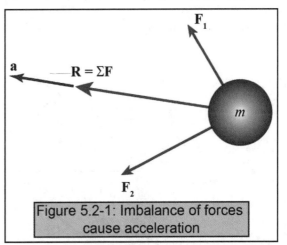

Figure 5.2-1: Imbalance of forces cause acceleration

applies when $\mathbf{a}$ = 0. Since Newton's second law is a vector equation, it is sometimes helpful to break up Equation 5.2-2 into its x- and y-components and apply each equation independently.

$$\sum F_x = ma_x \qquad\qquad \sum F_y = ma_y$$

Example 5.2-3

Answer the following questions about Newton's second law of motion.

a) Is Newton's second law a scalar or vector equation?

b) Looking at the equation representing Newton's second law of motion, what direction will the acceleration always be in?

c) The equation resulting from the application of Newton's second law is called the body's _____.

Conceptual Example 5.2-4

A go-kart is being pulled by two forces as shown in the figure. The wheels on the go-kart allow it to move freely in any direction. List in order, from largest to smallest, the magnitude of the acceleration of the go-kart.

Largest: _____ Next: _____ Next: _____ Smallest: _____

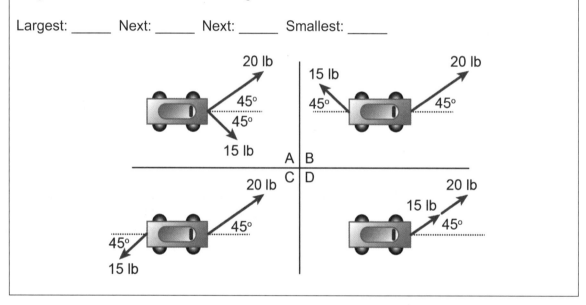

Conceptual Example 5.2-5

If no external forces are applied, the space shuttle shown will drift in a straight line from point A to C. However, halfway between the two points (point B), the engine engages. This produces a constant thrust that is perpendicular to the line drawn between points A and C. Which of the following trajectories most accurately represents the motion of the space shuttle?

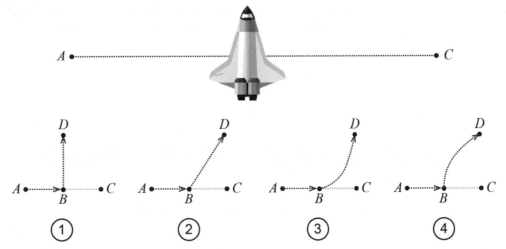

As the space shuttle moves from point B to D, its speed is
    a) constant.
    b) increases at a constant rate.
    c) increases for a short time and then remains constant.
    d) decreases at a constant rate.
    e) decreases for a short time and then remains constant.

At point D, the engine completely shuts off. Which of the trajectories most accurately represents the motion of the space shuttle after the engine is shut down?

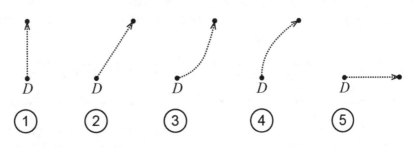

Beyond point D, the space shuttle's speed is
    a) constant.
    b) increases at a constant rate.
    c) increases for a short time and then remains constant.
    d) decreases at a constant rate.
    e) decreases for a short time and then remains constant.

Conceptual Example 5.2-6

A truck drives along a horizontal road with a constant velocity. Its motion is retarded and eventually halted by a set of crash barrels. The amount of barrels and sand within the barrels are arranged in such a way to decelerate the truck at a constant rate. Four test runs are performed where the velocity and mass of the truck are varied. In each of the following situations, the barrels stop the truck in the same amount of time. Rank the force required to stop the truck for the following situations from greatest to least.

Greatest: _____   Next: _____   Next: _____   Least: _____

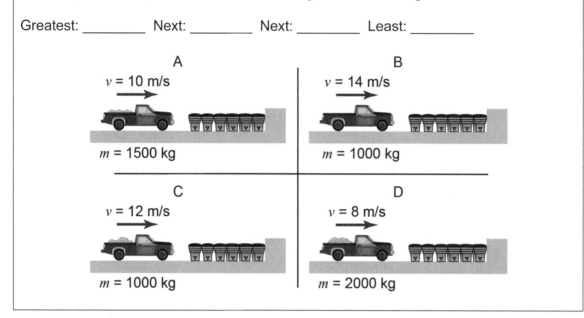

A	B
$v = 10$ m/s	$v = 14$ m/s
$m = 1500$ kg	$m = 1000$ kg
C	D
$v = 12$ m/s	$v = 8$ m/s
$m = 1000$ kg	$m = 2000$ kg

5.2.3) NEWTON'S THIRD LAW

The third and final of Newton's laws states that when a body A applies a force on a second body B, the body B applies an equal force back on body A. Furthermore, these two forces act on the same line (though in opposite directions). This law, when written in paragraph form, is most likely familiar.

> **Newton's Third Law:** For every force of action, there is an equal and opposite reaction.

An example of Newton's third law is a block simply sitting on the ground as shown in Figure 5.2-2. In this instance, the force due to the weight ($m\mathbf{g}$) of the block pushes down on the ground and the ground pushes back on the block with an equal and opposite force. This reaction force is generally referred to as the normal force ($\mathbf{N}$). These two reaction forces constitute an interaction pair. In this instance, $\mathbf{N}$ will equal $m\mathbf{g}$ because the block is not accelerating in the y-direction and there are no other externally applied forces.

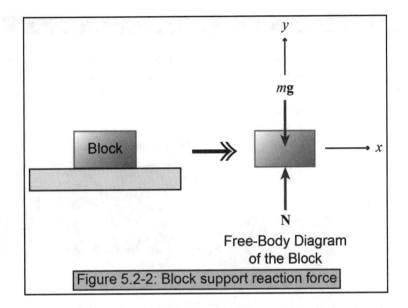

Free-Body Diagram
of the Block

Figure 5.2-2: Block support reaction force

Now consider a tractor towing a trailer, as shown in Figure 5.2-3, as another example of Newton's third law. Consider the tractor and the trailer as separate systems attached together by a hitch in the middle. There is a tension force from the tractor pulling the trailer forward and an equal and opposite force of the trailer pulling back on the tractor.

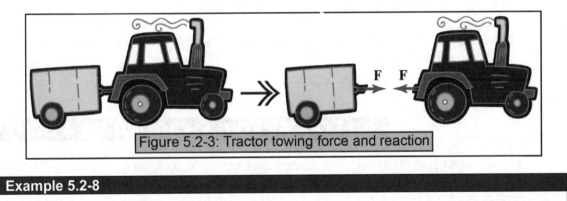

Figure 5.2-3: Tractor towing force and reaction

Example 5.2-8

Consider a uniform beam resting on two supports. What are the support reactions?

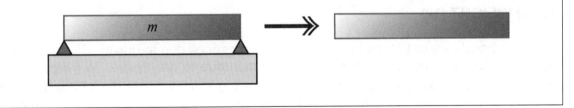

Conceptual Example 5.2-9

A large truck breaks down on the road and receives a push back into town by a small compact car as shown. While the car, still pushing the truck, is speeding up, the amount of force that the car applies to the truck is **equal to, smaller than** or **greater than** the force with which the truck pushes back on the car.

For the car and truck above, which is the correct statement of the situation?

a) The truck moves forward because the car pushes slightly harder on the truck than the truck pushes back on the car.
b) The car cannot move the truck because the truck pushes back on the car with the same force with which the car pushes on the truck.
c) The car can only push the truck forward if the truck weighs less than the car.
d) The car can only move the truck if the car's tractive force (i.e. the force generated between the wheels and the road) is greater than the force with which the truck pushes back on the car.

5.3) TYPICAL FORCES ACTING ON A BODY

Kinetic analysis deals with the forces that act on a body and how those forces create motion. There are forces that always act on a body such as its *weight* and forces that typically act on a body such as *normal* and *friction* forces. Below is a discussion of forces that you will commonly encounter when analyzing the motion of a body.

5.3.1) WEIGHT (W)

A body's **weight** is the force the body experiences due to gravity. Weight has a magnitude equal to its mass times the acceleration due to gravity (i.e. $W = mg$) and is directed straight down. More precisely, it is directed toward the center of the earth or another celestial body. Be careful to distinguish between a body's weight and its mass. A 100-lb body has a mass of m = 100 lb / 32.2 ft/s^2. The force of a body's weight is applied to its center of gravity. This will become a much more important concept when we start dealing with rigid bodies. Since particles theoretically have no size, the force of a body's weight can be placed anywhere. However, it is good practice to place it near the body's center. Figure 5.3-1 shows an apple sitting on a table. The apple has mass and, therefore, weight. In order to apply Newton's laws to this apple, we need to represent the apple's weight as a vector force on the figure. The weight will act

downward and equals the apples mass times the acceleration due to gravity as shown in
Figure 5.3-1. The weight force is also applied at the apples center of gravity (CG). However, as mentioned before, this does not become important until we start dealing with rigid bodies.

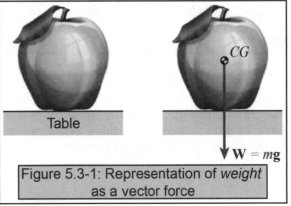

Figure 5.3-1: Representation of *weight* as a vector force

5.3.2) NORMAL FORCE (N)

The reaction of a surface to its contact with a body is the **normal force** (N). The normal force has a direction that is perpendicular (i.e. normal) to the surface the body is contacting. Often, a normal force is the reaction of a surface to a body resting on it. Figure 5.3-2 shows the apple represented in Figure 5.3-1. This time the table has been removed and replaced with a normal force. This normal force is the force that the table was applying to the apple that prevented the apple from falling to the ground.

We now know the direction of a normal force and, at least conceptually, the magnitude of the normal force. But, where along the bottom of the apple should the normal force be applied? It is good practice to draw the normal force in the middle of the contacting surface of the body. However, the real answer is, "It doesn't matter, at least for now." We are currently analyzing particles. A particle has no significant size, therefore, the normal force cannot create a moment no matter where you place it. On the other hand, when analyzing rigid bodies, the point of application of the normal force is important and will affect your analysis.

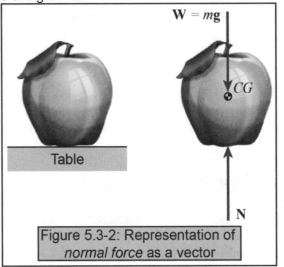

Figure 5.3-2: Representation of *normal force* as a vector

5.3.3) TENSION FORCE (T)

A **tension force** results from pulling on a cable, rope or chain and always acts along the line of the supporting medium. We usually idealize these components and say that their mass is negligible. In addition, we say that the rope, for example, cannot be stretched or that it is inextensible. In this idealized situation, the force applied at one end of the rope is transmitted perfectly along its entire length. This means that no matter where you cut the rope, you will see the same tension force. Figure 5.3-3 shows a crate suspended by a rope looped over a hook. The rope is transmitting the forces that are holding the crate. If we replace the rope with two tension forces, the forces will

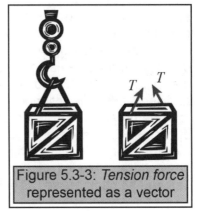

Figure 5.3-3: *Tension force* represented as a vector

be directed along the length of the rope. If we neglect friction at the hook, then the tension forces in both sides of the rope will be equal.

Conceptual Example 5.3-1

An elevator is being lifted up a shaft at a constant speed by a steel cable as shown. All frictional effects are negligible. In this situation, is the upward force of the cable **greater than, equal to** or **smaller than** the downward force of gravity?

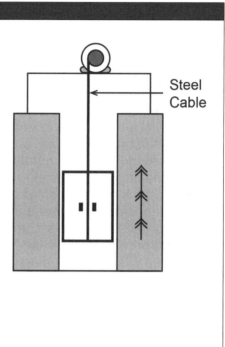

Steel Cable

What if the elevator is accelerating upward? Is the upward force of the cable **greater than, equal to** or **smaller than** the downward force of gravity.

5.3.4) FRICTION FORCE ($\mathbf{F}_f$)

Friction occurs between two contacting surfaces. If the surfaces are moving relative to each other the friction force is classified as **kinetic friction** and acts to retard or slow down the motion. The energy of kinetic friction dissipates in the form of heat. The heat generated while striking a match, ignites the chemicals as shown in Figure 5.3-4. If the surfaces are not moving relative to each other, then there must be a force applied to at least one of the bodies before a friction force can develop. If an external force is slowly applied, the friction force will oppose the applied force and the body will not move until the applied force is great enough to create an imbalance of forces. In the case of surfaces that aren't

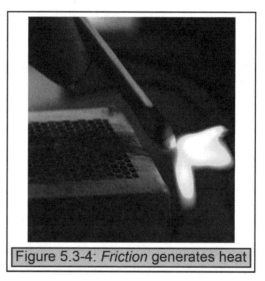

Figure 5.3-4: *Friction* generates heat

moving relative to each other, the frictional force is called **static friction** and it acts to prevent the intended motion. Friction is a very important and complicated concept. Later in the chapter, we will devote an entire section to friction.

Example 5.3-2

Consider a block being pushed along a flat horizontal surface.

- What forces are acting on the block?

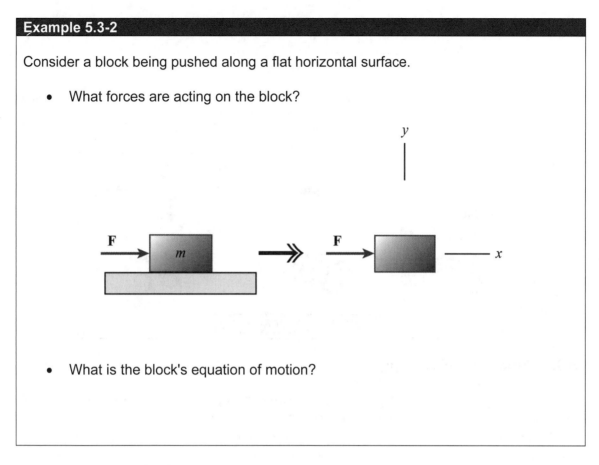

- What is the block's equation of motion?

5.3.5) SPRING FORCE (F_s)

The force generated by a linear spring is often described by **Hooke's law** $F_s = kx$, where k is a constant representing the spring's stiffness and x is the deformation of the spring. Hooke's law states that the force generated by a spring is linearly proportional to its deformation. All real world springs generate a force that varies nonlinearly with deformation. This is true, in particular, as the spring moves farther away from its equilibrium position. If you work within a narrow deflection band, Hooke's law often will provide a relatively accurate model of the spring force.

Whether the spring is compressed or stretched, the direction of the force always opposes the direction of deformation. Figure 5.3-5 shows a Jack-in-the-box. Initially, Jack is in the box and is straining to get out. When the lid opens, Jack pops out because the spring is compressed and pushes Jack out of the box. Jack will eventually pass the un-stretched length of the spring (l_o). This is the length of the spring where no spring forces are generated. In other words, this is the equilibrium position of the spring. As Jack passes the un-stretched length, the spring will generate a force to return the spring to its un-deformed state and thus Jack will proceed back towards the box. This push-pull action of the spring will continue until friction, air resistance and material hysteresis eventually bring Jack to a stop.

Spring force: 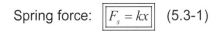   (5.3-1)

F_s = spring force
k = spring coefficient
x = compression or extension of the spring beyond its un-stretched length

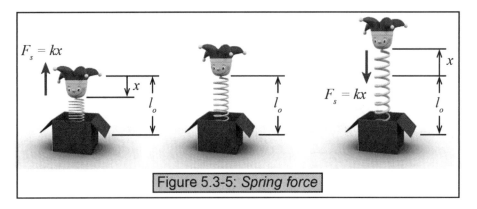

Figure 5.3-5: *Spring force*

5.3.6) DAMPING FORCE ($\mathbf{F}_d$)

The force generated by a **damper**, like a dashpot (i.e. a type of hydraulic or mechanical damper), can often be modeled as $F_d = c\dot{x}$, where the force is proportional to the body's speed ($\dot{x}$) and c is a constant representing the amount of damping. The damping force acts in a direction that opposes the velocity.

A common type of damper is the shock absorber on your car. A shock absorber smoothes out or dampens shock impulses from the road. Shock absorbers slow down and reduce the magnitude of vibratory motions by turning the kinetic energy of suspension movement into heat energy that is dissipated through the hydraulic fluid. A shock absorber is basically an oil pump placed between the frame of the car and the wheels. Figure 5.3-6 shows a twin-tube shock absorber. This is one of the most common types of shock absorbers. As the shock absorber is either compressed or extended, the hydraulic fluid is forced to flow through small orifices. This slows down the relative motion between the two ends of the shock absorber.

Damping force:    (5.3-2)

F_d = damping force
c = damping coefficient
$\dot{x}$ = time rate of change of the compression or extension of the damper

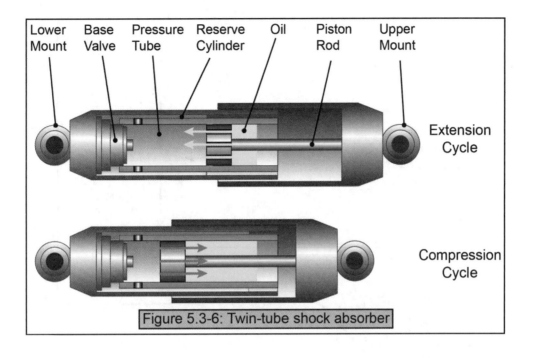

Figure 5.3-6: Twin-tube shock absorber

5.4) FRICTION

Modeling friction is very complex. **Friction** derives from the molecular interaction of two contacting surfaces. A detailed examination of friction is beyond the scope of this book. There are many theories that model the phenomenon of friction. These theories capture the dependence of the friction force on various conditions, such as, the surfaces involved, the normal force and the relative velocities between the surfaces. A few examples of different types of models are viscous friction, adhesion friction, delamination friction and Coulomb friction, though there are others. Each theory has its place and should be used only after all assumptions, loading and lubrication conditions are well understood.

A Coulomb model of friction is a very simple theory. This model states that, under static conditions, the friction force is equal and opposite to the component of the applied resultant force that is parallel to the contacting surfaces. If the applied force increases, the static friction force will also increase until the applied force reaches a level that overcomes the maximum static friction force and causes the bodies to slide relative to each other. Once motion is induced, the friction is classified as kinetic. The kinetic friction force is modeled as being proportional to the normal force and opposes the direction of the body's motion. Figure 5.4-1 illustrates the Coulomb friction force versus the pushing force for the block shown.

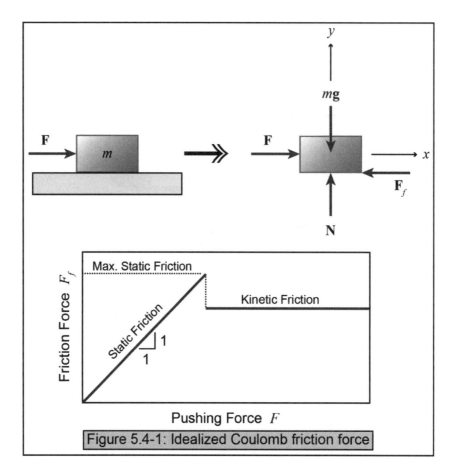

Figure 5.4-1: Idealized Coulomb friction force

A Coulomb model of friction consists of two distinct parts; the *static friction* portion and the *kinetic friction* portion. The **static friction** force occurs when there is no relative motion between the contacting surfaces. As a force is applied to one of the bodies to initiate motion, the static friction will increase until it reaches a maximum (see Figure 5.4-1). The **maximum static friction force** is the amount of friction that must be overcome in order to initiate motion.

The static friction force must be calculated by applying Newton's first law to the system (i.e. $\Sigma \mathbf{F} = 0$). The maximum static friction force may be calculated using Equation 5.4-1. The magnitude of the maximum static friction force is modeled as proportional to the normal force. The proportionality constant is called the **coefficient of static friction** (μ_s) which is dependent on the types of surfaces that are in contact.

The **kinetic friction** force occurs when the two surfaces in contact are moving relative to each other. The magnitude of the kinetic friction force is proportional to the normal force. The proportionality constant is called the **coefficient of kinetic friction** (μ_k) which, like the coefficient of static friction, is dependent on the types of surfaces that are in contact. In general, the magnitude of the kinetic friction force is less than the maximum static friction force (see Figure 5.4-1). Once the body begins to slide relative to the surface that it is touching, the magnitude of the friction is calculated using Equation 5.4-2. The kinetic friction force acts in a direction opposite to the relative velocity of the two contacting surfaces.

Maximum static friction force: $\boxed{F_{fs,max} = \mu_s N}$ (5.4-1)

Kinetic friction force: $\boxed{F_{fk} = \mu_k N}$ (5.4-2)

$F_{fs,\,max}$ = maximum static friction force μ_s = coefficient of static friction
F_{fk} = kinetic friction force μ_k = coefficient of kinetic friction
N = normal force

There are advantages and disadvantages to using a Coulomb model of friction. The most obvious advantage is the simplicity of its equations. However, the relationship between the kinetic friction force and the normal force is often not truly constant. Because of this fact, the Coulomb model of friction is often only accurate over a limited velocity range. It is also not accurate if the two contacting surfaces are extremely adhesive. One example is the friction between very soft tires and the road (e.g. drag racing tires). Coulomb friction theory assumes that the two surfaces slide over each other and don't adhere. For this type of problem, the adhesion theory of friction would work better.

The viscous model of friction (like with the dampers discussed earlier) captures the dependence of the friction force on the relative velocity of the contacting surfaces. Like the Coulomb model of friction, the viscous model has its limitation. There are many theories or models for friction but none, so far, are universal. In this book, we will assume a Coulomb friction model unless otherwise stated.

What is the difference between the *static* and *kinetic* friction forces?

What is the difference between the *static* and *maximum static* friction forces?

Example 5.4-1

Answer the following questions about friction.

   a) If **F** is not large enough to initiate motion, what type of friction occurs?

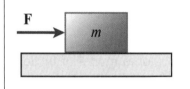

   b) Is the static friction force usually **equal**, **less than** or **greater than** $\mu_s N$?

   c) Is the kinetic friction force usually **larger** or **smaller** than the maximum static friction force?

5.4.1) FRICTION DEFINITION SUMMARY

Friction: Friction resists the relative motion of contacting surfaces. The friction force always opposes the direction of intended relative motion between the two contacting surfaces.

Coulomb Friction: Coulomb friction can be viewed as the tangential component of the contact force which is generated when two dry surfaces slide or are intending to slide relative to each other. The Coulomb friction force can be categorized as being static or kinetic.

Static Friction Force: The friction force that occurs under static conditions (i.e. no relative motion between the contacting surfaces). The static friction force is determined by applying the equilibrium equations to the system.

Maximum Static Friction Force: The largest static friction force that can occur before relative motion between the contacting surfaces begins. The maximum static friction force is determined by applying the equation $F_{fs,\max} = \mu_s N$, where N is the normal force between the contacting surfaces and μ_s is the static friction coefficient.

Kinetic Friction Force: The friction force that occurs when there is relative motion between the contacting surfaces. The kinetic friction force is determined by applying the equation $F_{fk} = \mu_k N$, where N is the normal force between the contacting surfaces and μ_k is the kinetic friction coefficient.

5.4.2) SOLVING A FRICTION PROBLEM

Once you start solving problems with friction using Newtonian mechanics, you will find that one of the biggest hurdles is determining whether or not the body can overcome the maximum static friction force and begin moving. If you are unsure how to approach the problem, try following these steps.
1. Assume that the friction force is static and that the given system is in equilibrium. Then, solve for the friction force F_{fs} from the force balance $\Sigma \mathbf{F} = 0$.
2. Check if $F_{fs} \le F_{fs,max}$, where $F_{fs,max} = \mu_s N$. If yes, then the friction force is static and you are done. Otherwise, proceed to the next step.
3. Since $F_{fs} > F_{fs,max}$, the system is not in equilibrium. It may now be assumed that the friction is kinetic and $F_{fk} = \mu_k N$. The problem must then be solved using Newton's second law $\Sigma \mathbf{F} = m\mathbf{a}$.

Example 5.4-2

Consider a block that is being pushed as shown in the figure below.

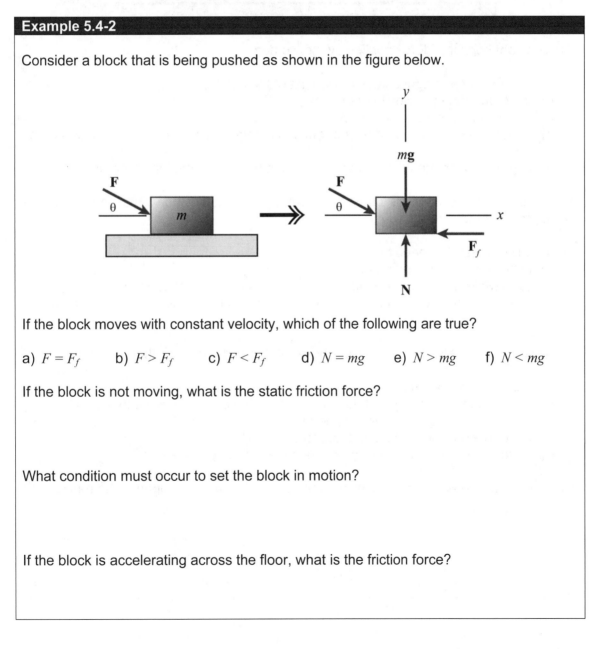

If the block moves with constant velocity, which of the following are true?

a) $F = F_f$ b) $F > F_f$ c) $F < F_f$ d) $N = mg$ e) $N > mg$ f) $N < mg$

If the block is not moving, what is the static friction force?

What condition must occur to set the block in motion?

If the block is accelerating across the floor, what is the friction force?

Conceptual Example 5.4-3

A woman exerts a constant horizontal force on a large box. As a result, the box moves across a horizontal floor at a constant speed v_o.

The constant horizontal force applied by the woman:

   a) has the same magnitude as the weight of the box.
   b) is greater than the weight of the box.
   c) is less than the weight of the box.
   d) has the same magnitude as the total friction force which resists the motion of the box.
   e) is greater than the total friction force which resists the motion of the box.

If the woman in the previous question doubles the constant horizontal force that she exerts, the box then moves:

   a) with a constant speed of $2v_o$.
   b) with a constant speed that is greater than v_o.
   c) for a while with an increasing speed, then with a constant speed thereafter.
   d) with a continuously increasing speed, assuming it is a super woman and is physically able to run with an infinite speed.

If the woman suddenly stops applying a horizontal force to the box, then the box will:

   a) immediately come to a stop.
   b) continue moving at a constant speed for a while and then slows to a stop.
   c) continue moving at a constant speed.
   d) slow continuously to an eventual stop.
   e) increase its speed for a while and then start slowing to a stop.

The above questions are theoretical and idealized, what would happen in the real world?

5.5) FREE-BODY DIAGRAMS

The solution of a kinetics problem should follow the same problem solving approach introduced in Chapter 1. This includes identifying the givens and unknowns and drawing a picture in order to help identify the approach needed for relating the givens and unknowns. The major difference in this section is the drawing. The most important step in applying the Newtonian mechanics approach of analysis is drawing the proper free-body diagram. A free-body diagram identifies all of the forces acting on a particle and their directions. Without a free-body diagram to guide you, solving a Newtonian mechanics problem is difficult and can lead to mistakes. All equations of motion come directly from the free-body diagram. Follow these steps to ensure that you will produce a proper free-body diagram (FBD).

1) Draw the body of interest separate from its supports and other bodies that it is contacting. The orientation of the drawn body should be the same as the orientation of the body in the context of the supports and other contacting bodies (see Figure 5.5-1).

2) Draw a coordinate system. If possible orient one axis in the direction of motion. The other (perpendicular) axis will then be oriented in a direction where there is no motion. This may not be possible for body-fixed coordinate systems such as a polar coordinate system (see Figure 5.5-2).

3) Draw all the forces that are acting on the body. This will include the weight, normal forces, externally applied forces, friction forces, etc. If you do not know if a force is positive or negative, you may assume a direction for the force. In the process of your analysis, if you solve for a force and it comes out to be negative, you know the force is actually in the direction opposite of what you assumed. If you, generally, don't know the angle a force makes in the plane, you may draw the force in terms of its components. Do not, however, draw redundant forces. For example, if an external force is applied on an angle, don't draw the external force and its x- and y-component forces on the same FBD. This can lead to confusion to anyone reading your FBD and to you later on. If you wish to include redundant forces, use a different color pen or pencil to differentiate the applied forces from the redundant ones (see Figure 5.5-3).

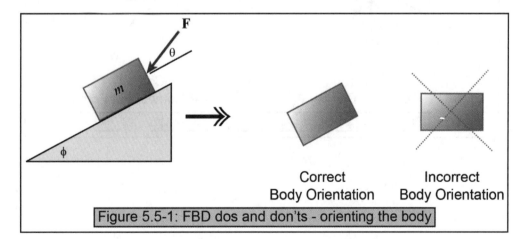

Figure 5.5-1: FBD dos and don'ts - orienting the body

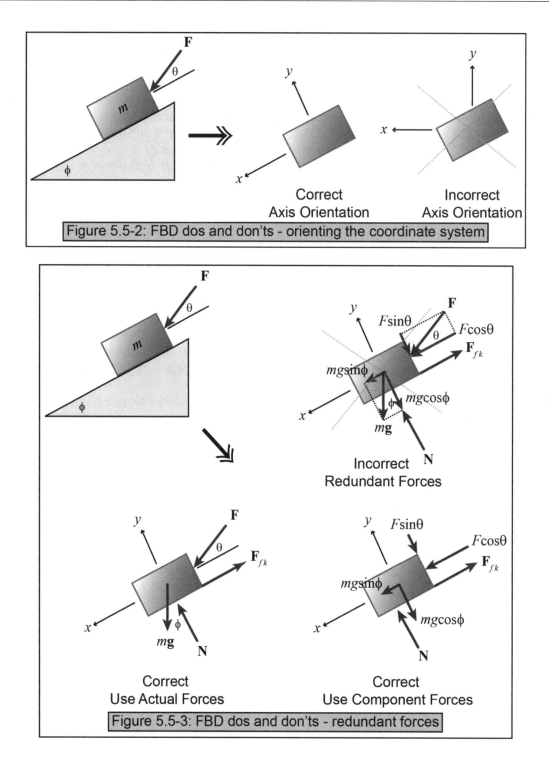

Correct
Axis Orientation

Incorrect
Axis Orientation

Figure 5.5-2: FBD dos and don'ts - orienting the coordinate system

Incorrect
Redundant Forces

Correct
Use Actual Forces

Correct
Use Component Forces

Figure 5.5-3: FBD dos and don'ts - redundant forces

Conceptual Example 5.5-1

A girl is getting ready to jump off of a diving board and into a swimming pool. Her legs are flexed and pushing on the board so that her body will accelerate up and forward.

    a) At the instant right before she loses contact with the diving board, draw a free-body diagram for both the diver's body and the board.

    b) Draw a free-body diagram of the girl once she is in the air.

Conceptual Example 5.5-2

A boy is running while flying a kite.

    a) Draw a free-body diagram of the kite.

    b) Draw a free-body diagram of the boy.

5.6) APPLYING NEWTON'S SECOND LAW

Solving Newtonian mechanics problems does not limit you to a particular coordinate system. However, the choice of coordinate frame can greatly affect the ease with which a solution is obtained. In problems where motion proceeds in a straight line (i.e. rectilinear motion) and there is only acceleration in a single direction, a rectangular coordinate frame is commonly employed. With general planar motion, the applied forces may be unbalanced in multiple directions. In this case, other coordinate frames may be preferred. Depending on the type of motion, this includes normal-tangential or polar coordinates. No matter the coordinate frame, Newton's second law can be applied in each orthogonal coordinate direction independently. This will also be the case for rectilinear motion, but in general the axes will be chosen such that there is motion in only one of the component directions.

Many problems in this chapter will require kinematic as well as kinetic analysis. Therefore, everything that we have learned in the previous chapters will still be useful here. Consider the following two examples combining kinematics and kinetics.

1. Solve a kinetics problem (e.g. $\Sigma\mathbf{F}=m\mathbf{a}$) to determine a particle's acceleration. Then, solve a kinematics problem to find some other quantity (e.g. velocity, displacement, time, etc.).
2. Solve a kinematics problem to determine a particle's acceleration. Then, solve a kinetics problem (e.g. $\Sigma\mathbf{F}=m\mathbf{a}$) to determine the required force.

The following is a summary of the different forms that Newton's second law can take, depending on the coordinate system chosen.

5.6.1) NEWTON'S SECOND LAW IN RECTANGULAR COORDINATES

Newton's second law can be applied in each orthogonal coordinate direction of the Cartesian coordinate system (i.e. the x- and y-directions) as shown in Equations 5.6-1 and 5.6-2. Recall, this type of coordinate system is especially useful when the orthogonal components of a body's acceleration are independent, as in projectile motion. Cartesian coordinates are also useful when the motion of the particle is exclusively in the x- or y-direction.

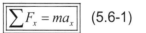 (5.6-1) $$\boxed{\sum F_x = ma_x}$$

 (5.6-2) $$\boxed{\sum F_y = ma_y}$$

F_x, F_y = forces acting on the body in the x- and y-directions respectively
m = mass of the body
a_x, a_y = acceleration of the body in the x- and y-directions respectively

Conceptual Example 5.6-1

A weightlifter squats motionless on a scale with a barbell held on her shoulders. First the weightlifter stands up and then she lifts the barbell up and over her head ultimately holding the barbell motionless. If the woman's weight is 120 lb and the barbell's total weight is 70 lb, answer the following questions.

a) What does the scale read when the weightlifter is initially squatting on the scale?

b) What does the scale read when the weightlifter, after standing, moves the barbell up with an acceleration of 9 ft/s^2?

c) What does the scale read when the weightlifter has the barbell up over her head?

Conceptual Example 5.6-2

The six figures shown depict various situations where a block is attached to a rope. List, from highest to lowest, the resulting tension in the rope. Note that for the situations where the body has velocity, this velocity is constant.

Highest: _____ Next: _____ Next: _____ Next: _____ Next: _____ Lowest: _____

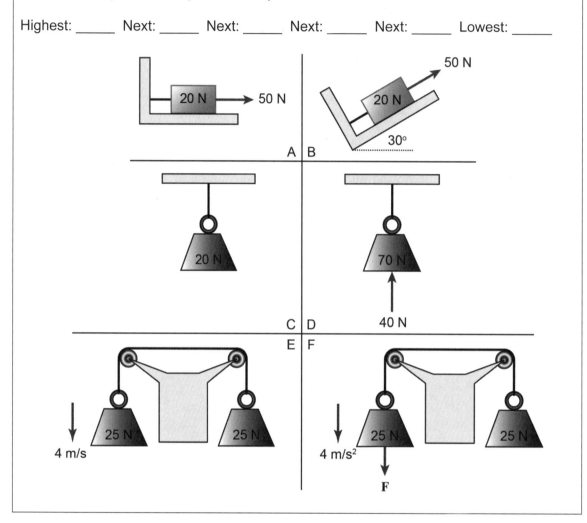

5.6.2) NEWTON'S SECOND LAW IN NORMAL-TANGENTIAL COORDINATES

Newton's second law can be applied in each orthogonal coordinate direction of the normal-tangential coordinate system (i.e. the n-, t- and b-directions) as shown in Equations 5.6-3, 5.6-4 and 5.6-5. Applying Newton's second law along the bi-normal axis, b, can lead to useful results. Therefore, this third axis is introduced here in this section. This type of coordinate system is useful if you are given or desire information about the path of the body (e.g. the radius of curvature) or the body's travel along the path (e.g. the speed).

The tangential acceleration (a_t) shown in Equation 5.6-3 is due to the body either speeding up or slowing down. Therefore, the derivative of the speed, with respect to time ($\dot{v}$), yields the tangential acceleration. This, in general, is different than the total acceleration that the body experiences. Also, recall that the velocity is always in the tangential direction.

The normal acceleration (a_n) is due to the body's direction changing. The resulting force in the normal direction (F_n) has a special name. It is called the centripetal force. For example, on a loopty loop roller coaster, this is the force that pushes you around the loop. It is your inertia that keeps you from falling out of the seat.

The normal and tangential coordinate system has a third axis called the bi-normal axis or b-axis. This axis is perpendicular to both the n- and t-axes. In the case of planar motion, the particle moves only in the t- and n-directions, therefore, a_b is equal to zero as shown in Equation 5.6-5.

$$\sum F_t = ma_t = m\dot{v} \quad \text{(5.6-3)} \qquad \sum F_n = ma_n = m\frac{v^2}{\rho} \quad \text{(5.6-4)} \qquad \sum F_b = 0 \quad \text{(5.6-5)}$$

F_t, F_n, F_b = forces acting on the body in the t-, n- and b-directions respectively

a_t, a_n = acceleration of the body in the t- and n-directions respectively

m = mass of the body
v = speed of the body
ρ = radius of curvature

5.6.2.1) Centripetal force

When dealing with general planar motion, there is sometimes the temptation to introduce a nonexistent force to explain what is being observed in a moving reference frame. For example, if you are sitting in a car rounding a sharp curve, there will appear to be some mysterious force pushing you into the side of the car. This mysterious force is often called the *centrifugal force*, but it is not real! What is really happening is that the car is rounding the curve, but from Newton's first law, you remain moving in a straight line. So really, the car door moved into you, not the other way around. Therefore, do not add this imaginary force to your free-body diagrams!

A similar force that is real is the *centripetal force*. Once you hit the car door, the reaction of the door on you that is directed toward the center of the curve is the **centripetal force**. This force pushes you around the curve. Another example of a centripetal force is the tension force in a string attached to a ball being swung around a circle. The tension force causes the ball to accelerate toward the center of the circle and to move in a circular path. If the string was to break, the ball would continue to move in a straight line with the tangential velocity it had at the instant the string broke.

5.6.3) NEWTON'S SECOND LAW IN POLAR COORDINATES

Newton's second law can also be applied in each orthogonal coordinate direction of the polar coordinate system (i.e. the r- and θ-directions) as shown in Equations 5.6-6 and 5.6-7. A polar coordinate frame is often useful when you are given or desire information about a body's motion with respect to a fixed reference frame (e.g. the radius and angle). This is especially true when the body also travels on a curved path.

$$\sum F_r = ma_r = m(\ddot{r} - r\dot{\theta}^2) \quad \text{(5.6-6)} \qquad \sum F_\theta = ma_\theta = m(r\ddot{\theta} + 2\dot{r}\dot{\theta}) \quad \text{(5.6-7)}$$

F_r, F_θ = forces acting on the body in the r- and θ-directions respectively

a_r, a_θ = acceleration of the body in the r- and θ-directions respectively

m = mass of the body
r = radial position
θ = transverse coordinate angle

Conceptual Example 5.6-3

A ball is shot at high speed into a frictionless channel that is in the shape of a circle segment with center at O. The channel is frictionless and sits on a horizontal table top. The figure shown is the top view of this situation. Which of the following force(s) are acting on the ball when it is within the frictionless channel at position B?

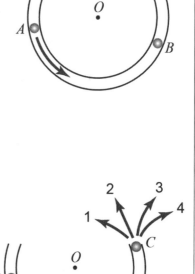

1) A downward force of gravity.
2) A force pointing from O to B.
3) A force in the direction of motion.
4) A force pointing from B to O.

Which path would the ball most closely follow after it exits the channel at C and moves across the frictionless table top?

Conceptual Example 5.6-4

A yo-yo swings from a string like a pendulum.
   a) Draw a free-body diagram of the yo-yo when it is at the bottom of its motion.
   b) Draw a free-body diagram of the yo-yo when it has swung up to an angle of 45 degrees?
   c) At an angle of 50 degrees, the string suddenly breaks. Draw a free-body diagram of the yo-yo the instant after the string breaks.

Equation Summary

Abbreviated variable definition list

a = acceleration of the body
c = damping coefficient
F = force acting on the body
F_s = spring force
F_d = damping force
$F_{fs,\,max}$ = maximum static friction force
F_{fk} = kinetic friction force
k = spring coefficient
m = mass of the body
N = normal force

r = radial position
v = speed
x = compression or extension of the spring beyond its un-stretched length
$\dot{x}$ = time rate of change of the compression or extension of the damper
μ_s = coefficient of static friction
μ_k = coefficient of kinetic friction
θ = transverse coordinate angle
ρ = radius of curvature

Spring and damping forces

$$F_s = kx \qquad\qquad F_d = c\dot{x}$$

Friction forces

$$F_{fs,\max} = \mu_s N \qquad\qquad F_{fk} = \mu_k N$$

Newton's second law

$$\sum F_x = ma_x \qquad\qquad \sum F_y = ma_y$$

$$\sum F_t = ma_t = m\dot{v} \qquad \sum F_n = ma_n = m\frac{v^2}{\rho} \qquad \sum F_b = 0$$

$$\sum F_r = ma_r = m(\ddot{r} - r\dot{\theta}^2) \qquad \sum F_\theta = ma_\theta = m(r\ddot{\theta} + 2\dot{r}\dot{\theta})$$

Example Problem 5.6-5

A particle m is suspended in a rigid container by two relatively light, inextensible strings.   A force is applied to the container so that its speed increases at a constant rate in the direction indicated in the figure.   Under this constant acceleration, $\mathbf{a}$, the tension in string CD is three times greater than the tension in string AB. Determine both the acceleration $\mathbf{a}$ and the tension in string AB.

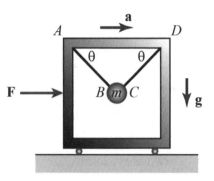

Given:

Find:

Solution:

Example Problem 5.6-6

The fighter airplane shown below has a mass of 30,000 kg and a maximum thrust-to-weight ratio of 1:1. If the airplane is cruising horizontally at 500 km/h when the engine engages its afterburners to generate its maximum possible thrust, how far does the airplane travel during the following 5 seconds? You may neglect the effect of air drag on the airplane. If you now consider air drag, how far must the plane travel to reach the same velocity it obtained when you ignored air drag? The air drag can be modeled as $D = Cv^2$, where C = 8 kg/m and v is the speed of the plane in meters per second.

Given:

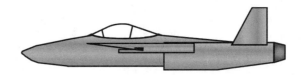

Find:

Solution:

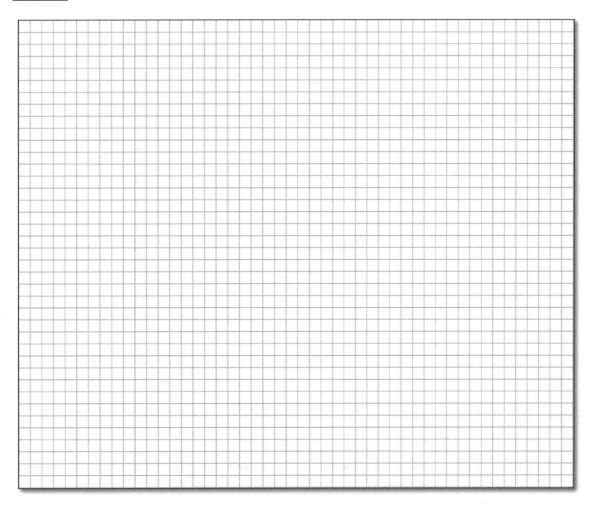

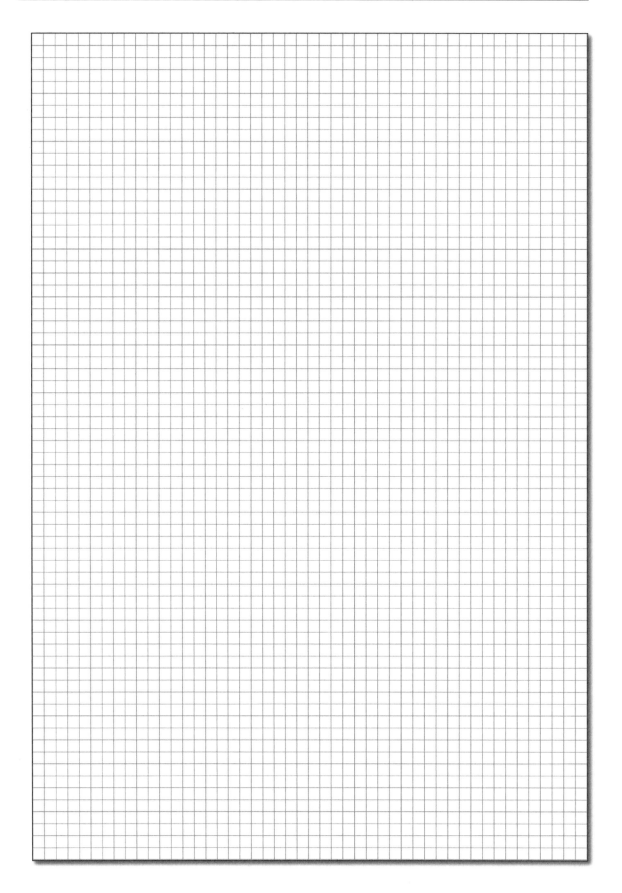

Example Problem 5.6-7

A 200-lb crate is placed on a θ = 30 degree incline and a horizontal force **F** is applied in an attempt to push the crate up the incline. Determine the resulting acceleration of the crate if a) F = 0, b) F = 150 lb and c) F = 400 lb. The crate is made from wood and the inclined ramp is made from metal. Estimate the friction characteristics as μ_s = 0.5 and μ_k = 0.3.

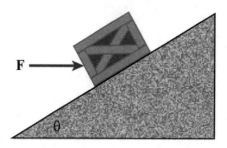

Given:

Find:

Solution:

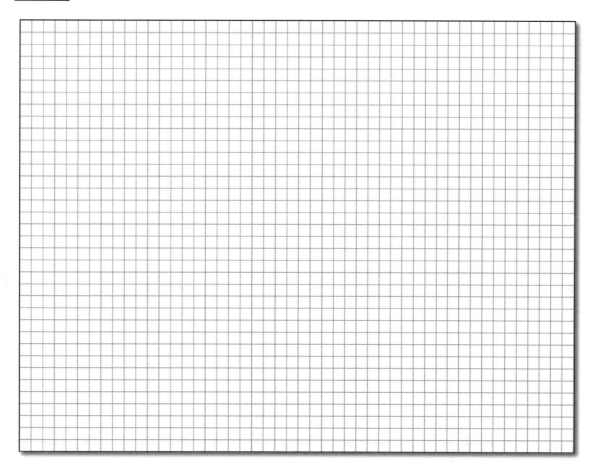

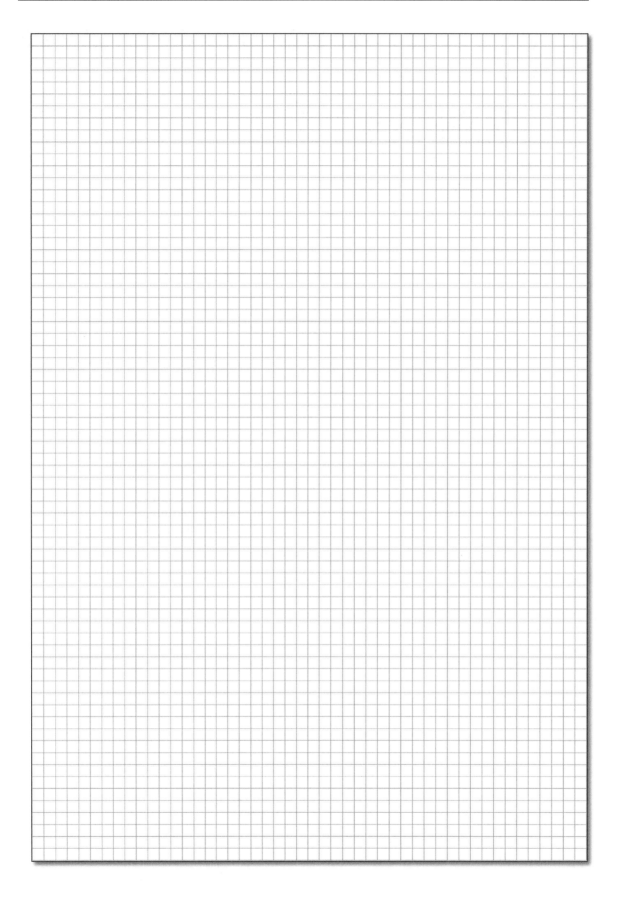

Example Problem 5.6-8

A passenger weighing 150 lb rides in a rollercoaster traveling at a constant 40 mph as it executes the loop shown. If the loop has a radius of 50 feet, determine the force exerted by the seat on the passenger at
    a) the bottom of the loop,
    b) half-way up the loop and
    c) at the top of the loop.

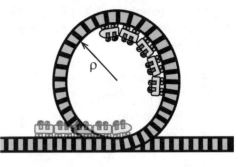

<u>Given:</u>

<u>Find:</u>

<u>Solution:</u>

Example Problem 5.6-9

A racecar weighing 3400 lb is racing in the Daytona 500. The track dimensions are as follows:

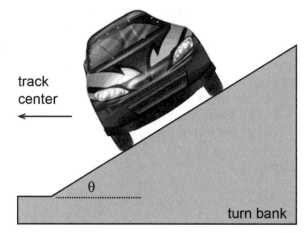

Superspeedway
- 2.5-mile tri-oval
- 40 feet wide
- 12- to 30-foot apron

Turns
- Banking: 31 degrees
- Length: 3,000 feet
- Radius: 1,000 feet

What is the range of speed that the race car can maintain through a turn without slipping on the track? Assume that the car maintains a constant speed through the turn and neglect the size of the car. Assume that the friction characteristics between the tire and track can be estimated as that between rubber and dry asphalt.

Given:

Find:

Solution:

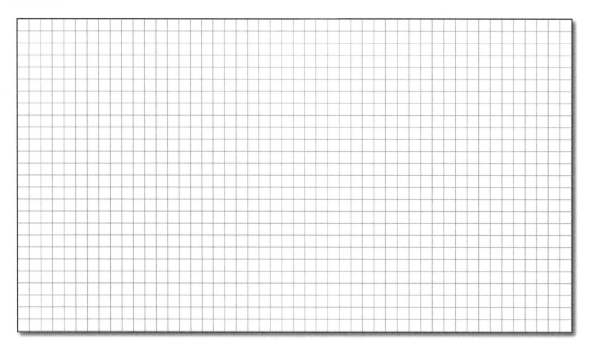

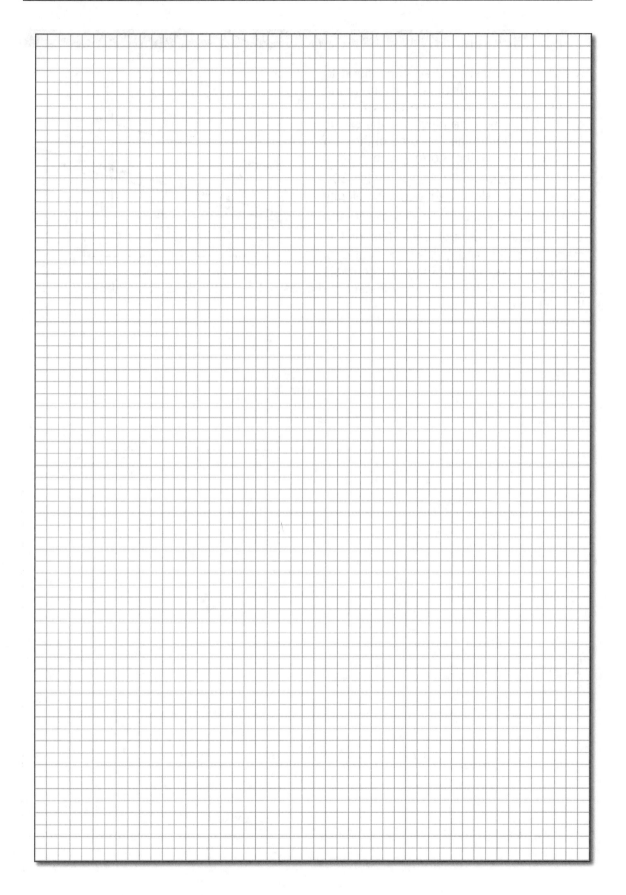

Example Problem 5.6-10)

A ball with a mass of 2 kg is confined to move along a vertical slot while being pushed by the rotating arm shown. The distance between the pivot and the slot is $L = 0.7$ m. Determine the force of the arm on the ball and the normal and friction forces of the slot on the ball when $\theta = 30°$. The rod is rotating with angular velocity and acceleration of $\dot{\theta} = 0.5$ rad/s and $\ddot{\theta} = 1$ rad/s^2, respectively. The ball is made of brass and the slot is made of mild steel. Assume that the ball contacts

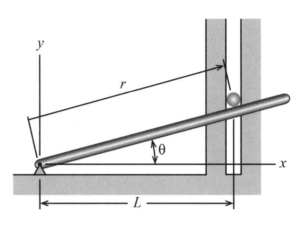

only one side of the slot at any instant and does not roll, but slides. See Appendix C for the friction characteristics.

Given:

Find:

Solution:

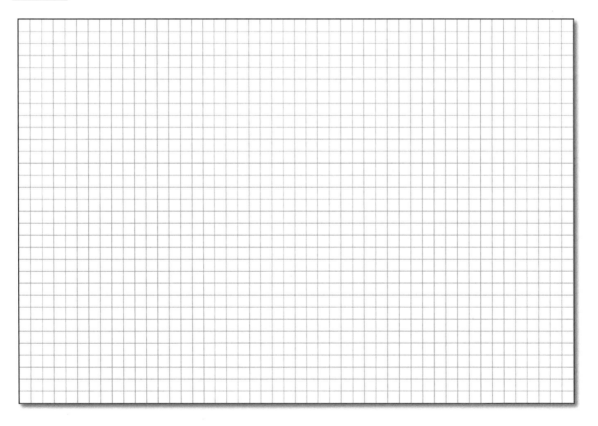

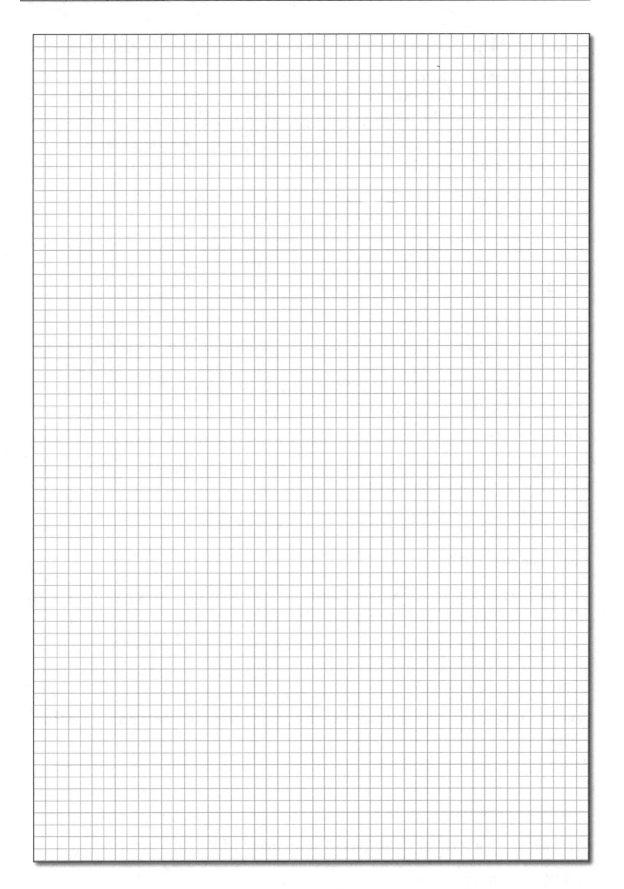

Solved Problem 5.6-11

A block, of mass m, is released from rest at $t = 0$ s from its position at the top of an incline of angle θ. If the static and kinetic friction coefficients between the two surfaces are μ_s and μ_k respectively, determine the condition for sliding. If the conditions are right for sliding, also determine the velocity of the block as a function of time.

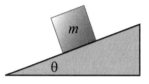

Given: mass = m $v_o = 0$ @ $t = 0$
       incline angle = θ friction characteristics = μ_s, μ_k

Find: conditions for sliding, v

Solution:

Free-body diagram

The first thing that we need to do when solving a Newtonian mechanics problem is draw a free-body diagram. Notice that we have oriented the axes in such a way that there is no motion in the y-direction.

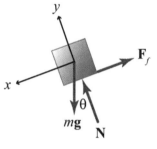

y-direction equation of motion

Let's apply Newton's second law in the y-direction. Because there is no motion in the y-direction, the acceleration is zero.

$$\sum F_y = 0 \qquad\qquad N - mg\cos\theta = 0 \qquad\qquad N = mg\cos\theta$$

Maximum static friction

Knowing the normal force, we can calculate the maximum static friction force. We will need this force to determine the conditions for sliding.

$$F_{fs,\max} = \mu_s N = \mu_s mg\cos\theta$$

x-direction equation of motion

Let's apply Newton's second law in the x-direction to determine the block's equation of motion.

$$\sum F_x = ma_x \qquad\qquad mg\sin\theta - F_f = ma_x$$

Conditions for sliding

To determine the condition for sliding, we will assume that the block is not moving but on the verge of moving. Therefore, the acceleration will be zero and the static friction force will have reached its maximum value.

$$mg \sin \theta - F_{fs,\max} \geq 0$$

The above equation states that the weight force needs to overcome the friction force in order for sliding to occur.

$$mg \sin \theta - \mu_s mg \cos \theta \geq 0 \qquad \boxed{\mu_s \leq \tan \theta}$$

Velocity

If the block slides, the friction will be kinetic. The kinetic friction force depends on the normal force and the nature of the contacting friction surfaces.

$$F_{fk} = \mu_k N = \mu_k mg \cos \theta$$

We will start with the x-direction equation of motion and determine the acceleration.

$$mg \sin \theta - F_{fk} = ma_x \qquad\qquad mg \sin \theta - \mu_k mg \cos \theta = ma_x$$

$$a_x = g(\sin \theta - \mu_k \cos \theta)$$

Notice that the acceleration is constant. Therefore, we can use the constant acceleration equations to determine the velocity.

$$v = v_o + a(t - t_o) \qquad\qquad \boxed{v = gt(\sin \theta - \mu_k \cos \theta)}$$

Example Problem 5.6-12

A rollercoaster car, starting from rest at the top of a hill, travels down the hill under the influence of gravity. The car's weight, including passengers, is **W**. The distance that the car travels along the track is measured by the variable s. The car's orientation (θ) is a function of s and equal to $\theta = bs$, where s is given in meters, b is a constant and θ is in radians. Determine the normal force between the track and car when the radius of curvature is equal to ρ. Neglect friction effects.

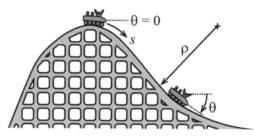

Given: $v_o = 0$ $\theta = bs$ rad

      ρ **W**

      Neglect friction

Find: N

Solution:

Free-body diagram

The given information should lead you to the understanding that the n-t coordinates would be the most practical system to use. The problem statement measures the distance that the car travels using the path coordinate s and gives a radius of curvature. We will draw the free-body diagram using the n-t coordinate system.

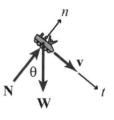

Equation of motion in the n-direction

Applying Newton's second law in the n-direction gives us an equation that contains the normal force. However, we need to determine the velocity of the car before we can determine N.

$$\sum F_n = ma_n = m\frac{v^2}{\rho} \qquad N - W\cos\theta = m\frac{v^2}{\rho} \qquad N = W\cos\theta + m\frac{v^2}{\rho}$$

Equation of motion in the t-direction

The velocity of the car is always in the tangential direction. We will apply Newton's second law in the t-direction to find the tangential acceleration and then use kinematics to find the speed.

$$\sum F_t = ma_t \qquad W\sin\theta = mg\sin\theta = ma_t \qquad a_t = g\sin\theta$$

Kinematics

Now that we have calculated the acceleration in the tangential direction, we may use kinematics to determine the speed of the car.

$$a_t \, ds = v \, dv \qquad \int_0^s g \sin\theta \, ds = \int_0^v v \, dv \qquad \int_0^s g \sin(bs) \, ds = \int_0^v v \, dv$$

$$-\frac{g}{b}\cos(bs)\Big|_0^s = \frac{v^2}{2}\Big|_0^v \qquad -\frac{g}{b}(\cos(bs)-1) = \frac{v^2}{2} \qquad v^2 = \frac{2g}{b}(1-\cos(bs))$$

Equation of motion in the n-direction

Now that we know the velocity we can calculate the normal force.

$$N = W\cos\theta + m\frac{v^2}{\rho} \qquad N = W\cos\theta + \frac{W}{g}\frac{2g(1-\cos(bs))}{b\rho}$$

$$\boxed{N = W\left(\cos(bs) + \frac{2(1-\cos(bs))}{b\rho}\right)}$$

Example Problem 5.6-13

The shape of the stationary cam shown in the given mechanism is described by the equation, $r = 1.5 + 0.5\sin(3\theta)$ inches. The deflection of the 6-lb/in spring holding the 1-lb cam follower, A, in place is $\Delta x = r - 1$ inches. If the slotted bar rotates with a constant angular velocity $\dot\theta = 5$ rad/sec, determine the normal force and the force that the arm applies to the follower as a function of θ.

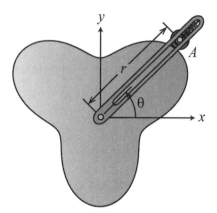

Given: $r = 1.5 + 0.5\sin(3\theta)$ inches
$\quad\quad \Delta x = r - 1$ in
$\quad\quad \dot\theta = 5$ rad/s = constant
$\quad\quad \mathbf{W} = 1$ lb
$\quad\quad k = 6$ lb/in

Find: N_{max}, F

Solution:

Free-body diagram

The problem statement indicates that the position of the follower is relative to a fixed coordinate system. Furthermore, the follower's position is given as a distance r and angle θ. This indicates that polar coordinates would be the best choice for this problem.

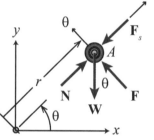

Equations of motion

To get an idea of what we need in order to calculate the unknown forces, let's write down the equations of motion in each coordinate direction.

$$\sum F_r = ma_r = m(\ddot r - r\dot\theta^2) \quad\quad \sum F_\theta = ma_\theta = m(r\ddot\theta + 2\dot r\dot\theta)$$

Time derivatives

We need the time derivatives of r and θ in order to calculate the unknown forces.

$$\dot\theta = 5\,\frac{\text{rad}}{\text{s}} \quad\quad\quad\quad \ddot\theta = \frac{d\dot\theta}{dt} = 0$$

$$r = 1.5 + 0.5\sin(3\theta)\ \text{in} \quad\quad \dot r = \frac{dr}{dt} = \frac{d(1.5 + 0.5\sin(3\theta))}{dt} = 1.5\dot\theta\cos(3\theta) = 7.5\cos(3\theta)\,\frac{\text{in}}{\text{s}}$$

$$\ddot r = \frac{d\dot r}{dt} = \frac{d(7.5\cos(3\theta))}{dt} = -22.5\dot\theta\sin(3\theta) = -112.5\sin(3\theta)\,\frac{\text{in}}{\text{s}^2}$$

Spring force

We also need to know what the spring force is at different locations. We will assume that the spring is linear and follows Hooke's law.

$$\Delta x = r - 1 = 1.5 + 0.5\sin(3\theta) - 1 = 0.5(1 + \sin(3\theta)) \text{ in}$$

$$F_s = k\Delta x = 0.5k(1 + \sin(3\theta)) = 3(1 + \sin(3\theta)) \text{ lb}$$

Equation of motion in the r-direction

$$\sum F_r = ma_r = m(\ddot{r} - r\dot{\theta}^2)$$

$$N - F_s - W\sin\theta = \frac{W}{g}(-112.5\sin(3\theta) - 25(1.5 + 0.5\sin(3\theta)))$$

$$N - 3(1 + \sin(3\theta)) - W\sin\theta = \frac{W}{g}(-112.5\sin(3\theta) - 25(1.5 + 0.5\sin(3\theta)))$$

$$\boxed{N = (\sin\theta - 0.88\sin(3\theta) + 1.84) \text{ lb}}$$

Equation of motion in the θ-direction

$$\sum F_\theta = ma_\theta = m(r\ddot{\theta} + 2\dot{r}\dot{\theta}) \qquad\qquad F - W\cos\theta = \frac{W}{g}75\cos(3\theta)$$

$$\boxed{F = (2.33\cos(3\theta) + \cos\theta) \text{ lb}}$$

5.7) NEWTON'S LAWS FOR SYSTEMS OF PARTICLES

The techniques presented in this chapter for analyzing the motion of individual particles can also be applied to a system of particles. Sometimes it is advantageous to analyze each particle of the system individually and sometimes it is better to look at the system as a whole. An advantage of analyzing the entire system rather than each particle individually is that some internal interaction forces cancel and do not need to be considered explicitly. This follows from Newton's third law as discussed earlier, "Every force of action has an equal and opposite reaction." A typical example is a system of two blocks. One block rests on top of the other as shown in Figure 5.7-1. If the bottom block is pulled with a force **F**, what will happen to the system? Will the system move as one or will the top block slide relative to the bottom block? This type of problem is usually analyzed as a system first and then each block individually. The system analysis would yield the acceleration that would occur if the top block does not slide. Then, using this acceleration, it is determined if the top block will slide by analyzing the blocks individually.

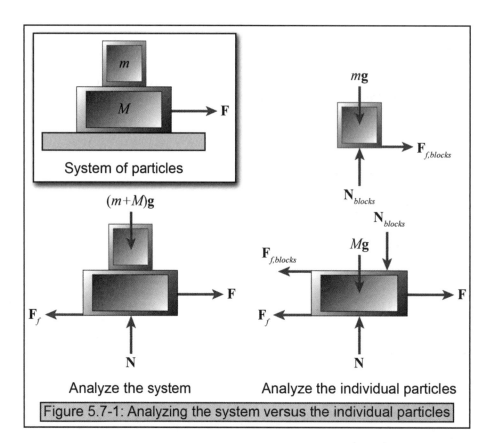

Figure 5.7-1: Analyzing the system versus the individual particles

As illustrated in Figure 5.7-1, if you analyze the system as a whole, the friction and normal forces acting at the block-block interface cancel out. When drawing free-body diagrams for each particle in a system (see Figure 5.7-2), there will be external forces $\mathbf{F}_i$ and internal forces $\mathbf{f}_i$ representing the interactions between the individual particles. If you sum the expressions for each particle in the system, the internal reaction force pairs will cancel, thereby, greatly simplifying the equations that must be solved. This relationship is especially useful if the accelerations of the individual particles are known. Equation 5.7-1 gives Newton's second law for a system of particles.

Newton's second law for systems of particles: $$\boxed{\sum \mathbf{F}_i + \sum \mathbf{f}_i = \sum \mathbf{F}_i = \sum m_j \mathbf{a}_j}$$ (5.7-1)

$\mathbf{F}_i$ = external forces acting on the body m_j = mass of the jth particle
$\mathbf{f}_i$ = internal forces acting on the body $\mathbf{a}_j$ = acceleration of the jth particle

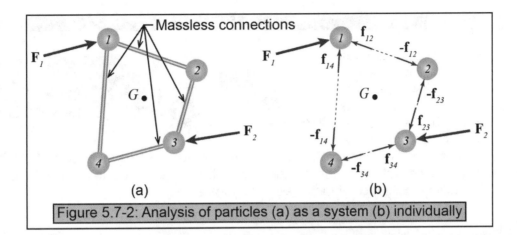

Figure 5.7-2: Analysis of particles (a) as a system (b) individually

Conceptual Example 5.7-1

A man clings to a rope that passes over a pulley. The man's weight is balanced by a weight attached to the other side. Both the man and the block are motionless. If the man starts to climb the rope, does the weight move? Why or why not?

Equation Summary

Variable definitions

F_i = external forces acting on the body
m_j = mass of the jth particle
a_j = acceleration of the jth particle

Newton's second law for a system of particles

$$\sum F_i = \sum m_j a_j$$

Example Problem 5.7-2

A 2000-kg pickup truck is pulling a series of two trailers as shown. The first trailer has a mass of 900-kg and the second trailer has a mass of 400-kg. If the traction force being generated at the road-tire interface of the front-wheel drive truck is 25000 N, determine the acceleration of the system as well as the forces in each of the two couplings. Rolling resistance at the wheels may be neglected.

Given:

Find:

Solution:

Example Problem 5.7-3

The coefficients of friction between block B and block A are $\mu_s = 0.5$ and $\mu_k = 0.3$ and the coefficients of friction between block A and the incline are $\mu_s = 0.62$ and $\mu_k = 0.4$. Block A weighs 30 lb and block B weighs 10 lb. Knowing that the system is released from rest in the position shown, determine the acceleration of B relative to A. The angle of the ramp is 40°.

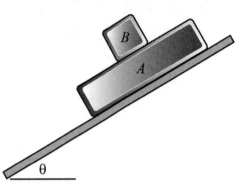

Given:

Find:

Solution:

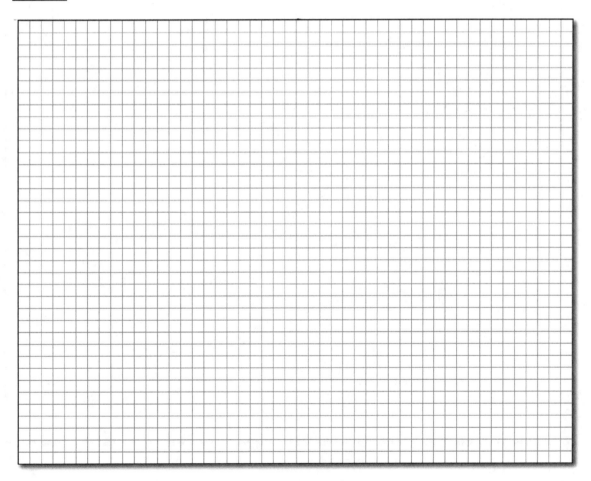

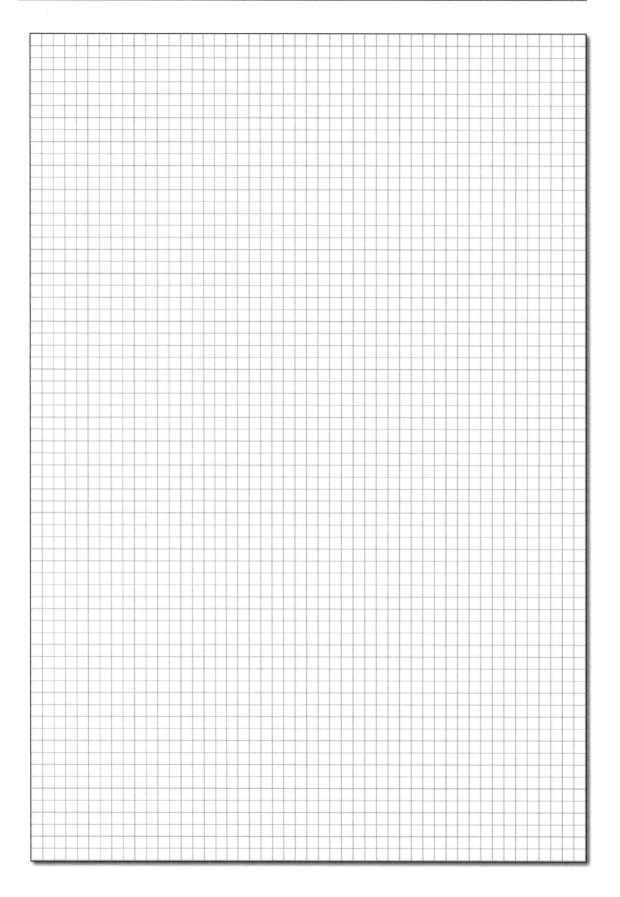

Solved Problem 5.7-4

An aluminum block A having a mass of m is released from rest when it is on top of a steel cart B and slides down its angled face. Cart B has a mass of $5m$ and a slant of θ = 35 degrees.   Determine the acceleration of the block and the acceleration of the cart. Assume that the wheels are frictionless and relatively light compared to the cart.   The friction characteristics may be obtained from Appendix C.

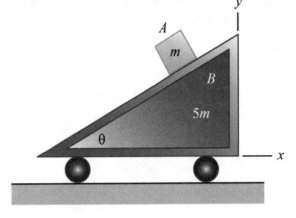

Given: $m_A = m$ $m_B = 5m$
              $\theta = 35^\circ$

Find: $\mathbf{a}_A, \mathbf{a}_B$

Solution:

Friction characteristics

The problem statement implies that the block will move. Therefore, we need to look up the kinetic friction coefficient. Appendix C indicates that the kinetic friction coefficient between steel and aluminum is 0.47.

μ_k = 0.47

Free-body diagram

The problem statement figure indicates the coordinate directions, however, these directions are chosen for a specific reason. Cart B has no motion in the y-direction. This will simplify the analysis of the problem.   Let's draw a free-body diagram for both particles. The forces that are common for both particles with be equal and opposite based on Newton's third law.

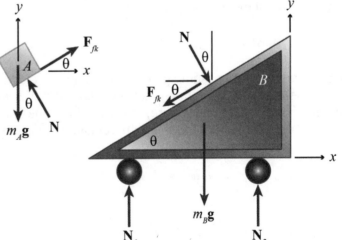

Cart B

Let's start by writing down the equation of motion for cart B. The cart will only move in the x-direction.

$$\sum F_x = m_B a_{Bx} \qquad\qquad N \sin\theta - F_{fk} \cos\theta = m_B a_{Bx}$$

Note that the kinetic friction force equals $F_{fk} = \mu_k N$, therefore, the above equation becomes

$$N(\sin\theta - \mu_k \cos\theta) = m_B a_{Bx} \qquad\qquad (1)$$

Block A

Block A moves in both the x- and y-directions. We will write equations of motion along each coordinate axis.

$$\sum F_x = m_A a_{Ax} \qquad\qquad -N \sin\theta + F_{fk} \cos\theta = m_A a_{Ax}$$

$$N(-\sin\theta + \mu_k \cos\theta) = m_A a_{Ax} \qquad\qquad (2)$$

$$\sum F_y = m_A a_{Ay} \qquad\qquad N \cos\theta + F_{fk} \sin\theta - m_A g = m_A a_{Ay}$$

$$N(\cos\theta + \mu_k \sin\theta) = m_A(a_{Ay} + g) \qquad\qquad (3)$$

Consider equations (1), (2) and (3). Note that we have three equations but four unknowns. We need to come up with another equation before we can solve this problem. This fourth equation will come from the geometry of the problem. There is a sliding constraint between the two bodies. We know that block A remains in contact with cart B and will move along its surface.

$$\mathbf{a}_{A/B} = a_{A/B}(\cos\theta \, \mathbf{i} + \sin\theta \, \mathbf{j})$$

Note that we are assuming that the relative velocity is going in the positive x- and y-directions. This is probably not the case, but it is easier than trying to figure out the actual directions. Furthermore, the true directions will become apparent when we solve the problem. Now let's break up the relative velocity into its component directions.

$$a_{A/B,y} = a_{A/B} \sin\theta = a_{Ay} - \cancel{a_{By}} = a_{Ay} \qquad\qquad a_{A/B} = \frac{a_{Ay}}{\sin\theta}$$

$$a_{A/B,x} = a_{A/B} \cos\theta = \frac{a_{Ay}}{\sin\theta}\cos\theta = \frac{a_{Ay}}{\tan\theta} = a_{Ax} - a_{Bx}$$

$$a_{Ay} = (a_{Ax} - a_{Bx})\tan\theta \qquad\qquad (4)$$

Now we have four equations and four unknowns. It is just a matter of substituting and solving for the accelerations. Dividing equation (2) with equation (3) we get

$$\frac{N(-\sin\theta+\mu_k\cos\theta)}{N(\cos\theta+\mu_k\sin\theta)}=\frac{m_Aa_{Ax}}{m_A(a_{Ay}+g)} \qquad\qquad \frac{(-\sin\theta+\mu_k\cos\theta)}{(\cos\theta+\mu_k\sin\theta)}=\frac{a_{Ax}}{(a_{Ay}+g)}$$

Substituting equation (4) into the above we get

$$\frac{(-\sin\theta+\mu_k\cos\theta)}{(\cos\theta+\mu_k\sin\theta)}=\frac{a_{Ax}}{((a_{Ax}-a_{Bx})\tan\theta+g)} \qquad\qquad (5)$$

Substituting equation (2) into (1) we get

$$a_{Bx}=\frac{m_Aa_{Ax}(\sin\theta-\mu_k\cos\theta)}{m_B(-\sin\theta+\mu_k\cos\theta)} \qquad\qquad (6)$$

Substituting equation (6) into (5) we can solve for a_{Ax}.

$$\frac{1}{(\cos\theta+\mu_k\sin\theta)}=\frac{5a_{Ax}}{((5a_{Ax}(-\sin\theta+\mu_k\cos\theta)-a_{Ax}(\sin\theta-\mu_k\cos\theta))\tan\theta+5g(-\sin\theta+\mu_k\cos\theta))}$$

$$a_{Ax}=\frac{5g(-\sin\theta+\mu_k\cos\theta)}{-5(-\sin\theta+\mu_k\cos\theta)\tan\theta+(\sin\theta-\mu_k\cos\theta)\tan\theta+5(\cos\theta+\mu_k\sin\theta)}$$

$$a_{Ax}=-1.48\,\frac{\text{m}}{\text{s}^2}$$

Plugging the result for a_{Ax} into equation (2) we can calculate the normal force as a function of the mass.

$N = 7.87m$ N

From equations (6) and (4) we can calculate the other two accelerations.

$$a_{Bx}=0.297\,\frac{\text{m}}{\text{s}^2} \qquad\qquad a_{Ay}=-1.24\,\frac{\text{m}}{\text{s}^2}$$

$$\boxed{\mathbf{a}_A=-1.48\mathbf{i}-1.24\mathbf{j}\,\frac{\text{m}}{\text{s}^2}}$$

$$\boxed{\mathbf{a}_B=0.297\mathbf{i}\,\frac{\text{m}}{\text{s}^2}}$$

RP5-1) What is equilibrium? Circle all that apply.

a) A particle is in equilibrium when it is not moving.
b) A particle is in equilibrium when it is accelerating.
c) A particle is in equilibrium when it is traveling at a constant velocity.
d) A particle is in equilibrium when it is traveling with constant speed around a curve.
e) A particle is in equilibrium when it is in free-fall.

RP5-2) What are Newton's first, second and third laws?

RP5-3) What is a body's equation of motion?

RP5-4) The velocity of a particle is given by the v-t (velocity-time) graph shown. Which of the given F-t (force-time) graphs best represent the corresponding net resultant force experienced by the particle as a function of time?

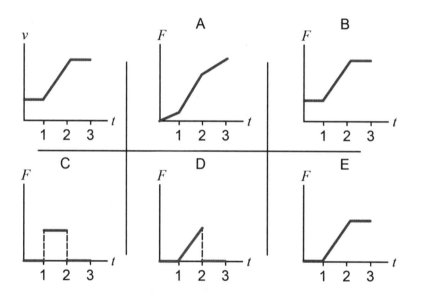

RP5-5) Is the normal force on a body always equal to its weight?

RP5-6) The normal force is always perpendicular or normal to what? Circle all that apply.

a) The contacting surface.
b) The weight force.
c) The friction force.

RP5-7) A boy swings on a rope, as shown in the figure. Which of the following forces are acting on the boy?

   a) A downward force of gravity.
   b) A force exerted by the rope pointing from A to O.
   c) A force in the direction of the boy's motion.
   d) A force exerted by the rope pointing from O to A.

RP5-8) A constant force pushes a block that slides on a frictionless surface. The block is initially at rest and the force (F) acts on the block for t_f seconds. The block reaches a final speed of v_f. To reach the same final speed with a constant force that is only half as big ($F/2$), the force would have to act on the block for _____ seconds.

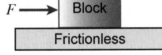

a) $4t_f$ b) $2t_f$ c) t_f d) $(1/2)t_f$ e) $(1/4)t_f$

Suppose that this time the block, instead of starting from rest, starts from an initial constant velocity of v_o before the force (F) is applied. After the force is applied for t_f seconds, the increase in the block's speed is...

a) $2v_o$ b) $4v_o^2$ c) $4v_o$ d) v_f e) Can't be determined.

Consider two blocks, initially at rest, similar to the block above. The first block is pushed by a constant force (F) and has a mass of m. It reaches a speed of v_f in t_f seconds. The second block is pushed by a constant force (F) and has a mass of $2m$. The speed of the second block is _____ if it is pushed for t_f seconds.

a) $(1/4)v_f$ b) $4v_f$ c) $(1/2)v_f$ d) $2v_f$ e) v_f

RP5-9) What is the difference between the static friction force and the limiting static friction force?

RP5-10) What is the equation that we use to calculate the kinetic friction force?

RP5-11) You are riding a bike along a flat horizontal road. Draw a free-body diagram for the following situations. Indicate the relative magnitudes of the forces by varying the length of the vectors.

   a) You are accelerating forward.
   b) You are traveling at a constant speed.
   c) You stop pedaling and apply the front brakes.

RP5-12) Imagine that you are swinging a bucket full of water in large circle within the vertical plane. What keeps the water from falling out?

RP5-13)   A car pulls a boat on a trailer. Starting from the same initial velocity, rank the situations shown from largest to smallest hitch force. The mass of the boat and the final velocity reached are given in the figure. In each situation, the car increases its speed at a constant rate and takes the same amount of time to reach its final velocity.

Largest: _____   Next: _____

Next: _____   Smallest: _____

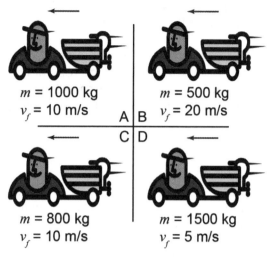

$m = 1000$ kg
$v_f = 10$ m/s

$m = 500$ kg
$v_f = 20$ m/s

A | B

C | D

$m = 800$ kg
$v_f = 10$ m/s

$m = 1500$ kg
$v_f = 5$ m/s

RP5-14) The 35-kg block shown is subjected to forces F_1 = 100 N and F_2 = 75 N. If the block is originally at rest, determine the distance it slides in 10 seconds. The coefficient of kinetic friction between the crate and the surface is μ_k = 0.4 and the forces are applied at angles of θ = 20° and ϕ = 35°.

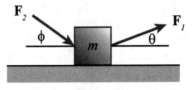

Given: m = 35 kg F_1 = 100 N
              F_2 = 75 N v_o = 0
              μ_k = 0.4 Δt = 10 s
              θ = 20° ϕ = 35°

<u>Find:</u> Δs

<u>Solution:</u>

Draw the free-body diagram.

<u>Equation of motion x–direction</u>

Determine the acceleration.

$a =$ _____

<u>Kinematics</u>

Determine velocity final.

<u>Equation of motion y–direction</u>

Determine the friction force.

$v_f =$ _____

Calculate the distance traveled.

$F_f =$ _____

$\Delta s =$ _____

RP5-15) The figure shows a conveyer belt system used to transport boxes. Determine the expression for the maximum speed a box may have in order to traverse the curve without slipping relative to the belt.   The design parameters are the radius of the curve and the static coefficient of friction between the box and the belt.

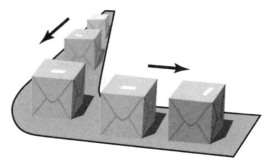

Given: ρ, μ_s

Find: v_{max}

Solution:

The solution will be simplest if the following coordinate system is used.

- Rectangular
- Normal-tangential
- Polar

Draw a free-body diagram.

Equations of motion

Determine the maximum possible velocity.

$v_{max} = $ _____

RP5-16)   Arm   *OA*   rotates counterclockwise with a constant angular velocity of 5 rad/s. As the arm passes the horizontal position, a 5-kg ball is placed at the end of the arm. As the arm moves upward, the ball begins to roll, with negligible rolling resistance, towards the pivot *O*. It is noted that at $\theta = 30°$ the ball is 0.8 meters from the pivot and moving towards *O* along the length of the arm. The ball moves with a speed of 0.5 m/s along the bar. What is the normal force that the arm applies to the ball at this instant?

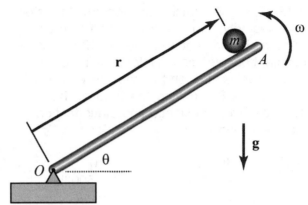

Given: $\theta = 30°$, $\omega = \dot{\theta} = 5$ rad/s, $\ddot{\theta} = 0$

        $m = 5$ kg

        $r = 0.8$ m, $\dot{r} = -0.5$ m/s

Find: *N*

Solution:

The solution will be simplest if the following coordinate system is used.

- Rectangular
- Normal-tangential
- Polar

Draw a free-body diagram of the ball.

Equations of motion

Determine the normal force.

$N = $ _____

RP5-17) A truck is forced to stop unexpectedly. If the truck is initially traveling at 40 mph and is able to stop in 180 ft with a constant deceleration. Determine if the 500-lb crate reaches the end of the flat bed if it is located d = 6 ft behind the front of the bed.

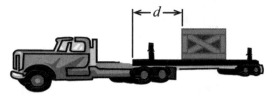

The coefficient of kinetic and static friction between the crate and the bed are 0.6 and 0.9 respectively.

Given: v_o = 40 mph $v_f = 0$
          s_{stop} = 180 ft W = 500 lb
          μ_k = 0.6 μ_s = 0.9

Find: Δs

Solution:

Draw a free-body diagram of the crate.

F_{fs} = _____

The crate will (slide, not slide).

Find the deceleration of the truck.

If the crate slides, determine the distance traveled by the crate relative to the truck bed.

a = _____

Determine if the crate will slide.

$\Delta s_{crate/truck}$ = _____

The crate (will, will not) reach the end of the flat bed.

$F_{fs,max}$ = _____

CHAPTER 5 PROBLEMS

P5.1) BASIC LEVEL X-Y COORDINATE PROBLEMS

P5.1-1)[fe] A car is climbing a hill of slope $\theta_1 = 20°$ at a constant speed v. If the slope decreases abruptly to $\theta_2 = 10°$ at point A, determine the acceleration a of the car just after passing point A. The driver does not change the gas pedal

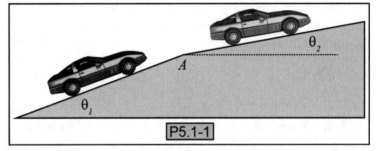

P5.1-1

position or shift into a different gear which means that the tractive force at the driving wheel remains constant. Any resistance forces (e.g. air drag and rolling resistance) may be neglected.

a) $a = 1.65$ m/s^2 b) $a = 0$ c) $a = 0.44$ m/s^2 d) $a = 5.42$ m/s^2

P5.1-2

P5.1-2) A RWD 2011 Corvette Coupe, starting from rest, accelerates from 0 to 60 mph in 4.5 seconds. The car's technical specifications are listed below. Assuming that the air drag is constant and equal to 100 lb, what is the car's traction force? Assuming that the car has enough engine power to accelerate at this same rate for 10 seconds, determine the

distance traveled by the car in 10 seconds. Assume that the front wheel rolling friction is negligible. The car's configuration is as follows: curb weight = 3175 lb, wheel base = 105.7 in, rear wheel drive and weight distribution = 51/49 f/r (%).

Ans: $F_f = 2028$ lb, $\Delta s = 978$ ft.

P5.1-3)[fe] The 250,000-kg jetliner shown is traveling in the x-direction with a speed of 200 m/s at the instant shown. If the jetliner experiences a lift force $L = 3000$ kN, a drag force $D = 100$ kN and a thrust force $T = 600$ kN in the directions shown, determine the magnitude

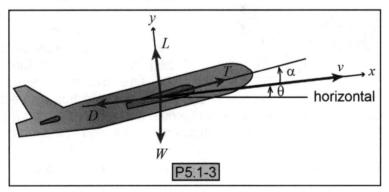

P5.1-3

of the jetliner's acceleration at this instant when the plane's attitude is $\theta = 10°$ and its angle of attack is $\alpha = 15°$.

a) $a = 1.73$ m/s^2 b) $a = 1.95$ m/s^2 c) $a = 5.64$ m/s^2 d) $a = 2.97$ m/s^2

P5.1-4) A man pushes a 100-lb wooden crate along a concrete floor. What is the acceleration of the crate if he applies a force of P = 100 lb at an angle of θ = 20°? Assume that the kinetic coefficient of friction is 80% that of the static coefficient of friction which may be found in Appendix C.

Ans: 8.8 ft/s^2

P5.1-4

P5.1-5) A particle, of mass m, is being pushed along a frictionless horizontal surface with a force $F = Ae^{-Bt} + C$. Determine the particle's position as a function of time and constants A, B and C, if the particle has the following initial conditions: at t_o = 0, s_o = 0 and v_o = 0.

Ans: $s = \dfrac{1}{m}\left[\dfrac{A}{B^2}e^{-Bt} + \dfrac{C}{2}t^2\right]$

P5.1-6)^{fe} A 55-kg block is being pulled up an inclined surface by a 320-N force through the rope and pulley system shown. The surface is inclined at a 20-degree angle and the end of the rope makes a 30-degree angle relative to the inclined surface. The coefficient of kinetic friction between the inclined surface and the block is 0.5. Determine the acceleration of the block.

a) a = 239.1 m/s^2

b) a = 109.5 m/s^2

c) a = 4.35 m/s^2

d) a = 9.1 m/s^2

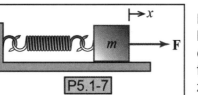

P5.1-6

P5.1-7)^{fe} Consider the block of mass m resting on a level surface and attached to a spring with spring constant k. Derive the differential equation of motion for this block in terms the block's position x, where x is zero from the point that the spring is undeformed. Consider that an external force F is being applied to the block in the direction shown and that the friction between the block and surface can be approximated by a viscous model that represents the friction force as proportional to the relative speed of the surfaces, $F_f = c\dot{x}$, and in a direction opposing the motion.

P5.1-7

a) $m\ddot{x} + k\dot{x} + cx = F$ b) $m\ddot{x} - c\dot{x} - kx = F$ c) $m\ddot{x} + c\dot{x} + kx = 0$ d) $m\ddot{x} + c\dot{x} + kx = F$

P5.2) INTERMEDIATE LEVEL X-Y COORDINATE PROBLEMS

P5.2-1) A father pulls his two kids on a sled with a force of $T = 100$ lb at an angle of θ. The combined weight of the sled and children is equal to 100 lb. What angle(s) does the father have to pull at for the sled to accelerate at 25 ft/s^2? Estimate the friction characteristics as that between waxed wood and dry snow (see Appendix C for coefficient of friction values).

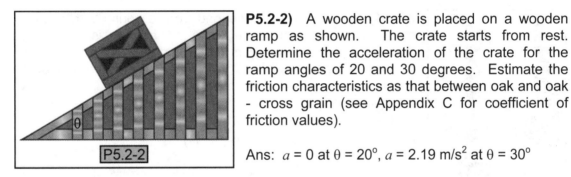

P5.2-1

Ans: $\theta = 37.6°$

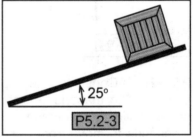

P5.2-2

P5.2-2) A wooden crate is placed on a wooden ramp as shown. The crate starts from rest. Determine the acceleration of the crate for the ramp angles of 20 and 30 degrees. Estimate the friction characteristics as that between oak and oak - cross grain (see Appendix C for coefficient of friction values).

Ans: $a = 0$ at $\theta = 20°$, $a = 2.19$ m/s^2 at $\theta = 30°$

P5.2-3) In order to estimate the static and kinetic coefficients of friction at the contact surface of a 35-kg crate and wooden board, a student performs the following experiment. The student places the crate on the board and with the crate at rest, the student slowly lifts the board until the crate begins to slide. The student then holds the board at this angle while timing the crate as it slides 4 m to the bottom of the board. The result of the experiment is that the crate begins to slide at an angle of 25° and it takes the crate 4 seconds to reach the bottom of the board. Estimate the coefficient of static friction μ_s and the coefficient of kinetic friction μ_k for these materials.

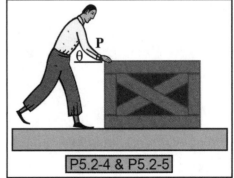

25°

P5.2-3

Ans: $\mu_s = 0.47$, $\mu_k = 0.41$

P5.2-4) A man pushes a 100-lb wood crate along a painted concrete floor. If he wishes to get the crate up to at least 3 mph starting from rest in a distance of 3 ft, what constant pushing force (P) is needed if it is applied at an angle of $\theta = 35°$? The kinetic and static coefficients of friction are 0.2 and 0.28, respectively.

Ans: $P = 42.6$ lb

P5.2-4 & P5.2-5

P5.2-5) A man pushes a 100-lb wood crate along a painted concrete floor (see figure above). If he wishes to accelerate the crate to at least 3 mph starting from rest in the span of 5 seconds, what constant pushing force (**P**) is needed if it is applied at an angle of θ = 35°? The kinetic and static coefficients of friction are 0.2 and 0.28, respectively.

Ans: P = 40.5 lb

P5.2-6) A horizontal force P = 20 N pushes a 20-kg block up a 20°-degree incline. The kinetic coefficient of friction is 0.3. If the block starts at a speed of 10 m/s, up the incline, how far does the block move?

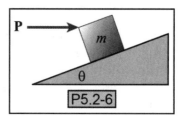

Ans: Δs = 11.05 m

P5.3) ADVANCED LEVEL X-Y COORDINATE PROBLEMS

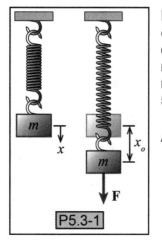

P5.3-1) A mass m is suspended from a spring with a spring constant k as shown in the figure. The mass is then pulled downward and released from rest. Find the acceleration of the mass as a function of x and then determine the velocity of the mass at x = 0 if the mass was initially pulled down 5 centimeters.

Ans: $\ddot{x} = (k/m)x$, $\dot{x}_{x=0} = \pm 0.05\sqrt{(k/m)}$

P5.3-2)   Consider the spring-mass system shown. The mass is set into motion and slides on a frictionless surface.   It takes the mass 3 seconds to move from position 1 to 2 to 3 and then back to position 1. If the spring constant is 7 N/m, determine the mass of the block.

Ans: m = 1.6 kg

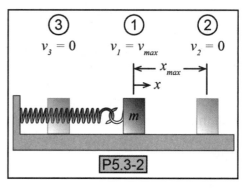

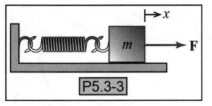

P5.3-3) A mass (m = 1 kg) is attached to a spring having a spring constant of k = 34 N/m as shown in the figure. A viscous friction model of $F_f = c\dot{x}$ is also employed, where c = 6 N-s/m and F_f opposes the direction of motion. Find the block's position x as a function of time (where x equals zero from the point the spring is undeformed) for (a) the case where the external applied force F equals zero, and (b) for the case where $F = \cos(2t)$ in Newtons. In both situations, assume that block has an initial position of x = 1 m and is initially at rest.

Ans: a) $x(t) = e^{-3t}(\cos(5t) + \dfrac{3}{5}\sin(5t))$ for $t \geq 0$,

b) $x(t) = e^{-3t}(0.971\cos(5t) + 0.577\sin(5t) + 0.0287\cos(2t) + 0.0115\sin(2t))$ for $t \geq 0$

P5.4) BASIC LEVEL N-T COORDINATE PROBLEMS

P5.4-1) A ski jump has the technical parameters listed below. The ski jump speed is rated at 94.3 km/h at takeoff. What normal force would an average 68-kg skier experience at take-off under these conditions?

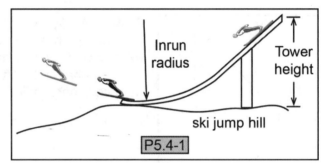

- Tower height: 60 m
- Inrun angle: 35°
- Take-off angle: 11°
- Inrun radius: 104 m

Ans: N = 1104 N

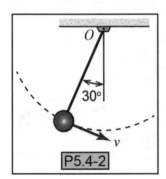

P5.4-2) The 10-lb bob of a pendulum has a velocity of 15 ft/s. If the cable supporting the bob is 5 feet long, determine the tension in the cable and the total acceleration of the bob at the instant shown.

Ans: $\mathbf{a} = 16.1\mathbf{e}_t + 45\mathbf{e}_n$ ft/s^2, T = 22.6 lb

P5.4-3)[fe] A bead slides with negligible friction down a circular hoop as shown in the figure. The bead is released from rest at the top of the hoop. The velocity of the bead varies with θ and is given by $v = \sqrt{2gr\sin\theta}$, where r is the radius of the hoop and θ is defined in the figure. Determine the normal force acting on the bead as a function of θ and m, where m is the mass of the bead.

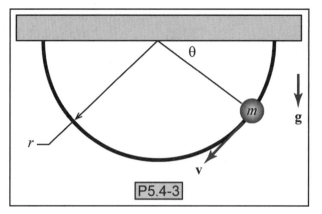

P5.4-3

a) $N = mg(\cos\theta + 2\sin\theta)$

b) $N = 3mg\cos\theta$

c) $N = 3mg\sin\theta$

d) $N = mg(\cos\theta - 2\sin\theta)$

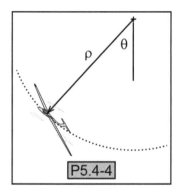

P5.4-4

P5.4-4) A 150-lb pilot, at one point, flies his 400-lb glider in a circular path having a radius of curvature of 200 ft. If the velocity of the glider is 40 mph and the glider is angled up $\theta = 35°$, what is the lift force acting on the glider? Note that lift force is perpendicular to the plane's flying direction.

Ans: $L = 16,884$ lb

P5.5) INTERMEDIATE LEVEL N-T COORDINATE PROBLEMS

P5.5-1) A 2500-lb car starts from rest at point A as shown and increases its speed at a constant rate of 2 ft/s^2. Determine the magnitude of the horizontal force the road applies to the car at
a) the inflection point B of the curve which is 1300 ft away from position A along the track and
b) position C which is an additional 900 ft along the track past point B.

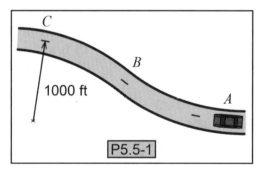

P5.5-1

Ans: F_B = 155.3 lb, F_C = 700.5 lb

P5.5-2)[fe] After being struck, a tether ball travels in a circular path around the pole. If the length of the rope is L = 6 ft, the weight of the ball is 10 oz and the angle that the rope makes with the pole is $\theta = 20°$, determine the velocity of the ball and the tension in the rope.

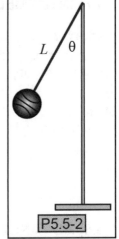

P5.5-2

a) T = 1.53 lb, v = 19.9 ft/s

b) T = 10.6 lb, v = 4.9 ft/s

c) T = 0.665 lb, v = 7.32 ft/s

d) T = 0.665 lb, v = 4.9 ft/s

P5.6) BASIC LEVEL POLAR COORDINATE PROBLEMS

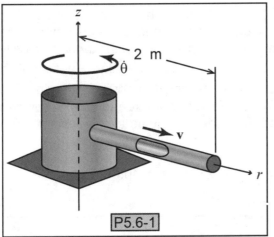

P5.6-1

P5.6-1)[fe] The gun turret shown rotates about its z-axis with a constant angular rate of $\dot\theta$ = 3 rad/s when a 0.3 kg shell is fired. If the speed of the shell relative to the barrel is 30 m/s as it exits the muzzle, determine the horizontal side-force the barrel exerts on the shell just before it emerges.

a) N = 34 N

b) N = 44 N

c) N = 54 N

d) N = 64 N

P5.6-2)[fe] A stationary cam lies flat on a horizontal table and its shape is described by the equation, $r = 0.15 + 0.05\sin(3\theta)$ in meters as shown. If the slotted bar rotates with constant angular velocity $\dot\theta$ = 3 rad/sec, determine the force exerted by the slotted bar on the 0.1-kg cam follower A when $\theta = 45°$.

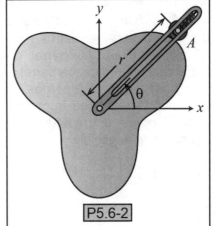

P5.6-2

a) N = 0.064 N

b) N = 0.191 N

c) N = 0.256 N

d) N = 0.549 N

P5.7) INTERMEDIATE LEVEL POLAR COORDINATE PROBLEMS

P5.7-1) A 5-kg block sits on a horizontal table a distance of 2 meters from the table's center O. The table is initially at rest and is subjected to a constant angular acceleration of $\ddot{\theta}$ = 0.5 rad/s^2 about its vertical axis. If the coefficient of static friction between the block and the table is 0.4, determine how long it takes before the block begins to slide.

Ans: t = 2.8 s

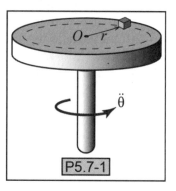

P5.8) ADVANCED LEVEL POLAR COORDINATE PROBLEMS

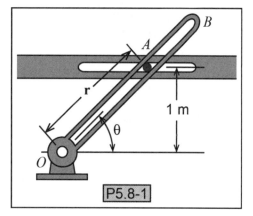

P5.8-1) Pin A has a mass of 0.5 kg and is constrained to slide along the horizontal guide shown. If arm OB rotates with angular velocity $\dot{\theta}$ = 3 rad/s and angular acceleration $\ddot{\theta}$ = -0.2 rad/s^2 when θ = 40°, determine the magnitudes of the forces exerted on pin A by arm OB and the horizontal guide at this instant. You may assume that friction is negligible.

Ans: F_{vert} = 36.23 N, F_{OB} = 40.89 N

P5.8-2) The gun turret shown rotates about its z-axis with a constant angular rate of $\dot{\theta}$ = 3 rad/s when a 0.3 kg shell is fired. If the shell is imparted with an initial speed of 30 m/s relative to the barrel, determine the horizontal side-force the barrel exerts on the shell just before it emerges. Assume that the shell is launched from an initial radial position of r_o = 0.1 m and that friction along the barrel is negligible.

Ans: F_θ = 55.1 N

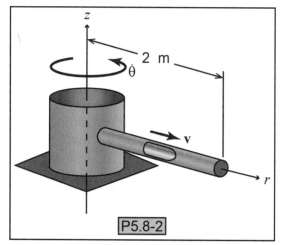

P5.9) BASIC LEVEL SYSTEM OF PARTICLES PROBLEMS

P5.9-1) A clamp holds three identical weights (W = 10 lb) with a force of N_C = 30 lb. What is the maximum upward clamp acceleration which may be attained without the blocks slipping relative to the clamp or each other? Also determine the frictional force between all surfaces at this acceleration. The coefficients of static friction between the block and between the blocks and clamp are μ_{sb} = 0.9 and μ_{sc} = 0.6, respectively.

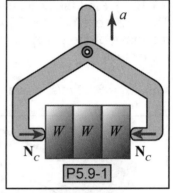

P5.9-2) The accompanying figure shows two pulley systems. Case (a) shows to blocks connected by a rope which rides along a frictionless and weightless pulley. Case (b) shows a similar system except that block A has been replaced by force **F**. In both cases, determine the acceleration of block B. The physical parameters of the system are as follows; weight of block A equals 50 lb, weight of block B equals 20 lb and force **F** is 50 lb.

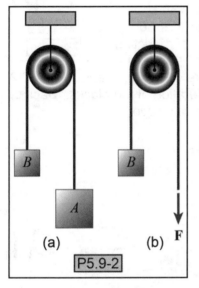

P5.9-3) Consider the accompanying figure of a truck pulling a trailer. The coupling between the truck and trailer is modeled as having stiffness k and damping c and the front-wheel drive truck is propelled forward by the force F at the road/tire interface. You may neglect rolling resistance, air drag, etc. The truck has mass m and the trailer has mass M, while their respective positions x and y are zero from the point that the spring representing the coupling is at its free length. Determine the differential equations of motion for both the truck and the trailer.

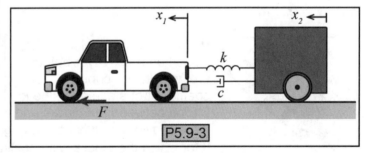

P5.10) INTERMEDIATE LEVEL SYSTEM OF PARTICLES PROBLEMS

P5.10-1) Find the acceleration of block A after the blocks are released. The mass of block A is 20 kg and the mass of block B is 10 kg. Neglect the mass of the pulleys and cables. Also, assume that the pulleys are frictionless.

Ans: a_A = 8.32 m/s^2 down

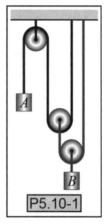

P5.10-1

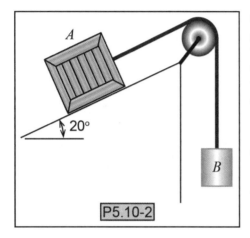

P5.10-2

P5.10-2) The crate and load shown are released from rest. The mass of crate A is 50 kg and the mass of the load B is 30 kg. The coefficients of friction between the crate and the inclined surface are μ_s = 0.3 and μ_k = 0.2. Determine the acceleration of the blocks. Back up your conclusion with calculations.

Ans: $a = 0$

P5.10-3) What is the velocity of mass B (30 kg) after 0.5 seconds if the pulley system shown is released from rest. The kinetic coefficient of friction between mass A (60 kg) and the sliding surface is 0.3.

Ans: v = 3.21 m/s

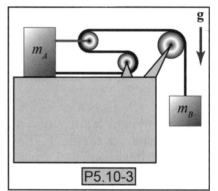

P5.10-3

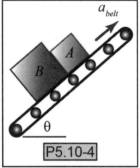

P5.10-4

P5.10-4) Two blocks are placed on a conveyer belt and are initially in contact. The coefficient of static and kinetic friction between the blocks and belt are 0.9 and 0.6 respectively. If the conveyer belt is not moving, determine the maximum angle θ for which the blocks will not slide. If the conveyer belt is set at an angle of 20°, determine the belt's maximum acceleration that can be achieved before the blocks slip relative to the belt. If the conveyer belt is accelerating at 20 ft/s^2, determine the acceleration of the blocks. If the coefficient of kinetic friction between block A and the belt is 0.7, determine the acceleration of the blocks.

P5.11) ADVANCED LEVEL SYSTEM OF PARTICLES PROBLEMS

P5.11-1) A 15-kg block rests on top of a 25-kg block which is subject to the constant 400-N force **F** as shown. If the coefficients of static and kinetic friction, between all surfaces, are 0.4 and 0.3 respectively, determine the acceleration of both blocks.

Ans: a_M = 9.53 m/s^2, a_m = 2.94 m/s^2

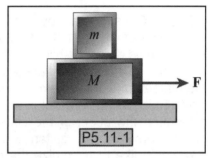

P5.11-1

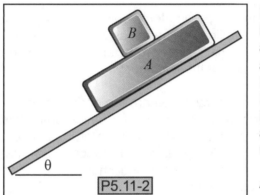

P5.11-2

P5.11-2) The coefficients of friction between block B and block A are μ_s = 0.5 and μ_k = 0.3 and the coefficients of friction between block A and the incline are μ_s = 0.62 and μ_k = 0.4. Block A weighs 30 lb and block B weighs 10 lb. Knowing that the system is released from rest in the position shown, determine the acceleration of B relative to A. The angle of the ramp is 35°.

Ans: $a_{B/A}$ = 10.55 ft/s^2

CHAPTER 5 COMPUTER PROBLEMS

C5-1) Consider a 1200-kg rear-wheel drive car subject to a drag force that is modeled as a quadratic function of the car's speed, $F_d = C_d v^2$, where the coefficient of drag C_d is 15 kg/m. Write the differential equation of motion for this car in terms of car's speed v, acceleration $\dot{v}$ and time t. You may neglect rolling resistance. The resulting differential equation is nonlinear in v due to the

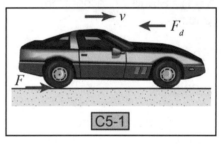

C5-1

squared term. Nonlinear differential equations are in general difficult, if not impossible, to find a closed-form solution. It is often necessary to approximate the solution to such a nonlinear differential equation numerically. Simulate the car (solve the equation of motion numerically) for 10 seconds assuming the traction force is given by the equation $F(t) = 2000t$ where F is in Newtons and t is in seconds and the car is initially at rest.

CHAPTER 5 DESIGN PROBLEMS

D5-1) Consider a toy car that is released from rest at the top of a hill. The car rolls down the hill until it reaches a plateau. The car continues to roll on the plateau until friction eventually brings it to a halt. Using Newtonian mechanics, determine the affect of the car's design parameters on the distance traveled. We would like to maximize the distance

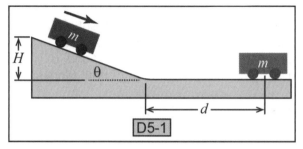

D5-1

traveled by the car (d). Consider the following design parameters: weight, aerodynamic design and rolling resistance. You may model the rolling resistance as $R_r = f_r W$, where f_r is the rolling resistance and depends on the tire material. The aerodynamic drag may be modeled as $R_a = Cv^2$, where C depends on the design of the car's body.

D5-2) Apply the results obtained in problem D5-1 to a toy car where the body is made of metal and the tires are made of rubber. Research and document information on metals and rubbers that you considered using to make the car. Also, research and document body shapes as they relate to aerodynamic drag. The goal is to produce a car that travels a maximum distance (d). Decide on the size, material and shape of the car. Give reasons for your choices and document all assumptions.

CHAPTER 5 ACTIVITIES

A5-1) Stand on a bathroom scale in an elevator and see what happens to the scale reading.

Group size: 1-2 students

Supplies

- Portable bathroom scale

Procedure

- Place the scale on the floor of a stationary elevator and note your weight.
- Ride the elevator up and note the maximum number attained on the scale.
- Ride the elevator down and note the minimum number attained on the scale.

Calculations and Discussion

- Determine the maximum acceleration and deceleration of the elevator.
- List all of the assumptions that were applied in your calculations.
- Discuss your results
- List possible sources of experimental error.

A5-2) Determine the static friction coefficient between contacting surfaces.

Group size: 2-3 students

Supplies

- Three different materials with flat surfaces.
- Protractor or ruler.

Procedure

1) Take two materials and place one on top of the other.
2) Angle the bottom material until the top material slides and note the angle that produced slip.
3) Repeat steps 1) and 2) for every combination of the materials.
4) Calculate the corresponding static friction coefficient.

Calculations and Discussion

- Calculate the static friction coefficient for each material combination.
- List all of the assumptions that were applied in your calculations.
- Discuss your results
- List possible sources of experimental error.

NOTES

PART III: KINETICS - NEWTONIAN MECHANICS

CHAPTER 6: RIGID-BODY NEWTONIAN MECHANICS

CHAPTER OUTLINE

CHAPTER SUMMARY

In the chapter on *Particle Newtonian Mechanics*, we applied Newton's second law to analyze how forces cause particles to translate. Forces applied to a rigid body may cause it to both translate and rotate. A rigid body has size as well as mass; therefore, an applied force may create a moment and cause the body to rotate.

In this chapter, using Newton's laws, we look at how forces and moments influence the motion of a rigid body. More specifically, we examine how forces and moments cause the rigid body to accelerate (both linear and angular acceleration). We will begin this chapter by looking at the simplified case of rigid bodies under pure translation and pure rotation. If we understand how a rigid body translates and rotates, we can combine that knowledge and use it to analyze the general planar motion of a rigid body.

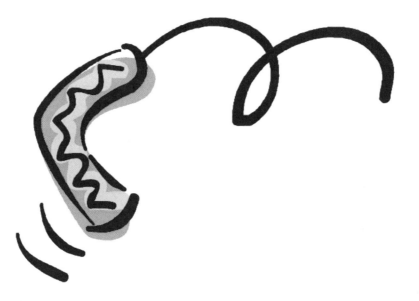

6.1) RIGID-BODY MOTION

6.1.1) RIGID-BODY MOTION

In the chapter on *Particle Newtonian Mechanics*, we looked at how forces influenced the motion of a particle. Particles, by definition, may only translate. A rigid body has both size and mass; therefore, we must be concerned with the body's rotation as well as its translation. Consider the picture of the rigid body given in Figure 6.1-1 where the force **F** is applied at a distance from the body's mass center (G). This force will cause the body to translate just as it would with a particle. Intuitively, we also know that this force will cause the body to rotate since the force is offset from the body's center of mass (G).

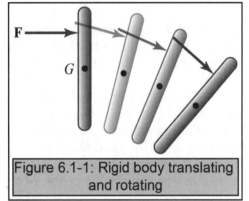

Figure 6.1-1: Rigid body translating and rotating

In this chapter, we will learn how to use Newton's laws to analyze the motion of a rigid body. In other words, how do forces and moments cause a body to accelerate? We will start simply by looking at pure translation and then move on to pure rotation. If we understand how a rigid body translates and rotates separately, we can combine that knowledge and use it to analyze the general planar motion of a rigid body.

Rigid bodies experience what two types of motion?

6.1.2) PREREQUISITE KNOWLEDGE

Before we begin analyzing how forces and moments influence the motion of a rigid body, there are a few topics that should be reviewed. As learned in the chapter on *Particle Newtonian Mechanics*, drawing free-body diagrams is an important part of applying Newton's laws. One force that is common to most free-body diagrams is a body's weight ($W = mg$). When analyzing particles, it did not matter at what point W was applied. However, the location of W is important when analyzing rigid bodies because all forces can now create moments. The weight force (W) is applied at the center of gravity/mass of the body when drawing its free-body diagram.

Since rigid bodies have size, a force applied to a body may generate a moment causing the body to rotate. Moments and center of gravity/mass are important concepts that must be understood when applying Newtonian mechanics to analyze the motion of a rigid body. We will devote some time reviewing centers of gravity/mass, mass moments of inertia, calculating moments and the rotational kinematic relationships.

6.2) CENTER OF MASS / GRAVITY

The *center of gravity* of a body in many instances coincides with its *mass center,* but they do not share the same definition. The **mass center** is the mean location of all the mass in a given body or system. The **center of gravity**, usually denoted as G, is the mean location of the gravitational force acting on the body. This is the point where you apply the $m\mathbf{g}$ force in your free-body diagram. You can also think of the center of gravity as a balancing point. If you balance an object on your finger, you are balancing it at its center of gravity.

The center of gravity and mass center are different concepts as illustrated by their definitions; however, in a uniform gravitational field they coincide. Therefore, they are often used interchangeably. Another concept that may get confused for the center of gravity is the *centroid*. The **centroid** of a body is the center of its volume. If the body has a uniform density, its centroid coincides with its center of mass. However, if the body is a composite or has varying density, its center of mass and its centroid may be in different locations.

How do the concepts of *center of mass* and *center of gravity* differ? When do they coincide?

Why do we need to know where the center of gravity of a body is located?

How do the concepts of *centroid* (center of volume) and *center of mass* differ? When do they coincide?

6.2.1) CENTER OF MASS FOR A SYSTEM OF PARTICLES

Consider a system of particles as shown in Figure 6.2-1. This system may be analyzed by considering the motion of each particle individually. However, if the system moves as one body, it may be more useful to analyze the motion of the overall center of mass for the system. The position of the center of mass for the system, $\mathbf{r}_G$, is the weighted average of each particle's position, $\mathbf{r}_i$. The weighting factors are the particle masses. This relationship is given in the following equation. Rearranging this equation allows us to solve for the overall mass center of the system as given in Equation 6.2-1.

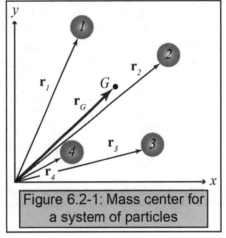

Figure 6.2-1: Mass center for a system of particles

$$m\mathbf{r}_G = \sum m_i \mathbf{r}_i$$

Mass center for a system of particles: $\boxed{\mathbf{r}_G = \dfrac{\sum m_i \mathbf{r}_i}{m}}$ (6.2-1)

$\mathbf{r}_G$ = location of the mass center for a system of particles
$\mathbf{r}_i$ = location of the ith particle

m = total mass of the system of particles
m_i = mass of the ith particle

6.2.2) CENTER OF MASS FOR A RIGID BODY

If we consider a rigid body as a collection of infinitesimally small particles as depicted in Figure 6.2-2, then Equation 6.2-1 can be expressed using integral notation as shown in Equation 6.2-2. Each incremental piece of the rigid body has mass dm and location $\mathbf{r}$.

Center of mass for a rigid body:

$$\boxed{\mathbf{r}_G = \frac{1}{m} \int \mathbf{r}\, dm}$$ (6.2-2)

$\mathbf{r}_G$ = location of the mass center for the rigid body
$\mathbf{r}$ = location of each incremental mass element dm
m = mass of the rigid body

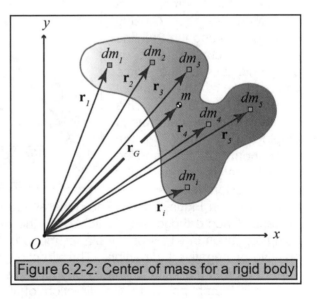

Figure 6.2-2: Center of mass for a rigid body

6.3) MOMENTS

It should be clear from the previous discussion that moments play an important role in rigid body motion. Therefore, let's take a moment (no pun intended) and review how a moment induced by a force is calculated. Moments are vectors. They have both

Units of Moments (force·length)
Metric Units:
  • Newton-meters (N-m)
US Customary Units:
  • pound-foot (lb-ft)

magnitude and direction. Unlike a force vector, the direction of a moment indicates the direction of the axis about which the moment is induced. The magnitude of the moment is equal to the perpendicular moment arm times the magnitude of the force creating the moment. We will present three methods for calculating a moment's magnitude and direction.

Does a moment have direction? If so, how do we indicate its direction?

For all of the methods described in the following sections, consider a force $\mathbf{F}$ that is applied at some position ($\mathbf{r}$) relative to a reference point O. Our job is to calculate the moment that $\mathbf{F}$ generates about point O.

6.3.1) PERPENDICULAR-MOMENT ARM METHOD

In this method, we calculate the moment that force $\mathbf{F}$ creates about O using only the component of the position vector $\mathbf{r}$ that is perpendicular to the force vector $\mathbf{F}$ (i.e. the moment arm). Alternatively, the moment may be determined employing only the component of the applied force $\mathbf{F}$ that is perpendicular to the position vector $\mathbf{r}$. This

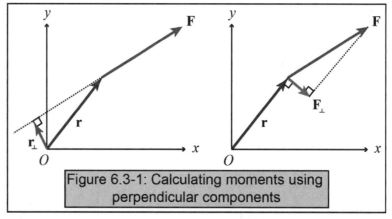

Figure 6.3-1: Calculating moments using perpendicular components

is illustrated in Figure 6.3-1. The general strategy should be to employ the method that is easiest. Once orthogonality is achieved, the magnitude of the moment is easily calculated using Equation 6.3-1. For x-y planar motion, the moment vector direction is either in the positive or negative $\mathbf{k}$-direction. The direction of the moment can be determined using the right-hand rule (i.e. roll the fingers of your right hand in the direction of the induced moment and your thumb points in the direction of the moment vector). The right-hand rule is explained further in Appendix B.

Moment about O: $\boxed{M_O = rF_\perp = r_\perp F}$ (6.3-1)

M_O = moment induced by $\mathbf{F}$ about O r = magnitude of $\mathbf{r}$

F = magnitude of $\mathbf{F}$ $r_\perp$ = component of $\mathbf{r}$ that is perpendicular to $\mathbf{F}$

$F_\perp$ = component of $\mathbf{F}$ that is perpendicular to $\mathbf{r}$

6.3.2) FORCE-COMPONENT METHOD

This method calculates the moment induced by force $\mathbf{F}$ about O using the x- and y-components of $\mathbf{F}$ as shown in Figure 6.3-2. It is usually easy to determine the x- and y-components of a force and position vector, making this method very useful. The direction of the moment is determined using the right-hand rule as in the previous method. The difference is that the right-hand rule is applied to each component separately. The magnitude of the moment may be calculated using an equation similar to Equation 6.3-2. Similar because the positive and negative signs of the terms may change depending on where the force is applied. Equation 6.3-2 specifically corresponds to Figure 6.3-2.

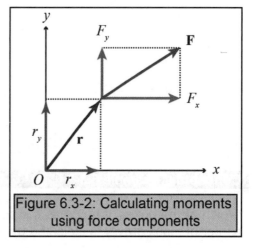

Figure 6.3-2: Calculating moments using force components

Moment about O: $\boxed{M_O = r_x F_y - r_y F_x}$ (6.3-2)

M_O = moment induced by $\mathbf{F}$ about O
F_x, F_y = x- and y-components of force $\mathbf{F}$ respectively
r_x, r_y = x- and y-components of position vector $\mathbf{r}$ respectively

6.3.3) CROSS-PRODUCT METHOD

If the perpendicular distances or x- and y-components of the force are inconvenient to determine, the moment may be calculated employing the cross product as shown in Equation 6.3-3. One big advantage of this method is that the resulting moment is given as a vector, that is, the magnitude and direction of the components are given directly. This is especially useful for analyzing cases of general three-dimensional motion as shown in Figure 6.3-3.

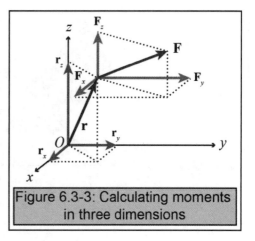

Figure 6.3-3: Calculating moments in three dimensions

$\boxed{\mathbf{M}_O = \mathbf{r} \times \mathbf{F}}$ (6.3-3)

$\mathbf{M}_O$ = moment induced by $\mathbf{F}$ about point O
$\mathbf{r}$ = position vector from point O to the point of application of force $\mathbf{F}$
$\mathbf{F}$ = the force creating the moment

Example 6.3-1

Calculate the moment induced by force $\mathbf{F} = 2\mathbf{i} - 0.5\mathbf{j}$ about the origin (O) using the cross product if the location of the force's point of application is given by $\mathbf{r} = 2\mathbf{i} + 3\mathbf{j}$.

$\mathbf{M}_O = \mathbf{r} \times \mathbf{F} =$

Conceptual Example 6.3-2

Which of the boxes, if any, has the possibility of tipping over? The location of the box's center of mass/gravity is indicated by point G.

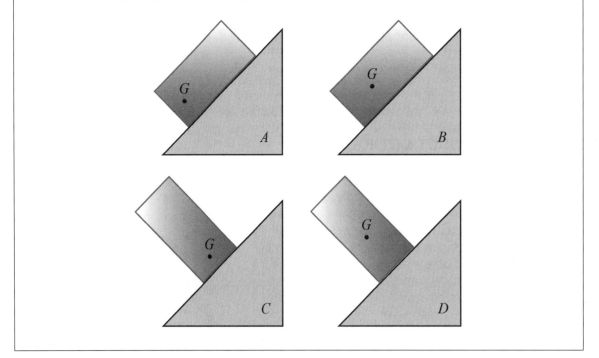

Conceptual Example 6.3-3

You need to loosen a very tight nut. Which arrangement will be the most effective? Rank them from best to worst.

Best: _____   Next: _____   Next: _____   Worst: _____

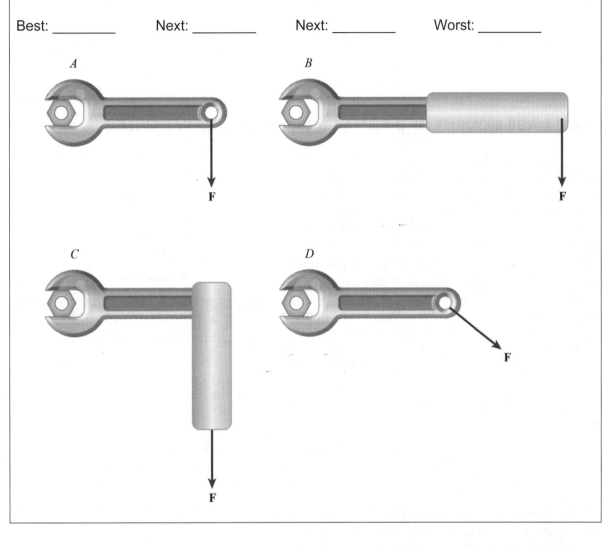

6.4) MASS MOMENTS

In order to analyze the angular motion of a rigid body, a property that accounts for the distribution of the body's mass with respect to a given axis is needed. This property is called the body's *mass moment of inertia*. The **mass moment of inertia** (I) is a measure of a body's resistance to angular acceleration about a given axis. A body's mass moment is its rotational inertia. Understanding the mass moment of inertia of a rigid body is an important part of analyzing its motion. A quantity that uses the same variable (I) as its nomenclature, is the area moment of inertia. The *area moment of inertia* is not the same as the *mass moment of inertia*. The **area moment of inertia** is a measure of a body's resistance to bending about a given axis. The area moment of inertia is used extensively in analyzing the strength of materials.

Is the *area moment of inertia* (I), used in strength of materials analysis, and the *mass moment of inertia* (I) the same? If not, what are their differences?

6.4.1) MASS MOMENT OF INERTIA FOR A SYSTEM OF PARTICLES

The mass moment of inertia for a single particle with respect to an axis perpendicular to its plane of motion is given by Equation 6.4-1. A representative particle is shown in Figure 6.4-1. Equation 6.4-1 states that the moment of inertia is proportional to the particle's mass m, and the squared distance r^2, where r is the perpendicular distance from the reference axis through O to the particle. Interpreting Equation 6.4-1 shows that the further the particle is away from O or the more massive the particle is, the larger I will become. The larger I becomes, the harder it will be to rotate the particle about the axis passing through O.

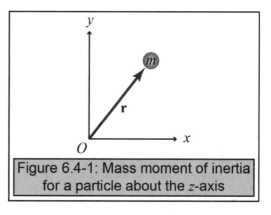

Figure 6.4-1: Mass moment of inertia for a particle about the z-axis

Mass moment of inertia for a particle: $\boxed{I_O = mr^2}$ (6.4-1)

I_O = mass moment of inertia about the axis through O
r = perpendicular distance from the axis passing through O to the particle
m = mass of the particle

For a system of particles like that shown in Figure 6.4-2, the overall mass moment of inertia with respect to an axis perpendicular to the plane of motion is simply the sum of the mass moments of the individual particles, as defined by Equation 6.4-2. Some further intuition for where this equation comes from will be provided when we introduce impulse and momentum.

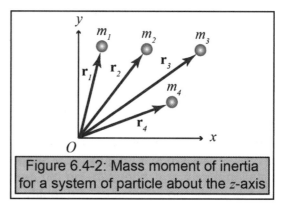

Figure 6.4-2: Mass moment of inertia for a system of particle about the z-axis

Mass moment of inertia for a system of particles: $\boxed{I_O = \sum m_i r_i^2}$ (6.4-2)

I_O = mass moment of inertia about the axis through O
r_i = perpendicular distance from the axis passing through O to the ith particle
m_i = mass of the ith particle

6.4.2) MASS MOMENT OF INERTIA FOR A RIGID BODY

As we did with the center of mass, we can consider a rigid body as a collection of infinitesimally small particles as shown in Figure 6.4-3. In this case, Equation 6.4-2 can be expressed using integral notation as shown in Equation 6.4-3. Each incremental piece of the rigid body has mass dm and perpendicular distance from the reference axis r.

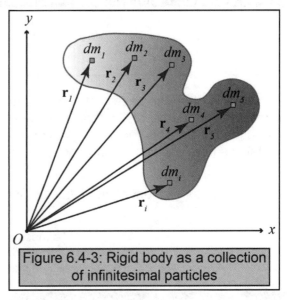

Mass moment of inertia for a rigid body:

$$I_O = \int r^2 dm \quad (6.4\text{-}3)$$

I_O = mass moment of inertia about the axis through O

r = perpendicular distance of each incremental mass element dm from an axis through O

Figure 6.4-3: Rigid body as a collection of infinitesimal particles

Mass moments of inertia for many simple rigid body shapes are given in Appendix A. Many of these inertias are given with respect to the axis that passes through the body's center of mass (G). In order to analyze rigid-body motion, we may need the mass moment of inertia of the body with respect to an arbitrary axis as shown in Figure 6.4-4.   The parallel-axis theorem (Equation 6.4-4) may be used to calculate the mass moment of inertia of the body with respect to an axis parallel to the axis passing through the body's mass center.   To illustrate how

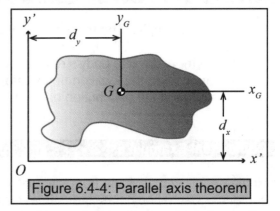

Figure 6.4-4: Parallel axis theorem

Equation 6.4-4 is applied, consider the body shown in Figure 6.4-4. If we want to know the mass moment of inertia about the x'-axis, but are only given the mass moment of inertia about the x_G axis (the axis through the body's center of mass) and the distance between the two axes, the equation would be $I_{x'} = I_{x_G} + md_x^2$.

Parallel-axis theorem: $$I' = I_G + md^2 \quad (6.4\text{-}4)$$

I' = mass moment of inertia with respect to an axis parallel to the axis passing through G

I_G = mass moment of inertia with respect to an axis passing through G

m = mass of the body

d = the perpendicular distance from the parallel axis to mass center G

Notice that the mass moment of inertia tables in Appendix A give values of the inertia with respect to a specific axis. What if the axis given is not the one we need? How do we calculate *I* with respect to a different axis?

It is often the case that a body's mass moment of inertia with respect to an axis is expressed in terms of its radius of gyration (k) with respect to that axis as shown in Equation 6.4-5. This radius of gyration can be thought of as the weighted average of the distances from each point of the body to the reference axis.

Mass moment of inertia/radius of gyration relationship: $\boxed{I = mk^2}$ (6.4-5)

I = mass moment of inertia
k = radius of gyration
m = mass of the body

Conceptual Example 6.4-1

Choose all that apply. The mass moment of inertia ....

a) is a measure of a body's resistance to angular acceleration.
b) depends on the location of the reference axis.
c) is larger if the distribution of the body's mass is moved away from the reference axis.

Conceptual Example 6.4-2

Rank the following objects from largest to smallest mass moment of inertia about their mass centers. Each body has the same mass. The circular bodies have the same outer diameters, while the rod has a length equal to the diameter of the other two bodies.

Largest: _____   Next: _____   Smallest: _____

Disk   Ring   Rod

Conceptual Example 6.4-3

If I_G is the mass moment of inertia of a rigid body with respect to an axis passing through its center of mass, then the mass moment of inertia I of the same body with respect to a parallel axis some distance away from the center of mass is ...

a) smaller than I_G.
b) larger than I_G.
c) the same as I_G.
d) Not enough information given to answer the question.

6.5) PURE TRANSLATION

One way to build intuition for the concepts and equations that we will be applying to rigid bodies is to first consider an object constructed of several discrete particles. Rigid bodies are not particles, but we may consider a rigid body as being made up of an infinite number of infinitesimally small particles.

6.5.1) TRANSLATION OF A SYSTEM OF PARTICLES

The techniques presented in the chapter on *Particle Newtonian Mechanics* for analyzing the motion of a system of particles will be summarized and expanded upon in this section. Previously, we learned that the internal interaction forces within a system of particles cancel, and hence, do not need to be considered. The is one reason why it may be advantageous to analyze a system of particles as a whole and not as individual particles. The force of gravity is not an internal interaction force and must be considered.

Newton's second law applies to both a single particle and a system of particles. For a single particle, Newton's second law states that the sum of all forces equals the mass of the particle multiplied by its acceleration.

$$\sum \mathbf{F} = m\mathbf{a}$$

The particle form of Newton's second law shown above may be slightly modified and applied to a system of particles (see Figure 6.5-1). If we sum all of the forces acting on a system of particles, then the internal interaction forces $\mathbf{f}$ cancel. This leaves us with only the externally applied forces $\mathbf{F}$. Since a system of particles is made up of several different masses, Newton's second law includes the accelerations of all the particle masses as shown in Equation 6.5-1. You may also analyze the system as a whole using Equation 6.5-2. This form of Newton's second law is completely analogous to the case of a single particle except now care is taken to recognize that the acceleration $\mathbf{a}_G$ is that of the system's mass center G.

Newton's second law for systems of particles: $\boxed{\sum \mathbf{F} + \sum \mathbf{f} = \sum \mathbf{F} = \sum m_i \mathbf{a}_i}$ (6.5-1)

Newton's second law for systems of particles: $\boxed{\sum \mathbf{F} = m\mathbf{a}_G}$ (6.5-2)

$\mathbf{F}$ = external forces acting on the system
$\mathbf{f}$ = internal interaction forces acting within the system
m_i = mass of the ith particle

m = total mass of the system of particles
$\mathbf{a}_i$ = acceleration of the ith particle
$\mathbf{a}_G$ = acceleration of the center of mass of the system of particles

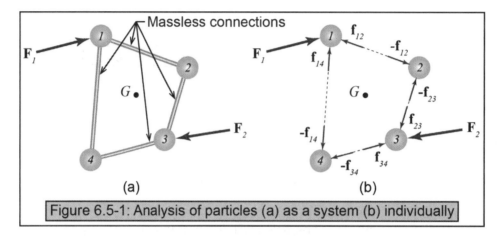

Figure 6.5-1: Analysis of particles (a) as a system (b) individually

Equation Derivation

Newton's second law as applied to a system of particles may take on two different forms as shown in the following equations. The first equation is used if you analyze the system using the motion of the individual particles. The second equation is used if you are analyzing the system as a whole. The latter case, is easier to implement if the system's mass center does not move relative to the individual particles (i.e. the particles are rigidly attached to each other). The following derivation proves that the two forms are equivalent.

$$\sum \mathbf{F} = \sum m_i \mathbf{a}_i \qquad \sum \mathbf{F} = m\mathbf{a}_G$$

We will start the derivation by substituting the second time derivative of a particle's position ($\mathbf{a} = d^2\mathbf{r}/dt^2$) for the acceleration of each individual particle into the first equation above. We then divide through by the total mass m as shown below. Note, the following equation is derived by assuming the masses of the individual particles are constant, the total mass of the system is m and the position of each particle is defined by $\mathbf{r}_i$.

$$\sum \mathbf{F} = \sum m_i \mathbf{a}_i = \sum m_i \frac{d^2 \mathbf{r}_i}{dt^2} = \frac{d^2}{dt^2} \left(\sum m_i \mathbf{r}_i \right)$$

$$\frac{\sum \mathbf{F}}{m} = \frac{d^2}{dt^2} \left(\frac{\sum m_i \mathbf{r}_i}{m} \right)$$

We can use the definition of the system's mass center $\mathbf{r}_G$, previously derived, to simplify the above equation.

$$\mathbf{r}_G = \frac{\sum m_i \mathbf{r}_i}{m} \qquad\qquad \frac{\sum \mathbf{F}}{m} = \frac{d^2}{dt^2}\left(\frac{\sum m_i \mathbf{r}_i}{m}\right) = \frac{d^2 \mathbf{r}_G}{dt^2} = \mathbf{a}_G$$

Rearranging the above equation gives us the second version of Newton's second law for a system of particles shown below.

$$\sum \mathbf{F} = m\mathbf{a}_G$$

Equation Summary

Abbreviated variable definition list

$\mathbf{F}$ = external forces acting on the system $\mathbf{a}$ = acceleration
m = total mass of the system of particles

Newton's second law

$$\sum \mathbf{F} = m\mathbf{a}_G \qquad\qquad \sum \mathbf{F} = \sum m_i \mathbf{a}_i$$

Example Problem 6.5-1

The figure shows a system that consists of three identical 5-kg spheres connected by rigid bars of negligible mass. If the system is initially at rest on a smooth horizontal surface, determine the velocity of the system's mass center after the system has been acted on by the two external constant forces (constant magnitude and direction) shown for a period of 4 seconds. Will the system of particles rotate as well as translate? Explain.

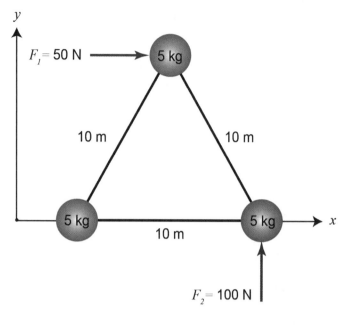

Given:

Find:

Solution:

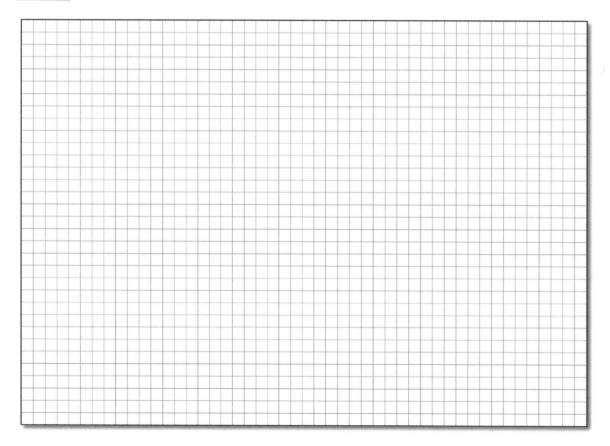

6.5.2) TRANSLATION OF A RIGID BODY

Analyzing a rigid body undergoing pure translation is a bit different than the analysis for a single particle. This is because a rigid body has size. Consider the ball and car shown in Figure 6.5-2. Both objects are accelerating to the right and are supported by the ground. The normal force that is applied to the ball by the ground acts at a single point. This is

*Note
In actuality, a car tire contacts the ground over a small area, the contact patch. The normal force is distributed throughout the contact patch. We are idealizing the normal force and modeling it as a single force acting at a single point.

because the ball is small and it can be modeled as a particle. In contrast, the car is larger and, in this case, is modeled as a rigid body supported by the ground in four places, a normal force at each tire*. Imagine that the normal forces that are applied to the front tires are taken away. What would happen to the car? The car would rotate because it is a rigid body and has size. The normal forces that support the front tires not only react to the weight force and prevent the car from translating vertically, but they also create a moment that prevents the car from rotating. What is the moral of this story? The forces that are applied to a rigid body undergoing pure translation not only make the body translate or prevent it from translating (in the case of support reactions), but also prevent the body from rotating.

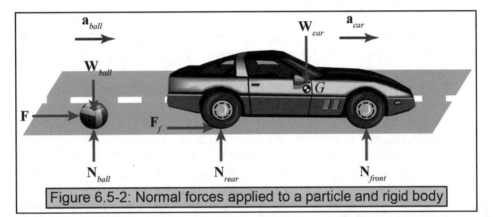

Figure 6.5-2: Normal forces applied to a particle and rigid body

Referring to Figure 6.5-2, why can the ball be modeled as a particle? Why does the car need to be modeled as a rigid body?

Referring to Figure 6.5-2, what would happen if N_{front} went away?

An imbalance of forces acting on a body causes it to accelerate. Because a rigid body has size, forces may be applied at some distance from each other. This may result in both an imbalance of forces and an imbalance of moments. These imbalances will cause the body to translate as well as rotate. Pure translation occurs when the moments balance, but the forces do not. This means that the body will translate, but not rotate. As was the case with particles, the translational motion of a rigid body can be described by Newton's second law given in Equation 6.5-3.

$$\text{Newton's second law:} \quad \boxed{\sum \mathbf{F} = m\mathbf{a}_G} \quad \text{(6.5-3)}$$

$\mathbf{F}$ = external forces acting on the rigid body
m = mass of the body
$\mathbf{a}_G$ = acceleration of the center of mass

Since the acceleration of each point on a rotating rigid body is not necessarily the same, the acceleration in Newton's second law refers to the acceleration of the center of mass (G). If, however, the rigid body is undergoing pure translation, then the acceleration of every point on the rigid body is equal.

In addition to Equation 6.5-3, an equation is needed to describe how the offset forces prevent the body from rotating. If you remember your course on statics, this would imply that the sum of the moments

> *Note
> A real car, while decelerating, pitches forward because it is not rigid. There is compliance in its suspension and tires.

equals zero, $\Sigma \mathbf{M} = 0$. In statics, however, the body under consideration is not accelerating like it may be in a dynamic situation. Do you think this fact changes this equilibrium equation? Consider the car shown in Figure 6.5-2. Imagine you are driving the car down a road with constant velocity. Under this situation, all of the moments are balanced. But what happens when you suddenly slam on the brakes? The car tips forward. Why does it tip forward if it was initially going in a straight line and the deceleration the mass center experiences is in that same line? It is because the moments are no longer in balance. As a result, the moments applied to a body that is both accelerating and undergoing pure translation will not balance ($\Sigma \mathbf{M} \neq 0$), unless the moments are taken about the body's center of mass (G).

A useful way to analyze a body that is translating and accelerating is to apply d'Alembert's principle. D'Alembert's principle states that the product of the mass m of a body with its acceleration is equal in every respect to the resultant of all forces acting on m. We have already seen, through Newton's second law, that the sum of all forces acting on a

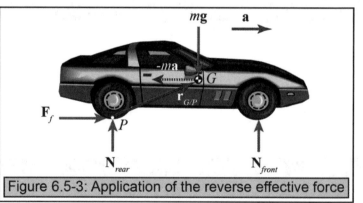

Figure 6.5-3: Application of the reverse effective force

body ($\Sigma \mathbf{F}$) is equal to its mass multiplied by its acceleration ($m\mathbf{a}$). The product of the mass of a body with its acceleration ($m\mathbf{a}$) is called the **effective force**. In other words, the *effective force* ($m\mathbf{a}$) is in equilibrium with the sum of all forces acting on the body ($\Sigma \mathbf{F}$). The word equilibrium may be a strange way to describe forces acting on an accelerating body, but the *effective force* allows us to look at a dynamic body in a different way. If we apply the *reverse effective force* ($-m\mathbf{a}$) to the body's center of mass, it transforms the system to static equilibrium as shown in the equation below. Figure 6.5-3 shows the free-body diagram of a car with the reverse effective force applied at the center of mass. You can see that Newton's second law and the equation resulting out of d'Alembert's principle, shown in the following, are equivalent. D'Alembert's principle only allows us to think of a dynamic system as static. The effective force is rarely applied when analyzing the forces, however, it is most useful when analyzing the moments.

Newton's second law: $\sum \mathbf{F} = m\mathbf{a}$

Static equilibrium applying the reverse effective force: $\sum \mathbf{F} - m\mathbf{a} = 0$

As stated previously, the effective force of a body ($m\mathbf{a}$) is equal to the resultant force acting on that body. The effective force acts like an applied force in every way. This means that it can also create a moment. The moments applied to a system in static equilibrium sum to zero or balance (i.e. $\Sigma \mathbf{M} = 0$). In general, the moments applied to a dynamic or accelerating body do not balance (i.e. $\Sigma \mathbf{M} \neq 0$). Just like the force analysis described above, the application of the reverse effective force ($-m\mathbf{a}$) at the body's center of mass reduces the system equations to the static case. Equation 6.5-4, in conjunction with Figure 6.5-3, expresses d'Alembert's principle using an arbitrary reference point P.

It is often considered strange to think of a dynamics system as static and to apply the reverse effective force instead of the actual effective force. Therefore, the second equivalent version of Equation 6.5-4, in conjunction with Figure 6.5-4, may be used if you wish to apply the effective force ($m\mathbf{a}$) instead of the reverse effective force. Remember that the effective force ($m\mathbf{a}$) is already captured in Newton's second law ($\Sigma F = m\mathbf{a}$). If a body is accelerating and

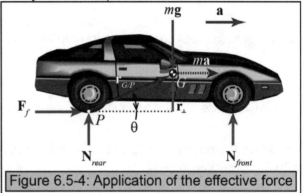

Figure 6.5-4: Application of the effective force

you choose to sum the moments about the body's center of mass (G), the effective force's moment arm is zero leading to Equation 6.5-5.

Sum of the moments about an arbitrary point P:
$$\sum \mathbf{M}_P - \mathbf{r}_{G/P} \times m\mathbf{a} = 0$$
$$\sum \mathbf{M}_P = \mathbf{r}_{G/P} \times m\mathbf{a}$$
(6.5-4)

Sum of the moments about mass center G: $\boxed{\sum \mathbf{M}_G = 0}$ (6.5-5)

$\mathbf{M}_P$, $\mathbf{M}_G$ = moments about an arbitrary point P and about mass center G respectively
$\mathbf{r}_{G/P}$ = the position of G relative to point P

m = mass of the body
a = acceleration of the body

The magnitude of the cross product in Equation 6.5-4 can be written in terms of the sine of the angle between the vectors (i.e $|\mathbf{r}_{G/P} \times m\mathbf{a}| = r_{G/P}\, ma\, \sin\theta = r_\perp ma$). It can also be written in terms of the perpendicular distance between point P and the acceleration vector drawn through the body's center of mass. The sign of the cross product can be determined using the right-hand rule.

Example 6.5-2

Consider a front wheel drive car that is accelerating down the road. What are the forces acting on the car?

Given the acceleration and mass of the car, attempt to determine all forces acting on the car using Newton's second law ($\sum \mathbf{F} = m\mathbf{a}_G$).

What other concept can we use to help us determine the forces?

Is $\sum \mathbf{M} = 0$ true like it was in statics? Let's see.

Calculate the moments created by all the forces about the contact point of the front tire with the ground (point A).

If $\sum \mathbf{M}_A = 0$, what is the relationship between N_{back} and N_{front}?

Does this relationship seem reasonable given that the car is accelerating?

Example 6.5-2 Continued

Calculate the moments created by all the forces about the center of mass G.

If $\sum \mathbf{M}_G = 0$, what is the relationship between N_{back} and N_{front}?

When the car is accelerating, which friction force is larger $F_{f,front}$ or $F_{f,back}$?

One of the above moment equations is correct ($\sum \mathbf{M}_A = 0$ or $\sum \mathbf{M}_G = 0$). Which one?

Why do we get a contradiction? What did we miss?

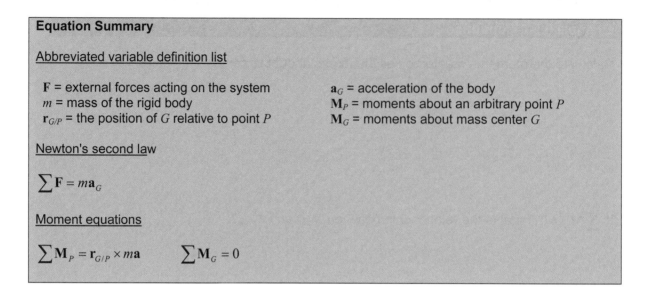

Equation Summary

<u>Abbreviated variable definition list</u>

$\mathbf{F}$ = external forces acting on the system
m = mass of the rigid body
$\mathbf{r}_{G/P}$ = the position of G relative to point P

$\mathbf{a}_G$ = acceleration of the body
$\mathbf{M}_P$ = moments about an arbitrary point P
$\mathbf{M}_G$ = moments about mass center G

<u>Newton's second law</u>

$$\sum \mathbf{F} = m\mathbf{a}_G$$

<u>Moment equations</u>

$$\sum \mathbf{M}_P = \mathbf{r}_{G/P} \times m\mathbf{a} \qquad \sum \mathbf{M}_G = 0$$

Example Problem 6.5-3

The 30,000-kg fighter aircraft shown has a primary engine that generates 200 kN of thrust during takeoff. Neglecting the effect of air drag, determine the acceleration of the aircraft and the reaction forces experienced by the front landing gear wheel and each of the two rear landing gear wheels while the aircraft is still traveling horizontally on the ground. Neglect the rolling friction of the wheels.

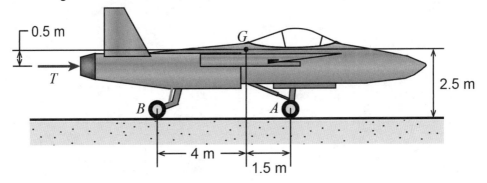

Given:

Find:

Solution:

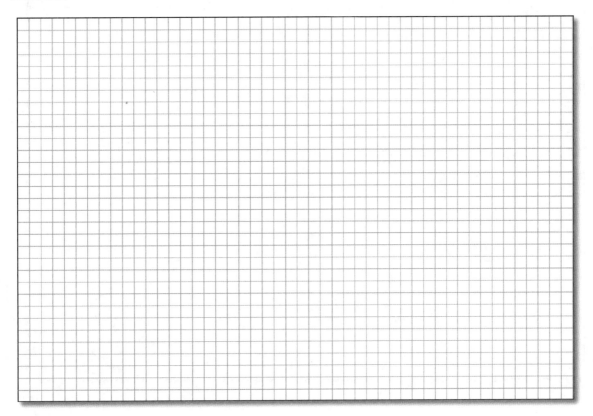

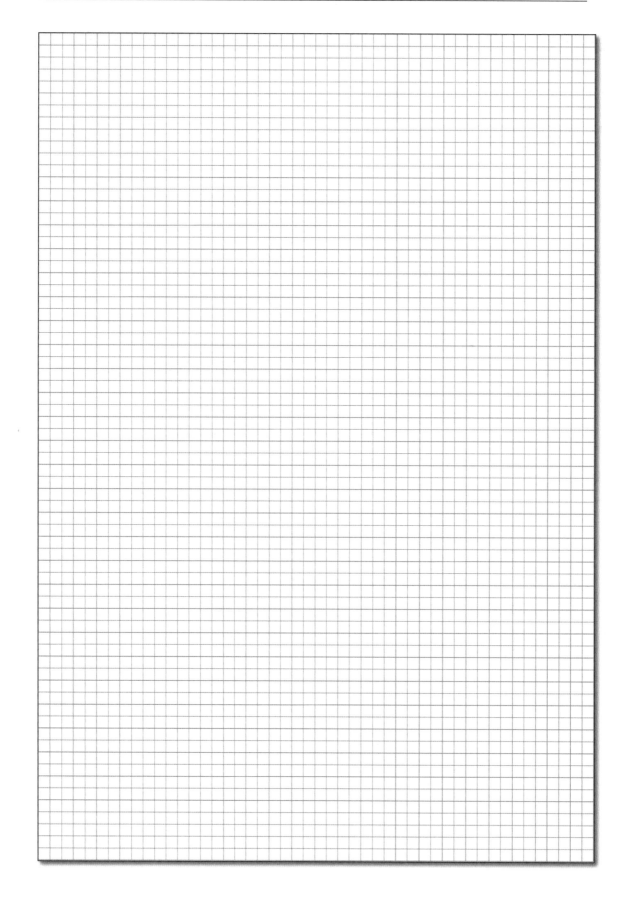

Example Problem 6.5-4

What is the horizontal acceleration a of the cart that will cause a steady-state deflection angle θ of the attached pendulum from the vertical? The slender rod has a length of l and a mass of m. The ball at the end of the pendulum has a mass of $2m$ and a radius of r. The motion of the pendulum is restricted by a torsional spring installed at the pivot O. The moment induced by the torsional spring has a spring constant k and is proportional to the pendulum's displaced angle θ from the vertical in radians.

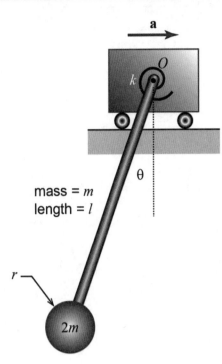

mass = m
length = l

$2m$

Given:

Find:

Solution:

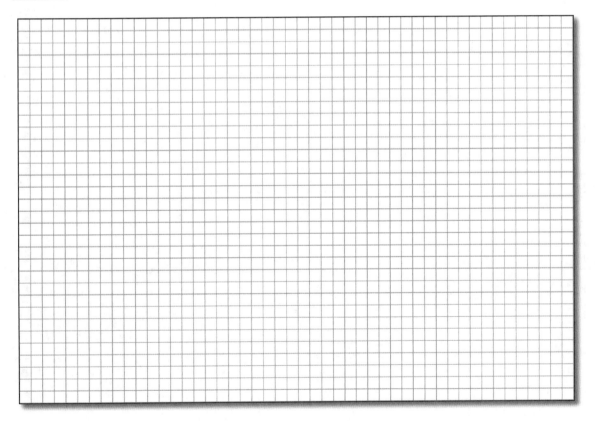

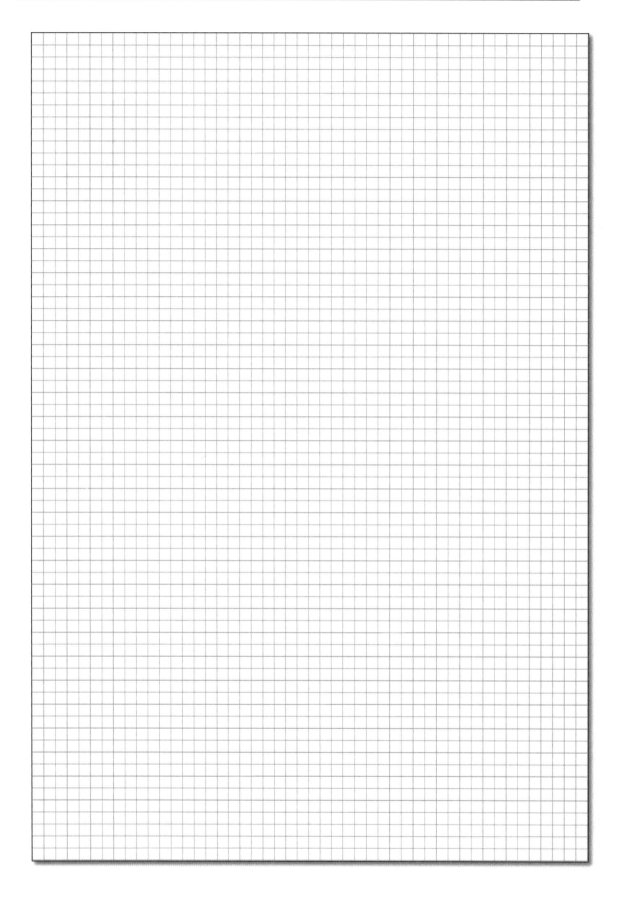

A Front wheel drive (FWD) pickup truck's technical specifications are listed below. What is the truck's maximum acceleration on level wet concrete when the driving wheels slip? The total rolling resistance is equal to $R = f_r W$, where the rolling resistance coefficient is $f_r = 0.02$. Neglect aerodynamic effects.

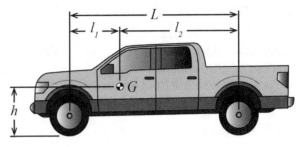

- Curb weight: $W = 4685$ lb
- Wheel base: $L = 133.3$ in (Distance between the axles.)
- CG height: $h = 39$ in
- Front wheel drive
- Weight distribution: 55/45 f/r (%)

Given: $R = f_r W$ $f_r = 0.02$ $W = 4685$ lb $L = 133.3$ in
        $h = 39$ in   55/45 f/r (%)   FWD

Find: a_{max}

Solution:

Free-body diagram

The first step in any Newtonian mechanics problem it to draw a free-body diagram. Notice the following things.

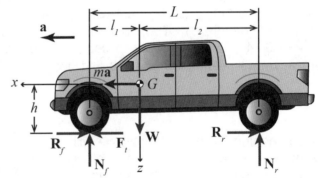

- All of the forces acting on the truck, with the exception of the drag force, are included.
- The effective force is included, however, it is drawn in a different color so that it is not confused with the applied forces.
- The coordinate axes may appear strange. The configuration shown with the z-axis directed downward is standard when analyzing vehicles.

Fore and aft position of the center of mass

The fore and aft position of the center of mass may be calculated using the weight distribution information (55/45 f/r (%)). The weight distribution data states that when the vehicle is still and on level ground, the front wheel bares

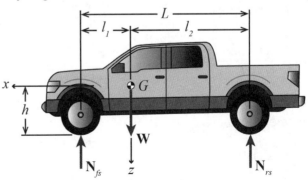

55% of the vehicle's weight. Therefore, the static normal forces on the truck's tires are.

$$N_{fs} = 55\%W = 2576.75 \text{ lb} \qquad\qquad N_{rs} = 45\%W = 2108.25 \text{ lb}$$

Using this information and the static FBD of the truck, we can calculate the fore and aft position of the center of mass (G). We will do this by summing the moments about the contact between the front tire and the ground.

$$\sum M_f = 0 \qquad N_{rs}L = Wl_1 \qquad l_1 = \frac{N_{rs}L}{W} = 59.985 \text{ in} \qquad l_2 = L - l_1 = 73.315 \text{ in}$$

Equations of motion

We need to find the acceleration of the truck (**a**). We will start by applying Newton's second law in the x-direction. The forces acting in the x-direction are the tractive force (F_t) and the rolling resistance ($R = R_f + R_r$).

$$\sum F_x = ma_x \qquad\qquad F_t - R = ma_x$$

We can calculate the rolling resistance using the equation given in the problem statement.

$$R = f_r W = 0.02 * 4685 = 93.7 \text{ lb}$$

We are told that the front wheel slips, therefore, the tractive force may be calculated using the following equation. In order to calculate the tractive force, we need to calculate the normal force acting on the front wheel. We also need to look up the coefficient of kinetic friction (Appendix C). The kinetic friction coefficient between rubber and wet concrete is 0.45 - 0.74. Since we want to conservatively estimate the acceleration capability of the truck, we will use the low end of the friction coefficient range and go with $\mu_k = 0.45$.

$$F_t = \mu_k N_f$$

To determine the normal force, we will sum the moments about the contact between the rear tire and the ground.

$$\sum \mathbf{M}_r = \mathbf{r}_{G/r} \times m\mathbf{a}$$

$$N_f L - W l_2 = -ma_x h$$

$$N_f = \frac{Wl_2 - ma_x h}{L}$$

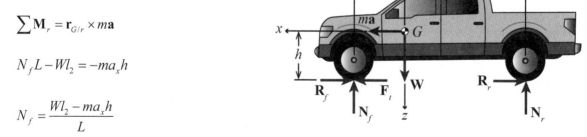

Now we can plug in the equation for N_f into Newton's second law and solve for the acceleration.

$$\sum F_x = ma_x \qquad\qquad F_t - R = ma_x \qquad\qquad \mu_k N_f - R = ma_x$$

$$\mu_k \frac{Wl_2 - ma_x h}{L} - R = ma_x \qquad a_x = \frac{\mu_k Wl_2 - RL}{m(L + \mu_k h)} \qquad \boxed{a_x = 6.47 \ \frac{ft}{s^2}}$$

6.6) ROTATION ABOUT A FIXED AXIS

Pure rotation of a rigid body (i.e. rotation about a fixed axis), as shown in Figure 6.6-1, results in all points on the body moving in circular paths around the central fixed axis. All points on the body, except for points along the fixed axis, may experience linear and angular acceleration. Recall from our previous kinematics discussion that the velocity and acceleration of any point on the rigid body can be determined based on the planar angular motion equations given below. In this chapter, we will assume planar motion unless otherwise stated.

$$\mathbf{v}_A = r_{A/O}\omega \mathbf{e}_t \qquad \text{or} \qquad \mathbf{v}_A = \boldsymbol{\omega} \times \mathbf{r}_{A/O}$$

$$\mathbf{a}_A = \dot{v}\mathbf{e}_t + \frac{v^2}{r_{A/O}}\mathbf{e}_n \qquad \text{or} \qquad \mathbf{a}_A = \boldsymbol{\alpha} \times \mathbf{r}_{A/O} - \omega^2 \mathbf{r}_{A/O}$$

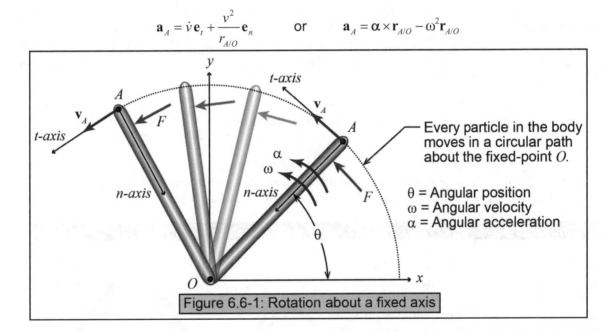

Every particle in the body moves in a circular path about the fixed-point O.

θ = Angular position
ω = Angular velocity
α = Angular acceleration

Figure 6.6-1: Rotation about a fixed axis

Previously, we used Newton's second law to analyze how forces affect the motion of a particle. This law holds true for both particles and rigid bodies. The difference arises when you consider the acceleration. Newton's second law states that the sum of all the forces acting on the body equals its mass times its acceleration. A particle, by the nature of its definition, has a single acceleration. For a rotating rigid body, every point on the body may have a different acceleration. Knowing that Newton's second law ($\Sigma \mathbf{F} = m\mathbf{a}$) contains an acceleration term, the question becomes, "Which acceleration do we use?" The answer is, "The acceleration of the *center of mass*." This answer was verified in the section on *Translation of a System of Particles*. The key to applying Newton's second law to a rigid body is to remember that the acceleration in Equation 6.6-1 is of the center of mass, $\mathbf{a}_G$. Other than this difference, Equation 6.6-1 is used in the same manner as when it is applied to a particle. For a rigid body in pure rotation, Equation 6.6-2 may be used to determine the acceleration of the body's center of mass.

Newton's second law: $\boxed{\sum \mathbf{F} = m\mathbf{a}_G}$ (6.6-1)

Acceleration of the center of mass: $\boxed{\mathbf{a}_G = \alpha \times \mathbf{r}_{G/O} - \omega^2 \mathbf{r}_{G/O}}$ (6.6-2)

$\mathbf{F}$ = external forces acting on the body
m = mass of the body
$\mathbf{a}_G$ = acceleration of the center of mass

$\mathbf{r}_{G/O}$ = position vector of the center of mass G relative to the fixed axis through O
ω = angular velocity of the body (rad/s)
α = angular acceleration of the body (rad/s^2)

A force applied to a body at a distance from the fixed axis will cause the rigid body to rotate. The moment induced by this applied force will accelerate the rate at which the body rotates in proportion to the body's rotational inertia. The relationship between the applied moment and the resulting angular acceleration is analogous to Newton's second law. The rotational equivalents of the variables found in Newton's second law (i.e. force ($\mathbf{F}$), acceleration ($\mathbf{a}$) and mass (m)) are the moment ($\mathbf{M}$), the angular acceleration (α) and the mass moment of inertia (I), respectively. The rotational analog of Newton's second law is a form of Euler's second law and is given in Equation 6.6-3. In this equation, the moments and the mass moment of inertia are calculated with respect to the fixed axis of rotation through O. Note that Equation 6.6-3 assumes planar motion, therefore, $\mathbf{M}_O$ and α are either in the clockwise or counterclockwise direction.

Sum of the moments about the fixed axis through O: $\boxed{\sum \mathbf{M}_O = I_O \boldsymbol{\alpha}}$ (6.6-3)

$\mathbf{M}_O$ = moments about an axis through O
I_O = mass moment of inertia about the axis through O
α = angular acceleration of the body (rad/s^2)

Example 6.6-1

Consider two bicycle wheels each having a mass of 1 kg. If both wheels start from rest and a force is applied as shown, how large must $\mathbf{F}_2$ be if both wheels are to have the same angular acceleration? Assume that both wheels rotate about a fixed axis passing through the hub and that most of the mass is concentrated on the outer rim of the wheel. ($I_{G,wheel} = mr^2$)

$\mathbf{F}_1 = 1$ N

$\mathbf{F}_2$

$m_1 = 1$ kg
$r_1 = 0.5$ m

$m_2 = 1$ kg
$r_2 = 1$ m

    a) 1/2 N
    b) 1 N
    c) 2 N
    d) 3 N

Conceptual Example 6.6-2

Consider a pendulum that is released from the position shown and falls under the influence of gravity. What affect do you think increasing the size of m_{sphere}, m_{bar}, L, and r have on the angular acceleration (α)? You can check your assertions with the book's website (see preface for the url).

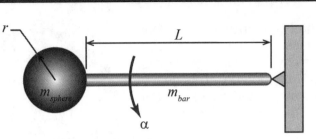

Parameter	α decreases	α increases	Notes
m_{sphere}			
m_{bar}			
r			
L			

Equation Summary

Abbreviated variable definition list

$\mathbf{F}$ = external forces acting on the system
m = mass of the rigid body
$\mathbf{a}_G$ = acceleration of the center of mass G
$\mathbf{r}_{G/O}$ = position vector of the center of mass G
    relative to the fixed axis through O

ω = angular velocity of the body (rad/s)
α = angular acceleration of the body (rad/s^2)
$\mathbf{M}_O$ = moments about an axis through O
I_O = mass moment of inertia about the axis
    through O

Newton's second law

$$\sum \mathbf{F} = m\mathbf{a}_G$$

Moment equation

$$\sum \mathbf{M}_O = I_O \boldsymbol{\alpha}$$

Acceleration

$$\mathbf{a}_G = \boldsymbol{\alpha} \times \mathbf{r}_{G/O} - \omega^2 \mathbf{r}_{G/O}$$

Example Problem 6.6-3

In the figure shown, gear A has a mass of 3 kg and is being driven by an engine generating 100 Nm of torque. This gear is then connected via a belt to gear B which has an effective mass of 1 kg and drives an accessory of the engine that has negligible mass and friction. If gear A has a radius of 10 cm and gear B has a radius of 3 cm, determine the angular acceleration of gear B. You may approximate the gears as disks and assume that the belt is inextensible with negligible mass.

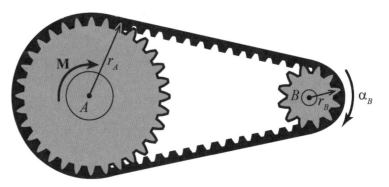

Given:

Find:

Solution:

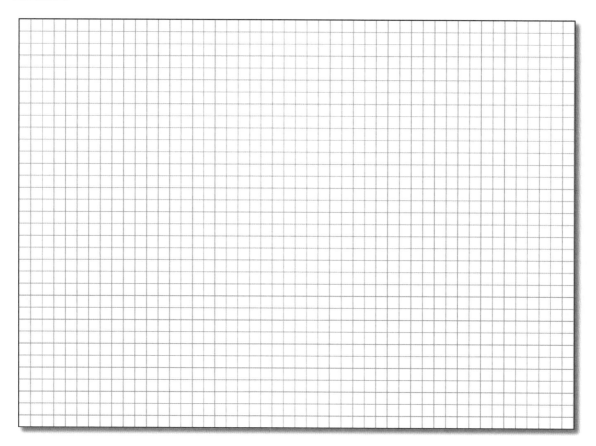

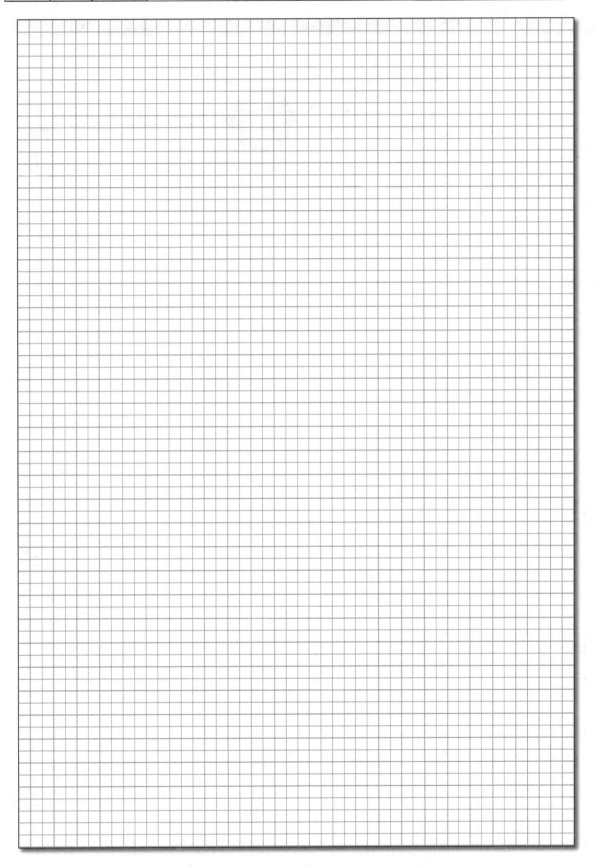

Example Problem 6.6-4

The pendulum shown consists of a 10-lb sphere (r = 2 in) and a 10-lb slender rod (L = 1.5 ft). Determine the reaction force at the support pin O as the pendulum passes the horizontal position with an angular velocity of ω = 6 rad/s.

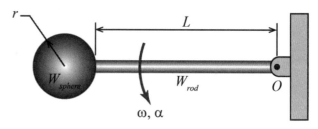

Given:

Find:

Solution:

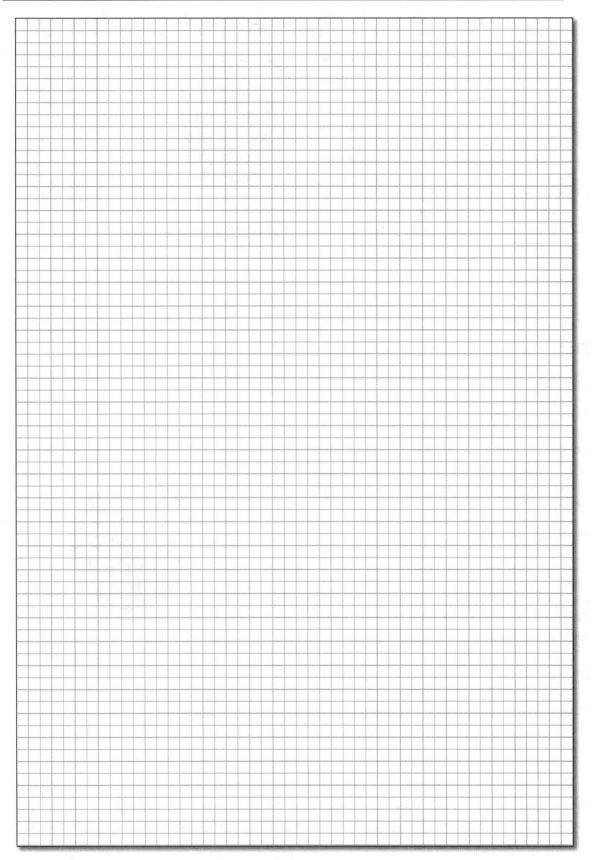

Example Problem 6.6-5

A disk with a mass imbalance rotates about fixed shaft at O. The mass imbalance is produced by an off-center cylinder. Both the disk and cylinder are made of 1020 steel with the dimensions shown. A motor applies a moment ($M = 5$ lb-ft) to the initially still disk with the cylinder directly above the shaft. The bearing that supports the disk is poorly lubricated and resists the disk's motion with a frictional torque ($M_f = 1$ lb-ft). Determine the bearing support force when the cylinder has completed 10 full revolutions.

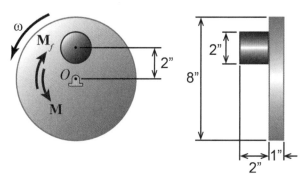

Given: $M = 5$ lb-ft, $M_{fric} = 1$ lb-ft $\Delta\theta = 10$ rev

$r_{disk} = 4$ in, $h_{disk} = 1$ in $r_{cylinder} = 1$ in,

$h_{cylinder} = 2$ in $d = 2$ in (cylinder offset)   1020 steel

Find: t_f, **O**

Solution:

Free-body diagram

To start we will draw a free-body diagram making sure that we draw the cylinder at an angle. This will introduce the angular position variable (θ) that we need for the support reaction results.

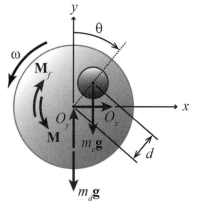

Equation of motion

We will use Newton's second law and a form of Euler's second law to determine the motion of the disk and the reaction forces. We will start with summing the moments. Since this is a case of pure rotation, we will use the fixed axis as a reference.

$$\sum \mathbf{M}_O = I_O \boldsymbol{\alpha}$$

When the motor is accelerating the disk, the angular acceleration is α.

$$M - M_{fric} - m_{cylinder}gd\sin\theta = I_O\alpha \qquad\qquad \alpha = \frac{M - M_{fric} - m_{cylinder}gd\sin\theta}{I_O} \qquad (1)$$

To find the value of α, we need to calculate the mass moment of inertia of the system with respect to O, the mass of each component and the angular position of the cylinder after 10 revolutions.

Angular position

The disk makes 10 full revolutions starting at a position of $\theta = 0$ as defined in the FBD.

$$\theta_f = \theta_o + 10 \text{ rev} = 0 + 10 \text{ rev} \frac{2\pi \text{ rad}}{\text{rev}} = 20\pi \text{ rad}$$

Mass

The mass of a body is its density (ρ) multiplied by its volume (V). The density of 1020 steel may be found in Appendix C.

$$\rho_{steel} = 0.2843 \frac{\text{lb}}{\text{in}^3} \left(\frac{1}{32.2 \text{ ft/s}^2} \right) = 0.00883 \frac{1}{\text{in}^3} \frac{\text{lb} \cdot \text{s}^2}{\text{ft}} = 0.00883 \frac{\text{slug}}{\text{in}^3}$$

$$V = \pi r^2 h \qquad\qquad m = \rho V$$

$$m_{disk} = \rho_{steel} \pi r_{disk}^2 h_{disk} = 0.4438 \text{ slug} \qquad\qquad m_{cylinder} = \rho_{steel} \pi r_{cylinder}^2 h_{cylinder} = 0.05548 \text{ slug}$$

Mass moment of inertia

$$I_{disk,O} = \frac{m_{disk} r_{disk}^2}{2} \qquad\qquad I_{cylinder,G} = \frac{m_{cylinder} r_{cylinder}^2}{2}$$

Note that the tables only give the mass moment of inertia for a cylinder about its center axis. Therefore, we will have to use the parallel axis theorem to calculate the inertia with respect to an axis parallel to the axis passing through O.

$$I_{cylinder,O} = I_{cylinder,G} + m_{cylinder} d^2 = \frac{m_{cylinder} r_{cylinder}^2}{2} + m_{cylinder} d^2$$

Now that both moments of inertia are about the same axis, they can be added.

$$I_O = I_{cylinder,O} + I_{disk,O} = \frac{m_{cylinder} r_{cylinder}^2}{2} + m_{cylinder} d^2 + \frac{m_{disk} r_{disk}^2}{2} = 3.8 \text{ slug} \cdot \text{in}^2$$

Equation of motion

From equation (1) we can calculate the final angular acceleration.

$$\alpha_f = 151.6 \frac{\text{rad}}{\text{s}^2}$$

Kinematics

In order to determine the linear acceleration of the cylinder, which is needed to calculate the reaction forces, we need to calculate the angular velocity.

$$\int_0^\theta \alpha \, d\theta = \int_0^\omega \omega \, d\omega \qquad\qquad \int_0^\theta (151.6 - 11.28 \sin\theta) \, d\theta = \int_0^\omega \omega \, d\omega$$

$$151.6\theta + 11.28 \cos\theta \Big|_0^\theta = \frac{\omega^2}{2} \qquad\qquad \omega = \sqrt{2(151.6\theta + 11.28(\cos\theta - 1))}$$

$$\omega_f = 138.02 \ \frac{\text{rad}}{\text{s}}$$

Because the disk is in pure rotation about its center of mass, this acceleration is zero.

$$\mathbf{a}_{G,disk} = 0$$

The acceleration of the cylinder's center of mass may be calculated using kinematic relations.

$$\mathbf{a}_{G,cylinder} = \mathbf{a}_O + \boldsymbol{\alpha} \times \mathbf{r}_{G/O} - \omega^2 \mathbf{r}_{G/O}$$

$$\begin{aligned}
\mathbf{a}_{G,cylinder} &= \alpha\mathbf{k} \times d(\sin\theta\,\mathbf{i} + \cos\theta\,\mathbf{j}) - \omega^2 d(\sin\theta\,\mathbf{i} + \cos\theta\,\mathbf{j}) \\
&= \alpha d(\sin\theta\,\mathbf{j} - \cos\theta\,\mathbf{i}) - \omega^2 d(\sin\theta\,\mathbf{i} + \cos\theta\,\mathbf{j}) \\
&= d\big((\alpha\cos\theta + \omega^2\sin\theta)\mathbf{i} + (\alpha\sin\theta + \omega^2\cos\theta)\mathbf{j}\big)
\end{aligned}$$

$$\mathbf{a}_{G,cylinder,f} = d\left(-\alpha_f\,\mathbf{i} - \omega_f^2\,\mathbf{j}\right) = -25.27\,\mathbf{i} - 3174.92\,\mathbf{j}\ \frac{\text{ft}}{\text{s}^2}$$

Equation of motion

We can now apply Newton's second law to determine the bearing reaction forces.

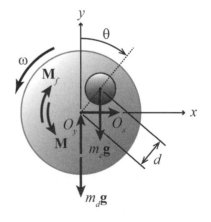

$$\sum \mathbf{F} = \sum m_i \mathbf{a}_{G,i}$$

$$\mathbf{O} - (m_{disk}g + m_{cylinder}g)\mathbf{j} = m_{disk}\mathbf{a}_{G,disk} + m_{cylinder}\mathbf{a}_{G,cylinder}$$

$$= m_{cylinder}\mathbf{a}_{G,cylinder}$$

$$\boxed{\mathbf{O}_f = -1.40\mathbf{i} - 160.17\mathbf{j}\ \text{lb}}$$

6.7) GENERAL PLANAR MOTION WITH A CENTER OF MASS REFERENCE

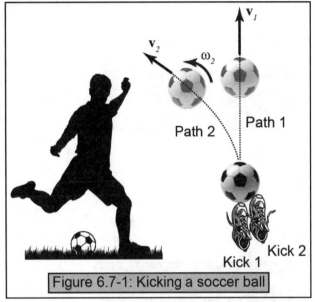

Figure 6.7-1: Kicking a soccer ball

Imagine that you are playing soccer and an empty goal is straight in front of you. The logical thing to do would be to kick the ball straight and score. Where do you kick the ball? Do you kick the ball in the middle or off to the side? This might seem like a silly question, but it illustrates an important concept. Most of us have experience kicking a ball of some sort or another. We know that if we kick the ball in the middle (i.e. at its center of mass) it will move straight out in front of you. If we kick the ball off to one side it will not go straight in front of you, it will veer off to one side or the other. You are not just changing the ball's trajectory, your are also causing the ball to rotate as shown in Figure 6.7-1. This concept is used by billiards (pool) players to perform various trick shots by putting spin (English) on the billiard balls. They do this by hitting the cue ball off-center.

It is often the case that an object will naturally want to rotate about its center of mass (G). This makes G a special point. You will see that when we analyze the motion of an object and choose G as the point to sum the moments about, we greatly simplify our calculations.

Under the influence of an imbalanced force and the moment it induces, an unconstrained rigid body will translate and rotate. If the point we use to sum the moments about is the center of mass G, the equations of motion take the form shown in Equations 6.7-1 and 6.7-2. Note that Equation 6.7-2 does not include the moment due to the effective force ($m\mathbf{a}$) discussed in the context of applying d'Alembert's principle. This is because the effective force is applied at the center of mass and does not create a moment about G. Note that Equation 6.7-2 assumes planar motion, therefore, $\mathbf{M}_G$ and α are either in the clockwise or counterclockwise direction.

Newton's second law: $\boxed{\sum \mathbf{F} = m\mathbf{a}_G}$ (6.7-1)

Sum of the moments about mass center G: $\boxed{\sum \mathbf{M}_G = I_G \alpha}$ (6.7-2)

$\mathbf{F}$ = external forces acting on the rigid body
m = mass of the body
$\mathbf{a}_G$ = acceleration of the center of mass

$\mathbf{M}_G$ = moments about mass center G
I_G = mass moment of inertia about the axis through mass center G
α = angular acceleration of the body (rad/s^2)

Conceptual Example 6.7-1

How is this block going to move under the various conditions shown?

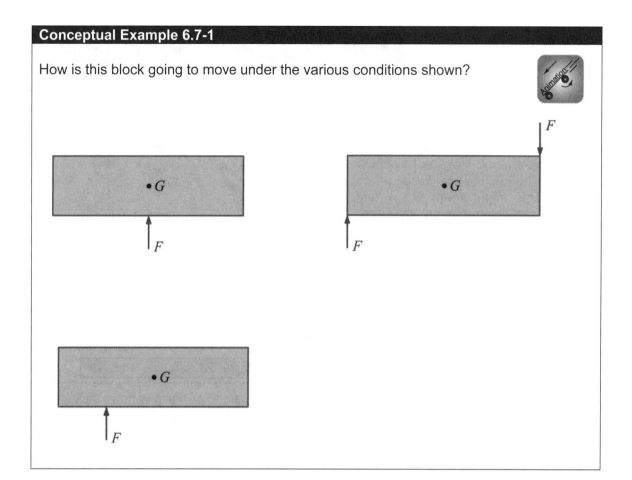

Example 6.7-2

The angular motion of a body undergoing general planar motion (i.e. translating and rotating) is described by the following equation, where P is an arbitrary reference point.

$$\sum \mathbf{M}_P = I_P \boldsymbol{\alpha} + \mathbf{r}_{G/P} \times m \mathbf{a}_P$$

Explain why, if G is the reference point, the body's angular motion is described by the following equation.

$$\sum \mathbf{M}_G = I_G \boldsymbol{\alpha}$$

6.7.1) KINETICS OF ROLLING

A common type of rigid-body motion that involves translation and rotation is rolling. The kinetics of rolling may not be immediately intuitive and deserves closer attention. Consider the rolling disk shown in Figure 6.7-2. The disk is rolling to the left. Drawing the free-body diagram, other than the force (**P**) that moves the disk, there is the weight (*mg*) of the disk acting downward at the body's center of mass and the normal force (**N**) of the floor reacting to the contact with the disk. There is, however, another reaction force: friction (**F**$_f$). As the disk is rolling counter-clockwise, the floor applies a horizontal friction force to the disk at its contact point. In this case, the disk would not rotate without the friction force. The magnitude of this reaction force is determined by the friction characteristics between the two surfaces. If the disk is rolling without slip, then recall from our kinematics analysis that the point of the disk in contact with the floor is an instantaneous center of zero velocity. Therefore, under the conditions of rolling without slip, the friction can be modeled as static. If the disk is slipping, then the friction must be modeled as kinetic. Even though it may not be apparent, the static friction case may be more difficult to analyze. This is because the static friction force may be one of many different values ($F_{fs} = 0$ to $\mu_s N$), whereas, the kinetic friction force has only one value ($F_{fk} = \mu_k N$). In cases where we do not know if a body is rolling without slip, we may employ the approach to solving problems with friction introduced in the chapter on *Particle Newtonian Mechanics*. First, solve the problem under the assumption that the friction is static (i.e. rolling without slip), then check the assumption to see if the required friction force is greater than the maximum possible static friction force ($F_{fs,max} = \mu_s N$). If the required force is not greater than the maximum, then we are done. If the friction force is greater than the maximum, then the problem must be solved for rolling with slip ($F_{fk} = \mu_k N$).

Since the value of the friction force can be difficult to determine in the case of rolling without slip, it is often advantageous to sum the moments about the point where the disk contacts the ground. Summing the moments in this manner eliminates the moment due to the friction force. Analyzing the rotation of a rigid body about an axis through a point that is neither fixed, nor the body's mass center, is discussed in the next section.

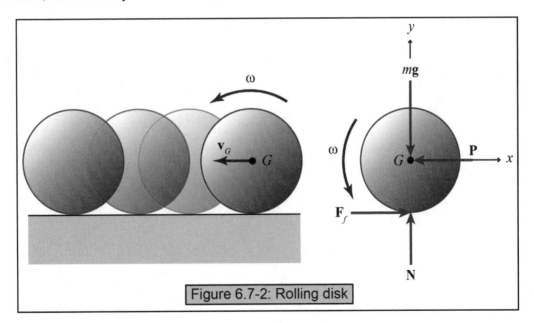

Figure 6.7-2: Rolling disk

Equation Summary

Abbreviated variable definition list

$\mathbf{F}$ = external forces acting on the system
m = mass of the rigid body
$\mathbf{a}_G$ = acceleration of the center of mass G

α = angular acceleration of the body (rad/s^2)
$\mathbf{M}_G$ = moments about an axis through G
I_G = mass moment of inertia about the axis through G

Newton's second law

$$\sum \mathbf{F} = m\mathbf{a}_G$$

Moment equation

$$\sum \mathbf{M}_G = I_G \alpha$$

Example Problem 6.7-3

When rockets rise to an altitude above the earth's atmosphere, their aerodynamic control surfaces are no longer able to generate the lift forces necessary to maneuver the vehicle. One approach for steering rockets under these conditions is called thrust vectoring and involves gimbaling the rocket's engine such that the direction of the thrust force can be controlled. This is the situation depicted in the figure of a 20,000-kg rocket directed 20 degrees away from vertical with its thrust vector gimbaled 1 degree away from the longitudinal axis of the rocket. If the rocket is experiencing 300 kN of thrust at an altitude where the acceleration due to gravity is approximately 8 m/s^2, determine the acceleration of the mass center G and the angular acceleration α of the rocket. The rocket has a centroidal radius of gyration of 5 m and the length L depicted is 10 m.

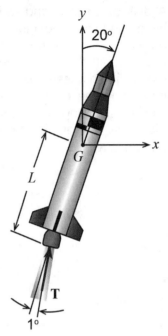

Given:

Find:

Solution:

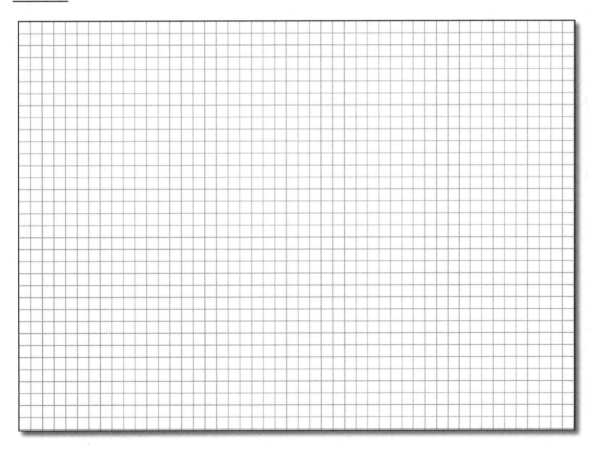

Example Problem 6.7-4

A disk of mass m and radius r rolls down an inclined hill as shown in the figure. A force P, which is applied parallel to the inclined surface, resists the motion of the disk. If the disk is released from rest, determine the angular acceleration of the disk if $P = 0$. Also, determine the minimum force P required to stop the disk from rolling down the incline. Assume that the disk rolls without slipping. The coefficient of static friction is μ_s.

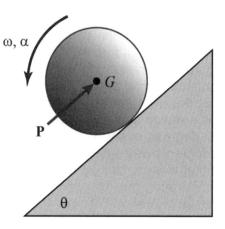

Given:

Find:

Solution:

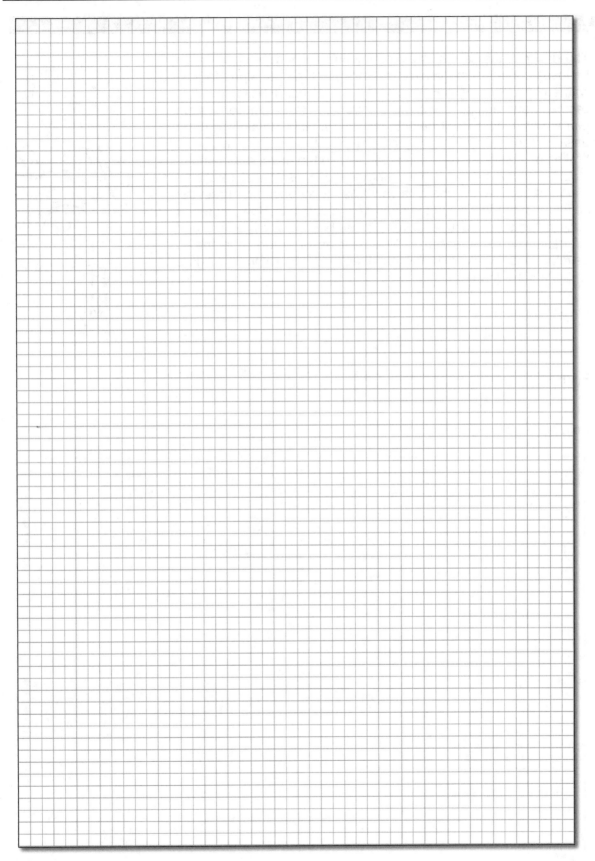

Example Problem 6.7-5

The driver of the car shown is attempting to round a curve when he loses control of the vehicle. At this point the car starts to spin with a clockwise angular velocity of 0.3 rad/sec. Fortunately, this car has been equipped with an electronic stability control system that is activated once it detects that the vehicle is skidding in an unsafe manner. The stability control system is able to apply the brakes of individual wheels in order to help the driver maintain the stability of the vehicle. The vehicle has a centroidal rotational inertia of 3500 kg-m^2 about its yaw axis. It is desired to arrest the spinning of the vehicle in the span of one second. How large of a constant force F must be developed by braking the left front and left rear wheels as shown? You may assume the road forces developed by braking have the directions shown in the figure where $\theta = 15$ degrees.

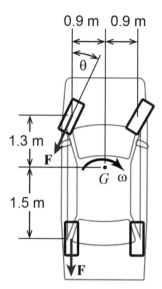

Given:

Find:

Solution:

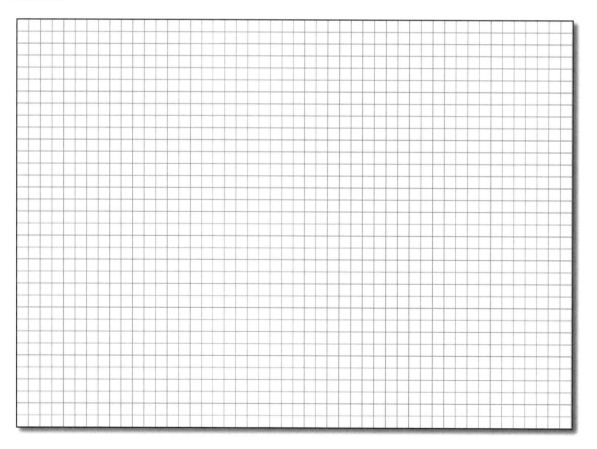

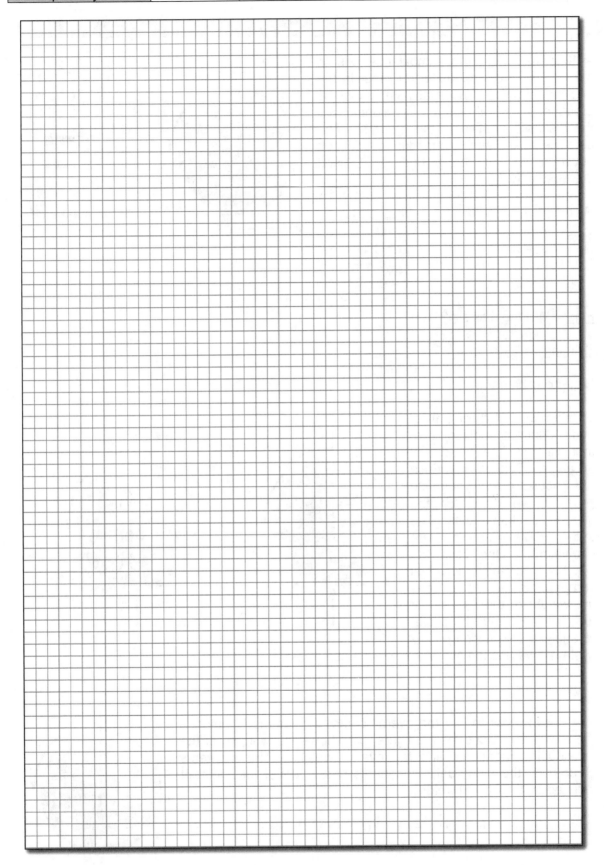

Example Problem 6.7-6

A 10-lb disk with a mass imbalance rolls on the ground. It is released from rest in the position shown. The mass imbalance is produced by an off-center 2-lb cylinder. Will the disk slip relative to the ground? The coefficient of static and kinetic friction between the disk and ground are 0.6 and 0.4, respectively.

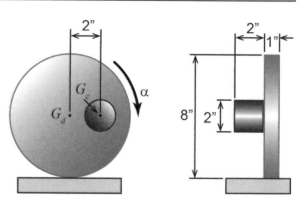

Given: $W_{cylinder}$ = 2 lb W_{disk} = 10 lb
        W = 12 lb ω = 0
        μ_s = 0.6 μ_k = 0.4
        $r_{cylinder}$ = 1 in r_{disk} = 4 in
        d = 2 in θ = 90°

Find: Will the disk slip?

Solution:

Free-body diagram

To start we will draw a free-body diagram making sure that we draw the cylinder at an angle. This will introduce the angular position variable (θ) that we need for the support reaction results. The two FBD's shown are equivalent. One uses the weights of the individual components and one uses the weight of the system. It turns out that it will be easier to analyze the motion of this system if we use the overall center of mass.

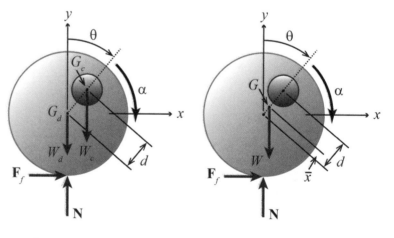

Center of mass

The easiest way to find the overall center of mass is to balance the moments. Consider the figure that shows the weights of the individual components and the overall weight. The moments created by the individual components has to equal the moment induced by the overall weight.

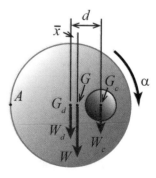

$$\sum M_A = W_{disk} r_{disk} + W_{cylinder}(r_{disk} + d) = W(r_{disk} + \bar{x})$$

$$\overline{x} = \frac{W_{disk}r_{disk} + W_{cylinder}(r_{disk} + d)}{W} - r_{disk} = 0.333 \text{ in}$$

Mass moment of inertia

$$I_{disk,Gd} = \frac{m_{disk}r_{disk}^2}{2} \qquad\qquad I_{cylinder,Gc} = \frac{m_{cylinder}r_{cylinder}^2}{2}$$

Note that the tables only give the mass moment of inertias for a cylinder and disk about their respective center axes. Therefore, we will have to use the parallel axis theorem to calculate the inertias with respect to an axis parallel to the axis passing through G.

$$I_{cylinder,G} = I_{cylinder,Gc} + m_{cylinder}(d - \overline{x})^2 = \frac{m_{cylinder}r_{cylinder}^2}{2} + m_{cylinder}(d - \overline{x})^2$$

$$I_{disk,G} = I_{disk,Gc} + m_{disk}\overline{x}^2 = \frac{m_{disk}r_{disk}^2}{2} + m_{disk}\overline{x}^2$$

Now that both moments of inertia are about the same axis, they can be added.

$$I_G = I_{cylinder,G} + I_{disk,G} = \frac{m_{cylinder}r_{cylinder}^2}{2} + m_{cylinder}(d - \overline{x})^2 + \frac{m_{disk}r_{disk}^2}{2} + m_{disk}\overline{x}^2 = 2.723 \text{ slug} \cdot \text{in}^2$$

Equation of motion

We will use Newton's second law and a form of Euler's second law to determine the motion of the disk and the reaction forces. We will start with summing the moments. Since this is a case of general planar motion we will use the center of mass of the system as a reference.

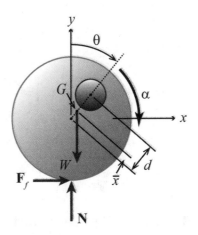

$$\sum \mathbf{M}_G = I_G \alpha \qquad\qquad -F_f r_{disk} + N\overline{x}\sin\theta = I_G \alpha$$

$$\alpha = \frac{-F_f r_{disk} + N\overline{x}\sin\theta}{I_G} = \frac{-F_f r_{disk} + N\overline{x}}{I_G} \qquad (1)$$

Kinematics

We don't, at this moment, know whether the friction is static or kinetic. Let's assume that there is no slip and then check our assumption at the end. If there is no slip, the acceleration of the center of the disk may be determined by the following equation.

$$a_{Gd,x} = r_{disk}\alpha$$

The acceleration of the overall center of gravity may be determined with the following equation. Remember that the system is released from rest (i.e $\omega = 0$) and the disk is in the $\theta = 90°$ position.

$$\mathbf{a}_G = \mathbf{a}_{Gd} + \boldsymbol{\alpha} \times \mathbf{r}_{G/Gd} - \omega^2 \mathbf{r}_{G/Gd}$$

$$= r_{disk}\alpha\mathbf{i} + \alpha(-\mathbf{k}) \times \overline{x}(\sin\theta\,\mathbf{i} + \cos\theta\,\mathbf{j}) - \omega^2\overline{x}(\sin\theta\,\mathbf{i} + \cos\theta\,\mathbf{j})$$

$$= r_{disk}\alpha\mathbf{i} + \alpha\overline{x}(-\sin\theta\,\mathbf{j} + \cos\theta\,\mathbf{i}) - \omega^2\overline{x}(\sin\theta\,\mathbf{i} + \cos\theta\,\mathbf{j})$$

$$= \alpha(\overline{x}\cos\theta + r_{disk})\mathbf{i} - \alpha\overline{x}\sin\theta\,\mathbf{j}$$

$$= \alpha r_{disk}\mathbf{i} - \alpha\overline{x}\,\mathbf{j}$$

$$(2)$$

Equation of motion

We can now apply Newton's second law to determine the friction force and normal force at the ground contact. We will use the angular acceleration given in Equation (1) and the acceleration given in Equation (2).

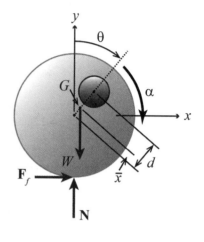

$$\sum\mathbf{F} = \sum m_i\mathbf{a}_{G,i}$$

x-direction

$$F_{fs} = \frac{W}{g}a_{Gx} = \frac{W}{g}[\alpha r_{disk}] = \frac{Wr_{disk}}{g}\left[\frac{-F_{fs}r_{disk} + N\overline{x}}{I_G}\right] = 0.1823N - 2.1898F_{fs} \qquad (3)$$

y-direction

$$N - W = \frac{W}{g}a_{Gy} = \frac{W}{g}[-\alpha\overline{x}] = \frac{W\overline{x}}{g}\left[\frac{F_{fs}r_{disk} - N\overline{x}}{I_G}\right] \qquad N\left[1 + \frac{W\overline{x}^2}{gI_G}\right] = W\left[1 + \frac{F_{fs}r_{disk}\,\overline{x}}{gI_G}\right]$$

$$N = W\left[\frac{gI_G + F_{fs}r_{disk}\,\overline{x}}{gI_G + W\overline{x}^2}\right] = 11.82 + 0.1796F_{fs} \qquad (4)$$

Substituting Equation (4) into Equation (3) we get an expression for the friction force and use that to calculate the normal force.

$$F_{fs} = (0.1823(11.82 + 0.1796F_{fs}) - 2.1898F_{fs}) \qquad F_{fs} = 0.68\text{ lb} \qquad N = 11.94\text{ lb}$$

Maximum static friction force

$$F_{fs,max} = \mu_s N = 7.17\text{ lb} > F_{fs}$$

Since the maximum static friction force is greater than the static friction force calculated, **the disk will not slip relative to the ground.**

6.8) GENERAL PLANAR MOTION

In the previous sections of this chapter, we looked at methods to analyze the rotational motion of a rigid body. The method chosen will depend on how the body moves or is constrained. Before applying the method discussed in this section, you should make sure that the other methods don't apply. In this section, we will explain a method for analyzing the general planar motion of a rigid body that allows you to use an arbitrary reference point P. The freedom to choose such an arbitrary point can be helpful in that it allows you the freedom to eliminate unwanted and/or unknown forces from the moment equation. For example, consider a ladder sliding down a wall as shown in Figure 6.8-1. From its free-body diagram, we can see that there is a normal force from the floor pushing on the ladder at point A and a normal force from the wall pushing on the ladder at point B. If we sum the moments about A we can eliminate the floor's normal and friction forces from the moment equation. If the wall and floor were frictionless, then point P would be a good choice. Point P is at the intersection of the lines of action of the two normal forces, thereby, eliminating the normal forces from the moment equation. The point you choose to sum the moments about should depend on what is given in the problem statement and what you need to find.

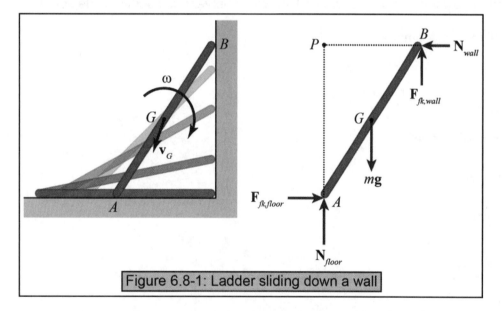

Figure 6.8-1: Ladder sliding down a wall

When using an arbitrary reference point P, the equations of motion are given by Equations 6.8-1 through 6.8-3. D'Alembert's principle is again employed where the effective force ($m\mathbf{a}$) is applied to the body's center of mass and, therefore, it creates a moment which is included as part of Equation 6.8-2 and 6.8-3. When analyzing the rotation of a rigid body, either Equation 6.8-2 or 6.8-3 may be used. These equations can be shown to be equivalent through the application of the parallel-axis theorem. However, there is a restriction on the choice of reference point P when applying Equation 6.8-2. When using Equation 6.8-2, the reference point P must be affixed to the body under consideration. Equation 6.8-3, on the other hand, allows you to choose a reference point P that does not necessarily have to move with the body. Note that Equations 6.8-2 and 6.8-3 assume planar motion, therefore, $\mathbf{M}_P$ and α are either in the clockwise or counterclockwise direction.

Newton's second law: $\boxed{\sum \mathbf{F} = m\mathbf{a}_G}$ (6.8-1)

Sum of the moments about an affixed point P: $\boxed{\sum \mathbf{M}_P = I_P\boldsymbol{\alpha} + \mathbf{r}_{G/P} \times m\mathbf{a}_P}$ (6.8-2)

Sum of the moments about an arbitrary point P: $\boxed{\sum \mathbf{M}_P = I_G\boldsymbol{\alpha} + \mathbf{r}_{G/P} \times m\mathbf{a}_G}$ (6.8-3)

$\mathbf{F}$ = external forces acting on the rigid body
m = mass of the body
$\mathbf{a}_P$, $\mathbf{a}_G$ = acceleration of point P and the acceleration of point G respectively
$\mathbf{M}_P$ = moments about P

I_P = mass moment of inertia about the axis through P
I_G = mass moment of inertia about the axis through G
α = angular acceleration (rad/s^2)
$\mathbf{r}_{G/P}$ = the position of G relative to point P

Example 6.8-1

Determine how the following bodies will move in the various conditions depicted.

How will this object move?

Equations of motion?

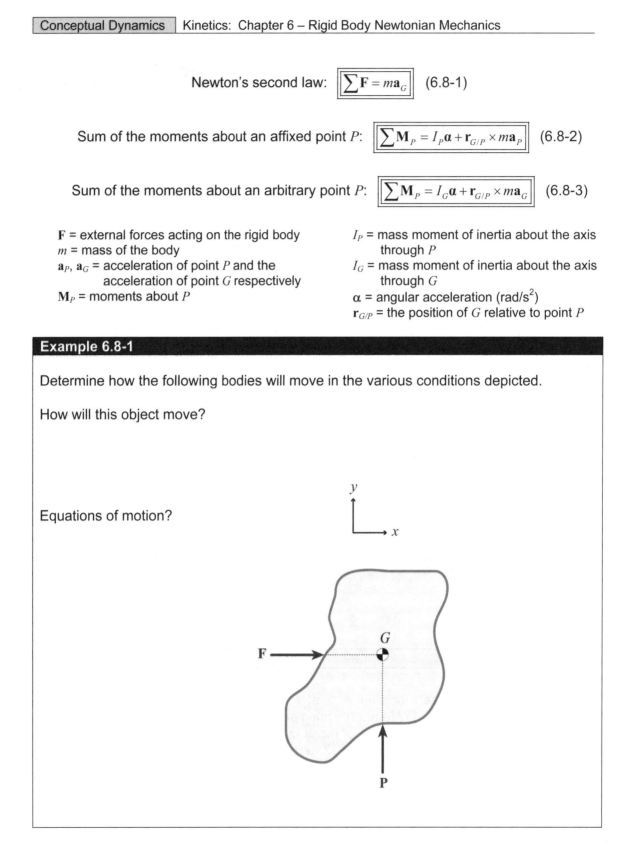

Example 6.8-1 Continued

How will this object move?

Equations of motion?

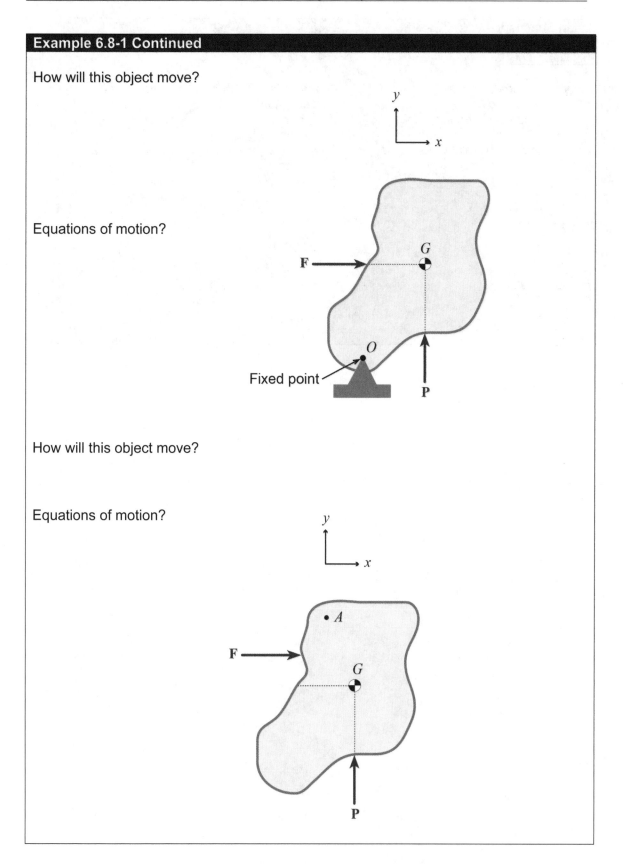

How will this object move?

Equations of motion?

Equation Summary

Variable definition

$\mathbf{F}$ = external forces acting on the system
m = mass of the rigid body
$\mathbf{a}_G$ = acceleration of the center of mass G
$\mathbf{a}_P$ = acceleration of point P
$\mathbf{r}_{G/P}$ = the position of G relative to point P

α = angular acceleration of the body (rad/s^2)
$\mathbf{M}_P$ = moments about P
I_G = mass moment of inertia about the axis through G
I_P = mass moment of inertia about the axis through P

Newton's second law

$$\sum \mathbf{F} = m\mathbf{a}_G$$

Moment equations

$$\sum \mathbf{M}_P = I_P \boldsymbol{\alpha} + \mathbf{r}_{G/P} \times m\mathbf{a}_P \qquad \sum \mathbf{M}_P = I_G \boldsymbol{\alpha} + \mathbf{r}_{G/P} \times m\mathbf{a}_G$$

Example Problem 6.8-2

A uniform slender bar (L = 1.5 m) rests on a smooth horizontal sheet of ice when a force **F** is applied normal to the bar's longitudinal axis at point P. Point P is observed to have an acceleration a_P of 10 m/s² in the direction shown at this instant, and the bar has a corresponding angular acceleration α = 9 rad/s² in the directions shown. Determine the distance d.

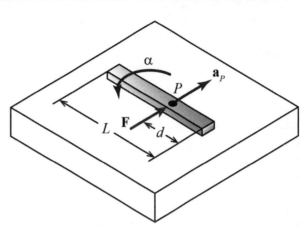

Given:

Find:

Solution:

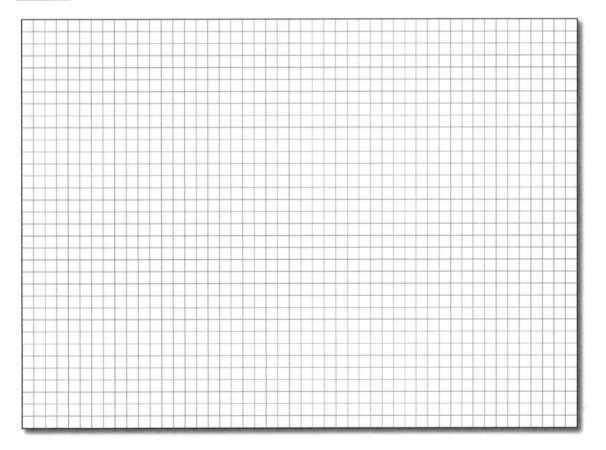

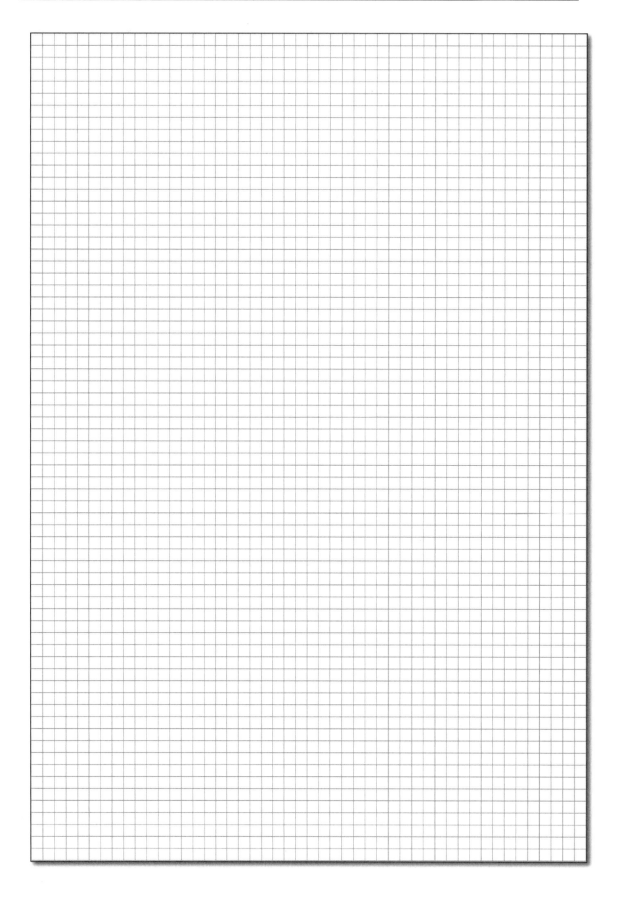

Example Problem 6.8-3

A 12-in ruler is perfectly balanced on a man's hand as shown in the figure. The man then accelerates his hand in the x-direction at a rate of 1 ft/s^2. If the acceleration remains constant, calculate the angular velocity ω of the ruler as it reaches the horizontal position just before it hits the man's arm. Assume that point O at the end of the ruler doesn't slide relative to the moving hand.

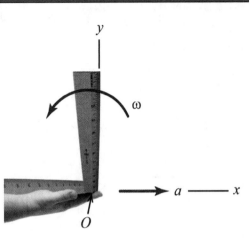

Given:

Find:

Solution:

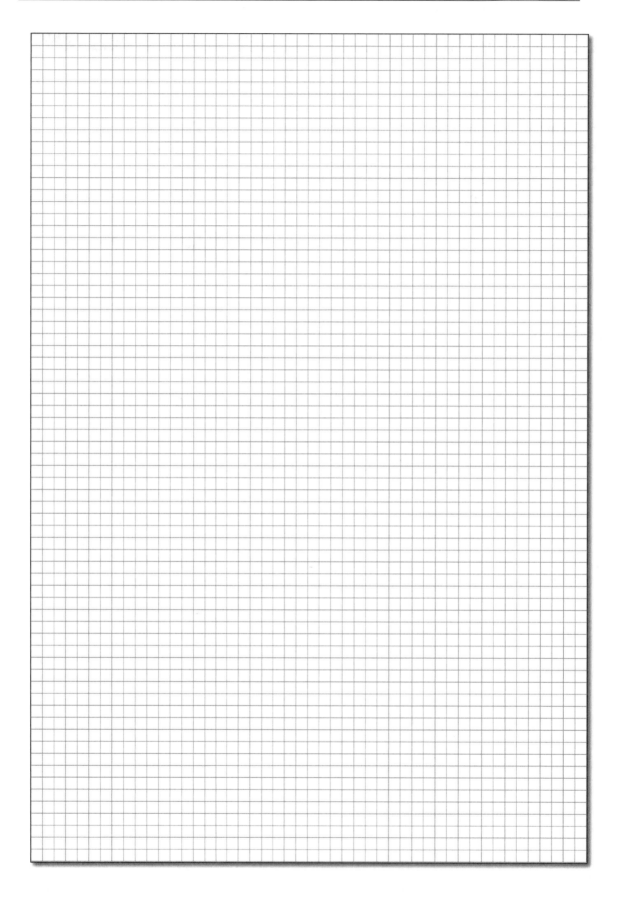

Solved Problem 6.8-4

Consider the 24-kg ladder shown which initially makes an angle of 30° with the wall it is leaning against. If the ladder is released from rest and the wall and floor can be assumed to be frictionless, determine the reaction forces at the wall and floor at the instant of release.

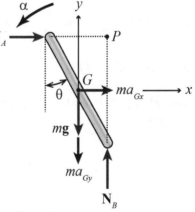

Given: m = 24 kg ω = 0
       θ = 30° L = 6 m
       Frictionless

Find: N_A, N_B

Solution:

Problem familiarization

The ladder both rotates and translates. Therefore, we have to decide which of the following equations should be used to analyze its motion. If we use Equation (1), we will end up with 3 unknowns (i.e. N_A, N_B, α). Newton's second law may be used to eliminate one of these unknowns, but this is still not enough. So our job will be to pick a point (P) that will eliminate some of the unknowns. If the point that we choose is not affixed to the body, we cannot use Equation (2). Therefore, we are left with using Equation (3).

$$\sum \mathbf{M}_G = I_G \alpha \quad (1) \qquad \sum \mathbf{M}_P = I_P \alpha + \mathbf{r}_{G/P} \times m\mathbf{a}_P \quad (2) \qquad \sum \mathbf{M}_P = I_G \alpha + \mathbf{r}_{G/P} \times m\mathbf{a}_G \quad (3)$$

Free-body diagram

Knowing which equation form we will be using helps us draw the necessary information on our FBD. Notice that the location of point P will eliminates the moments created by the normal forces.

$$\sum \mathbf{M}_P = I_G \alpha + \mathbf{r}_{G/P} \times m\mathbf{a}_G$$

$$mg\frac{L}{2}\sin\theta = I_G\alpha + ma_{Gy}\frac{L}{2}\sin\theta + ma_{Gx}\frac{L}{2}\cos\theta \quad (4)$$

Mass moment of inertia

The mass moment of inertia about the mass center for a bar is given by

$$I_G = \frac{mL^2}{12}$$

Kinematics

In order to solve Equation (4), we need to determine the acceleration of the ladder's mass center. Apply the relative acceleration equation between point B and the mass center. Remember that the angular speed at this instant is zero.

$$\mathbf{a}_G = \mathbf{a}_B + \boldsymbol{\alpha} \times \mathbf{r}_{G/B} - \dot{\phi}^2 \mathbf{r}_{G/B} = \mathbf{a}_B + \alpha \mathbf{k} \times \frac{L}{2}(-\sin\theta\,\mathbf{i} + \cos\theta\,\mathbf{j}) = \mathbf{a}_B + \alpha\frac{L}{2}(-\sin\theta\,\mathbf{j} - \cos\theta\,\mathbf{i})$$

Knowing that the acceleration of point B is only in the x-direction, we can determine the y-component acceleration of the mass center.

$$\mathbf{a}_{Gy} = -\alpha\frac{L}{2}\sin\theta\,\mathbf{j} \quad (5)$$

Following similar logic for point A, where A moves only in the y-direction.

$$\mathbf{a}_G = \mathbf{a}_A + \boldsymbol{\alpha} \times \mathbf{r}_{G/A} - \dot{\phi}^2 \mathbf{r}_{G/A} = \mathbf{a}_A + \alpha \mathbf{k} \times \frac{L}{2}(\sin\theta\,\mathbf{i} - \cos\theta\,\mathbf{j}) = \mathbf{a}_A + \alpha\frac{L}{2}(\sin\theta\,\mathbf{j} + \cos\theta\,\mathbf{i})$$

$$\mathbf{a}_{Gx} = \alpha\frac{L}{2}\cos\theta\,\mathbf{i} \quad (6)$$

The directions for the mass center accelerations agree with what we have drawn in the FBD. Substitute Equations (5) and (6) into Equation (4) and solve for α. Note that the positive value for a_{Gy} is used. This is because we assumed a_{Gy} to be negative when writing the moment equation.

$$mg\frac{L}{2}\sin\theta = \frac{mL^2}{12}\alpha + m\alpha\frac{L^2}{4}\sin^2\theta + m\alpha\frac{L^2}{4}\cos^2\theta \qquad \alpha = \frac{6g\sin\theta}{L(1 + 3\sin^2\theta + 3\cos^2\theta)} = 1.23\,\frac{\text{rad}}{\text{s}}$$

Newton's second law

We will apply Newton's second law in the x- and y-directions to determine the support reactions.

$$\sum F_x = ma_{Gx} \qquad N_A = ma\frac{L}{2}\cos\theta \qquad \boxed{N_A = 76.7\text{ N}}$$

$$\sum F_y = ma_{Gy} \qquad N_B - mg = -ma\frac{L}{2}\sin\theta \qquad \boxed{N_B = 191.16\text{ N}}$$

CHAPTER 6 REVIEW PROBLEMS

RP6-1) What is the difference between a body's centroid, center of mass and center of gravity? Select the quantity that matches the definition.

- A body's center of volume. (centroid, center of mass, center of gravity)
- The average location of a body's mass. (centroid, center of mass, center of gravity)
- The average location of the gravitational force on a body. (centroid, center of mass, center of gravity)
- The centroid and center of mass of a body always have the same location. (true, false)
- The center of mass and the center of gravity of a body always have the same location. (true, false)

RP6-2) What is the difference between a body's area moment of inertia and a body's mass moment of inertia? Select the quantity that matches the definition.

- A measure of a body's resistance to angular acceleration about a given axis. (area moment, mass moment)
- A measure of a body's resistance to bending about a given axis. (area moment, mass moment)
- The area moment of inertia and mass moment of inertia are both identified by the variable I. (true, false)
- The area moment of inertia and the mass moment of inertia always have the same value. (true, false)

RP6-3) Explain the physical significance of the $\mathbf{r}_{G/P} \times m\mathbf{a}_P$ term in the moment equation $\sum \mathbf{M}_P = \mathbf{r}_{G/P} \times m\mathbf{a}_P$ as employed for the case of pure translation.

RP6-4) One form of Euler's second law states that $\sum \mathbf{M}_O = I_O \boldsymbol{\alpha}$. What must be true about point O for this equation to hold true? Circle all that apply.

a) Point O is a fixed point.
b) Point O is moving with constant velocity.
c) Point O is accelerating.
d) Point O coincides with the mass center of the body.

RP6-5) The forklift shown is carrying a load of sand bags weighing 700 lb (318 kg). The forklift itself weighs 5000 lb (2268 kg). Determine the maximum safe upward acceleration of the load if the forklift is not moving and when the forklift is moving forward with an acceleration of 5 m/s². The forklift dimensions are as follows: c = 2 m, b = 1 m, d = 1.7 m, h = 1.5 m and h_L = 5 m.

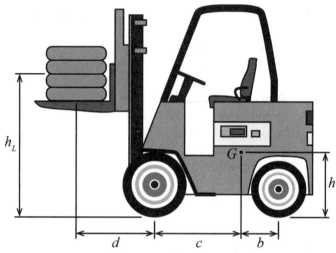

Given: m_{Fork} = 2268 kg
m_{Load} = 318 kg b = 1 m
c = 2 m d = 1.7 m
h = 1.5 m h_L = 5 m

Find: a_{max} if a_{Fork} = 0, a_{max} if a_{Fork} = 5 m/s²

Solution:

Draw the free-body diagram.

Find the maximum upward load acceleration if the forklift is not moving.

a_{max} = _____

Find the maximum upward load acceleration if the forklift is moving forward at 5 m/s².

Equations of motion

What point should we sum the moments about to simplify the calculations?

Sum the moments about _____

a_{max} = _____

RP6-6) The articulating robot arm shown has a total of 4 motors controlling each joint. Determine the minimum size required (in terms of torque) of the motor controlling the 2nd axis if it is desired that the end affecter (approximated as joint 4) is capable of reaching a speed of 1 m/s from rest in 1 seconds in a straight are position (i.e. the arms are in line and the motors at the 3rd and 4th joint are not being operated.) The arm parameters are as follows: m_{arm1} = 25 kg, m_{arm2} = 15 kg, L_{arm1} = 1.5 m, L_{arm2} = 2 m, and the mass of the motors controlling the 3rd and 4th axes are m_{m3} = 3 kg and m_{m4} = 2 kg, respectively. You may assume that each arm is a slender rod and that each motor is a particle. Note that the highest demand on the 2nd axis motor will be when all sections of the arm are horizontal.

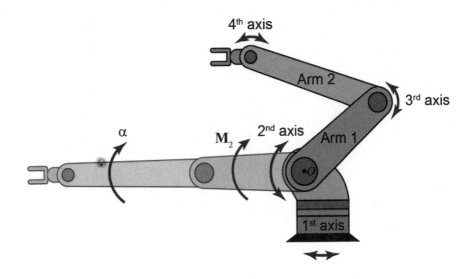

Given: $v_o = 0$ v_f = 1 m/s in t = 1 s
     m_1 = 25 kg m_2 = 15 kg
     L_1 = 1.5 m L_2 = 2 m
     m_{m3} = 3 kg m_{m4} = 2 kg
     Arms = slender rod

Find: M_2

Solution:	Kinematics
Draw the free-body diagram.	Calculate the angular acceleration of the outstretched robot arm. First you need to calculate the tangential acceleration of the 4th axis.
	a_{4t} = _____
	α = _____

Mass moment of inertia

Calculate the total mass moment of inertia. (Note: This will include both arms and motor 3 and 4.)

Equation of motion

Determine the minimum motor torque.

$M_2 =$ _____

$I_o =$ _____

RP6-7) A 15-inch car tire starting from rest rolls down a 30° ramp. How long will it take for the tire's center to travel 20 meters along the ramp? The coefficients of static and kinetic friction are μ_s = 0.9 and μ_k = 0.6, respectively. The tire's radius of gyration about its center is 1.2 ft.

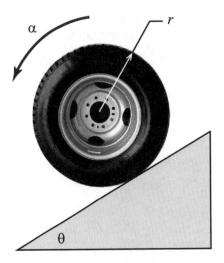

Given: W = 50 lb k_G = 1.2 ft
           r = 15 in μ_s = 0.9
           μ_k = 0.6   released from rest
           θ = 30°

Find: t at s = 20 ft

Solution:

Draw the free-body diagram.	Assume no slip and find α as a function of the static friction force by summing the forces.
	α = _____
Mass moment of inertia	Calculate the static friction force.
I_G = _____	
Equations of motion	
Sum the moments and calculate α as a function of the friction force.	F_{fs} = _____
	Is the no-slip assumption valid?
	Yes No
α = _____	

Calculate the angular acceleration and the acceleration of the center of mass.

Kinematics

Determine the time is takes to reach 20 ft.

$\alpha =$ _____

$\mathbf{a}_G =$ _____

$t =$ _____

RP6-8) A slender bar (m = 9 kg, L = 2 m) rests on a smooth block of ice. A force (**F**) is applied at a distance d = 0.75 m from the center of mass (G) in the direction shown. The resulting acceleration of point A (the point where the force is applied) at this instant is 1 m/s^2. Determine the acceleration of point B and the magnitude of force **F** required to generate this motion.

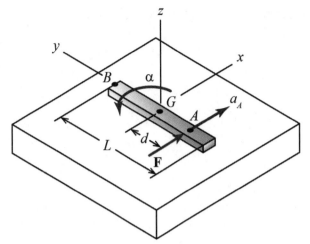

Given: Slender Bar L = 2 m
        d = 0.75 m m = 9 kg
        a_A = 1 m/s^2 ω_o = 0

Find: a_B, F

Solution:

Which reference point should be used?

A B G

Calculate the mass moment of inertia for the bar using the reference point chosen above.

$I_? =$ _____

Equation of motion

Determine the angular acceleration of the bar using the reference point.

$\alpha =$ _____

Kinematics

Determine the acceleration of point B.

$a_B =$ _____

Determine the acceleration of point G.

Equation of motion

Sum the forces to determine the magnitude of F.

$a_G =$ _____

$F =$ _____

P6.1) BASIC LEVEL PURE TRANSLATION PROBLEMS

P6.1-1)[fe] A bicyclist is riding his bike down a 10-percent grade. He realizes that his speed has reached an unsafe level so he applies the front brakes. Determine the maximum deceleration that the biker can attain without tipping over his front wheel. The location of the combined center of mass is shown in the figure.

a) $a = 1.19$ m/s^2 b) $a = 1.39$ m/s^2

c) $a = 1.59$ m/s^2 d) $a = 1.69$ m/s^2

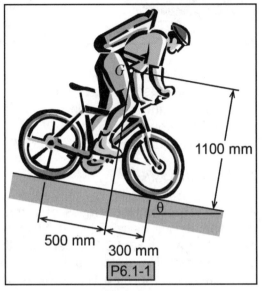

P6.1-1

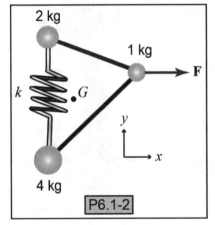

P6.1-2

P6.1-2)[fe] The three spheres shown are connected by a spring, with spring constant $k = 1000$ N/m that is initially compressed 10 cm, and two flexible cords. If the spheres lie in the plane of a frictionless table and a 10-N force is applied in the direction shown, determine the resulting acceleration of the system's center of mass G. You may assume that the spring and cords are massless. Note, not all given information is needed.

a) $\mathbf{a}_G = (10/6)\mathbf{i}$ m/s^2 b) $\mathbf{a}_G = (10/7)\mathbf{i}$ m/s^2

c) $\mathbf{a}_G = (10/2)\mathbf{i}$ m/s^2 d) $\mathbf{a}_G = (10/4)\mathbf{i}$ m/s^2

P6.1-3)[fe] A car is traveling at a constant speed around a curve of radius ρ. The track of the car is b and the height of its center of mass G from the ground is h. The center of mass is located in the middle of the track. Determine the maximum speed v that the car can traverse the curve without tipping.

a) $v = \sqrt{b\rho g / 2h}$ b) $v = \sqrt{b\rho g / h}$

c) $v = \sqrt{b\rho / 2h}$ d) $v = \sqrt{2b\rho g / h}$

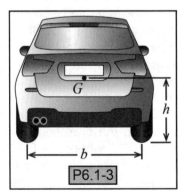

P6.1-3

P6.1-4

P6.1-4) A RWD 2011 Corvette Coupe's technical specifications are listed below. Assuming that the rolling resistance coefficient is f_r = 0.014, what is the car's maximum acceleration on level ground? Estimate the friction characteristics as that between rubber and dry asphalt and note that the total rolling resistance is equal to $R_r = f_r W$.

- Curb weight: W = 3175 lb
- Wheel base: L = 105.7 in (Distance between the axles.)
- CG height: h = 19.8 in
- Rear wheel drive
- Weight distribution: 51/49 f/r (%)

Ans: a = 16.54 ft/s^2

P6.1-5) The combined mass of a man and the personal transport device he is riding is 110 kg. The wheel of the device has a radius of 0.3 meters and the center of mass G of the man-device system is 1.1 meters from the center of the wheel O. The man wishes to accelerate forward at a constant 3 m/s^2 while maintaining a constant incline. Determine the incline angle θ and the traction force F that will provide the conditions desired by the rider.

Ans: F = 330 N, θ = 21.6°

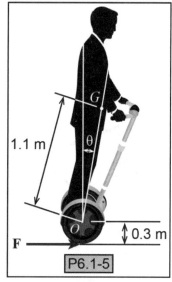

P6.1-5

P6.2) INTERMEDIATE LEVEL PURE TRANSLATION PROBLEMS

P6.2-1) A woman pushes a 60-lb lawn mower with a force of P in the direction shown in the figure (θ = 51°). Determine the maximum acceleration that the lawn mower can attain before the front wheel lifts off the ground and the associated value of P. The rolling resistance between the tires and the grass is $R_r = f_r W$, where f_r = 0.1 and W is the weight of the lawn mower. Neglect the rotational inertia of the wheels.

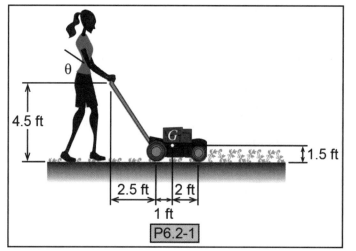

P6.2-1

Ans: a = 421.1 ft/s^2, P = 1256.7 lb

P6.2-2) An automobile of weight W is accelerating in a straight line with acceleration a. The wheel base of the automobile is L and its center of mass is located l_1 behind the front axle and h above the ground. Determine the dynamic load transfer. This is the amount of load that is transferred from the front tire to the rear tire during acceleration.

Ans: Dynamic load transfer = Wha/gL

P6.2-3) A RWD 2011 Corvette Coupe's technical specifications are listed below. Assuming that the air drag is constant and equal to 100 lb and the rolling resistance coefficient is $f_r = 0.014$, what is the car's maximum acceleration on level ground and driving uphill on a road with a 10% grade? Estimate the friction characteristics as that between rubber and dry asphalt and note that the total rolling resistance is equal to $R_r = f_r W$.

P6.2-3

- Curb weight: W = 3175 lb
- Wheel base: L = 105.7 in (Distance between the axles.)
- CG height: h = 19.8 in
- Rear wheel drive
- Weight distribution: 51/49 f/r (%)

Ans: a = 12.24 ft/s^2 at 10% grade, a = 15.5 ft/s^2 at 0% grade

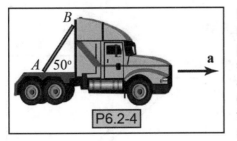

P6.2-4

P6.2-4) A truck is accelerating forward with a beam resting against the bed and the cab of the truck at points A and B, respectively. If the coefficient of static friction between the bed and the beam is 0.5 and the surface of the cab is very smooth, determine the maximum horizontal acceleration the truck may experience before the beam begins to slip.

P6.3) ADVANCED LEVEL PURE TRANSLATION PROBLEMS

P6.3-1) A RWD 2011 Corvette Coupe's technical specifications are listed below. What is the car's maximum acceleration on level ground and driving uphill on a road with a 10% grade at 30 mph? Note that the total rolling resistance is equal to $R_r = f_r W$ and the total aerodynamic drag is equal to $R_a = (\rho / 2)C_D A_f v_r^2$, where ρ is the density of air and

P6.3-1

v_r is the speed of the car relative to the wind. Assume still air. Assume that the aerodynamic drag force is applied at the height of the CG. Estimate the friction characteristics as that between rubber and dry asphalt.

- Curb weight: W = 3175 lb
- Wheel base: L = 105.7 in (Distance between the axles.)
- CG height: h = 19.8 in
- Rear wheel drive
- Weight distribution: 51/49 f/r (%)
- Rolling resistance coefficient: f_r = 0.014
- Aerodynamic drag coefficient: C_D = 0.29
- Frontal area: A_f = 22.3 ft^2

Ans: a = 16.39 ft/s^2 at 0% grade, a = 13.1 ft/s^2 at 10% grade

P6.4) BASIC LEVEL PURE ROTATION PROBLEMS

P6.4-1)[fe] A 10-kg uniform slender rod of length 0.8 m is released from rest at an angle of θ = 40°. Determine the angular acceleration of the rod at θ = 40° and θ = 90°.

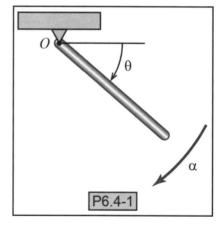

P6.4-1

a) α = 56.36 rad/s^2 at 40°, α = 0 at 90°
b) α = 25.01 rad/s^2 at 40°, α = -3.15 rad/s^2 at 90°
c) α = 14.09 rad/s^2 at 40°, α = 0 at 90°
d) α = 46.26 rad/s^2 at 40°, α = 1.25 rad/s^2 at 90°

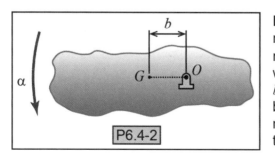

P6.4-2

P6.4-2)[fe] A non-uniform body is released from rest in the position shown in the figure. The resulting angular acceleration is $\alpha = g/b$, where g is the acceleration due to gravity and b is the distance from the fixed axis to the body's center of mass. Determine the reaction force of the support at O in terms of the body's weight W.

a) $O = 0$ b) $O = W$ c) $O = 2W$ d) $O = 3W$

P6.4-3) A slender rod, pinned at O, is released from rest in the position shown. Determine the distance of the pinned joint O from the center of mass G that would result in an angular acceleration of $\alpha = 3g^2 / L^2$.

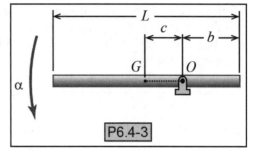

P6.4-3

Ans: $c = L^2 / 6g$

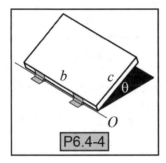

P6.4-4

P6.4-4) The trapped door shown is installed in a horizontal floor. What angle θ must the trapdoor be released from in order for the acceleration of its center of mass to be $g/2$, where g is the acceleration due to gravity?

Ans: $\theta = 48.2°$

P6.5) INTERMEDIATE LEVEL PURE ROTATION PROBLEMS

P6.5-1) The pendulum impact tester shown consists of a pendulum that rotates freely about the pivot O. The pendulum is made up of a sender rod of length L = 800 mm and mass 5 kg and a block at the end of the rod. The block has the dimensions: a = 250 mm and b = 200 mm with a mass of 30 kg. Plot the reaction forces at O and the velocity of the block's center of mass as a function of θ from pendulum release ($\theta = 0$ and $\omega = 0$) to impact ($\theta = 90$ degrees).

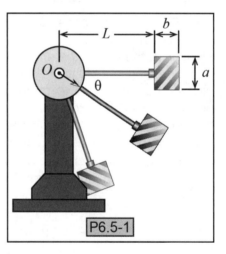

P6.5-1

Ans: $O_t = 20.2\cos\theta$ N, $O_n = 989.4\sin\theta$ N,
$v_{Gb} = 4.25(\sin\theta)^{0.5}$ m/s

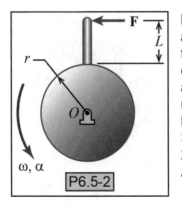

P6.5-2

P6.5-2) A disk rotates about its center under the influence of a force applied to a handle attached to the disk as shown in the figure. This force causes the body to rotate with a constant angular acceleration of 0.2 rad/s^2. Both the disk and handle are made of 6061 aluminum. The disk has a radius of 1 foot and a thickness of 5/8 inch. The handle may be approximated with a rod 11 inches long with a diameter of 3/8 inch. If, at the instant shown, the angular velocity is 2 rad/s, determine the support reaction at O. See Appendix C for density values.

Ans: O = 27.51 lb

P6.5-3) The 2-kg uniform slender rod in the given figure is initially at rest in the position shown when a horizontal force $F = 6$ N is applied to the end of the rod. Determine the distance d that will result in the horizontal component of the reaction force at O being zero. Also determine the associated angular acceleration of the rod for these conditions.

Ans: $d = (1/6)$ m, $\alpha = 18$ rad/s^2 ccw

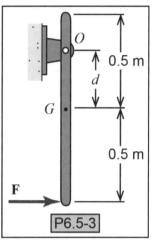

P6.5-3

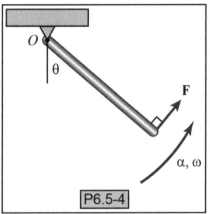

P6.5-4

P6.5-4) A 10-kg uniform slender rod of length 0.8 m is at rest when $\theta = 0°$ when a 100-N force is applied to the end of the rod. The force remains perpendicular to the rod throughout its motion. Determine the support reaction force at O when $\theta = 40°$.

Ans: $O = 257.7$ N

P6.5-5) A control pedal for an airplane is modeled as shown. Consider the lever as a massless shaft and the pedal as a lumped mass m at the end of the shaft. A spring of stiffness k and a damper with damping c are attached to the lever as shown to provide the pilot proper "feel" and to return the pedal to its equilibrium vertical position ($\theta = 0$) where the spring is unstretched. You may assume that the displacement θ is small such that the spring and damper can be approximated as moving horizontally. Find the differential equation of motion for this system in terms of θ and its derivatives for the case that an external horizontal force F is applied to the pedal.

Ans: $ml^2\ddot{\theta} + cl_2^2\dot{\theta}\cos^2\theta + kl_1^2\theta\sin\theta\cos\theta + mgl\sin\theta = Fl\cos\theta$

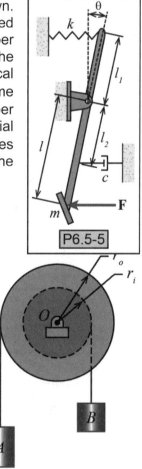

P6.5-5

P6.5-6) A 7-lb wheel consisting of an inner and outer hub is mounted on a shaft. A 5-lb load B is attached to the inner hub and a 15-lb load A is attached to the outer hub through an inextensible cable as shown. The bearing that allows the wheel to rotate on the shaft is at the end of its useful life and creates a constant resistive friction moment of 3 lb-ft. Find the angular acceleration of the wheel. The wheel properties are: $r_o = 1$ ft, $r_i = 0.75$ ft and $I_o = 0.1$ slug-ft^2.

Ans: $\alpha = 0$

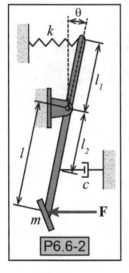

P6.5-7

P6.5-7) A flywheel rotates freely on a shaft located at its center (point O). The flywheel has a mass moment of inertia about its mass center of $I_o = 0.2$ slug-ft^2 and is attached to two masses as shown in the figure. Mass A ($m_A = 50$ kg) is attached to the inner radius ($r_i = 10$ cm) of the flywheel though an inextensible rope. Mass B ($m_B = 20$ kg) slides along a rough horizontal surface ($\mu_k = 0.3$) and is attached to the flywheel's outer radius ($r_o = 20$ cm) through an inextensible rope. If the flywheel is released from rest, calculate the wheel's angular velocity after mass B has moved 10 cm.

P6.6) ADVANCED LEVEL PURE ROTATION PROBLEMS

P6.6-1) A 10-kg uniform slender rod of length 0.8 m is released from rest at an angle of $\theta = 40°$. Determine the support reaction at O when the rod is at $\theta = 40°$ and $\theta = 90°$.

Ans: $O = 65.8$ N at $40°$, $O = 150.7$ N at $90°$

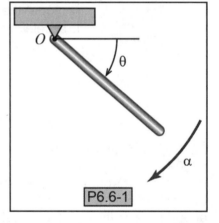

P6.6-1

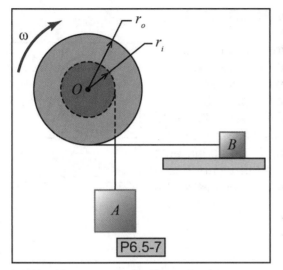

P6.6-2

P6.6-2) A control pedal for an airplane is modeled as shown. The lever may be treated as a massless shaft and the pedal as a lumped mass $m = 0.2$ kg at the end of the shaft where $l = 13$ cm. A spring of stiffness $k = 50$ N/m and a damper with damping $c = 20$ Ns/m are attached as shown where $l_1 = 6$ cm and $l_2 = 7$ cm. The spring is unstretched when the lever is in the vertical position ($\theta = 0$). The differential equation that governs this system is nonlinear because of the presence of $\sin\theta$ and $\cos\theta$ terms. Nonlinear differential equations are very difficult to solve analytically. In order to approximate the governing equation as a linear differential equation, assume that $\sin\theta \approx \theta$ and $\cos\theta \approx 1$. This approximation is accurate for small angles. Find the linearized version of the differential equation for this system and determine the angular position of the pedal as a function of time $\theta(t)$ when the pilot pumps the peddle with a horizontal force described by the function $F(t) = \sin(t)$ where F is in Newtons and the peddle is initially vertical and at rest.

Ans: $\theta(t) = 0.069e^{-5.47t} - 0.0038e^{-23.53t} + 0.287\sin(t) - 0.065\sin(t)$

P6.7) BASIC LEVEL GENERAL MOTION PROBLEMS

P6.7-1) Consider the 5-kg disk shown rolling to the right under the influence of a pulling force (P = 10 N). Typically when a wheel is rolling on a dry surface without being driven by an external torque, we can neglect slip at the contact surface and the friction force $\mathbf{F}_f$ will resist the disk's motion. Using this fact, estimate the angular acceleration α of the disk as well as the acceleration of its mass center $\mathbf{a}_G$. Assume that the coefficients of friction are μ_s = 0.45 and μ_k = 0.35 and that the wheel has a radius of 1.0 m. Verify that the disk does not slip relative to the ground.

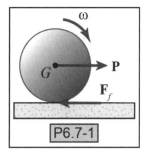

P6.7-1

Ans: α = 1.33 rad/s^2, a_G = 1.33 m/s^2

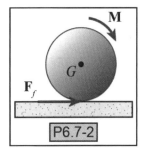

P6.7-2

P6.7-2) Consider the 5-kg disk shown rolling to the right under the influence of a moment (M = 50 Nm). When driven in this manner, the disk typically will experience slip and the friction force $\mathbf{F}_f$ will propel the wheel forward. Using this fact, estimate the angular acceleration α of the disk as well as the acceleration of its mass center $\mathbf{a}_G$. Assume that the coefficients of friction are μ_s = 0.45 and μ_k = 0.35 and that the wheel has a radius of 1.0 m. Very that the disk slips relative to the ground.

Ans: α = 13.13 rad/s^2, a_G = 3.43 m/s^2

P6.7-3)   Consider the rear-wheel drive, 1200-kg vehicle shown.   It has been determined that the mass moment of inertia of the rear wheels with drive train is I_C = 5 kg-m^2 about the axis through the wheel centers. The wheel radius equals 35 cm. If the vehicle is accelerating forward at 5 m/s^2 on level ground and the rear wheels are assumed to roll with

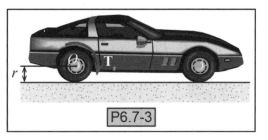

P6.7-3

negligible slip, determine the shaft torque $\mathbf{T}_s$ being applied to the rear wheels. You may assume vehicle air drag and the rolling friction at the front wheels is negligible.

Ans: T_s = 2171.4 N

P6.8) INTERMEDIATE GENERAL MOTION PROBLEMS

P6.8-1) A wood spool has a rope wrapped around its inner hub and rests on a cast iron track as shown in the figure. The spool has a radius of gyration about its mass center of 3.2 ft and a weight of 100 lb. The spool, starting from rest, has its rope pulled by a

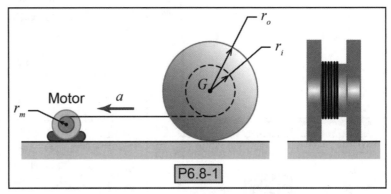

motor at 1 ft/s^2. Determine the tension in the rope and the frictional force between the spool and the track at this instant. Estimate the frictional characteristics as that between oak and cast iron and the kinetic coefficient is 80% of the static coefficient of friction. The physical parameters of the spool are as follows: r_o = 3 ft and r_i = 1.5 ft. The rope unwinds without slipping.

Ans: F_f = 27.4 lb, T = 29.5 lb

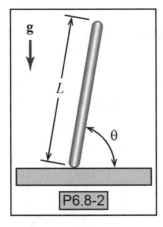

P6.8-2) A uniform, rigid rod of mass m is placed on a frictionless surface. The rod is released from rest from the shown position at t = 0. Determine the rod's equation of motion and the rod's initial angular acceleration if L = 1 m, $\theta(0) = 80^o$ and $\omega(0) = 0$. Assume that the end of the rod stays in contact with the ground.

Ans: α_o = 4.94 rad/s^2

P6.8-3) The 20-lb unbalanced disk shown has a center of mass G that is offset from its geometric center O. The wheel has a centroidal radius of gyration about an axis through G of k_G = 1 ft. If the wheel is released from rest from the position shown, determine the angular acceleration of the disk at this instant. The coefficients of static and kinetic friction between the disk and the rolling surface are μ_s = 0.45 and μ_k = 0.35, respectively.

Ans: α = 6 rad/s^2 ccw

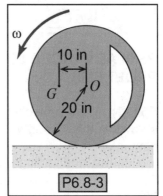

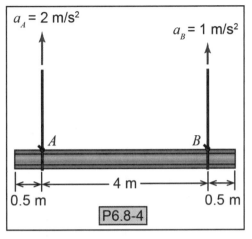

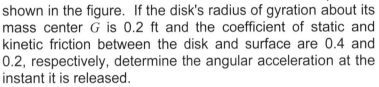

P6.8-4

P6.8-4) A 40-kg beam is being lifted from rest by two cables with accelerations as shown. Determine the tension in each of the two cables for the instant shown.

Ans: T_A = 231.4 N, T_B = 221.0 N

P6.8-5) A 35-lb non-uniform disk of radius 0.5 ft rests on a horizontal surface. The disk is released from rest in the position shown in the figure. If the disk's radius of gyration about its mass center G is 0.2 ft and the coefficient of static and kinetic friction between the disk and surface are 0.4 and 0.2, respectively, determine the angular acceleration at the instant it is released.

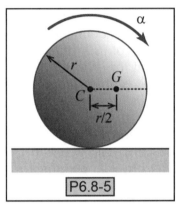

P6.8-5

Ans: α = 15.3 rad/s^2 ccw

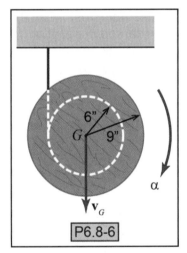

P6.8-6

P6.8-6) The 20-lb spool shown is hung by the ceiling with a rope that wraps around its inner radius. The spool is released from rest. Find the angular acceleration of the spool if it has a radius of gyration about its mass center of 4 in. Assume that the rope remains vertical and it unwinds without slipping.

Ans: α = 44.64 rad/s^2

P6.9) ADVANCED LEVEL GENERAL MOTION PROBLEMS

P6.9-1) The connecting rod AB of the piston-crank assembly shown weighs 2.4 lb with a radius of gyration about the mass center of 2.2 in. The piston assembly that is attached to the top of the connecting rod has a combined weight of 3.6 lb. Determine the piston pin reaction force at B for $\theta = 0$ and 20 degrees if link OA rotates at a constant angular velocity of 2700 rpm.

Ans: B_x = 0 and B_y = 4465.3 lb at $\theta = 0$,
  B_x = 1049.4 lb and B_y = 3994.5 lb at $\theta = 20°$

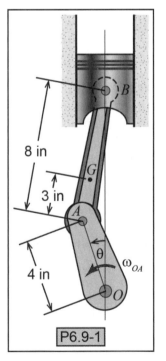

P6.9-1

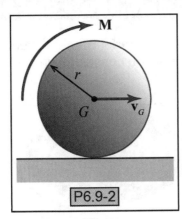

P6.9-2

P6.9-2) A 30-lb cylinder of radius 0.5 ft rolls along a horizontal surface under the influence of a moment. The moment causes the cylinder to roll with an angular speed described by $\omega = 4t^2$. Determine the time at which the cylinder slips if the coefficient of static friction between the cylinder and the surface is 0.4.

Ans: $t = 3.21$ s

P6.9-3) Consider the rear-wheel drive, 1200-kg vehicle shown. It has been determined that the mass moment of inertia of the rear wheels with drive train is $I_C = 5$ kg·m² about the axis through the wheel centers. The vehicle is resting in a simulation laboratory with its rear wheels in contact with a large steel drum. The drum is free to rotate about its central axis through O. The wheel and drum radii are $r_w = 35$ cm and $r_d = 90$ cm,

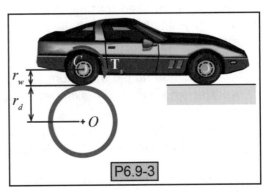

P6.9-3

respectively. The drum is employed to emulate the inertia of the vehicle when translating so that for a given shaft torque $\mathbf{T}_s$ the acceleration of the vehicle's wheels will closely match that experienced on the actual road. Determine the mass moment of inertia of the drum I_O needed to simulate the car accelerating at 5 m/s² on level road assuming there is no slip between the car wheels and the drum. The inertia of the drum may be adjusted by the addition of weights mounted symmetrically within the drum.

Ans: $I_O = 971.2$ kg-m²

CHAPTER 6 COMPUTER PROBLEMS

C6-1) The 20-lb spool shown is hung by the ceiling with a rope that wraps around its inner radius. The spool is released from rest. Find the tension in the rope and the speed of the spool's center of mass 10 seconds after it has been released. The spool has a radius of gyration about its mass center of 4 in. Assume that the rope remains vertical and unwinds without slipping.

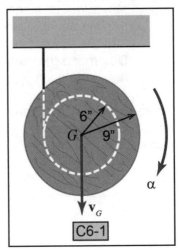

C6-1

CHAPTER 6 DESIGN PROBLEMS

D6-1) Locate a product that uses a motor driven mechanism to perform its function. Design the motor that drives the mechanism. List everything that you considered in your calculations such as material, dimensions and assumptions.

D6-2) Consider the "tea-cup" amusement park ride shown. Each cup has an occupancy limit of three riders. The cups are rotated with respect to the platform by the riders through a wheel at the center of the teacup. The platform itself is turned by a large electric motor. Use your knowledge of dynamics to address the following issues; a) it is desired that the sustained acceleration of a rider should be less than $3g$ and b) the peak acceleration experienced by a rider should be less than $5g$.

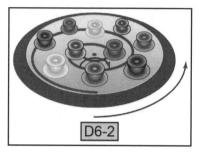

D6-2

- Recommend the maximum angular speed with which the platform can rotate.
- Recommend the maximum angular acceleration with which the platform can rotate.
- Recommend the value of the torque required by the electric motor driving the platform considering your previous two recommendations.

Note: There may be information that you need to complete your design which you do not have. Explain how you would obtained estimates of any missing information that you need.

D6-3) The design of the satellite shown is nearing completion. A few design decisions that must still be made are listed below. Use your knowledge of dynamics to address the following issues. Keep in mind that the life of the satellite is limited by the amount of fuel that can be carried on board the vehicle at the time of launch.

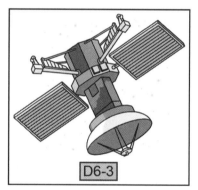

D6-3

- Recommend where, within the satellite structure, to place a very heavy battery pack.
- Recommend where, on the surface of the satellite structure, to place a series of thrusters for (a) moving the position of the satellite within its orbit, and (b) for reorienting the attitude of the satellite.
- Determine the amount of thrust necessary from each thruster in order to achieve any orientation of the satellite in less than 60 seconds.

Note: There may be information that you need to complete your design which you do not have. Explain how you would obtained estimates of any missing information that you need.

D6-4) Consider that you are designing a hybrid-electric bicycle. The design of the bicycle itself is set, but you need to size the motor/generator for the drive system. Assume that the bicycle frame and components weigh 20 lb, while the 26-in diameter front wheel weighs 5 lb. The rear-wheel assembly has the same diameter as the front, but weighs 10 lb because of the additional weight of the motor/generator system. The centroidal radius of gyration of the rear-wheel assembly is estimated to be approximately 0.2 slug-ft^2. It is required that the bike be able to achieve a top speed of 15 mph from rest in a span of 15 seconds under entirely electric propulsion on level ground. Estimate the average sustained torque the motor must be able to generate. Assume that the bicycle must be able to accommodate a 240-lb rider and that the motor applies torque directly to the wheel (no gearing). If there is any information that is missing that you may need, describe how you might obtain that information.

CHAPTER 6 ACTIVITIES

A6-1) Explore the concept of the center of gravity using household items.

Group size: 1 student

Supplies:

- Drinking glass
- 2 forks
- 1 coin (preferably a dollar coin or a quarter)

Procedure:

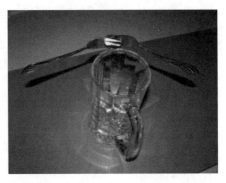

1. Place the glass on a level surface.
2. Link the tines of the forks onto the coin.
3. Place the edge of the coin on the glass as shown in the figure.
4. Move the ends of the forks in and out until the coin balances on the edge of the glass by itself.

Calculations and discussion

- Draw a schematic of the setup.
- On you drawing identify the weights and centers of gravity for each item.
- Determine, mathematically, where the location of each items' center of gravity is. Does this match what you found experimentally?
- Label the system's center of gravity on your schematic. The system includes the two forks and coin.
- List all of the assumptions that were applied in your calculations.
- List possible sources of experimental error.

CHAPTER 6 REFERENCES

[1] Horace Lamb, "*Dynamics*", Cambridge University Press, 1960.

[2] Wong, "*Theory of Ground Vehicles, 4th ed.*", Wiley, 2008, ISBN: 978-0-470-17038-0

PART IV: KINETICS - WORK AND ENERGY

CHAPTER 7: PARTICLE WORK AND ENERGY

CHAPTER OUTLINE

CHAPTER SUMMARY

Previously, we used Newtonian mechanics to analyze the motion of particles and rigid bodies. Newton's laws gave an instantaneous relationship between the resultant force on a body and its acceleration. If it is desired to examine what happens to the body over time or distance, then kinematic relationships were employed. In this chapter, an alternative approach to Newtonian mechanics will be introduced. This approach is referred to as the *work-energy* method. It examines the cumulative effect of the resultant force over a distance. Specifically, this chapter addresses the work-energy method as it applies to particles. The subsequent chapter will apply this method to rigid bodies.

For some situations, the work-energy method can efficiently determine the displacement and speed of a body. However, the work-energy method is not very useful for determining a body's acceleration or the direction of its velocity. The main idea with an energy approach is to analyze how a body's energy changes as it moves. The amount and form of a body's energy will change due to the body either doing work or due to work being done on the body.

Work is a concept that has meaning in our everyday lives. Everyone has done "work." We "work" for a living, we "work"out and we "work" hard. This is not exactly what we will be talking about here. If you are doing physical "work" like pushing or lifting something, then you are doing the type of work that we will learn how to calculate in this chapter. In the study of dynamics, work has a very specific mathematical definition that is given by Equation 7.1-1. This equation gives the most general equation for work; however, work in its most basic form is simply a *force* applied over a *distance* ($U = Fd$). So if you push a box across the floor from the living room to the kitchen, then you are doing work. However, if the box is very heavy and you push and push and sweat and sweat and the box does not move, then you have done no work. Even though you are exhausted, the box did not move and, therefore, the force you applied to the box did not do any work.

Work: $$U_{1\text{-}2} = \int_{\mathbf{r}_1}^{\mathbf{r}_2} \mathbf{F} \cdot d\mathbf{r}$$ (7.1-1)

$U_{1\text{-}2}$ = work done by $\mathbf{F}$ from $\mathbf{r}_1$ to $\mathbf{r}_2$ (work is a scalar)
$\mathbf{F}$ = force vector
$\mathbf{r}$ = position vector

Work is the amount of energy transferred by a force acting through a distance.

One definition of **work** is "The amount of energy transferred by a force acting through a distance." What does that mean in the context of dynamics? If a force is applied to a particle and the force causes that particle to move through a distance, the force has done work. This also means that the force has transferred some energy to (or from) the particle.

Units of Work
SI units:
- Newton–meter [N-m]
- Joule [J = 1 N-m]
- erg [erg = 1 x 10^{-7} J]

US customary units:
- foot–pound [ft-lb = 1.3558 J]
- kilocalorie [kcal = 4187 J]
- British thermal unit [Btu = 778.16 ft-lb = 1055 J]

Energy is the capacity of a particle to do work, which makes sense. If you put energy into a particle, then that particle now has the ability to do work. A particle can possess different forms of energy. The two forms related to dynamics that we will be discussing are the energy of motion (kinetic energy) and the energy of position (potential energy). So what is the difference between work and energy? Work and energy possess the same units (Joules) and they are related in that work is the process of a force transferring energy, while energy is a measure of the ability of a system to do work.

Work doesn't always have to add energy to a system, it can also take energy from a system. Consider that you are driving a car down the road at 70 mph. The car has the energy of its motion (kinetic energy). A deer crosses your path and you slam on the brakes causing you to skid. The friction force that develops between your tires and the road slows your car to an eventual stop. The friction force is applied over a distance (the stopping distance), therefore, it does work. The work done takes energy out of the car, which means that the work is negative.

Now that we have a physical understanding of what work is, let's go back and look at Equation 7.1-1 and attempt to gain a mathematical understanding of the work equation. Equation 7.1-1 states that the work U done by an external force $\mathbf{F}$ being applied to the particle is equal to the integral of the dot product of $\mathbf{F}$ with the differential change in position $d\mathbf{r}$. This most general form of the equation allows for the force to be a function of position and applied at an angle to the displacement. The force $\mathbf{F}$ in Equation 7.1-1 may be a single force or the summation of several forces. If you don't recall what a dot product is, refer to Appendix B. One interpretation of the dot product $\mathbf{F} \cdot d\mathbf{r}$ is that it is the product of the magnitude of $d\mathbf{r}$ with the magnitude of the vector $\mathbf{F}$ projected onto vector $d\mathbf{r}$. This explanation will make more sense after considering a couple of examples. Note that the result of a dot product of two vectors is a scalar; therefore, the work U done by a force is also a scalar.

7.1.1) WORK DONE BY A CONSTANT FORCE

Equation 7.1-1 gives the most general equation for calculating the work done by a force. To gain a better understanding of how to use this equation, we will examine several special cases. We will begin with an examination of work done by a constant force $\mathbf{F}$. This force will be constant in both magnitude and direction. This special case will help provide some insight into the notion of work. Since force $\mathbf{F}$ is constant, it can be factored out of the integral in Equation 7.1-1 to arrive at Equation 7.1-2. Equation 7.1-2 may be used whenever the force under consideration does not change as a function of position either in direction or magnitude.

$$U_{1-2} = \int_{\mathbf{r}_1}^{\mathbf{r}_2} \mathbf{F} \cdot d\mathbf{r} = \mathbf{F} \cdot \int_{\mathbf{r}_1}^{\mathbf{r}_2} d\mathbf{r} = \mathbf{F} \cdot (\mathbf{r}_2 - \mathbf{r}_1)$$

Work done by a constant force: $\boxed{U_{1-2} = \mathbf{F} \cdot \Delta\mathbf{r}_{1-2}}$ (7.1-2)

$U_{1\text{-}2}$ = work done by $\mathbf{F}$ from $\mathbf{r}_1$ to $\mathbf{r}_2$ (work is a scalar)
$\mathbf{F}$ = force vector
$\Delta\mathbf{r}_{1\text{-}2}$ = displacement vector

The expression for work given in Equation 7.1-2 may also be expressed in terms of the angle between the force and the direction of motion. The known angle allows you to project the force onto the displacement vector. Consider Figure 7.1-1 where force $\mathbf{F}$ pushes a block along a horizontal surface. We learned previously that work only occurs if a force manages to

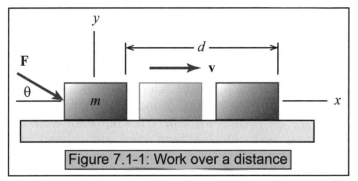

Figure 7.1-1: Work over a distance

move an object. The force shown in Figure 7.1-1 is moving the block, but the block is constrained to move only in the x-direction and has no motion in the y-direction. Therefore, only the x-component of force $\mathbf{F}$ does work. In other words, only the component of the force in the direction of the displacement does work. We can compute the work done by force $\mathbf{F}$ using only the component of the force that is in the direction of the displacement ($F \cos\theta$) as shown in Equation 7.1-3. The component $F \cos\theta$ can also be thought of as the force projected onto the direction of the displacement. Based on this logic, if a force is perpendicular to the direction of motion, then it does not do work and the work done by a force that is opposite the displacement is negative.

Work done by a constant force: $\boxed{U_{1-2} = (F\cos\theta)\, d_{1-2}}$ (7.1-3)

U_{1-2} = work done by $\mathbf{F}$ from position 1 to 2 (work is a scalar)
F = force magnitude
θ = angle between $\mathbf{F}$ and the direction of motion
d_{1-2} = displacement magnitude

7.1.2) WORK DONE BY A NON-CONSTANT FORCE

When determining the work done by a force that is not constant with respect to the position $\mathbf{r}$, then the force cannot be factored out of the integral and the integral must be evaluated. One way to simplify this type of calculation is to integrate the individual coordinate directions separately. In general, the dot product can be performed first so that the work can be calculated by carrying out multiple scalar integrals. For example, consider the following where the force and the displacement are expressed in rectangular coordinates.

$$U_{1-2} = \int_{\mathbf{r}_1}^{\mathbf{r}_2} \mathbf{F}\cdot d\mathbf{r} = \int (F_x\mathbf{i} + F_y\mathbf{j})\cdot(dx\mathbf{i} + dy\mathbf{j}) = \int (F_x\,dx) + (F_y\,dy) = \int_{x_1}^{x_2} F_x\,dx + \int_{y_1}^{y_2} F_y\,dy$$

7.1.3) WORK IN THE n-t COORDINATE SYSTEM

Another interesting example is when the force and the displacement are expressed in n-t coordinates. In this situation, the displacement s of the particle along the path is always in the tangential direction. Therefore, the normal component of the force does not do work and there is only a single scalar integral that must be evaluated.

$$U_{1-2} = \int_{\mathbf{r}_1}^{\mathbf{r}_2} \mathbf{F}\cdot d\mathbf{r} = \int_{s_1}^{s_2} (F_n\,\mathbf{e}_n + F_t\,\mathbf{e}_t)\cdot(ds\,\mathbf{e}_t) = \int_{s_1}^{s_2} F_t\,ds$$

Example 7.1-1

Choose the correct work equation for each of the following situations.

$U_{1-2} = \int_{\mathbf{r}_1}^{\mathbf{r}_2} \mathbf{F}\cdot d\mathbf{r}$ $\qquad$ $U_{1-2} = \int_{r_1}^{r_2} F\,dr$ $\qquad$ $U_{1-2} = \mathbf{F}\cdot\Delta\mathbf{r}_{1-2} = (F\cos\theta)\,d_{1-2}$ $\qquad$ $U_{1-2} = F\,d_{1-2}$

Work Equation	F is constant	F is in the direction of r
	Yes	No
	Yes	Yes
	No	No
	No	Yes

Example 7.1-2

Consider a block being pushed along a frictionless surface from position x_1 to position x_2.

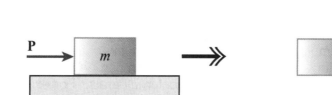

Answer the following questions about this problem.
- Draw a free-body diagram of the block.

Answer the following questions about this problem.
- What is the resultant force $\mathbf{F}$ that acts on the particle?
- What is the work done by $\mathbf{F}$ on the particle?

Which of the following statements are true?

a) The work done by a force $\mathbf{F}$ acting at an angle to the displacement $\Delta\mathbf{x}$ is equal to $U = F\Delta x$.

b) The work done by a force $\mathbf{F}$ pointing in the direction of the displacement $\Delta\mathbf{x}$ is equal to $U = F\Delta x$.

c) The work done by a force $\mathbf{F}$ that is perpendicular to the direction of the displacement $\Delta\mathbf{x}$ is equal to zero.

d) The work done by a force $\mathbf{F}$ pointing in the direction of the displacement $\Delta\mathbf{x}$ is negative.

Example 7.1-3

Consider a block being pulled up a rough slope from position x_1 to position x_2.

Answer the following questions about this problem.
- Draw a free-body diagram of the block.

Answer the following questions about this problem.
- What is the resultant force **F** in the direction of the displacement? What is the work done by **F** on the block?

Which of the following statements are true?

a) The weight force of an object never does work.
b) Only the component of the force in the direction of the displacement does work.
c) The work done by a force **F** pointing in the opposite direction of the displacement Δx is negative.

After completing Examples 7.1-2 and 7.1-3, the following notes about the calculation of work are now apparent.

> **Useful facts about *work***
> - If there is no displacement, there is no work done.
> - Only the component of the force that is parallel to the displacement does work.
> - A force that is perpendicular to the displacement doesn't do any work.
> - If the force and the displacement are in the same direction, then the resultant work is positive.
> - If the force and the displacement are in opposing directions, then the resultant work is negative.

Some useful terminology regarding forces related to work include: forces that do work are often called **active forces** and forces that don't do work are often called **constraint forces**.

Conceptual Example 7.1-4

Consider the following situations where a constant force **P** is applied to a block over a distance d. Rank the following situations in terms of the work done by force **P** from greatest to least.

Greatest: _____ Next: _____ Next: _____ Next: _____ Next: _____ Least: _____

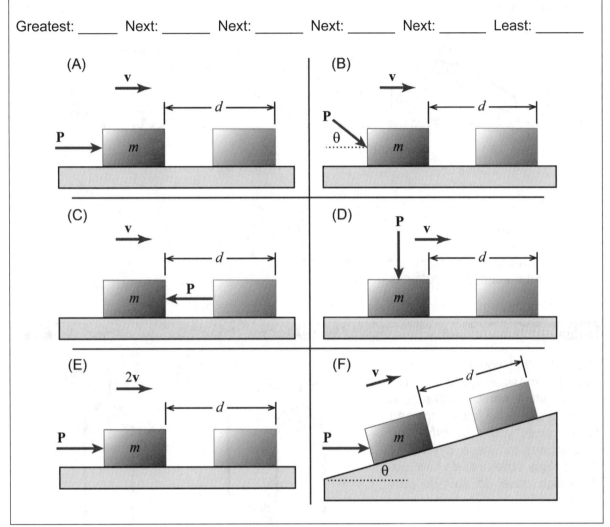

Conceptual Example 7.1-5

You would like to lift a block of mass m from the floor to a height d at a constant rate by applying a constant force. You have the opportunity to use a simple machine (a frictionless ramp) (Case (a)), or you can simply lift the block straight up (Case (b)). In which case do you do less work?

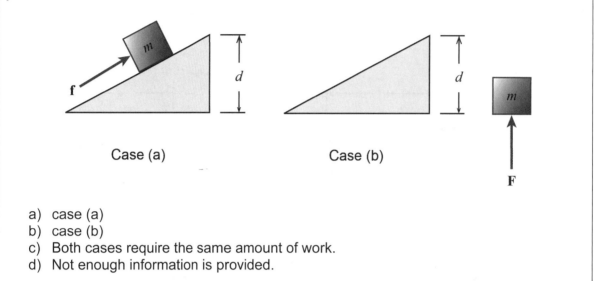

Case (a)   Case (b)

a) case (a)
b) case (b)
c) Both cases require the same amount of work.
d) Not enough information is provided.

Conceptual Example 7.1-6

Consider that you again have a block of mass m that you would like to lift a vertical distance d using a constant force. Instead of a ramp, you now have the opportunity to use a system of frictionless pulleys to do this job. In which case do you do less work?

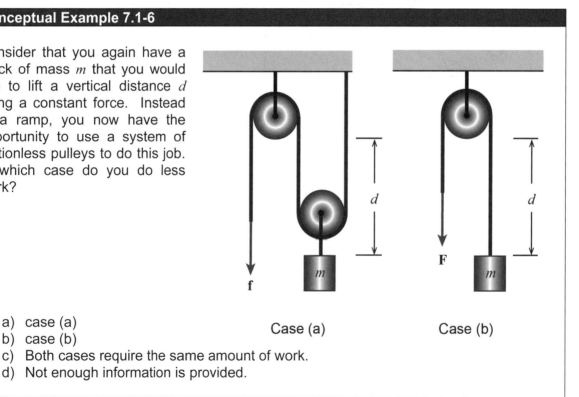

Case (a)   Case (b)

a) case (a)
b) case (b)
c) Both cases require the same amount of work.
d) Not enough information is provided.

Equation Summary

Abbreviated variable definition list

U_{1-2} = work done by $\mathbf{F}$ from $\mathbf{r}_1$ to $\mathbf{r}_2$ (work is a scalar)
$\mathbf{F}$ = force
$\mathbf{r}$ = position
d_{1-2} = displacement magnitude
θ = angle between $\mathbf{F}$ and the direction of motion

Work

$$U_{1-2} = \int_{\mathbf{r}_1}^{\mathbf{r}_2} \mathbf{F} \cdot d\mathbf{r} \qquad U_{1-2} = \mathbf{F} \cdot \Delta \mathbf{r}_{1-2} \qquad U_{1-2} = (F \cos\theta)\, d_{1-2}$$

Example Problem 7.1-7

A bushing slides on a shaft that forms a 20°
angle with respect to the x-axis. A constant
force $\mathbf{F} = 5\mathbf{i} + 4\mathbf{j}$ N acts on the bushing.
Determine the work the force performs on the
bushing if it moves a total distance of 2 m
along the shaft.

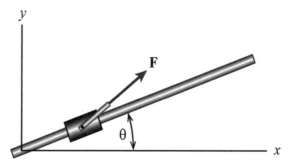

Given:

Find:

Solution:

Example Problem 7.1-8

A cart of mass m is attached to a spring, that is initially un-stretched, and is pulled by force **F** as shown in the figure. The mass moves through a distance d along the incline. Determine the expression for the work done all of the forces acting on the cart. Assume a linear spring with spring constant k and neglect friction.

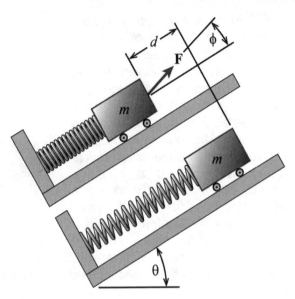

Given:

Find:

Solution:

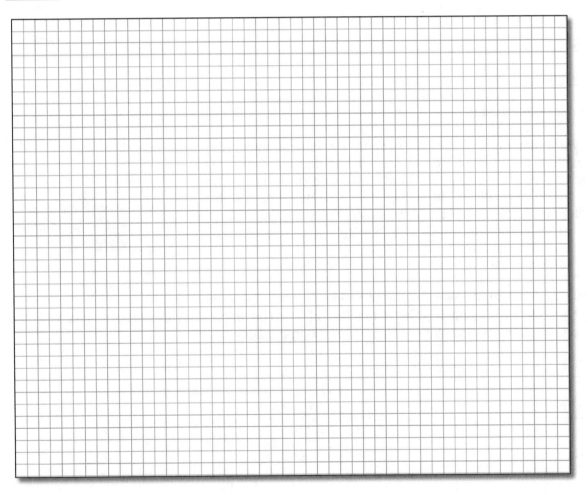

Solved Problem 7.1-9

A 500-lb crate is pulled along the ground for a distance of 10 ft. A pulley and rope system is used to pull the crate as shown in the figure. The angle (θ) that the rope makes with the horizontal changes as the crate moves along the floor. If the pulling force, **P**, is applied such that the crate travels along the floor at a constant speed, determine the work done by the tension force and the work done by the friction force between the crate and the floor. The crate is made of wood and it slides along a concrete floor. Estimate the kinetic friction coefficient as being 80% of the static value found in the appendix. The starting distances depicted in the figure are: d = 25 ft and h = 15 ft.

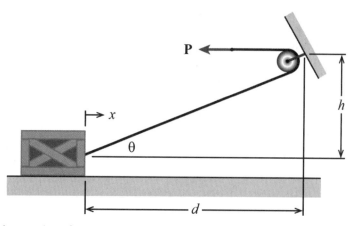

Given: W = 500 lb Δx = 10 ft
         d = 25 ft h = 15 ft
         μ_k = 0.5 (This value is obtained from Appendix C.)
         v = constant

Find: U_P, U_{Ff}

Solution:

Free-body diagram

When determining the work done by forces acting on a particle, it is a good idea to draw a free-body diagram. This allows you to visually see which forces do or do not perform work on the particle. In addition, we also need the free-body diagram to determine the value of **P**. Note that the tension in the rope is constant throughout its length.

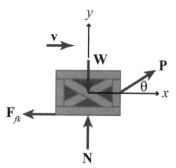

Newton's second law

We need to apply Newton's second law in both the x- and y-direction to determine P as a function of θ. The angle that the rope makes with the horizontal will change as the crate moves along the floor.

$$\sum F_y = 0 \qquad\qquad N = W - P\sin\theta \qquad\qquad F_{fk} = \mu_k N = \mu_k(W - P\sin\theta)$$

$$\sum F_x = ma_x = 0 \qquad P\cos\theta = F_{fk} = \mu_k(W - P\sin\theta)$$

$$P = \frac{\mu_k W}{\cos\theta + \mu_k \sin\theta}$$

Geometry

Since the work equation is expressed in terms of the distance traveled and not an angle, we need to determine a relationship between the angle θ and the distance x.

$$\sin\theta = \frac{h}{(h^2+(d-x)^2)^{1/2}} \qquad\qquad \cos\theta = \frac{d-x}{(h^2+(d-x)^2)^{1/2}}$$

Work

Since the pulling force is a function of displacement and not parallel to the displacement, we need to use the most general form of the work equation.

$$U_{1-2} = \int_{r_1}^{r_2} \mathbf{F}\cdot d\mathbf{r}$$

The crate only moves in the x-direction, therefore, the work equation may be simplified to the following.

$$U_P = \int_{x_1}^{x_2} F_x\, dx = \int_0^{\Delta x} P_x\, dx = \int_0^{\Delta x} P\cos\theta\, dx = \int_0^{\Delta x} \frac{\mu_k W(d-x)}{(d-x)+\mu_k h}\, dx$$

The above integral solution may be obtained from a calculus textbook or from a mathematical computer software package.

$$\int \frac{a-bx}{c-dx}\, dx = \frac{(bc-ad)\ln(c-dx)+bdx}{d^2}$$

$$U_P = \int_0^{\Delta x} \frac{\mu_k Wd - \mu_k Wx}{\mu_k h + d - x}\, dx = (\mu_k W(\mu_k h+d)-\mu_k Wd)\ln((\mu_k h+d)-x)+\mu_k Wx\Big|_0^{\Delta x}$$

$$\boxed{U_P = 1810.5 \text{ lb-ft}}$$

We could use the work equation to determine the work done by the friction force. However, if we realize that the crate is in equilibrium, we know that the amount of work being put into the system has to equal the amount of work being taken out of the system.

$$\boxed{U_{F_{fk}} = -1810.5 \text{ lb-ft}}$$

7.2) KINETIC ENERGY

7.2.1) ENERGY

Energy is defined as the capacity for doing *work*. It is, therefore, a scalar that has the same units as work and is directly related to work. Specifically, if a particle does work, its energy decreases by the amount of work it has performed. Conversely, if work is done on the particle, then its energy increases. Consider a block on the smooth level ground shown in Figure 7.2-1. Force **P** does positive work and increases the energy of the block by increasing its speed. This specific type of energy, the energy of motion, is referred to as **kinetic energy**.

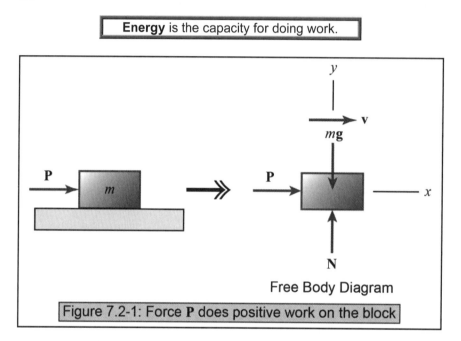

Energy is the capacity for doing work.

Free Body Diagram

Figure 7.2-1: Force **P** does positive work on the block

Example 7.2-1

Consider a block being pushed along a smooth surface as shown in Figure 7.2-1.

Answer the following questions.

1. The applied force P pushes the block causing the block to move, therefore, the force does work on the block. What is the work done by force P if the block travels through a distance of Δx?

2. The work done by the force on the block gets converted to _____.

Example 7.2-1 continued

3. Using Newton's second law (the equation of motion), determine the acceleration of the block.

4. Using the kinematic relationship $\int a\,dx = \int v\,dv$, determine the work done by force P on the block.

7.2.2) KINETIC ENERGY

Kinetic energy is the energy due to an object's motion. If a body has mass and is moving, then it has kinetic energy. The body's motion can be used to perform work. If a body is at rest, then it has no kinetic energy. The mathematical definition of kinetic energy is given by Equation 7.2-1.

Kinetic energy: $$T = \frac{1}{2}mv^2$$ (7.2-1)

T = kinetic energy of a particle
m = particle's mass
v = particle's speed

Kinetic energy is the energy possessed by a body due to its motion.

There are a few things that we can surmise by looking at Equation 7.2-1. First, this equation is not a vector equation, therefore, kinetic energy is a scalar. It has no direction. Second, kinetic energy depends on both mass and speed. Consider two identical balls. One ball is red and the other blue. The red

Units of Kinetic Energy
SI units:
- Newton–meter [N-m]
- Joule [J = 1 N-m]
- erg [erg = 1 x 10^{-7} J]

US customary units:
- foot–pound [ft-lb = 1.3558 J]
- kilocalorie [kcal = 4187 J]
- British thermal unit [Btu = 778.16 ft-lb = 1055 J]

ball is rolling with twice the speed of the blue one. Which one has the higher kinetic energy? The red one is moving faster, therefore, it has more energy. Now let's look at the case where the balls are moving with the same speed, but now the red ball has twice the mass as the blue ball. Which ball has the higher kinetic energy in this case? The red ball has more mass and, therefore, has more energy (i.e. a greater capacity to do work) even though both balls are moving at the same speed.   Note that kinetic energy is always positive. Mathematically, this is due to the fact that mass cannot be negative and that the speed, in Equation 7.2-1, is squared.   More intuitively, a moving body can always perform work regardless of the direction of its velocity. It is also interesting to note that the kinetic energy of a particle captures a particle's current state and not the particle's history. A particle's kinetic energy is based on the particle's current speed and mass, without regard to the force that brought the particle to that state.

To give you a sense of scale, let's compare two familiar things and the kinetic energy that they may possess. A 2010 Volkswagen Jetta sedan (3300 lb) moving at 60 mph possesses 538 kJ of kinetic energy. On the other hand, a flying mosquito has approximately 1.6 erg (i.e. 1.6×10^{-7} J) of kinetic energy.

Work and kinetic energy are related. An alternate definition of **kinetic energy** is "The work needed to accelerate a body of given mass from rest to its current speed." Kinetic energy also can be thought of as the amount of work that can be extracted from a moving particle of given mass in the process of bringing the particle from its current speed to rest.

Equation derivation

If a block is being pushed along a smooth surface by a force **P** as shown in the figure, the work done by force **P** moving the block between x_1 and x_2 can be shown to be equal to the change in kinetic energy. Two relationships are used in the following derivation: Newton's second law $\Sigma F_x = ma_x$ and the kinematic relationship $a_x\,dx = v\,dv$.

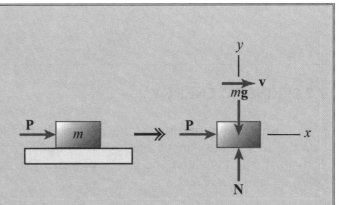

Let's start with the general equation for work and break it up into its x- and y-components. The y-component of the work is zero because the block does not move in the y-direction.

$$U_{1-2} = \int_{r_1}^{r_2} \sum \mathbf{F} \cdot d\mathbf{r} = \int_{x_1}^{x_2} \sum F_x\,dx + \cancel{\int_{y_1}^{y_2} \sum F_y\,dy}$$

We can substitute Newton's second law (i.e $\Sigma F_x = ma_x$) in for the sum of the forces. The only force acting in the x-direction is **P**.

$$U_{1-2} = \int_{x_1}^{x_2} P\,dx = \int_{x_1}^{x_2} ma_x\,dx$$

The acceleration-displacement term may be expressed in terms of the kinematic relationship $a_x\,dx = v\,dv$. Taking the integral gives us an expression that relates work and kinetic energy.

$$U_{1-2} = \int_{x_1}^{x_2} ma_x\,dx = \int_{v_1}^{v_2} mv\,dv = \frac{1}{2}mv_2^2 - \frac{1}{2}mv_1^2 = T_2 - T_1$$

7.3) WORK-KINETIC ENERGY BALANCE

Energy is subject to the law of *conservation of energy*. The law of **conservation of energy** states that energy cannot be created or destroyed, it can only change forms. We have talked about two different quantities that have energy units: work and kinetic energy. We will consider work as the mechanism by which the form of energy is transformed and the process by which energy is transferred between bodies. There are other mechanisms by which energy is transformed, but we will only consider mechanical forms of energy. For example, consider a boy pushing a block along a horizontal surface. The boy doing the pushing does work on the block, decreasing his energy. Where did the energy go and where did the boy get it in the first place? In the process of doing work, the boy converted some energy stored in food and body fat in terms of chemical energy into the motion of his body (i.e. kinetic energy). The kinetic energy of his body was then transferred to the block. A portion of the energy was transferred to the block in the form of kinetic energy and the rest went into overcoming the negative work of friction. The work done against the friction force caused some of the energy to be dissipated as heat and created microscopic changes in the materials at the sliding interface.

The relationship between work and kinetic energy is captured in Equation 7.3-1. In words, this equation states that the work done by all of the forces acting on a particle over a distance will equal the total change in kinetic energy of the particle.

$$\boxed{U_{1\text{-}2} = T_2 - T_1} \quad (7.3\text{-}1)$$

Work-kinetic energy balance:

T_1 = kinetic energy of particle at state 1
T_2 = kinetic energy of particle at state 2
$U_{1\text{-}2}$ = work done by forces acting on the particle between state 1 and 2

Another way to look at the work-kinetic energy balance is from a state point-of-view. That is, the amount of kinetic energy that the particle has at state 1 plus any work done on the particle is equal to the amount of kinetic energy that the particle has at state 2. This version of the work-energy equation is given in Equation 7.3-2 which states that the initial kinetic energy plus the work done by all forces equals the final kinetic energy of the particle. Note that Equation 7.3-1 and 7.3-2 are equivalent. Which equation you choose to use depends on how you would like to view the problem.

$$\boxed{T_1 + U_{1\text{-}2} = T_2} \quad (7.3\text{-}2)$$

Work-kinetic energy balance:

Equations 7.3-1 and 7.3-2 are very useful when solving problems that give or ask for speed, force or distance. These are all quantities that naturally appear in these equations. If the problem asked for acceleration, you may want to consider using Newtonian mechanics. Another advantage of employing a work-energy approach is that only those forces that do work need to be considered explicitly. For example, if a block is being pushed along a curved ramp, the normal force applied to the block will be changing as the block slides along the ramp. In some instances, it may be difficult to determine this normal force along the entire length of the ramp. For example, if the exact curvature of the ramp is unknown. Since the normal force is always perpendicular to the motion of the block, the normal force does not do work. Therefore, with a work-energy approach, you wouldn't need to determine the normal force to assess the block's change in speed.

When applying a work-energy approach to analysis, it is still recommended that you draw a free-body diagram of the body as we did with a Newtonian mechanics approach to analysis. Identification of the various forces acting on a body will help you to identify which forces do work and which don't. In other words, it helps you to identify which forces are active and which are constraints.

Conceptual Example 7.3-1

Consider two pucks resting on an air hockey table with the air on as shown in the figure. The two pucks are pushed across the table with a constant and equal force **F**. Neglecting friction, which puck will have the greater kinetic energy upon reaching the goal line?

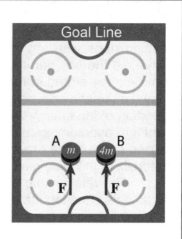

a) puck A
b) puck B
c) Both pucks will have the same kinetic energy.
d) Not enough information is given to answer the question.

Which puck will reach the goal line first?

a) puck A
b) puck B
c) Both pucks will reach the goal line at the same time.
d) Not enough information is given to answer the question.

Conceptual Example 7.3-2

A truck drives along a horizontal road with a constant velocity. Its motion is retarded and eventually halted by a set of crash barrels. The amount of barrels and sand within the barrels are arranged in such a way to decelerate the truck at a constant rate. Four test runs are performed where the velocity and mass of the truck are varied. In each of the following situations, the barrels stop the truck in the same distance. Rank the force required to stop the truck for the following situations from greatest to least.

Greatest: _____ Next: _____ Next: _____ Least: _____

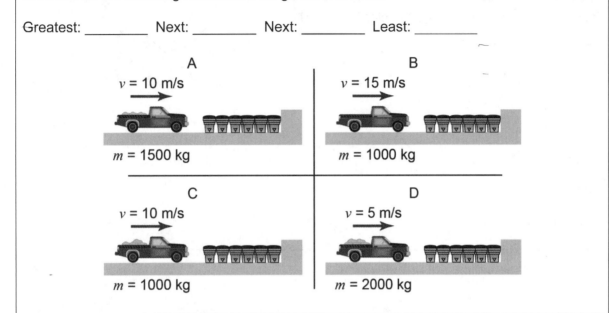

A
$v = 10$ m/s
$m = 1500$ kg

B
$v = 15$ m/s
$m = 1000$ kg

C
$v = 10$ m/s
$m = 1000$ kg

D
$v = 5$ m/s
$m = 2000$ kg

Equation Summary

Abbreviated variable definition list

T = kinetic energy of a particle
m = particle's mass
v = particle's speed
U_{1-2} = work done by forces on the particle between state 1 and 2

Kinetic energy

$$T = \frac{1}{2}mv^2$$

Work-kinetic energy balance
$$T_1 + U_{1-2} = T_2$$

Example Problem 7.3-3

A 40-lb box has a speed of 10 ft/s when it is at the top of the inclined surface shown, where $\theta = 40°$. The box is acted on by a constant force $P = 5$ lb that acts at the angle $\phi = 10°$ with respect to the inclined surface. Determine the speed of the box after it slides $d = 10$ ft down the inclined surface. The coefficient of kinetic friction between the crate and the plane is $\mu_k = 0.4$.

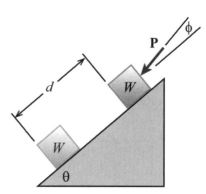

Given:

Find:

Solution:

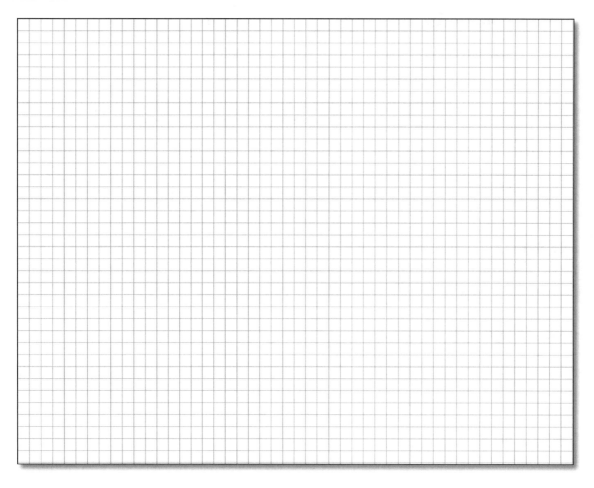

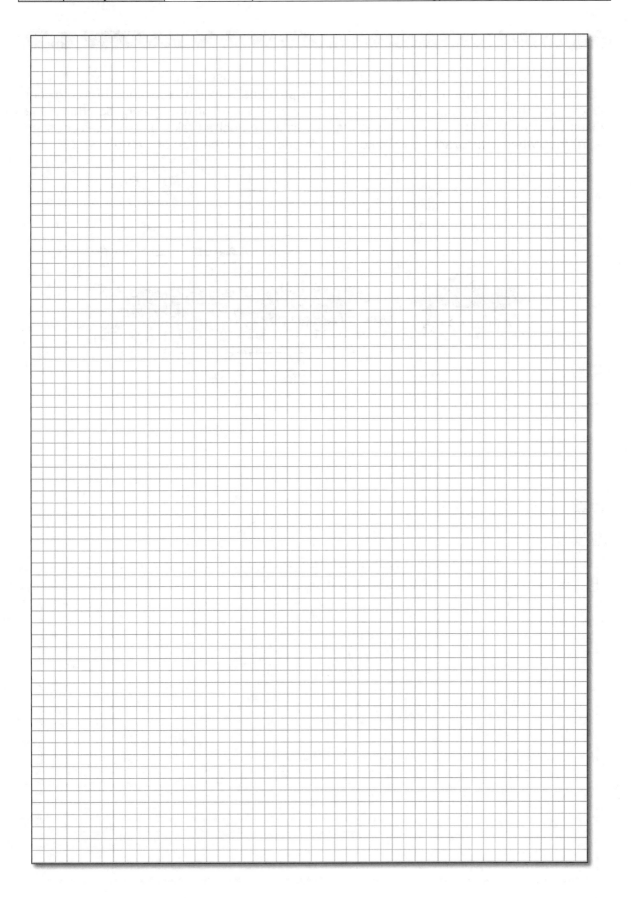

Example Problem 7.3-4

The motion of a pickup truck is arrested using a bed of loose sand AB and a set of crash barrels BC. If experiments show that the sand provides a rolling resistance of 120 lb per wheel and the crash barrels provide a resistance as shown in the graph, determine the distance x the 3120-lb truck penetrates the barrels if the truck's brakes fail and it is coasting at 70 mph when it approaches A. Take d = 70 ft and neglect the size of the truck.

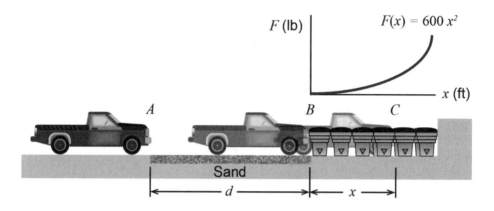

Given:

Find:

Solution:

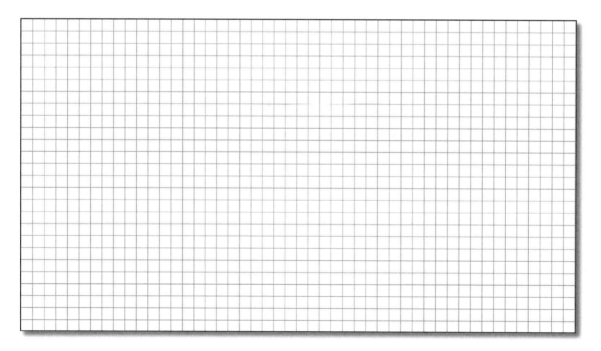

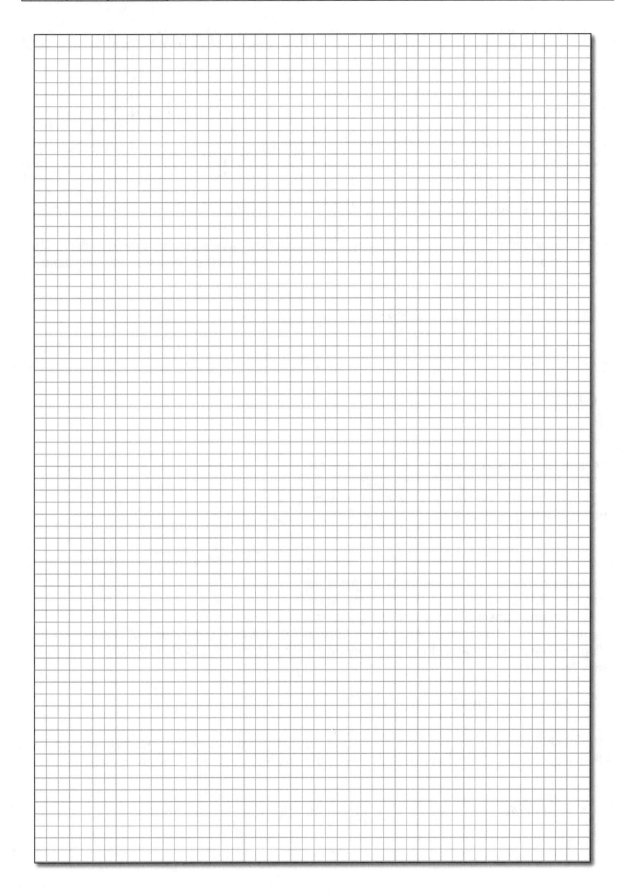

Solved Problem 7.3-5

A box (m = 1 kg), initially at rest, is pushed along a rough floor (μ_k = 0.4). A force $\mathbf{F}(t)$ pushes the box from its initial position at state 1 for 3 meters until it reaches state 2. The box is then allowed to coast to a stop at state 3. Determine the speed of the block at state 2 and the total distance, D, that the box travels before coming to rest. The force profile is given in the figure where F_o = 30 N and x_2 is the distance between state 1 and 2.

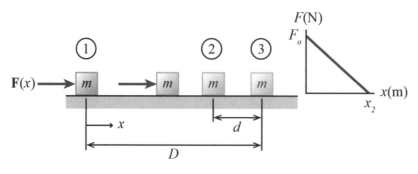

Given: m = 1 kg μ_k = 0.4 F_o = 30 N
   v_1 = 0 v_3 = 0 $x_2 = D\text{-}d$ = 3 m

Find: v_2, D

Solution:

Free-body diagram

Let's draw a free-body diagram of the box. This will enable us to visually determine which forces acting on the box do work and whether they do positive or negative work.

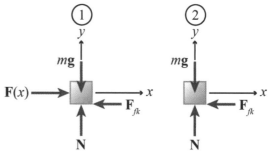

Work

Through inspection of the free-body diagram we can see that mg and $\mathbf{N}$ do not perform work, $\mathbf{F}(x)$ performs positive work and $\mathbf{F}_{fk}$ performs negative work. The friction force acts over the full distance (D), does not depend on the displacement and is parallel to the displacement, therefore, the simple work equation below may be used.

$$U_{F_{fk},1\text{-}3} = -F_{fk}D = -\mu_k mgD$$

The pushing force $\mathbf{F}(x)$ is in the same direction as the displacement but depends on the distance traveled. Therefore, we will have to use the integral form of the work equation.

$$U_{F,1\text{-}2} = \int_0^{x_2} F(x)\,dx = \int_0^{x_2} F_o\left(1 - \frac{x}{x_2}\right)dx = \int_0^{x_2} F_o\left(1 - \frac{x}{x_2}\right)dx = F_o\left(x - \frac{x^2}{2x_2}\right)\Bigg|_0^{x_2} = 0.5F_o x_2$$

Work-kinetic energy balance

We will apply the work-kinetic energy balance equation between state 1 and 3 to determine the total distance traveled.

$$\cancel{T_1} + U_{1-3} = \cancel{T_3} \qquad\qquad U_{F,1-2} + U_{F_{fk},1-3} = 0 \qquad\qquad 0.5F_o x_2 - \mu_k mgD = 0$$

$$D = \frac{0.5F_o}{\mu_k mg}x_2 = \frac{0.5(30)}{0.4(1)9.81}3 \qquad\qquad \boxed{D = 11.47 \text{ m}}$$

We will apply the work-kinetic energy balance equation between state 1 and 2 to determine the speed of the box at state 2.

$$\cancel{T_1} + U_{1-2} = T_2 \qquad\qquad U_{F,1-2} + U_{F_{fk},1-2} = T_2 \qquad\qquad 0.5F_o x_2 - \mu_k mgx_2 = \frac{1}{2}mv_2^2$$

$$v_2^2 = x_2\left(\frac{F_o}{m} - 2\mu_k g\right) = 3\left(\frac{30}{1} - 2(0.4)9.81\right) \qquad\qquad \boxed{v_2 = 8.15\ \frac{\text{m}}{\text{s}}}$$

7.4) CONSERVATIVE & NON-CONSERVATIVE FORCES

A **conservative force** is a force that conserves mechanical energy. This means that the work done by a conservative force is completely reversible and may be recovered at a later time. For example, consider a rock rolling with a speed v when it encounters a hill. The rock has enough kinetic energy to roll from ground level to the top of the hill. The force of gravity on the rock did negative work since the '$m\mathbf{g}$' force opposed the direction of the displacement. This negative work decreased the rock's kinetic energy; however, the kinetic energy could be recovered by letting the rock roll back down the hill. In the process of rolling down the hill, the gravitational force does positive work on the rock increasing its kinetic energy. While the rock was at the top of the hill, energy was stored in the rock due to its position in the gravitational field. This energy was ready to be converted back into kinetic energy when the rock rolled down the hill. The work done by gravity is recoverable; therefore, the gravitational force is classified as a conservative force. Examples of other conservative forces include elastic forces and electromagnetic forces.

When a **non-conservative force** does work that transfers energy, this energy is not stored, at least not in a form that can be readily recovered. A good example of a non-conservative force is friction. Consider a box sliding across a level floor. As the box slides the friction force will do negative work on the box and cause the box to slow down. The overall energy level of the box decreases. So, where did this energy go? It was dissipated into heat and complex material interactions. The energy converted to these forms cannot be employed to perform useful work. Therefore, the work done by the friction force cannot be recovered easily and hence the friction force is classified as non-conservative. Common examples of non-conservative forces are friction, externally applied forces, aerodynamic drag and non-elastic material stresses.

To summarize, the class of forces that conserve mechanical energy for reuse are called *conservative forces*. The class of forces that do not conserve mechanical energy for reuse are called *non-conservative forces*. If there is a question of whether or not a force is conservative or non-conservative, there is an easy test. The work done by a conservative force does not depend on the path taken by the particle between its initial point and its final point. Conversely, the work done by a non-conservative force depends on the path taken by the particle. If we can identify the conservative forces in our system, then we know which forces will do work that stores energy in a form that can be recovered (i.e. conservative work). We will learn in the next section that conservative work can be considered as being stored as *potential energy*. Knowing how to identify these forces will enable us to simplify our calculations.

Example 7.4-1

Is friction a conservative or a non-conservative force?

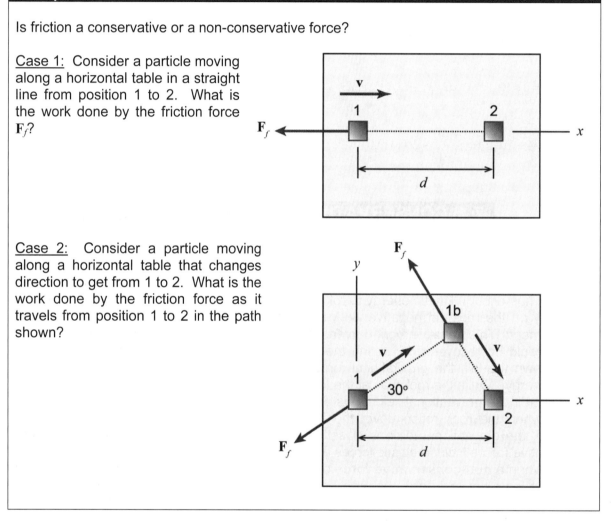

Case 1: Consider a particle moving along a horizontal table in a straight line from position 1 to 2. What is the work done by the friction force F_f?

Case 2: Consider a particle moving along a horizontal table that changes direction to get from 1 to 2. What is the work done by the friction force as it travels from position 1 to 2 in the path shown?

Example 7.4-2

Is the force of gravity a conservative or non-conservative force?

Case 1: Consider a particle that falls from position 1 to 2 in a straight line. What is the work done by **W** as it travels from point 1 to point 2?

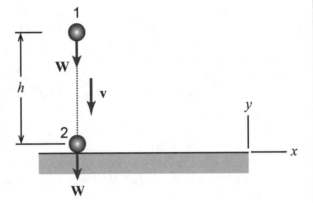

Is the force of gravity a conservative or non-conservative force?

Case 2: Consider a particle that falls from position 1 to 2 in an angled path. What is the work done by **W** from point 1 to 2?

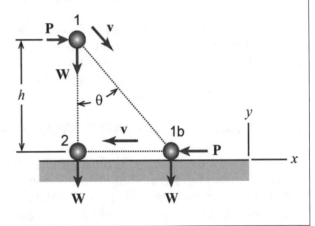

Mathematically and physically, why is friction a non-conservative force and gravity a conservative force?

7.5) POTENTIAL ENERGY

7.5.1) POTENTIAL ENERGY

There are types of energy other than kinetic energy introduced previously. Other types of energy include chemical, thermal, electrical and other forms of mechanical energy. In dynamics, we focus on mechanical energy which we split into two categories: kinetic energy (i.e. energy due to a body's motion)

> Units of Potential Energy
> SI units:
> - Newton–meter [N-m]
> - Joule [J = 1 N-m]
> - erg [erg = 1 x 10^{-7} J]
>
> US customary units:
> - foot–pound [ft-lb]
> - kilocalorie [kcal = 4187 J]
> - British thermal unit [Btu = 778.16 ft-lb = 1055 J]

and potential energy (i.e. energy due to a body's position). **Potential energy** comes from the work done by a conservative force which cause energy to be stored such that it can be recovered at a later time.

> **Potential Energy** is energy stored due to a body's position.

Since energy is the capacity to do work, one can imagine that a body falling in a gravitational field could be used to do work or a compressed spring could be released to do work. Therefore, these systems must have some energy stored in them. The specific amount of energy stored in these situations (i.e. potential energy) will be equal to the amount of work that can be done by the falling particle or the released spring.

7.5.2) GRAVITATIONAL POTENTIAL ENERGY

Gravity exerts a force on all bodies near the earth's surface. It can be shown that the work done by gravity does not depend on the path the particle takes as it falls. The work only depends on the initial and final points of the particle's path, in other words, only on the height h through which the particle falls. For example, Figure 7.5-1 shows three different paths going from point 1 to point 2. The gravity force does the same amount of work for each of the paths. Recall, this property of path-independence classifies the weight force as conservative. A force whose work depends on the path taken is referred to as a non-conservative force. One previously mentioned example of a non-conservative force is friction.

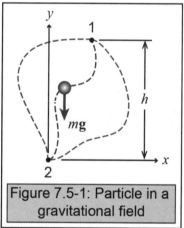

Figure 7.5-1: Particle in a gravitational field

The energy stored in a particle in a gravitational field can be calculated based on the work that its weight force can do. Since weight, in most instances, can be considered to have constant magnitude and direction, the work that it can perform is calculated as follows, where $\mathbf{W} = m\mathbf{g}$ and the height h is the component of the particle's displacement $\mathbf{r}$ that is in the direction of the weight force.

$$U_g = \int \mathbf{W} \cdot d\mathbf{r} \qquad \text{(The weight is constant and in the negative } \mathbf{j}\text{-direction)}$$

$$= \int -W\,\mathbf{j} \cdot (dx\,\mathbf{i} + dy\,\mathbf{j} + dz\,\mathbf{k}) = -W\int_h^0 dy$$

$$= Wh = mgh$$

Based on the above expression for the work done by gravity on a falling particle, the potential energy stored by a particle in a gravitational field is given by Equation 7.5-1. Note that the height is measured relative to a datum where the potential energy is defined as zero.

Relative gravitational potential energy: $\boxed{V_g = mgh}$ (7.5-1)

V_g = gravitational potential energy
m = mass of particle
g = acceleration due to gravity
h = height relative to a chosen datum where $V_g = 0$

> **Gravitational potential energy** is the energy a body possesses by virtue of its position in a gravitational field.

In the case where the particle falls through a very large distance, the weight force cannot be considered constant and the gravitational potential energy can be shown to have the more general form given in Equation 7.5-2, where the datum is considered to be when the particle is at a distance infinitely far from the earth such that $V_g = 0$. Note, that in this case, the weight force is still considered to be conservative.

Absolute gravitational potential energy: $\boxed{V_g = -\dfrac{mg_e R_e^2}{r}}$ (7.5-2)

V_g = gravitational potential energy
m = mass of particle
g_e = acceleration due to gravity at surface of the earth
R_e = radius of earth
r = distance of particle from the center of the earth

It is interesting to note that mathematically, the datum line or coordinate axis used to measure the potential energy may be placed anywhere. In Figure 7.5-1 the datum is placed at point 2, but it could just as easily have been placed at point 1. Make sure that in a given analysis, all potential energy terms use the same datum line. Choosing a non-intuitive datum may produce strange energy values, but could help simplify calculations. For instance, if the object is below the datum line, the potential energy will be negative, which may not truly represent its ability to do work.

Example 7.5-1

Can the gravitational potential energy $V_g = mgh$ be negative?

What is the potential energy of the particle at positions 1, 2 and 3?

$V_g = mgh$

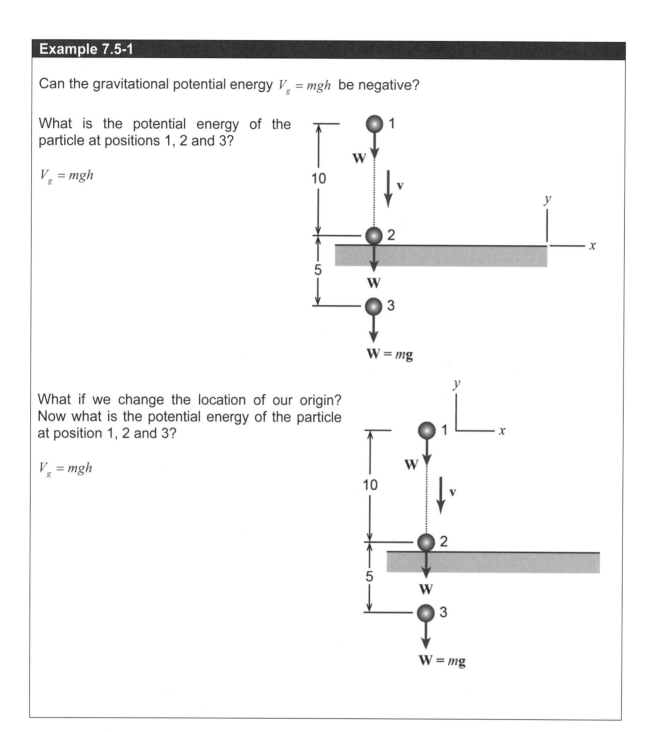

What if we change the location of our origin? Now what is the potential energy of the particle at position 1, 2 and 3?

$V_g = mgh$

7.5.3) ELASTIC POTENTIAL ENERGY

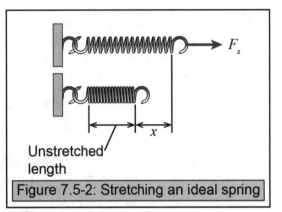

Figure 7.5-2: Stretching an ideal spring

The amount of energy stored in an elastic member that has been deformed can also be determined based on the amount of work it can perform. Consider the spring shown in Figure 7.5-2. It is being stretched through a distance x by the force F_s, which is equal to the force generated by the spring, assuming that the spring is massless. In addition, we will assume that the spring generates the force according to Hooke's law, $F_s = kx$, where x is the distance the spring deforms from its relaxed position and k is a constant that represents the spring's stiffness. The force generated by the spring is a function of the displacement x, therefore, the work done by the spring has to be calculated by evaluating the integral expression for work. Note that the work done by a spring as it is released is calculated as follows. The spring force and displacement will be in the same direction whether the spring has been stretched or compressed.

$$U_e = \int_{r_1}^{r_2} \mathbf{F} \cdot d\mathbf{r} \quad \text{(The force and displacement are in the same direction.)}$$

$$= \int_0^x F_s \, dx = \int_0^x kx \, dx = \frac{1}{2}kx^2$$

Based on the above expression for the work done by a spring, the potential energy stored in a spring is given by Equation 7.5-2.

Elastic potential energy: $\boxed{V_e = \dfrac{1}{2}kx^2}$ (7.5-2)

V_e = elastic potential energy
k = spring constant
x = distance the spring has been stretch or compressed

Equation 7.5-2 indicates that the potential energy stored in a spring is always greater than or equal to zero, whether the spring has been stretched or compressed. This makes sense because the work performed by the spring will be positive whether it is stretched or compressed. This is because the spring force is in the same direction as the displacement it produces. Also, note that since the work done by the elastic member depends only on the starting and ending points and not on the specific path taken, the force is conservative.

Example 7.5-2

Consider a spring that is being stretched. Answer the following questions.

- What is the work done by $F_s = k\,x$ as it moves the spring from position 1 to 2? If the elastic potential energy is the spring's ability to do work, then what is the elastic potential energy of the spring at position 2?

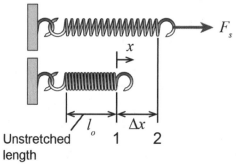

Example 7.5-3

What is the elastic potential energy of the spring shown at position 2 given that the un-stretched length of the spring is 15 cm, the final length is 25 cm and

$V_e = \dfrac{1}{2}kx^2$?

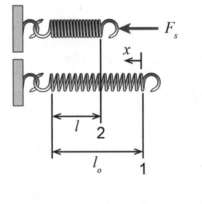

What is the elastic potential energy of the spring at position 1 in the above figure?

Can elastic potential energy ($V_e = \dfrac{1}{2}kx^2$) be negative?

What is the elastic potential energy of the spring shown below at position 2 given that the unstretched length of the spring is 25 cm, the final length is 15 cm and

$V_e = \dfrac{1}{2}kx^2$?

Conceptual Example 7.5-4

Consider two identical springs. This means that the two springs are made of the same wire, have the same diameter and number of coils and have the same un-stretched length which is equal to 10 cm. During operation, both springs work within their linearly elastic range. Spring A operates between 10 cm and 20 cm. Spring B operates between 15 cm and 25 cm. Which spring stores the larger amount of energy at its longest length?

    a) spring A b) spring B c) same energy

Which spring has the larger change in energy moving between its extreme operating points?

    a) spring A b) spring B c) same energy

7.6) WORK-ENERGY BALANCE & CONSERVATION OF ENERGY

7.6.1) THE WORK-ENERGY BALANCE EQUATION

In previous sections, the relationships between work and kinetic energy and work and potential energy were derived. In the section on the *Work-Kinetic Energy Balance*, a relationship between work and kinetic energy was given by the following Equation 7.6-1.

Work-kinetic energy balance: $\boxed{T_1 + U_{1-2} = T_2}$ (7.6-1)

T_1, T_2 = kinetic energy at state 1 and state 2 respectively
U_{1-2} = work done by forces on the particle between state 1 and 2

This equation states that the initial kinetic energy of the particle at state 1 plus the work done on the particle between state 1 and state 2 equals the final kinetic energy of the particle at state 2. In the section on *Potential Energy,* we learned that the work done on a particle by a conservative force is equal to the corresponding change in the particle's potential energy. Therefore, we can break up the work U_{1-2} into conservative work (i.e. the change in potential energy (V)) and non-conservative work ($U_{1-2,non}$) to generate a new version of the work-energy balance equation shown in Equation 7.6-2. This work and energy balance states that the initial kinetic and potential energy of the particle at state 1 plus the work done by non-conservative forces on the particle between state 1 and state 2 equals the final kinetic and potential energy of the particle at state 2. In Equation 7.6-2, V represents the gravitational potential energy together with the elastic potential energy of the particle and $U_{1-2,non}$ is the work done by all non-conservative forces acting on the particle.

Work-energy balance: $\boxed{T_1 + V_1 + U_{1-2,non} = T_2 + V_2}$ (7.6-2)

T_1, T_2 = kinetic energy at state 1 and state 2 respectively
V_1, V_2 = potential energy at state 1 and state 2 respectively
$U_{1-2,non}$ = work done by non-conservative forces acting on the particle between state 1 and 2

Another form of the work-energy balance equation rearranges the terms and is given in Equation 7.6-3. This equation essentially reads that the work done by the non-conservative forces acting on a body equals the change in the body's kinetic and potential energy.

Work-energy balance: $\boxed{U_{1-2,non} = \Delta T + \Delta V}$ (7.6-3)

ΔT = change in kinetic energy between state 1 and state 2
ΔV = change in potential energy between state 1 and state 2
$U_{1-2,non}$ = work done by non-conservative forces acting on the particle between state 1 and 2

7.6.2) CONSERVATION OF ENERGY

The principle of conservation of energy states that the total amount of energy in an isolated system does not change. In the context of analyzing dynamic systems, we will define a conservative system as having work done on it by only conservative forces. In a conservative system, the system's energy may change forms, but the total amount of energy remains constant. In other words, the energy may change between forms of kinetic and potential energy, but no work is done to transfer energy into or out of the system. If the system under consideration is conservative, the work-energy balance (Equation 7.6-2) may be simplified to give the conservation of energy equation shown in Equation 7.6-4. This conservation of energy equation states, that the sum of the kinetic and potential energies of the particle at the initial state equals the sum of the kinetic and potential energies of the particle at the final state.

Conservation of energy: $\boxed{T_1 + V_1 = T_2 + V_2}$ (7.6-4)

T_1, T_2 = kinetic energy at state 1 and state 2 respectively
V_1, V_2 = potential energy at state 1 and state 2 respectively

Physically, how can we interpret Equation 7.6-4? Consider a ball being tossed up in the air as shown in Figure 7.6-1. After the ball leaves your hand (state 1), the ball has significant speed and, hence, significant kinetic energy and little potential energy (relative to a datum at the level of your hand). As the ball rises, it begins to slow down and its kinetic energy decreases until it reaches its apex (state 2). At the apex, the ball's speed and kinetic energy are zero and the ball has obtained its maximum potential energy. Assuming no losses to air drag, the loss in kinetic energy as the ball travels from state 1 to state 2 is exactly offset by the increase in potential energy as seen by the increase in height h.

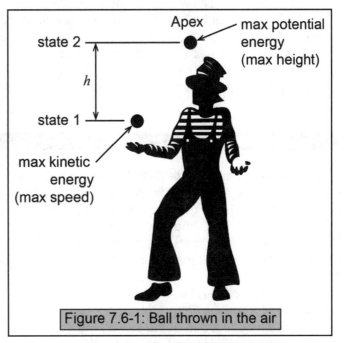

Figure 7.6-1: Ball thrown in the air

As the ball begins to fall, it speeds up again as the potential energy is converted back into kinetic energy.

If we can classify a system as being conservative, the analysis of the system can be simplified considerably. This exemplifies another advantage of a work-energy approach of analysis. When considering conservative systems (and conservative forces in general), we do not need to analyze the system along its entire path of travel, rather, we only need to know the system's initial and final state. The question then becomes, how do we know a system is conservative just by inspection? When trying to determine if a system is conservative or non-conservative, ask yourself the following questions. If you answer "no" to all of the questions, then your system is likely a conservative system.

1. Is there friction? If the system has friction and the friction is acting at a point where the contacting surfaces have relative velocity, then the system is not conservative.

2. Is there air drag or drag from another fluid? If the body experiences drag, then the system is not conservative. Two common examples of this would be a car traveling at high speeds and a boat traveling in water. Sometimes drag may be neglected at slow speeds.

3. Is there an externally applied force? If there is an externally applied force that does work (i.e. a force that isn't a constraint force), then the system is not conservative. These external forces do not include the particle's weight or spring forces.

4. Are there any energy losses due to material deformation? Most likely, we will not encounter this situation when working on basic dynamics problems, however, in a real life situation, this may be a factor.

Conceptual Example 7.6-1

Two metal balls have the same size but one weighs twice as much as the other. The balls are dropped from the roof of a single story building at the same instant of time. Assuming no air resistance, identify which of the following are true.

    a) The lighter ball will reach the ground with more kinetic energy.
    b) The heavier ball will reach the ground with more kinetic energy.
    c) Both balls will reach the ground with the same amount of kinetic energy.
    d) The lighter ball will reach the ground first.
    e) The heavier ball will reach the ground first.
    f) Both balls will reach the ground at the same time.

Conceptual Example 7.6-2

Two identical balls are released from rest at the top of inclines that are identical except for the bumps shown in the accompanying figure. Assuming air drag and friction are negligible, identify which of the following statements are true as the balls arrive at the bottom of the incline.

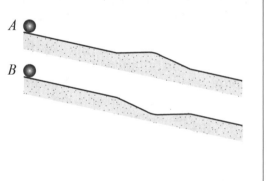

    a) Ball A arrives first.
    b) Ball B arrives first.
    c) Both balls arrive at the same time.
    d) Ball A arrives with higher speed.
    e) Ball B arrives with higher speed.
    f) Both balls arrive with same speed.

Conceptual Example 7.6-3

The carts shown initially move on a horizontal surface and then up a short incline. Assuming negligible friction, rank the carts from greatest to least achieved vertical height given the weights and velocities shown.

Greatest: _____ Next: _____ Next: _____ Least: _____

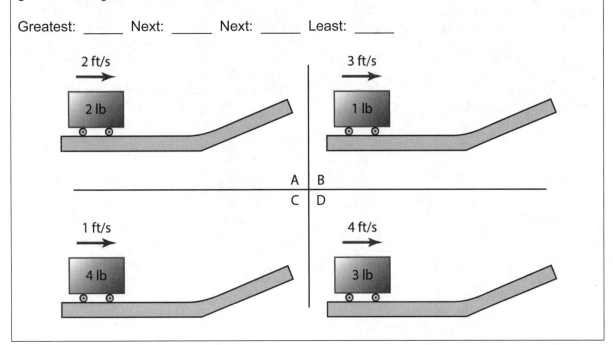

Equation Summary

Abbreviated variable definition list

U = work

$\mathbf{F}$ = force

$\mathbf{r}$ = position

$d_{1\text{-}2}$ = displacement magnitude

θ = angle between $\mathbf{F}$ and the direction of motion

T = kinetic energy of a particle

m = particle's mass

v = particle's speed

V = potential energy

g = acceleration due to gravity

g_e = acceleration due to gravity at surface of the earth

h = height relative to a chosen datum where $V_g = 0$

R_e = radius of earth

r = distance of particle from the center of the earth

k = spring constant

x = distance the spring has been stretch or compressed

Work

$$U_{1-2} = \int_{\mathbf{r_1}}^{\mathbf{r_2}} \mathbf{F} \cdot d\mathbf{r} \qquad U_{1-2} = \mathbf{F} \cdot \Delta \mathbf{r}_{1\text{-}2} \qquad U_{1-2} = \left(F \cos\theta \right) d_{1\text{-}2}$$

Kinetic energy

$$T = \frac{1}{2} m v^2$$

Potential energy

$$V_g = mgh \qquad V_g = -\frac{mgR_e^2}{r} \qquad V_e = \frac{1}{2} k x^2$$

Work-energy balance

$$T_1 + V_1 + U_{1-2,non} = T_2 + V_2$$

Conservation of Energy

$$T_1 + V_1 = T_2 + V_2$$

Example Problem 7.6-4

A 7-kg block compresses a spring 20 cm and is
released from rest. The spring has a stiffness
of 11 kN/m. Determine the velocity of the block
after it has traveled up the slope for 1.2 meters.
The coefficient of kinetic friction is 0.4 and
$\theta = 30°$.

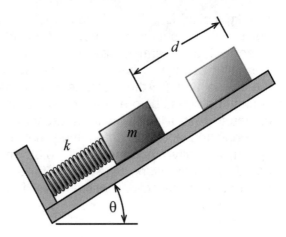

Given:

Find:

Solution:

Example Problem 7.6-5

A 4-kg collar is released from rest from point A on the curved rod shown. If friction is neglected, determine (a) the velocity of the collar at point B just prior to engaging the spring and (b) the distance x that the spring deforms before the collar again comes to rest at point C. If an experiment is conducted and it is determined that the spring actually deforms $x = 0.05$ m, calculate (c) the work done by friction over the length of the collar's entire travel from point A to point C.

Given:

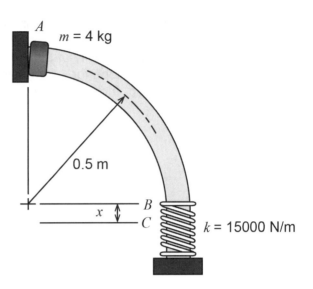

Find:

Solution:

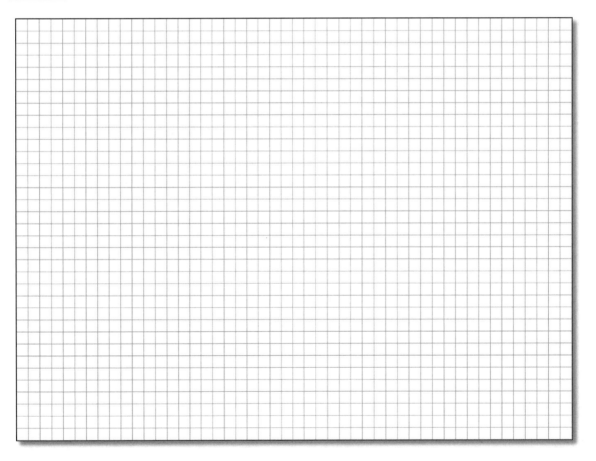

Solved Problem 7.6-6

The 2-kg mass starts from rest at position 1. At position 1, the mass compresses a 5000 N/m spring by 4 cm. After the mass is released, it slides down a frictionless slope into a curve of radius 10 cm. It then proceeds up a rough incline (μ_k = 0.5). If h = 12 cm and θ = 45°, determine how far up the rough incline the mass proceeds before coming to a stop and the normal force at position 2 when the mass is still on the curved surface. Note that position 2 may be considered to reside on the curve immediately before the rough incline.

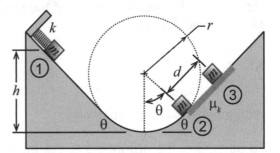

Given: m = 2 kg $v_1 = 0$ $v_3 = 0$
      k = 5000 N/m Δx = 0.04 m r = 0.1 m
      h = 0.12 m θ = 45° μ_k = 0.5

Find: N_2, d

Solution:

Getting familiar with the problem

The first thing we should do is to determine if the system is conservative or non-conservative. The system is conservative between state 1 and 2 and, because of the friction, is non-conservative between state 2 and 3.

We need to find the normal force at state 2. This will involve applying Newton's second law. Because the normal force points in the n-direction and the forces in that direction depend on the speed, the most convenient coordinate system to use is the n-t coordinate system.

$$\sum F_n = ma_n = m\frac{v^2}{r}$$

We will apply the conservation of energy equation between state 1 and 2 to determine the speed of the block at state 2. We will then apply the work-energy balance equation between state 1 and 3 to determine the distance that the mass slides up the hill.

Conservation of energy

When applying the work-energy balance equation or the conservation of energy equation, it is necessary to choose a datum where $V_g = 0$. Let's choose position 1 as our datum.

$$\cancel{T_1} + V_1 = T_2 + V_2 \qquad\qquad \frac{1}{2}k\Delta x^2 = \frac{1}{2}mv_2^2 - mg(h-(r-r\cos\theta))$$

$$v_2^2 = \frac{k\Delta x^2}{m} + 2g(h-(r-r\cos\theta)) = \frac{5000(.04)^2}{2} + 2(9.81)(0.12-0.1(1-\cos45)) \qquad v_2 = 2.4\frac{m}{s}$$

Newton's second law

Let's draw a free-body diagram of the mass at state 2 and apply Newton's second law in the n-direction.

$$\sum F_n = ma_n = m\frac{v^2}{r} \qquad N_2 - mg\cos\theta = m\frac{v_2^2}{r} \qquad \boxed{N_2 = 129.5\ \text{N}}$$

Work-energy balance equation

To determine the distance that the mass slides up the hill, we will apply the work-energy balance equation between state 1 and 3. We could apply this equation between states 2 and 3, however, the energies are simpler to determine at state 1.

$$T_1 + V_1 + U_{1-3} = T_3 + V_3 \qquad \frac{1}{2}k\Delta x^2 + U_{F_{fk}} = -mg(h - (r - r\cos\theta) - d\sin\theta)$$

The work done by the friction force is negative and equal to the friction force multiplied by the distance the friction acts over.

$$U_{F_{fk}} = -F_{fk}d = -\mu_k Nd$$

To determine the normal force we will consult the free-body diagram. This time we will switch to the x-y coordinate system. This is because on the incline, the mass does not accelerate in the y-direction.

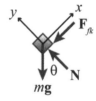

$$\sum F_y = 0 \qquad N = mg\cos\theta$$

$$U_{F_{fk}} = -F_{fk}d = -\mu_k Nd = -\mu_k dmg\cos\theta$$

Back to the energy-balance equation.

$$\frac{1}{2}k\Delta x^2 - \mu_k dmg\cos\theta = -mg(h - (r - r\cos\theta) - d\sin\theta) \qquad d = \frac{2mg(h - (r - r\cos\theta)) + k\Delta x^2}{2mg(\sin\theta + \mu_k\cos\theta)}$$

$$\boxed{d = 28\ \text{cm}}$$

7.7) WORK-ENERGY FOR A SYSTEM OF PARTICLES

As we learned in the discussion of Newtonian mechanics, a system of particles can be analyzed by either examining the particles individually, or by analyzing the system as a whole. One advantage of defining your system to include multiple particles is that sometimes the effect of internal forces may cancel. Hence, internal forces do not always need to be explicitly included in the analysis.

7.7.1) WORK

A force is a force whether it is applied to a single particle or a system of particles. This means that the definition of work does not change. **Work** is still the amount of energy transferred by a force acting through a distance. The equation for work (Equation 7.7-1) and all of its forms are the same for a particle as

Units of Work
SI units:
• Newton–meter [N-m]
• Joule [J = 1 N-m]
• erg [erg = 1 x 10^{-7} J]
US customary units:
• foot–pound [ft-lb = 1.3558 J]
• Kilocalorie [kcal = 4187 J]
• British thermal unit [Btu = 778.16 ft-lb = 1055 J]

they are for a system of particles. However, one thing to look out for when dealing with systems of particles is the distance used in the work equation. The distance used in this calculation is the distance traveled by the force's point of application. This distance may be different for each force being applied to the system.

Work: $$\boxed{U_{1-2} = \int_{r_1}^{r_2} \mathbf{F} \cdot d\mathbf{r}}$$ (7.7-1)

U_{1-2} = work done by $\mathbf{F}$ from $\mathbf{r}_1$ to $\mathbf{r}_2$ (work is a scalar)
$\mathbf{F}$ = force vector
$\mathbf{r}$ = position vector

As mentioned previously, the internal forces do not always need to be included in the analysis. For example, the work done by two internal forces in an interaction pair will cancel. This is because the forces are equal and opposite and applied over the same path. This simplification requires that the overall body be rigid or that the particles be connected by ideal frictionless connections.

7.7.2) KINETIC AND POTENTIAL ENERGY

Energy methods are very useful for analyzing systems of particles. This is mainly due to the fact that energy is additive. In a system of multiple particles (or multiple springs), the overall kinetic and potential energy for a system of particles is simply equal to the sum of the energies of the individual

Units of Energy
SI units:
• Newton–meter [N-m]
• Joule [J = 1 N-m]
• erg [erg = 1 x 10^{-7} J]
US customary units:
• foot–pound [ft-lb = 1.3558 J]
• Kilocalorie [kcal = 4187 J]
• British thermal unit [Btu = 778.16 ft-lb = 1055 J]

components. This fact is expressed in Equation 7.7-2 through 7.7-4.

Kinetic energy for a system of particles: $$T = \sum \frac{1}{2} m_i v_i^2$$ (7.7-2)

Gravitational potential energy for a system of particles: $$V_g = \sum m_i g h_i$$ (7.7-3)

Elastic potential energy for a system of particles: $$V_e = \sum \frac{1}{2} k_j x_j^2$$ (7.7-4)

T = kinetic energy of the system
V_g, V_e = gravitational and elastic potential energy of the system, respectively
m_i = mass of the ith particle
v_i = speed of the ith particle
h_i = height of the ith particle relative to a chosen datum where $V_g = 0$
k_j = spring constant of the jth spring
x_j = distance the jth spring has been stretch or compressed
g = acceleration due to gravity

7.7.3) WORK-ENERGY BALANCE

The equation that relates work and energy remains the same for a system of particles as it was for a single particle. The only thing to remember is that the energy of the system is the sum of the energies of the individual particles and that the work term is the sum of all the work done on each particle.

The notion of conservation of energy (Equation 7.7-6) can also be applied to the analysis of systems of particles. Energy is conserved for a system of particles if there is no work done by any external or internal forces that are not gravitational or elastic. Be careful to consider any non-conservative internal forces.

Work-energy balance: $$T_1 + V_1 + U_{1-2,non} = T_2 + V_2$$ (7.7-5)

Conservation of energy balance: $$T_1 + V_1 = T_2 + V_2$$ (7.7-6)

T_1, T_2 = total kinetic energy of the system at state 1 and state 2, respectively
V_1, V_2 = total potential energy of the system at state 1 and state 2, respectively
$U_{1-2,non}$ = total non-conservative work done on the system

Equation Summary

Abbreviated variable definition list

U = work
$\mathbf{F}$ = force
$\mathbf{r}$ = position
$d_{1\text{-}2}$ = displacement magnitude
θ = angle between $\mathbf{F}$ and the direction of motion
T = kinetic energy of the system

m = mass
v = speed
V = potential energy
g = acceleration due to gravity
h = height relative to a chosen datum where $V_g = 0$
k = spring constant
x = distance the spring has been stretch or compressed

Work

$$U_{1\text{-}2} = \int_{\mathbf{r}_1}^{\mathbf{r}_2} \mathbf{F} \cdot d\mathbf{r} \qquad U_{1\text{-}2} = \mathbf{F} \cdot \Delta \mathbf{r}_{1\text{-}2} \qquad U_{1\text{-}2} = (F \cos\theta)\, d_{1\text{-}2}$$

Kinetic energy

$$T = \sum \frac{1}{2} m_i v_i^2$$

Potential energy

$$V_g = \sum m_i g h_i \qquad V_e = \sum \frac{1}{2} k_j x_j^2$$

Work-energy balance

$$T_1 + V_1 + U_{1\text{-}2,non} = T_2 + V_2$$

Conservation of Energy

$$T_1 + V_1 = T_2 + V_2$$

Example Problem 7.7-1

A ball (m_1 = 4 kg) is attached to a light rod which is pinned at O. The rod is free to rotate in the vertical plane. It is desired to determine the velocity of the ball when θ is 90 degrees when the system is released from rest at position A. Calculate the velocity for the two cases shown in the figure. The first case has a block (m_2 = 10 kg) attached to the end of the rod. The second case includes a constant force ($F = m_2g$) that pulls down on the end of the rod. The force remains vertical throughout the motion. The total length of the rod is 450 mm, where l_1 = 300 mm and l_2 = 150 mm.

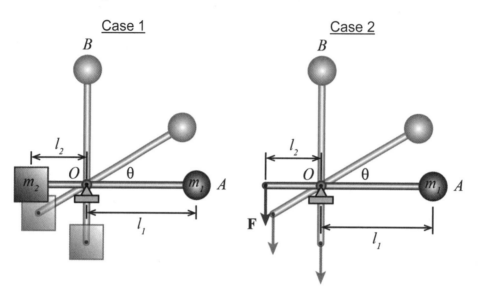

Given:

Find:

Solution:

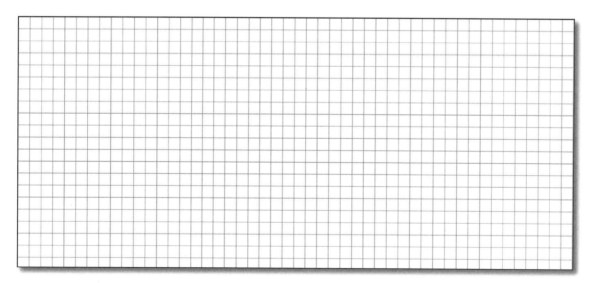

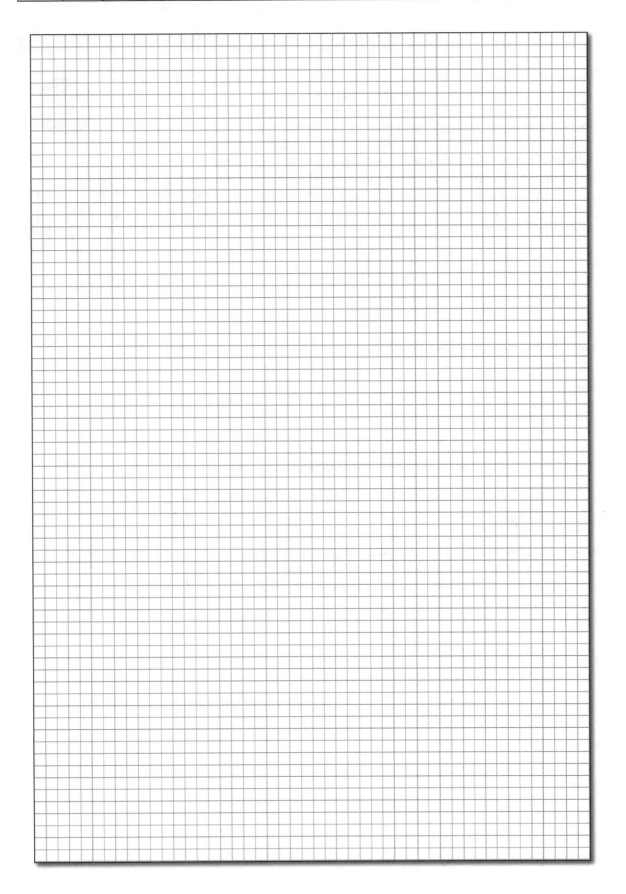

The system shown consists of two cylindrical masses (m_A = 10 kg, m_B = 8 kg) connected by a rope and suspended over a light frictionless pulley. Mass A is constrained to move within a tube that applies a constant pressure. Taking into account the surface area of mass A, this pressure translate into a 20-N force. The static and kinetic friction coefficient between mass A and the tube is 0.6 and 0.5, respectively. If the system is released from rest in the position shown, determine a) if the system moves and b) if it moves, how far will mass A compress the spring shown (k = 100 N/m) before it first comes to rest. The bottom of mass A is initially h = 0.8 m above the spring.

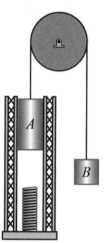

Given: m_A = 10 kg m_B = 8 kg
         N = 20 N k = 100 N/m
         h = 0.8 m μ_s = 0.6
         μ_k = 0.5

Find: Does the system move?, Δs

Solution:

Free-body diagram

The first step is to determine if the system moves. We will need to analyze the free-body diagrams of mass A and mass B.

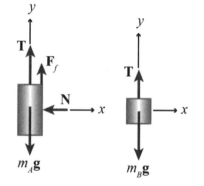

Newton's first law

We will assume that the system is static and solve for the static friction force. Then we will compare this force to the maximum static friction force to see if the system moves.

Mass B: $\sum F_y = 0$ $T = m_B g$

Mass A: $\sum F_y = 0$ $T + F_{fs} = m_A g$ $F_{fs} = (m_A - m_B)g = 19.62$ N

$F_{fs,\max} = \mu_s N = 0.6(20) = 12$ N $F_{fs} > F_{fs,\max}$ Therefore, the system will move.

Work-energy balance

Let's identify the two energy states of interest and then apply the work-energy balance equation to determine the spring's compression. We will choose the datum where $V_g = 0$ to correspond to the initial position of mass A (i.e. state 1).

$$\cancel{T_1} + \cancel{V_1} + U_{1-2} = \cancel{T_2} + V_2 \qquad U_{F_{fk}} = V_2$$

$$-F_{fk}(h + \Delta s) = -m_A g(h + \Delta s) + m_B g(h + \Delta s) + \frac{1}{2}k\Delta s^2$$

$$-\mu_k N(h + \Delta s) = (m_B - m_A)g(h + \Delta s) + \frac{1}{2}k\Delta s^2$$

$$\Delta s^2 + \frac{2}{k}\left(g(m_B - m_A) + \mu_k N\right)\Delta s + \frac{2}{k}\left(g(m_B - m_A) + \mu_k N\right)h = 0$$

$$\Delta s^2 + \frac{2}{100}\left(9.81(-2) + 10\right)\Delta s + \frac{2}{100}\left(9.81(-2) + 10\right)0.8 = 0$$

$$\Delta s^2 - 0.1924\Delta s - 0.15392 = 0 \qquad \Delta s = \frac{0.1924 \pm \sqrt{0.1924^2 - 4(-0.15392)}}{2}$$

$$\boxed{\Delta s = 0.50 \text{ m}}$$

7.8) POWER AND EFFICIENCY

7.8.1) POWER

In a previous section, we discussed the concept of *work*. A force does work if it is applied over a distance. You may have noticed that there was no mention of time. If a force moves a particle over a distance d in one minute or one day, it still does the same amount of work. The quantity that takes time into account is called *power*. **Power** is, by definition, the rate at which work is done or the rate at which energy is used. In other words, power is the derivative of work performed with respect to time and is given by Equation 7.8-1.

Power: $\boxed{P = \dfrac{dU}{dt}}$ (7.8-1)

P = power
U = work
t = time

Power is the rate at which work is being performed.

Power can also be calculated as the dot product of a force **F** that is applied to a body and the velocity **v** of the point of application of the force as shown in Equation 7.8-2.

Power of constant force: $\boxed{P = \mathbf{F} \cdot \mathbf{v}}$ (7.8-2)

P = power
F = force vector
v = velocity vector

Equation derivation

The derivation of Equation 7.8-2 use the fact that $\mathbf{r} = \mathbf{v}dt$ and that differentiation is the inverse operation of integration.

$$U = \int \mathbf{F} \cdot d\mathbf{r}$$

$$P = \frac{dU}{dt} = \frac{d\left(\int \mathbf{F} \cdot d\mathbf{r}\right)}{dt} = \frac{d\left(\int \mathbf{F} \cdot \mathbf{v}dt\right)}{dt} = \mathbf{F} \cdot \mathbf{v}$$

Machines are generally designed to perform work, but do much more than that. They are designed to perform work at a specified speed or over a range of speeds. Have you ever noticed that many purchased machines come with a power curve? A power curve is a graph of power versus speed. Figure 7.8-1 shows a typical power curve for a gasoline engine in a car. You can see that the power changes depending on the engine speed. The shape of this curve is a function of differences in the combustion process, the changing amount of friction, etc. It is also interesting to note that every

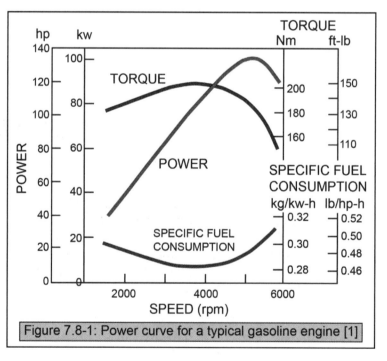

Figure 7.8-1: Power curve for a typical gasoline engine [1]

engine has a sweet spot, or an engine speed where the generated power is maximum. Be careful, however, this is not necessarily where you will get your maximum acceleration because power is a function of the speed as well as the force.

7.8.2) EFFICIENCY

We all know that there is no such thing as a perpetual motion machine. This is because there are always losses. So, what does that mean in terms of power? It means that you will never get the same amount of power out of a machine as you put into the machine. The ratio of the power out P_{out} to the power in P_{in} is called the efficiency of a machine and is given by Equation 7.8-3. Efficiency can be informally thought of as what you get compared to what you pay. Since any real machine will have losses due to things like friction, efficiency is always less than 1. These energy losses often take the form of mechanical energy being converted into heat. For example, internal combustion engines convert only about 20-25% of the fuel energy into useful motion.

Efficiency: $$\boxed{\varepsilon = \frac{P_{out}}{P_{in}}}$$ (7.8-3)

ε = efficiency
P_{in}, P_{out} = power in and power out, respectively

Equation Summary

Abbreviated variable definition list

P = power
U = work
t = time

$\mathbf{F}$ = constant force vector
$\mathbf{v}$ = velocity vector
ε = efficiency

Power

$$P = \frac{dU}{dt}$$ $$P_{ave} = \frac{\Delta U}{\Delta t}$$ $$P = \mathbf{F} \cdot \mathbf{v}$$

Efficiency

$$\varepsilon = \frac{P_{out}}{P_{in}}$$

Example Problem 7.8-1

A 16,000-lb truck climbs a 12% grade with a constant speed of 50 mph as shown.   During a 1 min time interval, determine the total tractive force produced by the wheels, the power developed by this force and the distance the truck travels.   Neglect rolling and air resistance.

Given:

Find:

Solution:

An 80-kg man runs up a flight of 12 stairs in 7 seconds with approximately constant speed. If each individual stair runs 20 cm and rises 22 cm, determine the average power the man delivered.

Given: m = 80 kg   12 stairs Δt = 7 s
           v = constant h = 22 cm (stair rise) d = 20 cm (stair run)

Find: P_{ave}

Solution:

Understanding the problem

Since this problem does not include a figure, it is good idea to draw out what the problem is stating.

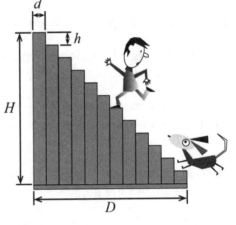

Power

Now let's write down the power equation to see what we need in order to calculate the power.

$$P_{ave} = \frac{\Delta U}{\Delta t}$$

We have the time it takes the man to climb the stairs, but we need to calculate the work.

Work

The work done by a force is the magnitude of the force multiplied by the distance it acts through. In this case there are many forces that vary significantly throughout the stair climbing process. Therefore, an easier way to determine the work is to use the work-energy balance equation. Since the problem statement states that the man runs at a constant speed, the change in kinetic energy will be zero (i.e. $\Delta T = 0$).

$$\cancel{T}_{bottom} + \cancel{V}_{bottom} + U_{b-t} = \cancel{T}_{top} + V_{top}$$

$$U_{b-t} = V_{top} = mgH = mgh(12) = 80(9.81)0.22(12) = 2071.9 \text{ J}$$

Now with the work determined, we can solve for the average power delivered by the man.

$$P_{ave} = \frac{\Delta U}{\Delta t} = \frac{2071.9}{7} \qquad \boxed{P_{ave} = 296 \text{ W}}$$

CHAPTER 7 REVIEW PROBLEMS

RP7-1) Work is energy. (True, False)

RP7-2) Is work a scalar or a vector?

RP7-3) Which of the following are non-conservative forces?

    a) Friction
    b) Weight
    c) Spring
    d) Aerodynamic drag
    e) Contact force

RP7-4) What makes a force conservative?

RP7-5) Does friction usually do positive or negative work?

RP7-6) A force (**F**) performs work on a particle. Given the following situations, choose the simplest work equation that may be employed to solve for the work.

    a) **F** is constant and parallel to the displacement.
    b) **F** is constant and not parallel to the displacement.
    c) **F** changes with and is not parallel to the displacement.

RP7-7) What is the work done by a force applied to a particle that is perpendicular to the displacement of the particle?

RP7-8) Kinetic energy... (Choose all that apply.)

    a) is the energy of motion.
    b) is the energy of position.
    c) may change as work is put into or taken away from a system.
    d) occurs only in a gravitational field.

RP7-9) Potential energy... (Choose all that apply.)

    a) is the energy of motion.
    b) is the energy of position.
    c) may change as work is put into or taken away from a system.
    d) occurs only in a gravitational field.

RP7-10) Can gravitational potential energy be negative?

RP7-11) Can elastic potential energy be negative?

RP7-12) Match the situation with the amount of work described.

Car: Your 2700-lb car broke down and you need to push it 1/2 mi back to town. How much work did you do? Assume that the road grade is fairly level and you are applying a constant horizontal force of 100 lb.

Package: You need to lift a 30-lb package from the floor to a shelf 6 feet off the ground. How much work did you do? Assume that you are applying a constant vertical force of 1.3 times the package weight.

Mosquito: A mosquito's maximum speed in air is 0.87 mph and has a typical mass of 2 milligrams. How much work is needed to stop a mosquito traveling at its maximum speed?

Cup: A ceramic coffee cup weighing 14 ounces is accidentally dropped from a height of 4.5 ft. How much work does the force of gravity perform on this cup?

Situation	Work performed
Car	234 ft-lb, 317 J
Package	3.94 ft-lb, 5.3 J
Mosquito	264000 ft-lb, 358 kJ
Cup	1.63×10^{-6} ft-lb, 2.21×10^{-6} J

RP7-13) A car on a roller coaster rolls down the track as shown in the figure. As the cart rolls past the point illustrated in the figure, what happens to its speed and acceleration?

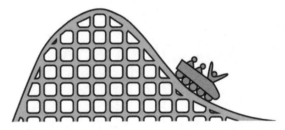

a) Both decrease.
b) Both increase.
c) Both remain constant.
d) The speed decreases and the acceleration increases.
e) The speed increases and the acceleration decreases.
f) Not enough information given to answer the question.

RP7-14)   A 2-kg collar is pulled along a shaft by force $P = 30$ N. The direction θ of this force varies such that $\theta = bx$, where x is in meters and b is 0.1 rad/m. Determine the work done by P from $x = 0$ to $x = 15$ m.

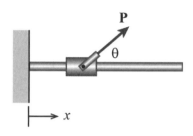

Given: $m = 2$ kg $P = 30$ N
              $\theta = bx$ $b = 0.1$ rad/m

Find: U_P

Solution:

Find the x-component of P as a function of x.

$P_x =$ _____

Find the work done by P from $x = 0$ to 15 m.

$U_P =$ _____

RP7-15) Packages are delivered to a smooth shoot via a conveyor belt. Once the packages exit the shoot they are deposited on to a rough floor (μ_k = 0.3). The packages travel along the conveyor belt at v_1 = 1 ft/s. The height of the shoot is h = 4 ft. Neglecting the size of the packages, determine the distance (d) that the packages travel along the rough level floor before coming to a stop.

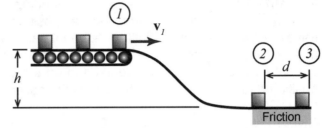

Given: v_1 = 1 ft/s v_3 = 0
             no friction 1 – 2 μ_k = 0.3 from 2 – 3

Find: d

Solution:

1 – 2

Determine the speed of the packages as they exit the shoot.

2 - 3

Determine the distance required to stop.

v_2 = _____

d = _____

RP7-16) A 12-oz collar travels along a smooth curved shaft (r = 1.2 ft). It starts from rest at position A and travels down the shaft until it encounters a spring (k = 11 lb/ft). Determine the maximum compression of the spring and the normal force exerted on the collar at position B (θ = 45°). Neglect the size of the collar.

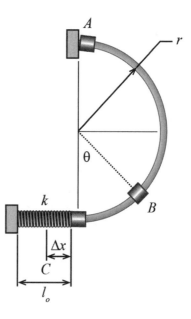

Given: W = 12 oz v_A = 0
        v_C = 0 k = 11 lb/ft
        r = 1.2 ft θ = 45°
        No Friction

Find: Δx, N_B

Solution:

Determine the maximum compression of the spring and the speed of the collar at state B.	Draw the free-body diagram of the collar at B.
	Determine the normal force at B.
Δx = _____	
v_B = _____	N_B = _____

RP7-17) A light flywheel rotates freely on a shaft located at its center (point O). The flywheel is attached to two masses as shown in the figure. Mass A (m_A = 50 kg) is attached to the inner radius (r_i = 10 cm) of the flywheel though an inextensible rope. Mass B (m_B = 20 kg) slides along a rough horizontal surface (μ_k = 0.3) and is attached to the flywheel's outer radius (r_o = 20 cm) through an inextensible rope. If the flywheel is released from rest, calculate the wheel's angular velocity after mass B has moved 10 cm.

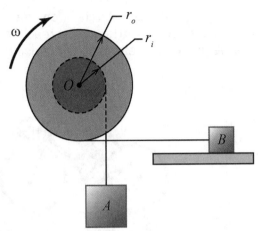

Given: m_A = 50 kg m_B = 20 kg
        r_i = 10 cm r_o = 20 cm
        μ_k = 0.3 d_B = 10 cm
        v_1 = 0

Find: ω

Solution:

Is this a conservative or non-conservative system?

Conservative   Non-conservative

Draw a free-body diagram of mass B.

Determine the friction force acting on mass B.

F_{fk} = _____

What is the relationship between the speed of mass A, the speed of mass B and the angular velocity?

v_A = _____ ω v_B = _____ ω

How far does mass A drop?

h_A = _____

Use the energy balance equation to determine the angular velocity of the flywheel.

ω = _____

RP7-18) A bicyclist, starting from rest on level ground, accelerates according to the power curve shown. The cyclist reaches his cruising speed in 1 minute. If the rider and bike weigh 160 lb, determine the cruising speed. Neglect air and rolling resistance.

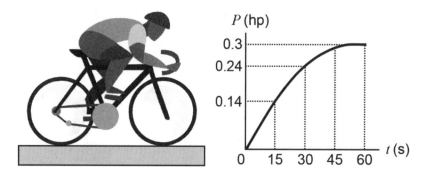

Given: W = 160 lb Δt = 1 min

Find: v

Solution:

Estimate the total work done by the cyclist for the first 1 minute.

Calculate the cruising speed.

U = _____ ft-lb

v = _____ mph

P7.1) BASIC LEVEL WORK PROBLEMS

P7.1-1)^{fe} A horizontal force pushes a particle along a horizontal surface. The force varies with the particle's displacement as shown in the graph. Determine the work done by this force.

a) $U = 12$ J   b) $U = 6$ J

c) $U = 3$ J   d) $U = 0$

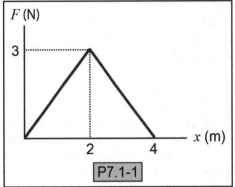

P7.1-1

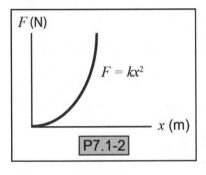

P7.1-2

P7.1-2)^{fe} A nonlinear spring is pulled by force F. The force required to deform the spring is given by $F = kx^2$, where x is the spring displacement beyond its unstretched length and k is the spring constant. Determine the work done by F as a function of spring displacement.

a) $U = kx^3/2$ b) $U = kx^3/3$

c) $U = kx^2$ d) $U = kx$

P7.1-3)^{fe} A 10-N force **F** pushes a 2-kg collar along the curved rod shown from position A to position B while maintaining a constant direction. Determine the work done by the force **F**. You may assume that the dimensions of the collar are sufficiently small to be neglected.

a) $U_F = 7$ J   b) $U_F = -7$ J

c) $U_F = 0.62$ J   d) $U_F = -0.62$ J

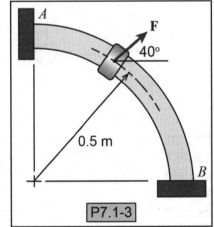

P7.1-3

P7.2) INTERMEDIATE LEVEL WORK PROBLEMS

P7.2-1) A 40-lb crate is displaced a distance of 2 ft up a 20-degree inclined surface by a constant 70-lb force F applied to a massless cord that passes over a frictionless pulley as shown. If the coefficient of kinetic friction between the crate and the inclined surface is 0.25, determine the total work done on the crate.

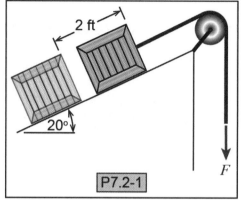

Ans: U_{1-2} = 93.8 lb-ft

P7.2-2)[fe] A pendulum is initially hanging straight down when a 1-N force **F** is applied to the bob of the pendulum. In one case the force **F** remains horizontal as it pushes the bob, while in another case the force **F** changes direction so as to always point in the direction of the bob's motion. Calculate the work done by the force **F** in moving the pendulum from its initial position to an angle of 60 degrees in each of the two cases.

(1)   a) U_F = 1 J   b) U_F = 1.21 J   c) U_F = 1.73 J   d) U_F = 2.4 J

(2)   a) U_F = 2 J   b) U_F = 2.09 J   c) U_F = 120 J   d) U_F = 122 J

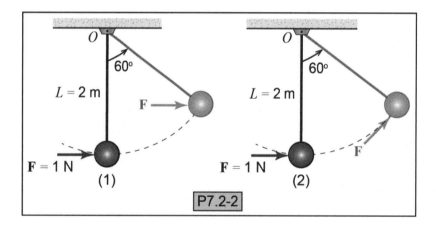

P7.2-3) The man shown applies a constant 100-N force to the end of the cord at B and lifts the 5-kg collar A up the smooth vertical shaft a distance of 2 m. If the distances shown in the figure at the beginning of the collar's motion are h = 8 m and D = 5 m, determine the total work done on the collar.

Ans: U_{1-2} = 64.3 J

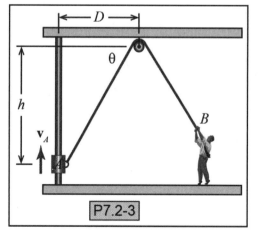

P7.3) BASIC LEVEL WORK-ENERGY PROBLEMS

P7.3-1)^{fe} A collar moves on a shaft under the action of force **F**. Force **F** remains horizontal throughout the motion of the collar. Determine an expression for the velocity of the collar as it reaches position B if it starts from rest at position A. Friction may be neglected.

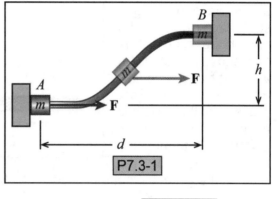

P7.3-1

a) $v_B = \sqrt{\dfrac{2(Fd - mgh)}{m}}$

b) $v_B = \sqrt{\dfrac{Fd - mgh}{m}}$

c) $v_B = \dfrac{2(Fd - mgh)}{m}$

d) $v_B = \sqrt{\dfrac{2(Fd + mgh)}{m}}$

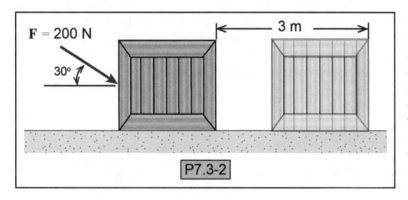

P7.3-2

P7.3-2)^{fe} A 50-kg crate is moved from rest by a 200-N force applied at an angle of 30 degrees as shown. If the coefficient of kinetic friction between the crate and the ground is 0.20, determine the velocity of the crate after it has moved 3 m to the right.

a) $v = 3.2$ m/s   b) $v = 3$ m/s   c) $v = 2.57$ m/s   d) $v = 2.12$ m/s

P7.3-3)^{fe} A 15,000-kg train car rolls up a 5%-grade track. Initially, the train car is traveling at 40 kph. After the car has coasted for a distance of $d = 50$ m, the brakes are applied. The braking force is constant and equal to 2% of the train car's weight. Neglecting rolling resistance, determine the total distance D that the train car travels before coming to a complete stop.

a) $D = 104.3$ m   a) $D = 110.5$ m   a) $D = 125.7$ m   a) $D = 132.9$ m

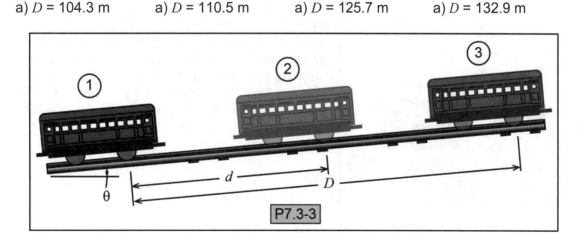

P7.3-3

P7.3-4) A 50-lb little girl is sliding down a slide as shown above. She starts from rest at the top of the slide. She descends a total of $h = 20$ ft before she flies off the end of the slide. The end of the slide has a radius of curvature of $\rho = 10$ ft. Determine the force exerted by the slide on the girl at the bottom. Neglect friction and air resistance.

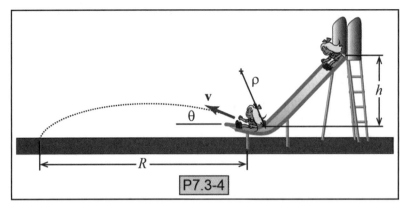

P7.3-4

Ans: $N = 150.1$ lb

P7.3-5

P7.3-5) A 60-lb child swings on a swing. The child reaches a maximum height when the swing chain creates an angle of $\theta = 60°$ with respect to the vertical. If the length of the swing chain is $L = 10$ ft, determine the force exerted by the swing on the child at the bottom of the swing motion.

Ans: $N = 120$ lb

P7.3-6)^{fe} Consider the pendulum shown where the bob is released from rest at position 1 and swings to position 2 after the string comes in contact with the vertical wall. Conservation of energy can be employed to analyze the motion of this pendulum. Explain why this is so. Now using the fact that energy is conserved, determine the minimum angle θ necessary for the pendulum to swing to position 2.

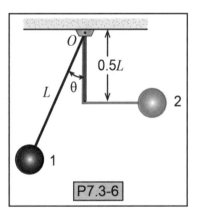

P7.3-6

a) $\theta = 20°$ b) $\theta = 40°$ c) $\theta = 60°$ d) $\theta = 80°$

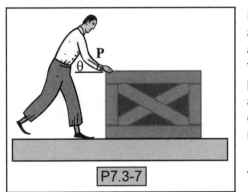

P7.3-7

P7.3-7) A man pushes a 100-lb wood crate along a painted concrete floor. If he wishes to accelerate the crate to at least 3 mph starting from rest in a distance of 3 ft, what constant pushing force (**P**) is needed if it is applied at an angle of $\theta = 35°$? The kinetic and static coefficients of friction are 0.2 and 0.28, respectively.

Ans: $P = 42.6$ lb

P7.3-8) A mass (m = 2 kg) is attached to two springs as shown in the figure. Initially, at state A, the springs are un-stretched. Force **F** pushes the mass from state A to state B, compressing/extending the springs by x_B = 5 cm. The mass is released from rest at state B. What is the speed of the mass as it passes its initial position at state A? The spring constants are k_1 = 200 N/m and k_2 = 300 N/m. The coefficient of kinetic friction is 0.3 and the springs do not support vertical weight.

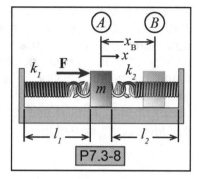

Ans: v_A = 0.58 m/s

P7.4) INTERMEDIATE LEVEL WORK-ENERGY PROBLEMS

P7.4-1) A 5-kg mass is attached to a linear spring and is free to move up and down between two supports. The mass is released from rest at position 1 where the spring is compressed a distance Δx. After the mass is released it reaches a height of h = 0.2 m above position 1 at which time it reverses its direction of motion. Determine the initial spring compression Δx if the spring has a stiffness of 3000 N/m. Also determine the speed of the mass at position 3.

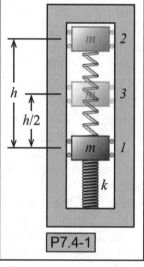

Ans: Δx = 0.116 m, v_3 = 245 m/s

P7.4-2) A pickup truck (W = 3200 lb) loses control and runs off of the road such that it ends up heading straight for a brick wall. When the truck is d = 25 ft from the wall and driving 60 mph, the driver slams on the brakes causing the truck to skid. What is the truck's velocity when it reaches the wall? What is the maximum amount of energy that the truck's bumper could possibly absorb during this crash? Estimate the friction characteristic between the tire and road as the high value of that between rubber and asphalt.

Ans: ΔE = 320,402 lb-ft, v = 80.3 ft/s

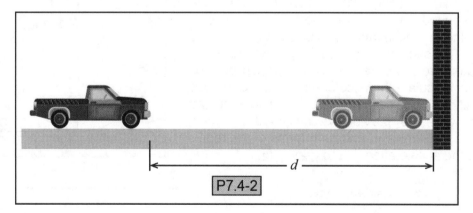

P7.4-3) A collar slides on a smooth circular shaft. The collar is released from rest at position A, slides down and then up the shaft under the influence of gravity until it reaches position E which is at the same height as position A. Determine the speed and the magnitude of the acceleration of the collar at positions A, B, C, D and E as a function of g, r and θ. Neglect the size of the collar.

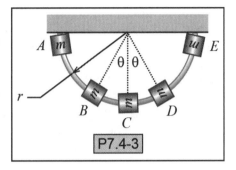

P7.4-3

Ans: $v_B = v_D = \sqrt{2gr\cos\theta}$, $v_C = \sqrt{2gr}$, $v_E = 0$, $\mathbf{a}_A = g\,\mathbf{e}_t$, $\mathbf{a}_C = 2g\,\mathbf{e}_n$,

$\mathbf{a}_B = g(\sin\theta\,\mathbf{e}_t + 2\cos\theta\,\mathbf{e}_n)$, $\mathbf{a}_D = g(-\sin\theta\,\mathbf{e}_t + 2\cos\theta\,\mathbf{e}_n)$, $\mathbf{a}_E = -g\,\mathbf{e}_t$

P7.4-4) A 7-kg mass slides down an inclined surface ($\theta = 40°$) until it hits and compresses a spring. The spring is a hard spring with a spring force that is defined by the equation $F_s = kx^2$, where k = 1000 N/m². The mass is released from rest at position 1 and slides down the inclined surface d = 2 m before it encounters the spring. If the coefficient of kinetic friction between the mass and inclined surface is

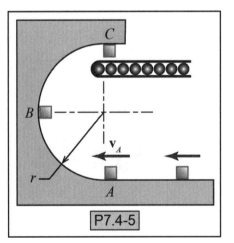

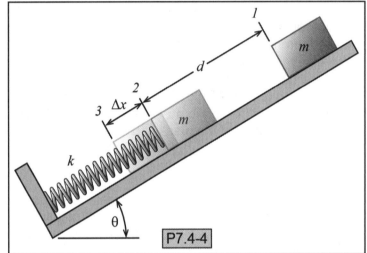

P7.4-4

0.62, determine the maximum compression of the spring Δx.

Ans: Δx = 0.44 m

P7.4-5) Packages travel on a smooth track that delivers them to a conveyor belt. Initially, the packages move along a flat section of the track and then proceed to a curved portion of radius r. Determine the minimum entry velocity to the curved portion (v_A) such that the packages will not start separating from the track until they reach position C. Also, calculate the acceleration of the packages at position B if the packages enter the curved section with the previously calculated minimum velocity. Determine v_A and $\mathbf{a}_B$ in terms of the radius (r) and the acceleration due to gravity (g).

Ans: $v_A = \sqrt{5rg}$, $\mathbf{a}_B = -g\mathbf{e}_t + 3g\mathbf{e}_n$

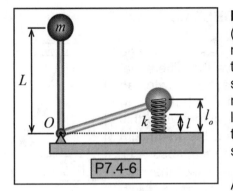

P7.4-6

P7.4-6)   A pendulum, consisting of a sphere ($m = 0.5$ kg) and a light rod pinned at O, starts from rest in the vertical position. The pendulum is allowed to fall in the clockwise direction until it compresses a spring ($k = 2000$ N/m) and, momentarily, comes to rest. If the spring is originally 12 centimeters in length, determine the shortest compressed length of the spring. The length of the rod is 1 m and the sphere may be treated as a particle.

Ans: $l = 5.18$ cm

P7.4-7) A 5-kg block compresses a nonlinear spring a distance of 0.5 m, where the spring generates a force according to the equation $F_s = 2100x^2$ where x is in meters and F_s is in Newtons. If the block is released from rest and the height of the frictionless ramp is $d = 1$m, determine the velocity with which the

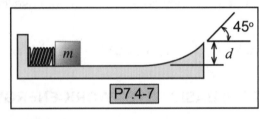

P7.4-7

block leaves the ramp and the maximum height reached by the block after it leaves the ramp. You may neglect air drag. If the ramp can no longer be considered frictionless, what further information would you need to solve this problem?

Ans: $v = 3.93$ m/s, $h = 1.39$ m

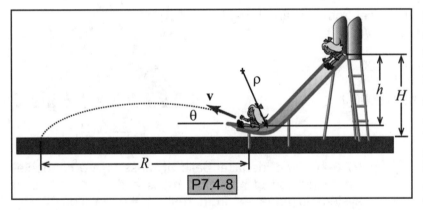

P7.4-8

P7.4-8) A little girl is sliding down a slide as shown. She starts from rest at the top of the slide.   She descends a total of $h = 18$ ft before she flies off the end of the slide. The end of the slide is curved causing her to leave the end of the slide at an angle of $\theta = 15°$ relative to the horizontal. Determine the horizontal distance she travels in the air. Neglect friction and air resistance.

Ans: $R = 23.7$ ft

P7.5) ADVANCED LEVEL WORK-ENERGY PROBLEMS

P7.5-1) Consider the pendulum shown where the bob has mass m and the light cord has length L and is inextensible. Derive the differential equation of motion for this system employing (a) a Newtonian mechanics approach and (b) a work-energy approach. Hint: derive an expression for the total energy of the pendulum then differentiate this expression with respect to time. Then, assuming that the pendulum swings through small angles (allows approximation $\sin\theta \approx \theta$, $\cos\theta \approx 1$), solve for the motion $\theta(t)$ of the pendulum for the case that the pendulum is released from rest from an angle of $\theta = 5$ degrees. With what frequency does the pendulum oscillate?

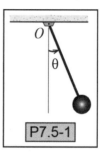

P7.5-1

Ans: $\ddot{\theta} + \dfrac{g}{L}\sin\theta = 0$, $\theta(t) = \dfrac{\pi t}{36}\cos\sqrt{\dfrac{g}{L}}$, $\omega = \sqrt{\dfrac{g}{L}}$ rad/s

P7.6) BASIC LEVEL WORK-ENERGY SYSTEM PROBLEMS

P7.6-1)^{fe} The Ferris wheel shown consists of 8 cars that each have a mass of approximately 500 kg when loaded with riders. The cars are located about the point O at a radius of 8 m. Neglecting the overall change in potential energy of the system and the mass of the Ferris wheel structure, estimate the work required to start the Ferris wheel from rest and bring it to an angular speed of 0.25 rad/sec.

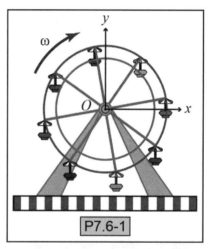

P7.6-1

a) $U = 16000$ J   b) $U = 500$ J

c) $U = 1000$ J   c) $U = 8000$ J

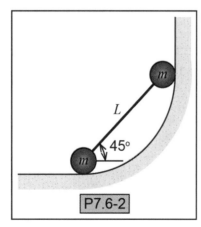

P7.6-2

P7.6-2)^{fe} Two identical balls of mass m are connected via a rigid rod of negligible mass and length L. If the system is released from rest at the position shown, determine the speed of the system when it reaches the bottom of the slope. Assume the sliding surface has negligible friction and the assembly slides in the vertical plane.

a) $v = 0.54\sqrt{gL}$ b) $v = 1.19\sqrt{gL}$

c) $v = 0.71\sqrt{gL}$ d) $v = 0.84\sqrt{gL}$

P7.7) INTERMEDIATE LEVEL WORK-SYSTEM ENERGY PROBLEMS

P7.7-1) Crate A with a mass of 50 kg is attached to load B of 30 kg by the massless, inextensible cord shown. If the system is released from rest, determine the speed of each mass after load B falls 2 m. The coefficient of kinetic friction between the crate and the inclined surface is 0.15 and the pulley can be considered massless.

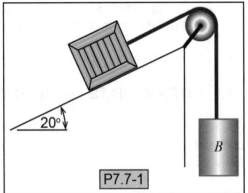

Ans: $v = 1.69$ m/s

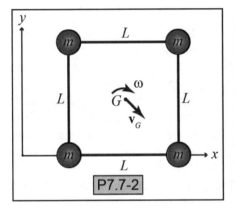

P7.7-2) A system consisting of four 5-kg spheres is connected by four 10 m long rigid, but light, bars. The structure is translating and rotating across the horizontal xy-plane in the direction shown. If the velocity of the mass center is $\mathbf{v}_G = 2\mathbf{i} - 2\mathbf{j}$ m/s and the system's overall kinetic energy is 100 N-m at the instant shown, determine the system's rotational speed ω.

Ans: $\omega = 0.2$ rad/s

P7.7-3) Consider the pendulum shown consisting of two 5-kg spheres connected by a rigid, but light, structure where $L = 2$ m. If the pendulum is rotating with an angular speed of 1 rad/sec counterclockwise at the instant shown, determine the angle the pendulum will swing through before momentarily coming to rest. Note: you may need to use a calculator/computer to help you solve the resulting equations.

Ans: $\theta = 11.55°$

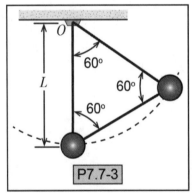

P7.8) BASIC LEVEL POWER PROBLEMS

P7.8-1)[fe] A women and her bike together weigh 150 lb. If the woman is riding up a 10%-percent grade with a constant speed of 10 mi/hr, determine the amount of power she is delivering.

a) $P = 0.368$ hp   b) $P = 0.473$ hp   c) $P = 0.528$ hp   d) $P = 0.691$ hp

P7.8-2) A 1200-kg electric vehicle is brought to rest from a speed of 55 km/h on level ground in the span of 10 seconds by its regenerative braking system. If the regeneration process has an efficiency of 60%, determine the average power that must be absorbed by the battery during this braking event.

Ans: P_{ave} = 8.40 kW

P7.9) INTERMEDIATE LEVEL POWER PROBLEMS

P7.9-1) A 1400-kg car starts from rest at the bottom of 5-percent grade and accelerates uniformly over a distance of 150 m to reach a speed of 80 km/h. Determine the power delivered by the car's drive wheels when it reaches this top speed.

Ans: P = 66.4 kW

P7.9-2) One limitation of existing electric vehicle battery technology is its limited power density (the amount of mass required to absorb energy at a particular rate). If a typical battery has a power density of 400 W/kg, determine the size of battery needed to absorb all of the energy generated by bringing a 2500-kg delivery truck to rest from a speed of 40 km/h in the space of 100 m on level ground. Assume that the braking event decelerates the truck uniformly and the efficiency of the regenerative braking system is 50%. An alternative to hybrid electric vehicles are hydraulic hybrid vehicles which store energy in pressurized fluid. Hydraulic hybrids have advantages in that they can be more efficient and typically have greater power density. Determine the mass of the system required for the same situation described above if the power density of the hydraulic regeneration system is 600 W/kg and the regeneration process is 70% efficient.

Ans: m_e = 21.5 kg for ε_e = 50%, m_h = 20 kg for ε_h = 70%

CHAPTER 7 COMPUTER PROBLEMS

C7-1) A roller coaster car is lifted to the top of the first peak of the roller coaster shown in the figure and allowed to fall from rest down the slope on the other side. The coaster parameters are as follows: $h = 30$ m, $R = 15$ m, $L = 15$ m, $R_1 = 12$ m, $R_2 = 10$ m, $m = 450$ kg (car and passengers) and $m_p = 70$ kg (passenger).

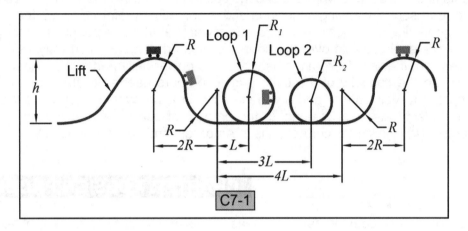

Assuming that the track is frictionless,
a) Calculate the work required to lift the car up to the first peak.
b) If the car is released from rest at the top of the first hill, plot the speed of the car versus distance traveled until the car reaches the final hill.
c) If the car is released from rest at the top of the first hill, plot the kinetic and potential energy of the car versus distance traveled until the car reaches the final hill.
d) Why is the speed of the car at the top of the second loop larger than that of the first loop?
e) Calculate the centripetal force that a passenger feels at the top of the first loop.
f) Calculate the centripetal force that a passenger feels at the top of the second loop.
g) Identify two reasons why the centripetal force of the second loop is larger than that of the first loop.
h) Determine the maximum radius of the first loop such that a passenger will remain in his/her seat at the top of the loop.

If friction is not neglected and equal to a constant 100 N,
i) If the car is released from rest at the top of the first hill, plot the speed of the car versus distance traveled until the car reaches the final hill.
j) If the car is released from rest at the top of the first hill, plot the kinetic and potential energy and the work performed on the car versus distance traveled until the car reaches the final hill.
k) Will the passengers fall out of their seats if the track has friction?

C7-2) The U.S. Environmental Protection Agency prescribes a few different drive cycles that it employs for testing vehicle emissions and fuel economy. More specifically, a drive cycle specifies a vehicle's velocity as a function of time. Automobile manufacturers also use these standard drive cycles for testing in order to enable comparison between various designs. Download the file for the Urban Dynamometer Drive Schedule (uddscol.txt) from http://www.epa.gov/nvfel/testing/dynamometer.htm. The first column represents time in seconds while the second column is speed in mph. This drive cycle is meant to represent city driving conditions and is sometimes referred to as "the city test." Employing this file, numerically estimate the power that must be supplied by the battery in a 4000-lb electric vehicle over the course of this drive cycle. You may assume that the efficiency of the vehicle from the battery to the power delivered at the wheels is 50%. Also estimate the capacity of the battery in amp-hours needed to traverse the course assuming a 150-Volt battery and a regenerative braking system that is 60% efficient. Recall that electrical power is equal to the product of current and voltage (i.e $P = IV$), where power (P) is in Watts, current (I) is in Amperes, and voltage (V) is in Volts.

CHAPTER 7 DESIGN PROBLEMS

D7-1) Design a rollercoaster that has at least 3 hills of different heights and at least one vertical loop. Research and take into account the following parameters.

- Design limitations put forth by code such as maximum height and maximum acceleration.
- The typical amount of resistance force that a car experiences as it travels down the track (e.g. bearing friction, aerodynamic drag).

Perform the following calculations.

- Verify that a rider will not fall out of the car at any point where this is a possibility.
- Verify that the rollercoaster car will make it to the end of the ride without assistance after being released from the initial hill.

A7-1) This activity gives students practice using the energy method to determine the resistance force of different surfaces.

Group size: 1 - 3 students

Supplies

- Construction materials (e.g. plywood, fasteners, glue)
- Tools (e.g. hammer, saw, screwdriver)
- 3 different surfaces (e.g. carpet, linoleum, towel, sandpaper, rubber)
- Toy car
- Hook and loop fasteners

Experimental setup

Construct a short ramp that allows different surfaces to be attached to it with hook and loop fasteners.

Procedure

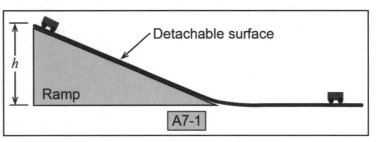

- Place the toy car at the top of the ramp and measure its height relative to the ground.
- Release the car from rest.
- Measure the total distance that the car travels before coming to rest.
- Using the concept of work and energy, determine the total force that resisted the motion of the car.
- Repeat steps 1-4 two more times on the same surface.
- Repeat steps 1-5 for the other two surfaces.

Calculations and Discussion

- Compare and contrast the results of the three surfaces giving a plausible explanation for the observed differences.
- List all of the assumptions that were applied in your calculations.
- Discuss your results
- List possible sources of experimental error.

CHAPTER 7 REFERENCES

[1] T. Gillespie, *"Fundamentals of Vehicle Dynamics"* , Society of Automotive Engineers Inc, March 1992, ISBN: 978-1560911999

PART IV: KINETICS - WORK AND ENERGY

CHAPTER 8: RIGID-BODY WORK AND ENERGY

CHAPTER OUTLINE

CHAPTER SUMMARY

In the chapter on *Particle Work and Energy*, we investigated a work-energy approach for analyzing the motion of particles. In this chapter, we revisit the work-energy method of analysis. This time we will apply it to rigid bodies. Again, the primary distinction with rigid bodies is that they have size and can rotate as well as translate. This distinction does not greatly affect the concepts of work and potential energy, but it does significantly change the manner in which we determine the kinetic energy of a rigid body. The kinetic energy of a rigid body must account for its translational motion as well as its rotational motion. This fact is motivated by first examining the kinetic energy of systems of particles, then using the results to provide intuition for the determination of the kinetic energy of rigid bodies.

8.1) WORK-ENERGY FOR SYSTEMS OF PARTICLES

In the chapter on *Particle Work and Energy,* we investigated a work-energy approach to analyzing the motion of particles. Here we address how to analyze a system of particles using the same approach. As we learned in the discussion of a Newtonian mechanics approach to analysis, a system of particles can be analyzed by either examining the particles individually or by analyzing the system as a whole. An advantage of defining your system to include multiple particles is that certain internal forces do not need to be explicitly included in the analysis. Our examination of the application of the work-energy method to a system of particles will also provide intuition on how to apply this method to rigid bodies.

8.1.1) WORK

In the chapter on *Particle Work and Energy,* we learned that **work** is the amount of energy transferred by a force acting through a distance and is calculated by using Equation 8.1-1.

> Units of Work
> SI units:
> - Newton-meter [N-m]
> - Joule [J = 1 N-m]
> - erg [erg = 1 x 10^{-7} J]
>
> US customary units:
> - foot-pound [ft-lb = 1.3558 J]
> - kilocalorie [kcal = 4187 J]
> - British thermal unit [Btu = 778.16 ft-lb = 1055 J]

Work:
$$U_{1\text{-}2} = \int_{\mathbf{r}_1}^{\mathbf{r}_2} \mathbf{F} \cdot d\mathbf{r} \quad (8.1\text{-}1)$$

$U_{1\text{-}2}$ = work done by $\mathbf{F}$ from $\mathbf{r}_1$ to $\mathbf{r}_2$ (work is a scalar)
$\mathbf{F}$ = force vector
$\mathbf{r}$ = position vector

> **Work** is the amount of energy transferred by a force acting through a distance.

Consider the system of particles shown in Figure 8.1-1. The system is acted upon by external forces $\mathbf{F}_i$ and internal forces $\mathbf{f}_i$. As mentioned previously, the system may be analyzed as a whole or as individual particles. The energy transferred to the whole system is calculated by only considering the external forces. The work done by the internal forces sum to zero if the system is rigid and connected by frictionless connections. This simplification follows from the fact that each pair of interaction forces have equal magnitudes, opposite directions and are applied over the same path. However, if the energy transferred to an individual particle is calculated, the work done by the internal as well as the external forces that act on the particle need to be included.

The work done by a force is calculated in the same manner whether the force is applied to a particle or a system of particles. One distinction is that a system of particles when analyzed as a whole has size. The external forces may be applied at a distance from the center of mass G. Therefore, the work done by a force depends on the displacement of the point of application of the force, not necessarily on the displacement of the body's mass center.

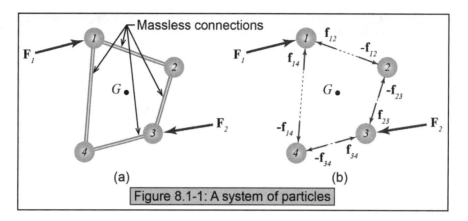

Figure 8.1-1: A system of particles

8.1.2) KINETIC ENERGY

In a system of multiple particles, the overall kinetic energy for the system is simply equal to the sum of the energies of the individual particles. The kinetic energy for a single particle is given by Equation 8.1-2. Summing the energies of the individual particles in a system gives us Equation 8.1-3.

Units of Kinetic Energy
SI units:
- Newton-meter [N-m]
- Joule [J = 1 N-m]
- erg [erg = 1 x 10^{-7} J]

US customary units:
- foot-pound [ft·lb = 1.3558 J]
- kilocalorie [kcal = 4187 J]
- British thermal unit [Btu = 778.16 ft-lb = 1055 J]

Kinetic energy for a single particle: $$T = \frac{1}{2}mv^2$$ (8.1-2)

Kinetic energy for a system of particles: $$T = \sum \frac{1}{2}m_i v_i^2$$ (8.1-3)

T = kinetic energy of the system
m_i = mass of the ith particle
v_i = speed of the ith particle

Kinetic energy is the energy possessed by a body due to its motion.

Since a system of particles has size, rotation is possible. When a system of particles is rotating, it may be inconvenient to calculate the individual speeds of each particle. Equation 8.1-4 may be used to analyze the kinetic energy of a rigid system of particles using the speed of its mass center G and its angular speed (ω). A system may be considered rigid if the positions of the individual particles relative to each other are constant. A physical interpretation of Equation 8.1-4 is as follows: the first term can be thought of as the kinetic energy due to the translation of the system of particles, and the second term can be thought of as the kinetic energy due to the rotation of the system.

Kinetic energy for a rigid system of particles: $\boxed{T = \dfrac{1}{2}mv_G^2 + \dfrac{1}{2}I_G\omega^2}$ (8.1-4)

T = kinetic energy of the system
m_i = mass of the ith particle
m = mass of the system
ω = system's angular speed

v_i = speed of the ith particle
v_G = speed of the system's center of mass G
I_G = the system's mass moment of inertia
    relative to the system's mass center G

Equation Derivation

The origins of Equation 8.1-3 should be clear based on what we learned about the kinetic energy of a single particle (Equation 8.1-2). However, intuition for the version of kinetic energy given in Equation 8.1-4 may be less obvious. To help understand the origins of this equation, a derivation is given below, beginning with Equation 8.1-3.

$$T = \sum \frac{1}{2}m_i v_i^2$$

The velocity of the individual particles $\mathbf{v}_i$ may be expressed in terms of their motion relative to the system's center of mass G (i.e. $\mathbf{v}_i = \mathbf{v}_G + \mathbf{v}_{i/G}$). Therefore, the above expression for kinetic energy T can be rewritten in terms of the velocity of the system's mass center $\mathbf{v}_G$ and the motion of the particles relative to the mass center $\mathbf{v}_{i/G}$, where m is the total mass of the system.

$$T = \sum \frac{1}{2}m_i(\mathbf{v}_G + \mathbf{v}_{i/G})\cdot(\mathbf{v}_G + \mathbf{v}_{i/G})$$

$$= \sum \frac{1}{2}m_i(v_G^2 + 2\mathbf{v}_G\cdot\mathbf{v}_{i/G} + v_{i/G}^2)$$

$$= \frac{1}{2}\left(\sum m_i\right)v_G^2 + \mathbf{v}_G\cdot\sum m_i\mathbf{v}_{i/G} + \frac{1}{2}\sum m_i v_{i/G}^2$$

Using the definition of the center of mass, we can show that $\sum m_i\mathbf{v}_{i/G} = 0$. This follows from the fact that $\sum m_i\mathbf{r}_{i/G} = 0$ as shown below, where $\mathbf{r}_G$ is the system's mass center.

$$\sum m_i\mathbf{r}_{i/G} = \sum m_i(\mathbf{r}_i - \mathbf{r}_G) = \sum m_i\mathbf{r}_i - \sum m_i\left(\frac{\sum m_i\mathbf{r}_i}{m}\right) = \sum m_i\mathbf{r}_i - \sum m_i\mathbf{r}_i = 0$$

Differentiating the expression $\sum m_i\mathbf{r}_{i/G} = 0$ with respect to time demonstrates that $\sum m_i\mathbf{v}_{i/G} = 0$. Hence, the kinetic energy for a system of particles may be expressed as the following.

$$T = \frac{1}{2}mv_G^2 + \frac{1}{2}\sum m_i v_{i/G}^2$$

If the system of particles is rigidly connected, then each particle can be thought of as rotating about any other point fixed to the system, for example, the mass center. Applying this idea, the above expression is modified as follows.

$$T = \frac{1}{2}mv_G^2 + \frac{1}{2}\sum m_i(r_{i/G}\omega)^2 = \frac{1}{2}mv_G^2 + \left(\frac{1}{2}\sum m_i r_{i/G}^2\right)\omega^2$$

Recalling the definition for the mass moment of inertia for a system of particles, the above expression leads to Equation 8.1-4 repeated below.

$$T = \frac{1}{2}mv_G^2 + \frac{1}{2}I_G\omega^2$$

8.1.3) POTENTIAL ENERGY

In a system of multiple particles (or multiple springs), the overall potential energy of the system is simply equal to the sum of the energies of the individual components. This fact is expressed in Equations 8.1-5 and 8.1-6. The gravitational potential energy of the system may also be expressed in terms of the position of the system's mass center G, as shown in Equation 8.1-7.

Units of Potential Energy
SI units:
- Newton-meter [N-m]
- Joule [J = 1 N-m]
- erg [erg = 1 x 10^{-7} J]

US customary units:
- foot-pound [ft-lb = 1.3558 J]
- kilocalorie [kcal = 4187 J]
- British thermal unit [Btu = 778.16 ft-lb = 1055 J]

Potential Energy is energy stored due to a body's position.

Gravitational potential energy for a system of particles: $\boxed{V_g = \sum m_i g h_i}$ (8.1-5)

Elastic potential energy for a system of particles: $\boxed{V_e = \sum \frac{1}{2}k_j x_j^2}$ (8.1-6)

Gravitational potential energy for a system of particles: $\boxed{V_g = mgh_G}$ (8.1-7)

V_g = gravitational potential energy of the system
V_e = elastic potential energy of the system
m_i = mass of the ith particle
m = total mass of the system
h_i = height of the ith particle relative to a chosen datum where $V_g = 0$

h_G = height of the system's mass center G relative to a chosen datum where $V_g = 0$
k_j = spring constant of the jth spring
x_j = distance the jth spring has been stretched or compressed
g = acceleration due to gravity

8.1.4) WORK-ENERGY BALANCE

The relationships between work and energy hold for systems of particles just as easily as they do for individual particles. The notion of conservation of energy can also be applied to the analysis of systems of particles. Specifically, energy is conserved for a system of particles if there is no work done by any external or internal forces that are not gravitational or elastic. Be careful to consider any non-conservative internal forces. The work-energy balance equations for a system of particles are given by Equation 8.1-8 through 8.1-10.

Work and kinetic energy balance: $\boxed{T_1 + U_{1-2} = T_2}$ (8.1-8)

Work-energy balance: $\boxed{T_1 + V_1 + U_{1-2,non} = T_2 + V_2}$ (8.1-9)

Conservation of energy: $\boxed{T_1 + V_1 = T_2 + V_2}$ (8.1-10)

T_1 = kinetic energy at state 1
T_2 = kinetic energy at state 2
V_1 = potential energy at state 1
V_2 = potential energy at state 2

U_{1-2} = work done by all forces acting on the system between state 1 and 2
$U_{1-2,non}$ = work done by non-conservative forces on the system between state 1 and 2

Equation Summary

Abbreviated variable definition list

U = work
$\mathbf{F}$ = force
$\mathbf{r}$ = position
T = kinetic energy
m = mass
v = speed
I = mass moment of inertia

ω = angular speed
V = potential energy
h = height relative to a chosen datum where $V_g = 0$
k = spring constant
x = distance the spring has been stretched or compressed
g = acceleration due to gravity

Work

$$U_{1-2} = \int_{\mathbf{r}_1}^{\mathbf{r}_2} \mathbf{F} \cdot d\mathbf{r}$$

Kinetic energy - Rigid system of particles

$$T = \frac{1}{2} m v_G^2 + \frac{1}{2} I_G \omega^2$$

Kinetic energy - System of particles

$$T = \sum \frac{1}{2} m_i v_i^2$$

Gravitational potential energy - System of particles

$$V_g = \sum m_i g h_i$$

Gravitational potential energy - Rigid system of particles

$$V_g = m g h_G$$

Elastic potential energy for a system of particles

$$V_e = \sum \frac{1}{2} k_j x_j^2$$

Work-energy balance

$$T_1 + V_1 + U_{1-2,non} = T_2 + V_2$$

Conservation of energy

$$T_1 + V_1 = T_2 + V_2$$

Solved Problem 8.1-1

The system shown consists of two 2-lb balls that may be affixed at any position along a light horizontal shaft. The horizontal shaft is rigidly attached to a light vertical supporting shaft that is driven by a motor that applies a constant torque of $M = 20$ lb-ft. The horizontal shaft extends out from the center of the vertical shaft by $L = 3$ feet in both directions. If the system starts at rest, determine the angular velocity of the system after the system has rotated 360 degrees if the balls are affixed to a position $L/2$ from the center of rotation. Compare these results to the results if the balls are affixed to

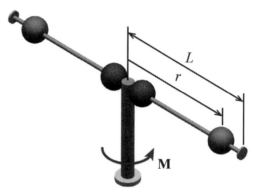

a position L from the center of rotation. Do you think there will be a difference in the results between the two cases?

Given: $W = 2$ lb
$M = 20$ lb-ft
$L = 3$ ft
$\theta_1 = 0$
$\theta_2 = 2\pi$ rad
$\omega_1 = 0$

Find: ω_2 when the balls are affixed $L/2$ and L from the center of rotation.

Solution

Getting familiar with the system

The moment applied to the system does work causing the balls to gain kinetic energy. Therefore, we will apply the work-energy balance to the system to determine the angular speed of the balls following a full rotation. The vertical and horizontal shafts are stated to be light, therefore, we will neglect the effect of their mass on the dynamics of the system.

Work

The general work equation for a system that has multiple forces is given by the following equation, where **r** is the position of the point of application of **F** not the constant radius shown in the figure.

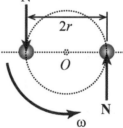

$$U_{1-2} = \int_{\mathbf{r}_1}^{\mathbf{r}_2} \sum \mathbf{F}_i \cdot d\mathbf{r}$$

The moment applied to the vertical shaft will cause the horizontal shaft to apply an equal normal force N to each ball in the directions shown in the figure. Since the shafts are treated as massless, the moments applied to the shafts must balance about the rotation center O. That is, the motor torque M will be balanced by the moment due to the normal forces, $2Nr$. Also, each normal force N travels through a distance of $s = r\theta$. Using these relationships and the fact that r is constant, the work equation becomes

$$U_{1-2} = \int_{r_1}^{r_2} \sum \mathbf{F}_i \cdot d\mathbf{r} = \int_{s_1}^{s_2} 2N \, ds = \int_{s_1}^{s_2} \frac{M}{r} \, ds = \int \frac{M}{r} \, d(r\theta) = \int_{\theta_1}^{\theta_2} M \, d\theta = M(\theta_2 - \theta_1) = M\theta_2$$

Work-energy balance

This system is not a conservative system due to the application of the motor torque. When calculating the kinetic energy of the system, note that both balls have the same mass and travel along the bar in a symmetric manner giving them equal speeds.

$$\cancel{T_1} + \cancel{V_1} + U_{1-2} = T_2 + \cancel{V_2} \qquad\qquad U_{1-2} = \frac{1}{2} \sum m_i v_{i,2}^2 = m v_2^2$$

The velocity of the balls may be written as $v = r\omega$.

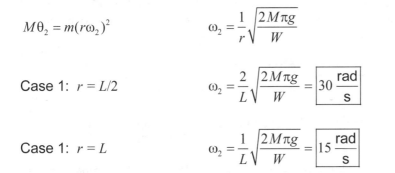

$$M\theta_2 = m(r\omega_2)^2 \qquad\qquad \omega_2 = \frac{1}{r}\sqrt{\frac{2M\pi g}{W}}$$

Case 1: $r = L/2$ $\qquad\qquad \omega_2 = \frac{2}{L}\sqrt{\frac{2M\pi g}{W}} = \boxed{30 \ \frac{\text{rad}}{\text{s}}}$

Case 1: $r = L$ $\qquad\qquad \omega_2 = \frac{1}{L}\sqrt{\frac{2M\pi g}{W}} = \boxed{15 \ \frac{\text{rad}}{\text{s}}}$

The distance that the balls are from the center of rotation affects the angular speed that the system achieves. If the balls are further out from the center, it takes more energy to accelerate them to an equal angular speed as compared to when the balls are closer to the center.

8.2) WORK

In the chapter on *Particle Work and Energy*, we learned that a force pushing or pulling an object through a distance does *work*. In the case of rigid bodies, this force can cause the body to translate as well as rotate. Despite the differences between particles and rigid bodies, the definition of work done by a force and its determination is the same whether that force is applied to a particle or a rigid body.

Units of Work
SI units:
• Newton-meter [N-m]
• Joule [J = 1 N-m]
• erg [erg = 1 x 10^{-7} J]
US customary units:
• Foot-Pound [ft-lb = 1.3558 J]
• Kilocalorie [kcal = 4187 J]
• British thermal unit [Btu = 778.16 ft-lb = 1055 J]

Because rigid bodies can rotate as well as translate, sometimes it will be inconvenient to determine the distance over which the force acts. Therefore, in some situations it is easier to determine the work done by a force based instead on the moment the force induces.

From this discussion we will also demonstrate how to determine the work done by a pure moment.

> **Work** is the amount of energy transferred by a force acting through a distance or a moment acting through an angle.

8.2.1) WORK DONE BY A FORCE

The **work** done by a force is the amount of energy transferred by a force acting through a distance. The amount of work the force does is calculated using Equation 8.2-1. Equation 8.2-1 may be rewritten and simplified for a few special cases as will be discussed.

Work done by a force: $$U_{1-2} = \int_{r_1}^{r_2} \mathbf{F} \cdot d\mathbf{r}$$ (8.2-1)

U_{1-2} = work done by **F** from position 1 to 2
F = force vector
r = position vector

If the force doing the work (**F**) is constant with respect to the position (**r**), the work equation may be simplified to give Equation 8.2-2. Positive work occurs when the force and displacement are in the same direction. Negative work occurs when the force is acting opposite to the displacement.

Work done by a constant force: $$U_{1-2} = (F\cos\theta)d_{1-2}$$ (8.2-2)

U_{1-2} = work done by **F** from position 1 to 2
F = force magnitude
θ = angle between **F** and displacement vector
d_{1-2} = displacement magnitude

As stated previously, the work done by a force is calculated in the same manner whether the force is applied to a particle or rigid body. One distinction that is helpful to recognize is that since rigid bodies have size, the force can be applied at a distance offset from the center of mass. Therefore, the work done by a force is determined based on the displacement of the point of application of the force, not necessarily based on the displacement of the body's mass center. Since rigid bodies can rotate as well as translate, it may be difficult to determine the path that a force moves through and/or the direction of the force relative to this path. If the orientation of the force relative to the body remains constant and the rigid body is undergoing pure rotation, work done by this force can be calculated using Equation 8.2-3. It can be shown that the work done by a force can be calculated using either Equation 8.2-1 or Equation 8.2-3. However, be careful not to calculate the work using both equations. Using both equations will result in the amount of work calculated being twice the actual work performed.

Work done by a force producing pure rotation: $$U_{1-2} = \int_{\theta_1}^{\theta_2} F_{\perp} r\, d\theta = \int_{\theta_1}^{\theta_2} M\, d\theta$$ (8.2-3)

r = moment arm
$F_{\perp}$ = force component that is perpendicular to the moment arm
M = moment induced by force **F**
θ = angular position

Equation Derivation

Consider for example the special case of rotation about a fixed axis where the direction of the force stays constant with respect to the body. The distance the force acts over is an arc length (s) which may be calculated by $s = r\theta$. If we start with the Equation 8.2-1, we can reduce this equation to the scalar form because the direction of the force is constant relative to the direction of motion.

$$U_{1-2} = \int_{\mathbf{r}_1}^{\mathbf{r}_2} \mathbf{F} \cdot d\mathbf{r} = \int_{s_1}^{s_2} F_\perp \, ds$$

If we substitute the arc length equation in for the displacement and recognize that $M = F_\perp r$, we arrive at Equation 8.2-3 which is repeated below.

$$U_{1-2} = \int_{\theta_1}^{\theta_2} F_\perp r \, d\theta = \int_{\theta_1}^{\theta_2} M \, d\theta$$

8.2.2) WORK DONE BY A MOMENT

Previously, we showed a special case where the work done by a force can be calculated directly using Equation 8.2-1 or based on the moment that the force induces (Equation 8.2-3). Equation 8.2-3 was derived for rotation about a fixed axis, but it also holds in the case of a pure moment that causes a rigid body to only rotate and not translate. This intuitively should make sense since a body may gain or lose energy due to the influence of this moment, hence, the moment must be doing work. The general vector form of Equation 8.2-3 is given in Equation 8.2-4. Here, the work done by a moment is calculated by multiplying the moment with the angular displacement over which it acts.

In the case of planar motion, the moment and the angular displacement both have parallel directions (i.e. positive or negative $\mathbf{k}$). Therefore, the expression for work can be written without the dot product as shown in Equation 8.2-5. As was the case with work done by a force, the work done by a moment is positive when it is in the same direction as the angular displacement and the work is negative when the moment is in the direction opposite of the angular displacement.

Work done by a moment: $$\boxed{U_{1-2} = \int_{\theta_1}^{\theta_2} \mathbf{M} \cdot d\boldsymbol{\theta}}$$ (8.2-4)

Work done by a moment (planar motion): $$\boxed{U_{1-2} = \int_{\theta_1}^{\theta_2} M \, d\theta}$$ (8.2-5)

U_{1-2} = work done by $\mathbf{M}$ from angular position 1 to 2
$\mathbf{M}$ = moment
θ = angular position

Example 8.2-1

Consider an unwinding spool fixed at its center being pulled by the rope that is wrapped around its inner hub.

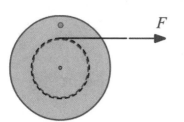

Describe the work done by force F.

Conceptual Example 8.2-2

Consider the following situations where a force of constant magnitude F is applied to a pendulum over an angular distance θ. Rank the following situations in terms of the work done from greatest to least.

Greatest _____

Next _____

Next _____

Least _____

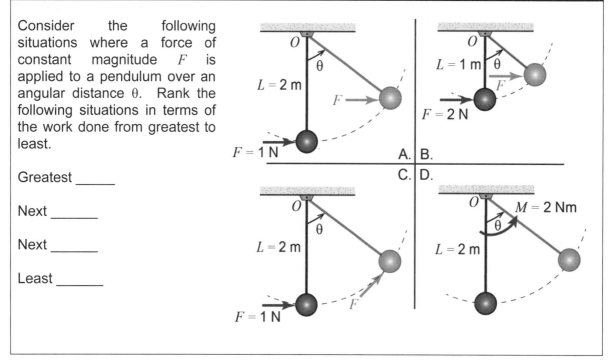

Equation Summary

Abbreviated variable definition list

U = work $\qquad\qquad\qquad\qquad\qquad\qquad\qquad\qquad$ $\mathbf{M}$ = moment
r = moment arm $\qquad\qquad\qquad\qquad\qquad\qquad\qquad$ θ = angular position
$F_{\perp}$ = force component that is perpendicular to the moment arm

Work done by a moment $\qquad\qquad\qquad\qquad$ Work done by a moment (planar motion)

$$U_{1-2} = \int_{\theta_1}^{\theta_2} \mathbf{M} \cdot d\boldsymbol{\theta}$$
$$U_{1-2} = \int_{\theta_1}^{\theta_2} M \, d\theta$$

Work done by a force producing pure rotation

$$U_{1-2} = \int_{\theta_1}^{\theta_2} F_{\perp} r \, d\theta = \int_{\theta_1}^{\theta_2} M \, d\theta$$

Example Problem 8.2-3

A 100-lb block is suspended from an inextensible cable, which is wrapped around the outer radius (r = 1.3 ft) of a flywheel. The flywheel rotates on a shaft supported by bearings. The bearings are poorly lubricated and apply a M_f = 30 lb-ft frictional moment to the flywheel. If the block drops h = 3 ft, determine the work done by the weight and the friction.

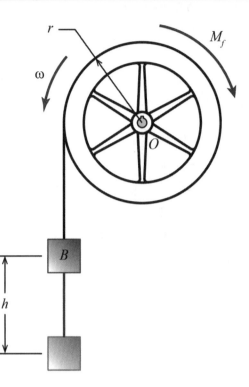

Given:

Find:

Solution:

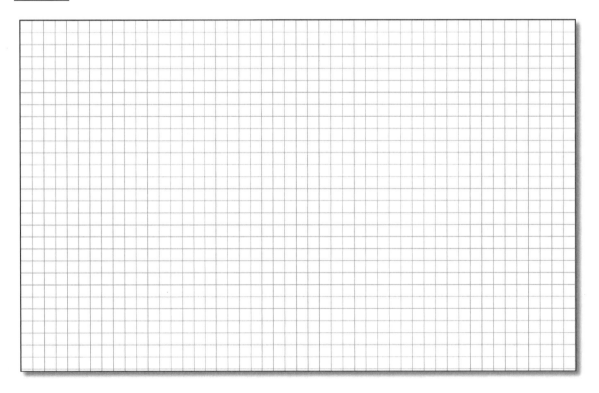

8.3) KINETIC ENERGY

In the chapter on *Particle Work and Energy*, we learned that kinetic energy is the energy of motion. This definition holds true for rigid bodies as well. The difference comes in the way that a rigid body moves. The only way that a particle can move is to translate. Therefore, a particle's kinetic energy is a function of its translational speed as shown in Equation 8.3-1.

> **Units of Kinetic Energy**
> SI units:
> - Newton-meter [N-m]
> - Joule [J = 1 N-m]
> - erg [erg = 1 x 10^{-7} J]
>
> US customary units:
> - Foot-Pound [ft-lb = 1.3558 J]
> - Kilocalorie [kcal = 4187 J]
> - British thermal unit [Btu = 778.16 ft-lb = 1055 J]

Kinetic energy of a particle: $$T = \frac{1}{2}mv^2$$ (8.3-1)

T = kinetic energy
m = mass
v = speed

| **Kinetic energy** is the energy possessed by a body due to its motion. |

A rigid body can rotate as well as translate. Therefore, you would expect the equation for the kinetic energy of a rigid body to account for its rotational motion in addition to its translational motion. A rigid body can translate so you would expect that it would still include a term that involves its linear speed (v), but it can also rotate so you may also expect that it would include a term that involves its angular speed (ω).

Whether or not the kinetic energy equation for a rigid body includes a term that involves only linear speed (v), only angular speed (ω), or both, depends on how the body is moving. In the coming sections, we will look at each case in turn and introduce the associated version of the kinetic energy equation.

Conceptual Example 8.3-1

Which bar has a higher kinetic energy during the motion?

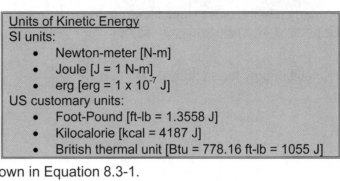

Conceptual Example 8.3-2

A body may possess kinetic energy due to its rotational motion. Using the kinetic energy for a particle as a reference, guess how the kinetic energy, due to rotation, is calculated.

Kinetic energy for a particle: $T = \dfrac{1}{2}mv^2$

Conceptual Example 8.3-3

Draw the motion of each body and then determine the form of the equation used for calculating its kinetic energy.

G

8.3.1) PURE TRANSLATION

If a rigid body is only translating, then its kinetic energy is only dependent on the linear speed of the body and is given by Equation 8.3-2. This equation is identical to the equation employed for calculating the kinetic energy of a particle. Note that in the case of pure translation, every point on the rigid body has the same velocity.

Kinetic energy for pure translation: $\boxed{T = \dfrac{1}{2}mv^2}$ (8.3-2)

T = kinetic energy
m = mass
v = speed

8.3.2) PURE ROTATION

If the rigid body is rotating about a fixed axis through point O, then its kinetic energy is determined only by the body's angular speed (ω) and is given by Equation 8.3-3. Note, this equation can also be used for a body that is rotating and translating when the point O is taken to be the instantaneous center of zero velocity for the body at that instant.

Kinetic energy for pure rotation: $\boxed{T = \dfrac{1}{2}I_O\omega^2}$ (8.3-3)

T = kinetic energy
I_O = mass moment of inertia about O
ω = angular speed (rad/s)

Equation Derivation

Some intuition for Equation 8.3-3 may be gained by considering the body as a rigid system of particles where the system has only a single angular speed ω. The kinetic energy for a system of particles is the sum of the energies of the individual particles. If the system is undergoing pure rotation, the velocity of any point on the system is given by $v = r\omega$. Substituting this relationship into the kinetic energy equation gives the following.

$$T = \sum \frac{1}{2}m_i v_i^2 = \sum \frac{1}{2}m_i(r_{i/O}\omega)^2 = \frac{1}{2}\left(\sum m_i r_{i/O}^2\right)\omega^2$$

If we consider a rigid body as a collection of an infinite number of particles, then its mass moment of inertia with respect to point O is given by $\sum m_i r^2_{i/O}$. Making this substitution gives us the final expression for kinetic energy given in Equation 8.3-3 which is repeated below.

$$T = \frac{1}{2}\left(\sum m_i r_{i/O}^2\right)\omega^2 = \frac{1}{2}I_O\omega^2$$

8.3.3) GENERAL PLANAR MOTION WITH CENTER OF MASS REFERENCE

In the most general case where a body is translating and rotating simultaneously, the body's kinetic energy is comprised of the energy due to its translational motion and the energy due to its rotational motion. Since the rigid body is rotating, the velocity of every point on the body may be different. Therefore, to calculate the translational portion of the kinetic energy we use the velocity of the center of mass of the body. If we choose the center of mass (G) as a reference point, the equation for the kinetic energy simplifies to the expression shown in Equation 8.3-4. Recall that even though different points on a rigid body may have different linear velocities, the body as a whole has only a single angular velocity since each point on the body sweeps out the same angle in the same amount of time.

Kinetic energy for general planar motion with the mass center G as a reference:

$$T = \frac{1}{2}mv_G^2 + \frac{1}{2}I_G\omega^2 \quad (8.3\text{-}4)$$

T = kinetic energy
m = mass
v_G = speed of the center of mass

I_G = mass moment of inertia about an axis passing through G
ω = angular speed (rad/s)

Conceptual Example 8.3-4

Consider the following situations where the same bar is translating and/or rotating with the given linear and angular speeds.   Rank the kinetic energy the bar in each situation from greatest to least.

Greatest _____

Next _____

Next _____

Least _____

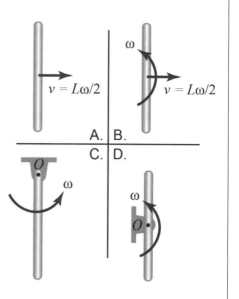

8.4) POTENTIAL ENERGY

As with a particle, the potential energy a rigid body possesses is due to its position, regardless of whether it is due to the body's position in a gravitational field or the position of the compression or expansion of a spring.

Units of Work
SI units:
- Newton-meter [N-m]
- Joule [J = 1 N-m]
- erg [erg = 1 x 10^{-7} J]
US customary units:
- foot-pound [ft-lb = 1.3558 J]
- kilocalorie [kcal = 4187 J]
- British thermal unit [Btu = 778.16 ft-lb = 1055 J]

Potential Energy is energy stored due to a body's position.

8.4.1) GRAVITATIONAL POTENTIAL ENERGY

Gravitational potential energy is the energy stored by a body due to its position in a gravitational field. Recall that for a particle, the gravitational potential energy is calculated using $V_g = mgh$.

If we look at the terms in the equation for the gravitational potential energy for a particle, what might change if we now consider a rigid body? Because a rigid body has size, each part of the rigid body has a different height. Therefore, some parts will have more potential energy than others. So where do we measure the height from? Recall that the center of mass (G) of a body captures the average location of all the "pieces" of the rigid body as weighted by their masses. Therefore, if we assume a constant gravitational field, the center of mass represents the average "height" for the entire body for the purposes of calculating the body's gravitational potential energy. This is captured in Equation 8.4-1.

$$\text{Gravitational potential energy:} \quad \boxed{V_g = mgh_G} \quad (8.4\text{-}1)$$

V_g = gravitational potential energy g = acceleration due to gravity
m = mass h_G = height of the body's center of mass (G) relative to a
                                                                                    chosen datum

8.4.2) ELASTIC POTENTIAL ENERGY

An elastic element does not care if the body it is attached to is a particle or a rigid body. The elastic potential energy equation does not change. It is still calculated using Equation 8.4-2.

$$\text{Elastic potential energy:} \quad \boxed{V_e = \frac{1}{2}kx^2} \quad (8.4\text{-}2)$$

V_e = elastic potential energy
k = spring constant
x = spring displacement relative to its unstretched length

Conceptual Example 8.4-1

Draw the motion of the object shown and discuss the change in potential energy of the pendulum after it moves.

8.5) WORK-ENERGY BALANCE

8.5.1) WORK-ENERGY BALANCE EQUATION

The relationship between work and kinetic energy for rigid bodies is the same as for a particle. The work-energy balance equation states that the initial kinetic energy plus the initial potential energy plus any non-conservative work performed on the body equals the final kinetic energy plus the final potential energy. The work-energy balance is given in Equation 8.5-1. The non-conservative work term ($U_{1\text{-}2,non}$) is the work done by all non-conservative forces and moments. Remember that the weight and elastic forces are conservative and are included in the potential energy. The work-energy principle may also be rewritten as Equation 8.5-2.

Work-energy balance: $\boxed{T_1 + V_1 + U_{1-2,non} = T_2 + V_2}$ (8.5-1)

Work-energy balance: $\boxed{U_{1-2,non} = \Delta T + \Delta V}$ (8.5-2)

T_1 = kinetic energy at state 1 V_1 = potential energy at state 1
T_2 = kinetic energy at state 2 V_2 = potential energy at state 2
ΔT = change in kinetic energy ΔV = change in potential energy
                                             $U_{1\text{-}2,non}$ = non-conservative work between state 1 and 2

8.5.2) WORK-ENERGY BALANCE FOR A SYSTEM OF RIGID BODIES

We have learned in several places that energies are additive. This means that the total energy of a system is just the sum of the individual energies of the bodies making up the system. We will not belabor this point in this section. A quick overview will be given along with some notes on what to watch out for.

Calculating work is covered in the previous work sections and will not be covered here. However, one thing to watch out for when dealing with a system of rigid bodies are internal forces. If there is significant internal friction coming from material hysteresis or poorly lubricated joints, then energy is not conserved. One interesting example of material hysteresis is the rolling friction (i.e. rolling resistance) of automobile tires. The major component of rolling friction does not come from the friction developed between the tire and the ground, it comes from material hysteresis. In this case, material hysteresis comes from the internal friction between the material's molecules. This, in turn, causes the system to lose energy through non-conservative work.

The kinetic and potential energy of a system of rigid bodies equals the sum of the energies of the individual bodies as shown in Equation 8.5-3 and 8.5-4. The individual kinetic energy equation used depends on each component of the system and how it is moving.

Kinetic energy for a system of rigid bodies: $\boxed{T = \sum T_i}$ (8.5-3)

Potential energy for a system of rigid bodies: $\boxed{V = \sum V_i}$ (8.5-4)

T = kinetic energy of the system
T_i = kinetic energy of the i_{th} rigid body

V = kinetic energy of the system
V_i = kinetic energy of the i_{th} rigid body

In the previous sections on *System of Particles*, the energy of the system could be obtained looking at the individual bodies or the system as a whole. In the case of a system of rigid bodies, it is usually very difficult to look at the system as a whole unless the system is rigid. If the system is not rigid (e.g. the system is connected through movable joints), then the components can move relative to each other. This will cause the center of mass to change locations relative to the components.

8.5.3) CONSERVATION OF ENERGY EQUATION

If there are no non-conservative forces or moments that do work, then energy is conserved for the system and Equation 8.5-5 may be applied. The great advantage of being able to employ conservation of energy is that you only need information about the body's initial and final state. You do not need any information regarding the path taken by the body between the two states.

Conservation of energy: $\boxed{T_1 + V_1 = T_2 + V_2}$ (8.5-5)

T_1 = kinetic energy at state 1
T_2 = kinetic energy at state 2

V_1 = potential energy at state 1
V_2 = potential energy at state 2

Even if there are non-conservative forces and moments applied to the body, they may not do any work. A force does no work if it has no component in the direction of the displacement. A moment does no work if there is no component in the direction of the angular displacement. With rigid bodies, the work done by a force is determined by the displacement of the point of application of the force, which may be different than the displacement of the body's mass center. An interesting example of this is rolling. Recall that when a body is rolling without slip, the point of contact of the body with the rolling surface is an instantaneous center of zero velocity. Since the friction experienced by the rolling body is also applied at this point of contact, the friction does no work. This is due to the fact that the point of application of the force has zero velocity. However, if a rolling body slips, then the friction force at the rolling contact will do work. Also, if the body is rolling without slip on a moving surface, such as a conveyer belt, then the friction force will do work.

Conceptual Example 8.5-1

A solid cylinder and a hollow tube roll without slip down an incline. Both the cylinder and tube have the same weight, radius, start from rest, and start at the same initial height. Which will reach the bottom of the incline first?

   a) the solid cylinder
   b) the hollow tube
   c) both will reach the bottom at the same time

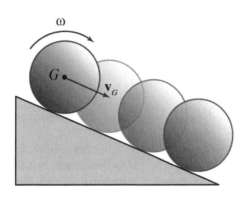 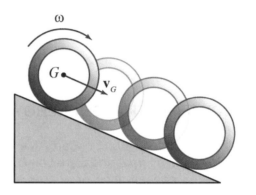

Reconsider the above problem. Which of the following has the potential to change your answer to the previous question?

   a) $m_{cylinder} > m_{tube}$
   b) $m_{cylinder} < m_{tube}$
   c) $r_{cylinder} < r_{tube}$
   d) $r_{cylinder} > r_{tube}$
   e) The tube will always be slower than the cylinder no matter what the relative values of m and r are.

Example Problem 8.5-2

Consider a 5-kg disk rolling to the right under the influence of (a) a pulling force (P = 10 N) and (b) a moment (M = 10 Nm). Assume that the disk does not slip relative to the ground and that the coefficient of static friction is μ_s = 0.45. The disk has a radius of 1.0 m. If the disk starts from rest, determine, in each case, the velocity of the disk's mass center after it has moved 5 m.

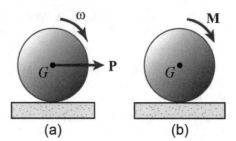

(a)   (b)

Given:

Find:

Solution

Example Problem 8.5-3

The drum shown has two blocks attached to it with inextensible rope, one to its inner hub and one to its outer hub. Block A has a mass of 50 kg and block B has a mass of 25 kg. The drum has a mass of 10 kg and a radius of gyration about the pin O of $k_O = 0.30$ m. Block A is initially 1.5 m above the floor and is moving downward at 3 m/s. To slow the decent of block A, a force of $F = 80$ N is applied to the brake arm in the location shown. Determine the speed at which Block A hits the floor.   The coefficient of kinetic friction at the brake pad is $\mu_k = 0.5$. ($L = 2$ m, $l_1 = 1.7$ m, $l_2 = 0.2$ m, $r_o = 0.3$ m, $r_i = 0.2$ m)

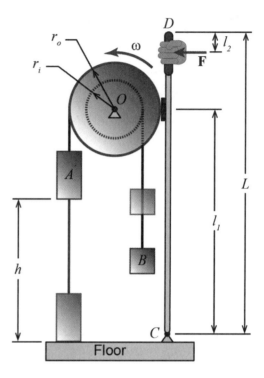

Given:

Find:

Solution:

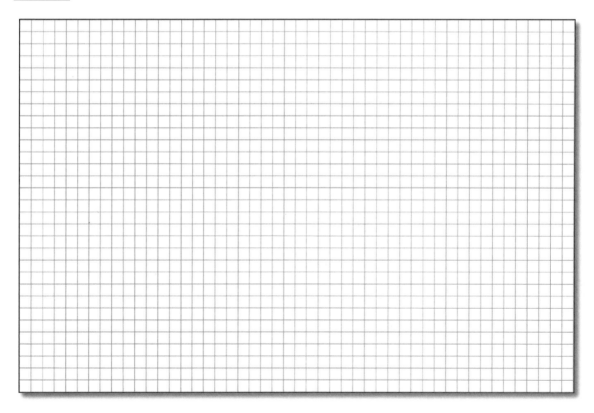

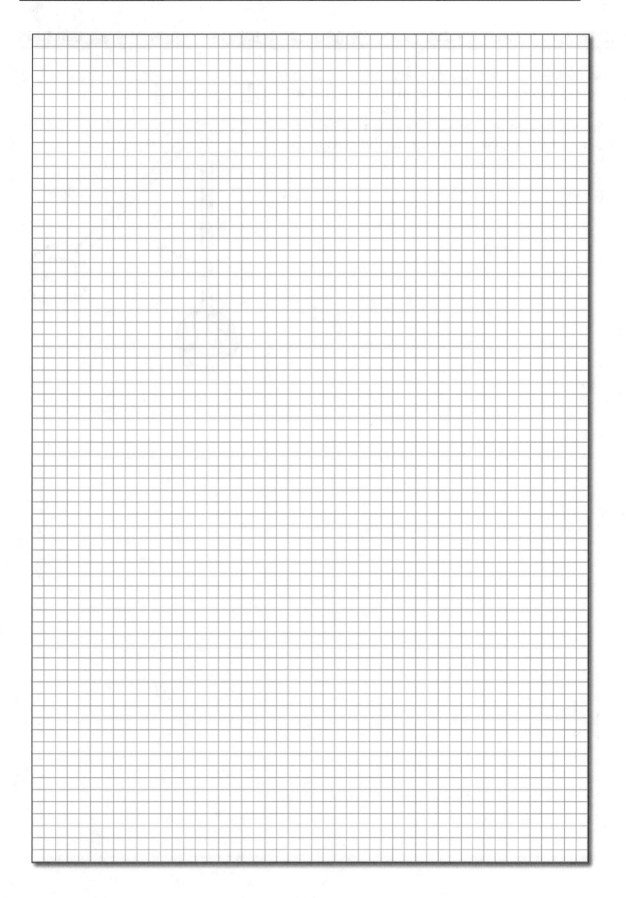

Example Problem 8.5-4

An 80-lb sphere ($r = 1.5$ in) is attached to a 20-lb slender rod ($L = 3$ ft) shown, which is pinned at point O. At the instant the pendulum is vertical, the rod has an angular velocity of $\omega = 0.5$ rad/s. Determine the angle θ to which the rod swings up to before it momentarily comes to rest.

Given:

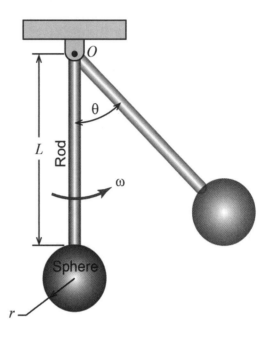

Find:

Solution:

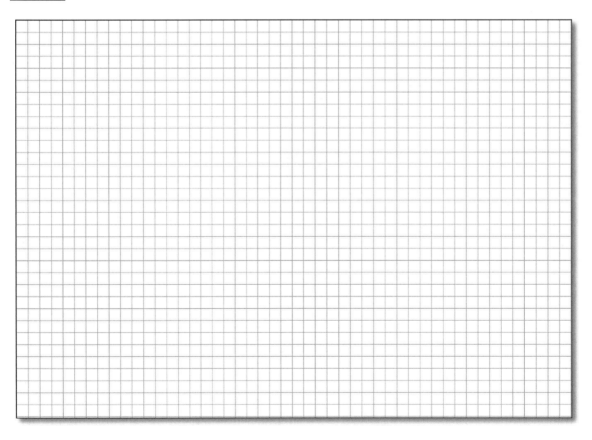

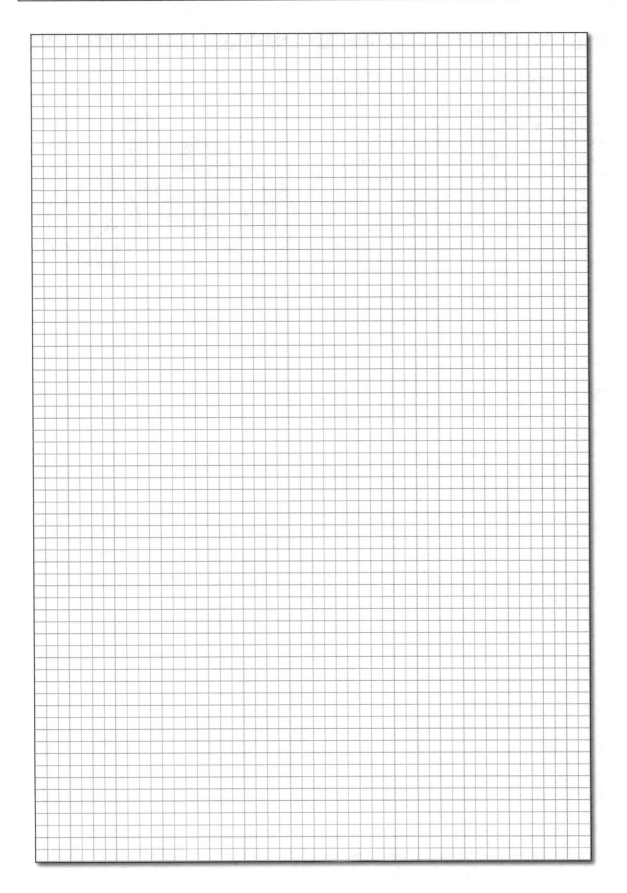

Example Problem 8.5-5

The braking system on a new concept bicycle is being tested. Both wheels of the bike have a diameter of 26 inches and weigh 1 lb. Most of that weight is concentrated on the circumference of each wheel. The bike frame weighs 2 lb and the rider weighs 140 lb. The braking system has the ability to apply a 10 lb-ft moment to each wheel when engaged. If the bike is traveling at 25 mph, what is the minimum stopping distance? You may assume that the wheels are rolling without slip.

Given:

Find:

Solution:

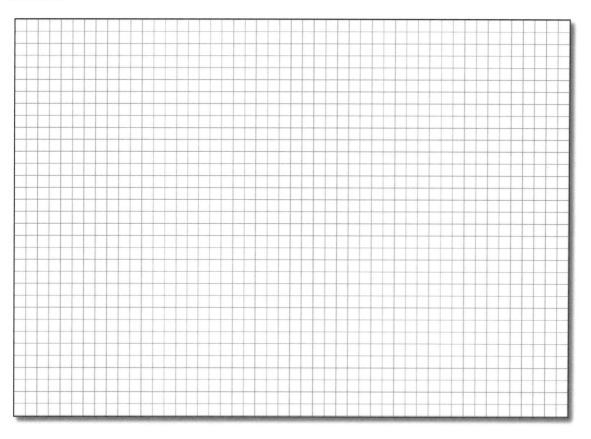

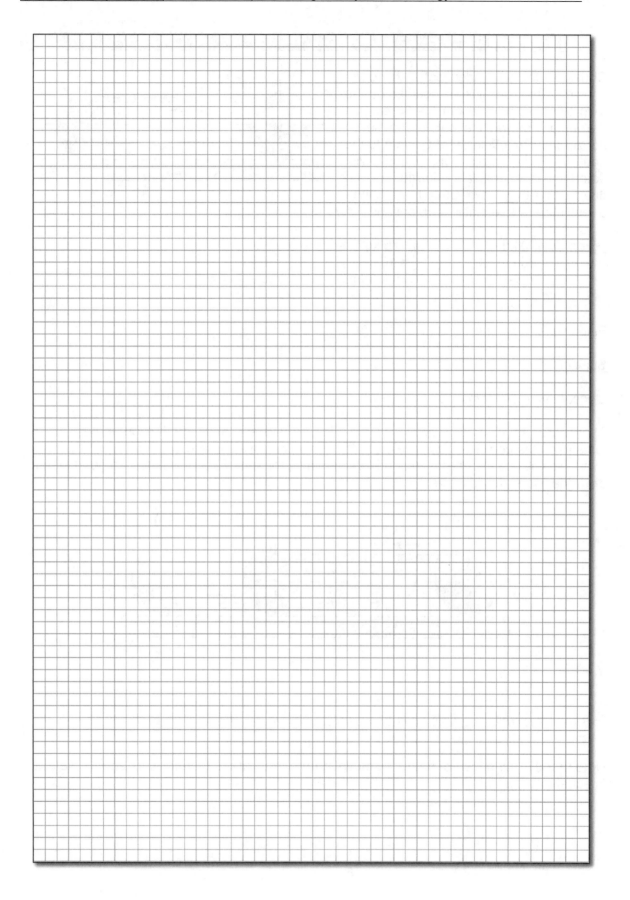

Solved Problem 8.5-6

Consider the two identical 2-kg disks shown. They are linked together by the 5-kg slender bar of length L = 30 cm. At the instant shown the bar is completely horizontal with a velocity of 0.5 m/s to the right.   If the disks roll without slip, determine the angular velocities of the disks when they have rotated one half of a revolution. Note that r_o = 10 cm and r_i = 7 cm.

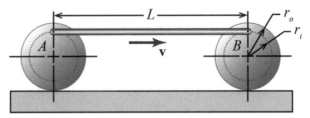

Given: m_d = 2 kg m_b = 5 kg
      L = 30 cm r_o = 10 cm
      r_i = 7 cm $\mathbf{v}_{b,1}$ = 0.5 **i** m/s
      No slip

Find: ω_2

Solution

Conservation of energy

There are no externally applied forces to the system and the friction forces act at instantaneous centers of zero velocity, therefore, this system is conservative. With this in mind, we will use the energy method to solve this problem. First, it is always a good idea to draw the two states of interest.

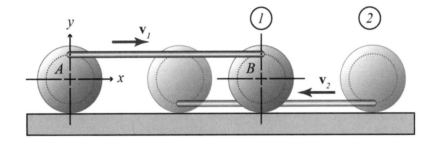

$$T_1 + V_1 = T_2 + V_2$$

$$\frac{1}{2}m_b v_{b,1}^2 + 2\left(\frac{1}{2}m_d v_{Gd,1}^2 + \frac{1}{2}I_{Gd}\omega_1^2\right) + m_b g r_i = \frac{1}{2}m_b v_{b,2}^2 + 2\left(\frac{1}{2}m_d v_{Gd,2}^2 + \frac{1}{2}I_{Gd}\omega_2^2\right) - m_b g r_i$$

Note that the bar does not rotate. It move in pure translation, therefore, its angular velocity is equal to zero.

Mass moment of inertia

The mass moment of a disk is given by

$$I_{Gd} = \frac{1}{2} m_d r_o^2$$

Kinematics

We need to relate all the linear speeds of the disk and bar to the angular speed of the disk.

$$v_{Gd} = r_o \omega$$

$$\mathbf{v}_b = \mathbf{v}_{Gd} + \boldsymbol{\omega} \times \mathbf{r}_{A/Gb} \qquad \mathbf{v}_{b,1} = v_{Gd,1}\mathbf{i} + \omega_1(-\mathbf{k}) \times r_i(\mathbf{j}) = (v_{Gd,1} + r_i\omega_1)\mathbf{i} = (r_o + r_i)\omega_1 \mathbf{i}$$

$$\mathbf{v}_{b,2} = v_{Gd,2}\mathbf{i} + \omega_2(-\mathbf{k}) \times r_i(-\mathbf{j}) = (v_{Gd,2} - r_i\omega_2)\mathbf{i} = (r_o - r_i)\omega_2 \mathbf{i}$$

Conservation of energy

$$\frac{1}{2}m_b v_{b,1}^2 + 2\left(\frac{1}{2}m_d v_{Gd,1}^2 + \frac{1}{2}I_{Gd}\omega_1^2\right) + m_b g r_i = \frac{1}{2}m_b v_{b,2}^2 + 2\left(\frac{1}{2}m_d v_{Gd,2}^2 + \frac{1}{2}I_{Gd}\omega_2^2\right) - m_b g r_i$$

$$\frac{1}{2}m_b v_{b,1}^2 + 2\left(\frac{1}{2}m_d r_o^2 \frac{v_{b,1}^2}{(r_o + r_i)^2} + \frac{1}{4}m_d r_o^2 \frac{v_{b,1}^2}{(r_o + r_i)^2}\right) + 2m_b g r_i = \frac{1}{2}m_b(r_o - r_i)^2\omega_2^2 + 2\left(\frac{1}{2}m_d r_o^2\omega_2^2 + \frac{1}{4}m_d r_o^2\omega_2^2\right)$$

$$v_{b,1}^2\left[\frac{1}{2}m_b + \frac{3m_d r_o^2}{2(r_o + r_i)^2}\right] + 2m_b g r_i = \omega_2^2\left[\frac{1}{2}m_b(r_o - r_i)^2 + \frac{3}{2}m_d r_o^2\right]$$

$$\boxed{\omega_2 = 15.5\ \frac{\text{rad}}{\text{s}}}$$

CHAPTER 8 REVIEW PROBLEMS

RP8-1) When calculating the work done on a body by a force, you need to add the work done by the force to the work done by the moment induced by the force. (True, False)

RP8-3) Consider a rigid body undergoing general planar motion. The gravitational potential energy is determined by the height of what point on the rigid body?

RP8-4) When calculating the kinetic energy of a rigid body rotating about a fixed-axis through point O you should use which of the following equations?

$(T = \frac{1}{2}I_O\omega^2$, $T = \frac{1}{2}mv_G^2 + \frac{1}{2}I_G\omega^2$, either)

RP8-5) When a rigid body is rolling <u>without</u> slip on a stationary surface, the frictional force does work. (True, False)

RP8-6) When a rigid body is rolling <u>with</u> slip the frictional force does work. (True, False)

RP8-7) When a rigid body is rolling <u>without</u> slip on a conveyor belt, the frictional force does work. (True, False)

RP8-8) If the external forces applied to a rigid body (or system of particles) sum to zero, then you are guaranteed that energy is conserved for the body (system). (True, False)

RP8-9) A two-line bucket carries a load of dirt. The combined weight of the bucket and dirt is 1500 lb. A winch, weighing 20 lb and having a radius of 4 inches, lifts the bucket by applying a torque of $M = 700$ lb-ft. If the bucket starts from rest and the winch has a radius of gyration about its center of mass of $k = 0.95$ ft, determine the speed of the bucket when it has been hoisted 10 ft.

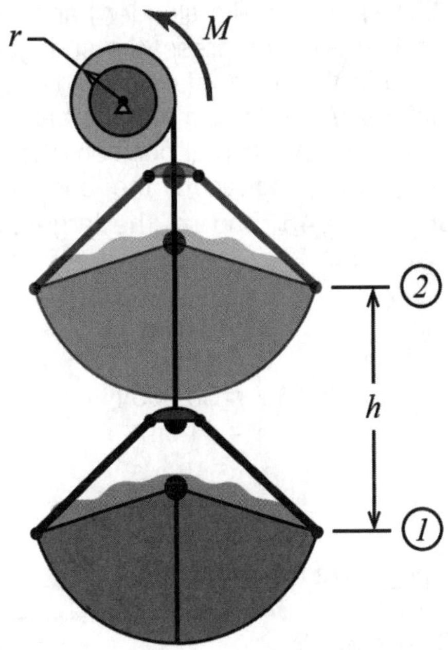

Given:
$W_{Bucket} = 1500$ lb
$W_{winch} = 20$ lb
$r = 4$ in
$M = 700$ lb-ft
$k = 0.95$ ft
$h = 10$ ft
$v_1 = 0$, $\omega_1 = 0$

Find:
v_2

Solution:

Determine the work done by M.

$U_M =$ _____

Determine the potential energy of the system at state 1 and state 2.

$V_1 =$ _____

$V_2 =$ _____

Determine the kinetic energy of the system at state 1 and state 2.

$T_1 =$ _____

$T_2 (v_2) =$ _____

Determine the speed of the bucket at state 2.

$v_2 =$ _____

RP8-9) The 50-kg slender bar shown has length $l = 1$ m and compresses a spring 11 centimeters when in the horizontal position. If the spring has stiffness $k = 30$ kN/m and the bar is released from rest in the horizontal position, then determine (a) the angular speed of the bar as it passes the angle $\theta = 45°$ and (b) the largest angle attained by the bar.

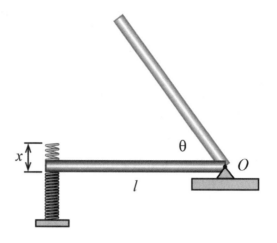

Given: $m = 50$ kg
                $l = 1$ m
                $k = 30,000$ N/m
                $x = 11$ cm
                $\omega_1 = 0$

Find: ω_2 at $\theta = 45°$, θ_{max}

Solution:

Determine the potential energy at all states.

$V_1 = $ _____

$V_2 = $ _____

$V_{max}(\theta_{max}) = $ _____

Determine the kinetic energy at all states.

$T_1 = $ _____

$T_2(\omega_2) = $ _____

$T_{max} = $ _____

Determine the angular velocity at $\theta = 45°$ and the maximum angle attained.

$\omega_2 = $ _____

$\theta_{max} = $ _____

RP8-10) A manual push mower consists of an array of rotating cutting blades connected to a set of wheels (r = 10 in). To initiate motion, the operator applies a constant horizontal pushing force of P = 30 lb. If the wheel/blade assembly has a combined mass of 15 lb and a radius of gyration of k_G = 0.5 ft and encounters a resistive moment of M = 10 lb-ft, determine the distance the mower travels before it reaches a velocity of 6 ft/s. The total weight of the mower is 50 lb. Assume that the handle angle remains constant and that the wheels roll without slip.

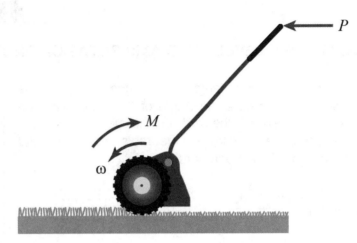

Given: W = 50 lb W_w = 15 lb
          r = 10 in P = 30 lb
          k = 0.5 ft M = 10 lb-ft
          v_1 = 0 v_2 = 6 ft/s

Find: d

Solution:

Determine the kinetic energy at both states.

$T_1 =$ _____

Determine the combined work done by the force and moment.

$U =$ _____

Determine the distance traveled.

$T_2 =$ _____

Determine the potential energy at both states.

$V_1 =$ _____

$V_2 =$ _____

$d =$ _____

CHAPTER 8 PROBLEMS

P8.1) BASIC LEVEL FIXED-AXIS ROTATION PROBLEMS

P8.1-1)[fe] A rod of length L is pinned at O. It is released from rest at an angle of θ. Determine the angular speed of the rod at the vertical position ($\theta = 90°$) as a function of the rod's mass (m) and length (L) and the angle of release (θ).

a) $\omega = \sqrt{\dfrac{3g(1-\cos\theta)}{L}}$　　　　b) $\omega = \sqrt{\dfrac{3g(1-\sin\theta)}{L}}$

c) $\omega = \sqrt{\dfrac{12g(1-\cos\theta)}{L}}$　　　　d) $\omega = \sqrt{\dfrac{12g(1-\sin\theta)}{L}}$

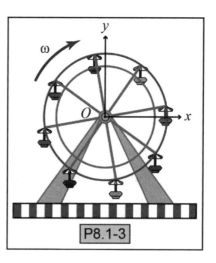

P8.1-1

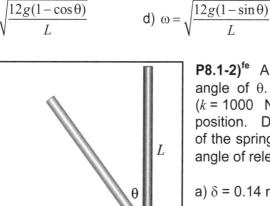

P8.1-2

P8.1-2)[fe] A 10-kg bar is released from rest at an angle of θ. It falls onto and contacts a spring ($k = 1000$ N/m) when it reaches a horizontal position. Determine the maximum displacement of the spring if the bar's length is $L = 1$ m and the angle of release is $\theta = 30°$.

a) $\delta = 0.14$ m　　　　b) $\delta = 0.24$ m

c) $\delta = 0.34$ m　　　　d) $\delta = 0.44$ m

P8.1-3)[fe] The Ferris wheel shown consists of 8 cars that each have a mass of approximately 500 kg when loaded with riders. The cars are located about the point O at a radius of 8 m. An experiment is performed to determine the amount of friction in the assembly. The procedure is as follows. The motor driving the Ferris wheel is used to accelerate the ride to an angular speed of 0.50 rad/sec. Power to the motor is then removed and the ride is allowed to slow to rest on its own. If the ride takes 10 revolutions to come to a rest, estimate the level of friction in the system. Model the frictional torque as constant. Also, neglect the overall change in potential energy of the system and the mass of the Ferris wheel structure.

a) $M_f = 120.6$ N-m　　b) $M_f = 509.3$ N-m　　c) $M_f = 52.8$ N-m　　d) $M_f = 356.9$ N-m

P8.1-4)[fe] Suppose that an electric vehicle battery can store 3 kWh of energy (equivalent to producing 3000 Watts of power for 1 hour). You would like to investigate using a very fast spinning disk (a flywheel) instead to store an equivalent amount of energy. If the disk cannot exceed a speed of 30,000 rpm for safety purposes and must have a radius of less than 1 m for packaging purposes, how much mass must the disk have?

a) m = 0.048 kg   b) m = 4.38 kg   c) m = 59.7 kg   d) m = 172.8 kg

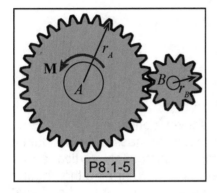

P8.1-5

P8.1-5) Consider gear A that is driven by a constant moment M = 0.1 N-m. Gear A has a radius r_A = 30 cm and gear B has a radius r_B = 10 cm. The respective moments of inertia of the two gears are I_A = 0.18 kg-m^2 and I_B = 0.02 kg-m^2. If the gears start from rest, estimate their angular velocities after gear B has turned 5 revolutions. Assume the friction in the system is negligible.

Ans: ω_A = 7.24 rad/s, ω_B = 21.7 rad/s

P8.2) INTERMEDIATE LEVEL FIXED-AXIS ROTATION PROBLEMS

P8.2-1) A 15-kg rod of length L = 1 m is pinned at O and is released from rest from the horizontal position (θ = 0°). The motion of the rod is retarded by a torsional spring (k = 10 N-m/rad) and a poorly lubricated bearing at O. The friction moment from the bearing is constant and equal to M_f = 2 Nm. The moment produced by the spring depends on the displaced angle and the spring constant ($M = k\theta$). The spring is un-deformed when the rod is in the horizontal position. Determine the angular speed of the rod when θ = 60°. Also, determine the maximum swept angle achieved by the rod before it, momentarily, comes to rest.

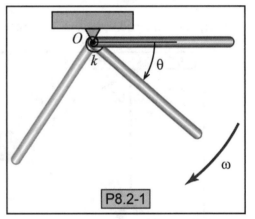

P8.2-1

Ans: ω = 4.74 rad/s, θ = 148.4°

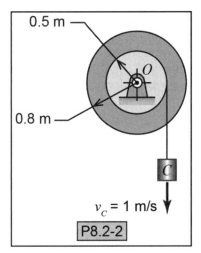

0.5 m

0.8 m

O

C

$v_c = 1$ m/s

P8.2-2

P8.2-2) A 5-kg cylinder initially has a downward velocity of 1 m/s. As the cylinder falls it turns a grooved drum by means of a cord wrapped around the inner diameter of the drum. If the cord unwraps without slipping and the drum has a mass of 15 kg with centroidal radius of gyration of 0.6 m, determine the speed of the cylinder after it has dropped 2 m. Assume that the frictional moment at O is a constant 5 N-m.

Ans: $\omega = 2.62$ rad/s

P8.2-3)   Consider the symmetric triangular pendulum shown consisting of two 5-kg masses connected by beams of length $L = 5$ meters and negligible mass. The pendulum starts from rest in the position shown where the left beam is completely vertical and a constant 200-N force F is applied to the left bob of the pendulum. Determine the angular velocity of the pendulum when the right beam of the pendulum becomes horizontal under the condition that (a) the force F remains perpendicular to the left beam of the pendulum through its entire trajectory, and (b) the force F remains horizontal through the entire trajectory of the pendulum.

Ans: a) $\omega = 1.72$ rad/s, b) $\omega = 1.66$ rad/s

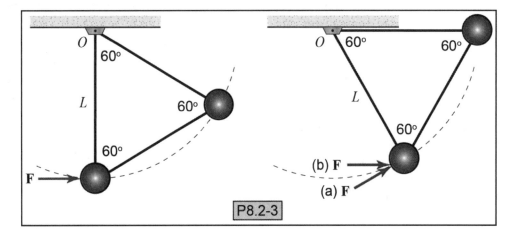

P8.2-3

P8.3) ADVANCED LEVEL FIXED-AXIS ROTATION PROBLEMS

P8.3-1) Consider the uniform slender rod pinned at O as shown. The rod has mass m and length L. Derive the differential equation of motion for this system employing (a) a Newtonian mechanics approach, and (b) a work-energy approach.   Hint: derive an expression for the total energy of the pendulum then differentiate this expression with respect to time. Then, assuming that the pendulum swings through small angles (allows approximation $\sin\theta \cong \theta$, $\cos\theta \cong 1$), solve for the motion $\theta(t)$ of the pendulum for the case

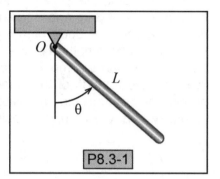

P8.3-1

that the pendulum is released from rest from an angle of θ = 5 degrees. With what frequency does the pendulum oscillate?

Ans: $\ddot{\theta} + \dfrac{3g}{2L}\sin\theta = 0$, $\theta(t) = \dfrac{\pi}{36}\cos\left(\sqrt{\dfrac{3g}{2L}}t\right)$, $\omega = \sqrt{\dfrac{3g}{2L}}$ rad/s

P8.4) BASIC LEVEL GENERAL PLANAR MOTION PROBLEMS

P8.4-1)[fe] A ball rolls up a hill that has a 5-ft vertical rise.   Determine the minimum speed (v_G) that the ball needs to make it up the hill. Assume that the ball rolls without slipping relative to the ground.

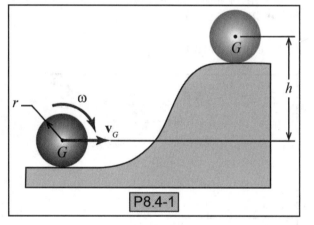

P8.4-1

a) v_G = 4.1 ft/s

b) v_G = 9.2 ft/s

c) v_G = 15.2 ft/s

d) v_G = 18.6 ft/s

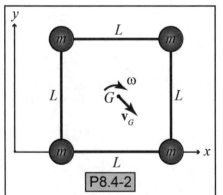

P8.4-2

P8.4-2)[fe] A system consisting of four 5-kg spheres connected by a rigid, but light, structure is translating and rotating across the horizontal xy-plane in the direction shown. If the velocity of the mass center is $\mathbf{v}_G$ = 2**i** − 2**j** m/s and the system's overall kinetic energy is 100 N-m at the instant shown, determine the system's rotational speed ω.

a) ω = 0.2 rad/s   b) ω = 0.3 rad/s

c) ω = 0.4 rad/s   d) ω = 0.5 rad/s

P8.4-3) A man and bicycle together have a mass of 100-kg, including the wheels. The wheels themselves each have a mass of 3-kg and a radius of 30 cm. If the cyclist descends from rest down a 10% grade for 300 m, estimate his velocity at the bottom of the hill. Solve for the velocity using a) a particle model and b) a rigid-body model including the rotation of the wheels. Treat the wheels as hoops and assume they roll without slip. Based on your results, which type of model would you employ?

P8.4-3

Ans: a) v = 24.2 m/s, b) v_G = 23.5 m/s

P8.5) INTERMEDIATE LEVEL GENERAL PLANAR MOTION PROBLEMS

P8.5-1) The 10-kg disk shown is free to rotate about a pin through its center which is attached to a spring of stiffness k = 1000 N/m. If the disk is released from rest from a position where the spring is stretched 0.2 m, determine the maximum angular speed achieved by the disk in the resulting motion. Assume the disk rolls without slipping.

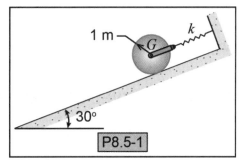

P8.5-1

Ans: ω = 1.23 rad/s

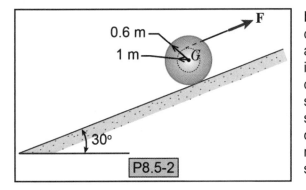

P8.5-2

P8.5-2) The 10-kg spool shown has a centroidal radius of gyration of 0.75 m and a cord wrapped securely about its inner radius. The cord is pulled by a constant force equal to F = 100 N. If the spool begins from rest, determine the speed v achieved by the spool's mass center G after the mass center has moved 2 m up the incline. Assume the spool rolls without slipping.

Ans: v_G = 5.33 m/s

P8.5-3) The 20-lb unbalanced disk shown has a centroidal radius of gyration of k_G = 12 in. The disk is released from rest from the position shown. If the disk rolls without slipping, determine the disk's angular velocity after it has rolled one quarter of a revolution.

Ans: ω = 5.63 rad/s

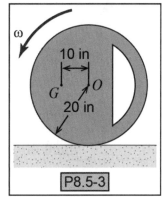

P8.5-3

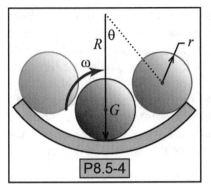

P8.5-4

P8.5-4) A disk rolls without slip on a concave surface as shown. The radius of the disk is r = 0.5 m and the radius of the concaved surface is R = 3 m. The angular velocity of the disk is 5 rad/s in the counterclockwise direction when it reaches the bottom. Determine the maximum angle θ that the disk attains on its way up the surface. If the disk were a ring of the same radius and mass, how would this affect your results?

Ans: θ_{disk} = 36.2°, θ_{ring} = 41.8°

P8.5-5) A cart, starting from rest, rolls without slip down a hill of angle θ. Determine the speed of the cart after it has traveled a distance of d. Determine the speed in terms of the mass of the cart body (M), the mass and radius of the front wheels (r_1, m_1), the mass and radius of the rear wheels (r_2, m_2), the distance traveled (d) and the angle of the slope (θ). Model the wheels as rings.

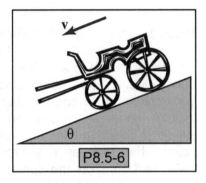

P8.5-6

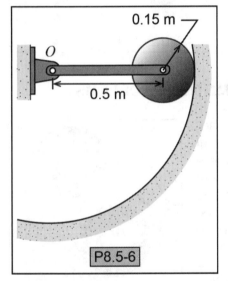

0.15 m

O

0.5 m

P8.5-6

P8.5-6) The 1-kg uniform slender bar and 2-kg disk shown are released from rest when the bar is horizontal. The disk rolls without slip on the curved surface shown with constant radius centered at O. Determine the angular velocity of the bar when the system reaches the position where the bar is completely vertical.

Ans: ω = 5.42 rad/s

P8.5-7)   Consider the mechanism shown that consists of two slender bars of mass 3 kg and length 1.5 m. If the system is released from rest from the position shown and moves in the vertical plane, determine the angular speed of the bars when they reach the ground. Assume there is negligible friction in the joints and between the sliding surfaces at point C.

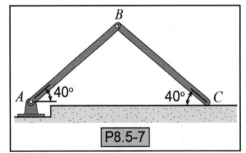

P8.5-7

Ans: ω = 3.55 rad/s

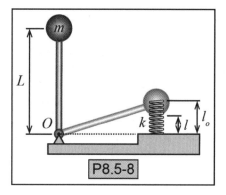

P8.5-8

P8.5-8) A pendulum consisting of a 0.5-kg sphere and a 0.5-kg rod is pinned at O. The system is released from rest in the vertical position. The pendulum is allowed to fall in the clockwise direction until it compresses a spring (k = 2000 N/m) and, momentarily, comes to rest. If the spring is originally 12 centimeters in length, determine the shortest compressed length of the spring. The length of the rod is 1 m and the radius of the sphere is 0.1 m.

Ans: l = 3.3 m

P8.5-9) The 5-kg collar shown below is released from rest at the position θ = 40 degrees and as it falls it pushes the 10-kg wheel shown below outward via the 7-kg bar. If the friction between the wheel-ground interface is sufficient to prevent slipping, determine (a) the velocity of the collar when the bar is horizontal and the collar first contacts the spring, and (b) the maximum deflection of the spring. Assume that the wheel is a disk.

Ans: v = 6.62 m/s, x = 0.131 m

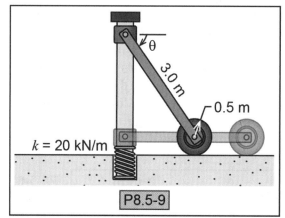

P8.5-9

P8.6) ADVANCED LEVEL GENERAL PLANAR MOTION PROBLEMS

P8.6-1) A sphere, starting from rest, rolls down a 60° incline. Assume that the sphere rolls down the hill without slipping. After the sphere's mass center G has traveled 3 ft, determine its speed. Determine the minimum coefficient of static friction between the sphere and the incline that still produces no slip. Based on the calculated static friction coefficient, determine the likelihood of the no slip assumption. If the sphere slips going down the incline, do you expect the speed of the sphere's mass center to be greater or less than the speed calculated above? Explain.

Ans: v = 10.9 ft/s

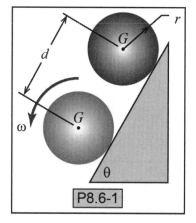

P8.6-1

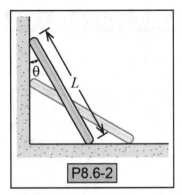

P8.6-2) A 24-kg ladder of length $L = 6$ m initially makes an angle of $\theta = 30°$ with the wall it is leaning against. If the ladder is released from rest and the wall and floor can be assumed to be frictionless, determine the angular velocity of the ladder when the angle the ladder makes with the wall has grown to $\theta = 60°$.

Ans: $\omega = 1.34$ rad/s

P8.6-3) Consider the 10-kg pulley shown that is free to rotate about a pin at its center G. The pin is attached to a spring of stiffness $k = 200$ N/m. A 100-N force F is applied to a cord wrapped tightly around the pulley such that the cord does not slip relative to the pulley. If the pulley begins from rest and at a position where the spring is unstretched, determine the resulting velocity of the pulley's center G after it has been displaced 0.5 m downward.

Ans: $v_G = 4.07$ m/s

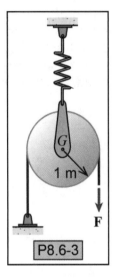

CHAPTER 8 COMPUTER PROBLEMS

C8-1) A rod of length L is pinned at O. It is released from rest at an angle of $\theta = 90°$. Plot the angular speed of the rod versus angular displacement (from $\theta = 90°$ to $\theta = 0°$). Assume the pendulum has a length $L = 1$ m.

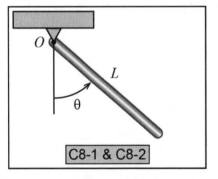

C8-2) Consider the pendulum shown in the figure. Determine the differential equation of motion for the pendulum and plot the angular position of the pendulum versus time for the case where the pendulum is released from rest from an angle of $\theta = 45°$ and also for the case where the pendulum is released from rest from an angle of $\theta = 5°$. Since this equation of motion is nonlinear, you will need to solve it numerically. Then, assuming that the pendulum swings through small angles (allows for the approximations $\sin\theta \approx \theta$, $\cos\theta \approx 1$), simulate the linearized model of the pendulum for both cases described above. How do the results compare? Explain any differences in agreement between the two cases.

CHAPTER 8 DESIGN PROBLEMS

D8-1) Consider that you are designing a hybrid-electric bicycle. The design of the bicycle itself is set, but you need to size the battery for the drive system. Assume that the bicycle frame and components have a mass of 10 kg, while the 30-cm radius front wheel has a mass of 3 kg. The rear-wheel assembly has the same radius as the front, but twice the mass because of the additional weight of the motor/generator system. The centroidal radius of gyration of the rear-wheel assembly is estimated to be approximately 0.4 kg-m^2. Without recharging the battery, the bicycle is required to have a range of 10 km, while achieving an elevation change of 2 km and an increase in its speed from rest to 8 m/s under purely electric propulsion. Estimate the required capacity of the battery in terms of Amp-hours. Assume that the bicycle must be able to accommodate a 100-kg rider. Also assume a 24-Volt battery and that the electric propulsion system has an overall efficiency of 50% (recall that electrical power is equal to the product of current and voltage (W = Ampere-Volt)). If there is any information that is missing that you may need, describe how you might obtain that information.

A8-1) This activity gives students practice calculating bearing friction.

Suggested group size: 3 students

Supplies

- Bicycle wheel
- Stop watch and tape measure or a bicycle speedometer
- Weight scale

Experimental setup

This experiment requires that the bicycle wheel is suspended so that it is vertical and free to rotate. This can be achieved by supporting the bicycle on a repair stand or by just holding up the bike with your hands. Feel free to build a fixture to hold the wheel.

Procedure

1. Measure the radius of the wheel and the radius of the hub.
2. Measure the weight of the wheel.
3. Research the relative weight of all the wheel components.
4. Suspend the bicycle wheel so that it is vertical and free to rotate.
5. Make a mark on the rim of the wheel to aid in your ability to count rotations.
6. Apply the moment to the wheel.
7. Measure the time the wheel takes to complete five full rotations. This may be difficult to do on the first rotation. It may be more accurate to start timing the second full rotation. Or, if you are using a speedometer, look at the speed for the first few rotations. This linear speed may be converted to angular speed if you know the radius of the tire.
8. Count the number of revolutions before the wheel comes to rest.
9. Repeat steps 6-8 two more times.

Calculations and Discussion

- Calculate the initial angular speed for each experimental run.
- Calculate the friction moment of the wheel for each experimental run.
- List all of the assumptions that were applied in your calculations.
- Discuss your results
- List possible sources of experimental error.

NOTES

PART V: KINETICS - IMPULSE AND MOMENTUM

CHAPTER 9: PARTICLE IMPULSE AND MOMENTUM

CHAPTER OUTLINE

CHAPTER SUMMARY

In this chapter, we study the last of the three approaches we will employ for analyzing the kinetics of particles: the *impulse-momentum* method. Previously, we studied *Newtonian mechanics* and the *work-energy* method. The impulse-momentum method involves examining the cumulative effect of forces and moments over time and how they affect a body's motion. There are two applications of this methodology, linear impulse-momentum and angular impulse-momentum. The linear impulse-momentum approach relates forces, time, linear velocities, and masses, while the angular impulse-momentum approach relates moments, time, angular velocities, and mass moments. Often, more than one of the three kinetic analysis methods: Newtonian mechanics, work-energy, and impulse-momentum can be used to analyze a specific situation. However, selecting the "best" approach can minimize the amount of calculation. It can also be the case that the required information is more readily available (i.e. can be measured or estimated more easily) for a specific analysis method.

9.1) LINEAR MOMENTUM

We have all used the term "momentum" in our everyday lives. You are working on a task and you don't want to stop because you will lose momentum. Or, a sports team is on a winning streak so they have momentum and they are going to be hard to stop. These two examples, in a sense, capture the meaning of the precise mechanics/physics definition of *momentum*. These examples give you a sense that you and the team have some sort of forward motion that will carry you through your next obstacle. The more momentum you have, the more difficult it will be for you to stop. Momentum, as defined in mechanics, depends on mass and velocity. If you have a lot of mass and velocity, you have a lot of momentum that will carry you forward.

Objects in motion have momentum. **Linear momentum** is defined as the product of the particle's mass and velocity as shown in Equation 9.1-1. Linear momentum is a vector quantity because it has both

Units of Linear Momentum
SI units:
• Kilogram-meter per second [kg-m/s]
• Newton-second [N-s]
US customary units:
• Pound-second [lb-s]

magnitude and direction. The direction derives from the particle's velocity. Because momentum is a vector, Equation 9.1-1 can be applied in each orthogonal coordinate direction. In many instances, *linear momentum* is referred to as just *momentum*. However, it is important to understand the difference between *linear momentum*, which is presented in this section, and *angular momentum* which is presented in a later section.

Momentum does not indicate the velocity of an object, although velocity (**v**) is used in the calculation of momentum (**G**). Momentum gives you a sense of the force of an object and its ability to do work*. Even if a particle has a small velocity, it could still have significant momentum. For instance,

*Note
Momentum is not equal to work. It only gives you a sense of an object's ability to do work.

consider a semi-truck. Its mass is very large, therefore, even if its velocity is small, it would still have a large amount of momentum. On the other hand, a bullet has a very small mass when compared to a semi-truck, but when fired from a gun, it also has a very large amount of momentum.

What quantity or quantities are a measure of how hard a moving object is to stop?

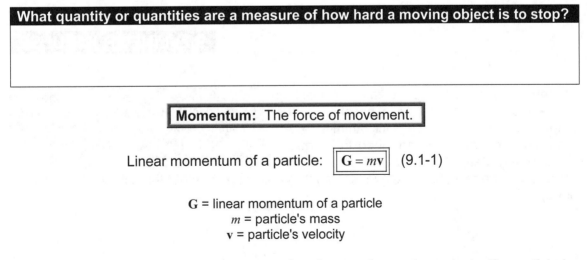

Momentum: The force of movement.

Linear momentum of a particle: $\boxed{\mathbf{G} = m\mathbf{v}}$ (9.1-1)

$\mathbf{G}$ = linear momentum of a particle
m = particle's mass
$\mathbf{v}$ = particle's velocity

It should be noted that momentum is reference-frame dependent. If a particle is moving relative to a reference frame, it has momentum in that reference frame.

However, it may have a different value of momentum relative to another reference frame. For example, if a reference frame is attached to and moves with the particle, then its momentum relative to this reference frame would be zero.

Equation Derivation

Linear momentum, Equation 9.1-1, relates to Newton's second law for a particle ($\sum F = ma$). This form of Newton's second law assumes that the particle's mass is constant. In fact, the more general form of Newton's second law equates the sum of the forces with the derivative of the particle's linear momentum. The following derivation shows how these two forms of Newton's second law are related.

$$\sum \mathbf{F} = m\mathbf{a}$$

$$= m\frac{d\mathbf{v}}{dt}$$

$$= \frac{d}{dt}(m\mathbf{v}) = \frac{d}{dt}(\mathbf{G}) = \dot{\mathbf{G}}$$

where $\mathbf{G} = m\mathbf{v}$

Conceptual Example 9.1-1

Make a guess as to the rank of the following situations in order of largest to smallest linear momentum.

   a) A semi-truck with no load traveling at 60 mph.
   b) A .32 caliber bullet just after firing.
   c) A 150-lb skydiver falling at terminal velocity.
   d) A cheetah running at its maximum speed.

Largest: _____   Next: _____   Next: _____   Smallest: _____

9.2) LINEAR IMPULSE

"Impulse" is often thought of as a fast acting force such as a bat hitting a baseball or the impact resulting from a jackhammer. An impulse, however, does not need to be quick. **Linear impulse** $(\mathbf{I})^{*}$ is a measure of the accumulated effect of a force over some interval of time (whether that interval be short or long).

*Note that linear impulse ($\mathbf{I}$) is not the same quantity as mass moment of inertia (I). In the scalar case, the two can be distinguished by the context of the problem and the attached subscripts.

Mathematically, the total linear impulse applied to a particle is the integral of the sum of forces acting on the particle with respect to time. Equation 9.2-1 gives the linear impulse equation. As shown in this equation,

Units of Linear Impulse
SI units:
• Kilogram-meter per second [kg-m/s]
• Newton-second [N-s]
US customary units:
• Pound-second [lb-s]

impulse is a vector quantity. The impulse direction derives from the resultant force and the impulse equation can be applied in the individual orthogonal coordinate directions.

What may be done to increase the linear impulse applied to a tennis ball by a tennis racket?

Impulse: The accumulated effect of forces acting over time.

Linear impulse:
$$\mathbf{I} = \int_{t_1}^{t_2} \sum \mathbf{F}\, dt \qquad (9.2\text{-}1)$$

$\mathbf{I}$ = linear impulse applied to the particle
$\sum \mathbf{F}$ = resultant of all the external forces acting on the particle
t = time

Note that integrating force with respect to time is equivalent to finding the area under the graph of force versus time. This fact can be helpful in estimating the total impulse produced by a force when a time-based equation for the force cannot easily be derived. For example, when you have empirical force data versus time from an experiment, you can estimate the associated impulse by numerically integrating the data.

Conceptual Example 9.2-1

A particle is acted upon by a time varying force $F(t)$. Rank the following force profiles from greatest to least applied linear impulse due to the action of the force F.

Greatest: _____

Next: _____

Next: _____

Least: _____

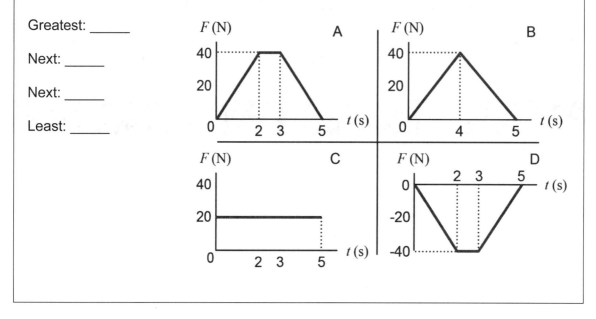

9.3) LINEAR IMPULSE-MOMENTUM PRINCIPLE

9.3.1) LINEAR IMPULSE-MOMENTUM PRINCIPLE

When a force is applied to a body over a period of time, the body's momentum changes. You've experienced this. You are standing still and someone comes up and pushes you. Now you are moving and have momentum. A small force applied over a long period of time can produce the same change in momentum as a large force applied over a short period of time. The quantity that affects a body's change in momentum is the product of the applied force and the time over which it acts (i.e. the linear impulse.) This is reflected in the *linear impulse-momentum principle* given in Equation 9.3-1.

> **Principle of linear impulse-momentum:** The impulse of the resultant force acting on a particle is equal to the change in the particle's linear momentum.

Linear impulse-momentum principle: $\mathbf{I} = \Delta\mathbf{G}$ $\displaystyle\int_{t_1}^{t_2}\sum\mathbf{F}\,dt = m\mathbf{v}_2 - m\mathbf{v}_1$ (9.3-1)

$\mathbf{I}$ = linear impulse applied to a particle
$\Delta\mathbf{G}$ = change in a particle's linear momentum
$\sum\mathbf{F}$ = resultant of all the external forces acting on the
        particle
m = mass

t = time
t_1 = time at state 1
t_2 = time at state 2
$\mathbf{v}_1$ = particle's velocity at state 1
$\mathbf{v}_2$ = particle's velocity at state 2

If you have ever taken a class in Martial Arts, especially Jujitsu, the Sensei spends a lot of time teaching students how to fall properly. One of the goals in falling properly is to extend the time of the impact with the ground by rolling or collapsing in some way. Why do you think this is important? Look at the impulse-momentum equation (Equation 9.3-1). When you fall, you start with a certain velocity and end with zero velocity. No matter what happens during your fall, the total impulse will remain constant. What then does increasing the time your body impacts the ground do? It decreases the reaction force from the ground experienced during the impact. This means that you are less likely to get hurt. The same principle is used when you are jumping from an elevated position and landing feet first on the ground. You want to bend your knees when you hit the ground to extend the duration of impact. If you land with straight and locked knees, the force of impact will be very high and can lead to injury.

It is important to remember that Equation 9.3-1 is a vector relationship and can be expressed in terms of its components. For example, in rectangular coordinates it can be written as the following set of equations. The impulse and momentum equations can be used in any coordinate frame. It is not restricted to the rectangular coordinate system.

$$\int_{t_1}^{t_2} \sum F_x \; dt = mv_{2x} - mv_{1x} \qquad \int_{t_1}^{t_2} \sum F_y \; dt = mv_{2y} - mv_{1y}$$

As with the previous approaches to kinetic analysis that were presented, it is helpful to draw a free-body diagram for the system under consideration. This process helps to identify the forces that apply impulse to the system and the directions in which the impulses are applied.

Equation Derivation

The principle of linear impulse and momentum, Equation 9.3-1, can be derived from Newton's second law as shown.

$$\sum \mathbf{F} = m\mathbf{a} = m\frac{d\mathbf{v}}{dt} = \frac{d}{dt}(m\mathbf{v}) = \frac{d\mathbf{G}}{dt}$$

The above equation is integrated over time to give the following relationship.

$$\int_{t_1}^{t_2} \sum \mathbf{F}\, dt = \int_{\mathbf{G}_1}^{\mathbf{G}_2} d\mathbf{G} = \mathbf{G}_2 - \mathbf{G}_1 = m\mathbf{v}_2 - m\mathbf{v}_1$$

In the above, the left-hand side of the equation is the linear impulse imparted on the particle by the resultant force. The linear impulse is the effect of all the forces acting on the particle accumulated over a given time interval. The right-hand side of the equation is the change in linear momentum of the particle. It makes intuitive sense that forces applied over an interval of time affect the particle's velocity and hence its linear momentum.

Conceptual Example 9.3-1

Consider the two shopping carts shown. Cart B carries groceries and has twice the mass of cart A. Neglecting friction, if both carts are pushed with a 50 lb force for 3 seconds, which cart experiences the larger change in momentum?

50 lb ⟶ Cart A   m

50 lb ⟶ Cart B   2m

a) cart A
b) cart B
c) They both experience the same change in momentum.
d) Not enough information is provided.

Which cart experiences the larger change in kinetic energy?

a) cart A
b) cart B
c) Both experience the same change in kinetic energy.
d) Not enough information is provided.

Conceptual Example 9.3-2

Suppose a billiard ball and a bowling ball are rolling toward you. By the time each ball reaches you, they have the same linear momentum. You stop each ball from rolling by applying equal forces to each. Assuming each ball may be treated as a particle and friction is negligible, which ball will take longer to stop?

$\mathbf{v}_{billiard}$

$\mathbf{v}_{bowling}$

a) billiard ball
b) bowling ball
c) Both balls take the same amount of time to stop.
d) Not enough information is provided.

Again, assuming that each ball may be treated as a particle, which ball will take a longer distance to stop?

a) billiard ball
b) bowling ball
c) Both balls take the same amount of distance to stop.
d) Not enough information is provided.

Conceptual Example 9.3-3

If you are in a head-on car crash with a stationary wall, from the perspective of impulse and momentum, are you better off being in a heavy or light car? Is it better to have a car that is stiff or a car that will crumple on impact?

9.3.2) LINEAR IMPULSE-MOMENTUM PRINCIPLE FOR A SYSTEM OF PARTICLES

The principle of linear impulse and momentum is not restricted to use on a single particle. In fact, it is very useful when it is applied to a system of particles. Consider the game of pool (billiards). The pool cue applies an impulse to the cue ball thus changing its momentum. The cue ball then collides with another pool ball and passes part of its momentum onto that ball. All of these components, the pool cue, the cue ball, and the other pool ball are all part of the system. The linear impulse and momentum equations may be applied to capture the impulse associated with each force and the momentum of each component in the system.

The total momentum of a system of particles is simply the sum of the momenta of the individual particles as expressed in Equation 9.3-2. The total impulse applied to the system can be calculated according to Equation 9.3-3. When using Equation 9.3-3, the forces internal to the system do not need to be explicitly considered. This is because each pair of internal interaction forces sum to zero. Each internal force has equal magnitude and opposite direction to its interaction partner and is applied over the same interval of time. This is one of the great advantages of the impulse-momentum approach as will be demonstrated in the subsequent section on conservation of linear momentum. The overall linear impulse-momentum principle for a system of particles is given in Equation 9.3-4.

Equations for a system of particles

Linear momentum: $\boxed{\mathbf{G} = \sum m_i \mathbf{v}_i}$ (9.3-2)   Linear impulse: $\boxed{\mathbf{I} = \int_{t_1}^{t_2} \sum \mathbf{F}\, dt}$ (9.3-3)

Linear impulse-moment principle: $\boxed{\mathbf{I} = \Delta \mathbf{G} \qquad \int_{t_1}^{t_2} \sum \mathbf{F}\, dt = \sum_i m_i \mathbf{v}_{i,2} - \sum_i m_i \mathbf{v}_{i,1}}$ (9.3-4)

$\mathbf{G}$ = linear momentum of a system of particles
$\Delta \mathbf{G}$ = change in linear momentum of the system
    of particles
$\mathbf{I}$ = linear impulse
m_i = mass of the ith particle
$\mathbf{v}_i$ = velocity of the ith particle

$\mathbf{v}_{i,1}$ = velocity of particle i at state 1
$\mathbf{v}_{i,2}$ = velocity of particle i at state 2
$\sum \mathbf{F}$ = external resultant force acting on
    all the particles in the system
t = time
t_1 = time at state 1
t_2 = time at state 2

9.3.3) CONSERVATION OF LINEAR MOMENTUM

With the work-energy method, there were certain conditions under which the energy of a particle would remain constant. Similarly, there are also conditions under which the linear momentum of a particle will remain constant. If the resultant force $\sum \mathbf{F}$ is zero for some interval of time, then the linear impulse is zero and hence the change in linear momentum is zero. This means the linear momentum remains constant over this time interval and is conserved. Therefore, the initial momentum equals the final momentum. For a particle, the principle of conservation of linear momentum is expressed by Equation 9.3-5.

Conservation of linear momentum: $\boxed{\mathbf{G}_1 = \mathbf{G}_2 \qquad m\mathbf{v}_1 = m\mathbf{v}_2}$ (9.3-5)

$\mathbf{G}_1$ = particle's linear momentum at state 1
$\mathbf{G}_2$ = particle's linear momentum at state 2
m = particle's mass

$\mathbf{v}_1$ = particle's velocity at state 1
$\mathbf{v}_2$ = particle's velocity at state 2

Note that since this linear momentum is a vector property, it may be conserved in one coordinate direction and not another. The property of conservation of linear momentum tends to be most useful when considering a system of particles. If the resultant force on a system of particles is zero during an interval of time, the system's linear momentum must remain constant. For a system of particles, the conservation of linear momentum is given by Equation 9.3-6.

Conservation of linear momentum for a system of particles:

$$\boxed{\sum m_i \mathbf{v}_{i1} = \sum m_i \mathbf{v}_{i2}}$$ (9.3-6)

m_i = mass of particle i
$\mathbf{v}_{i1}$ = velocity of particle i at state 1
$\mathbf{v}_{i2}$ = velocity of particle i at state 2

As stated above, the concept of conservation of linear momentum is very useful when considering a system of particles. In particular, a case where there is the collision of multiple particles. By defining the system to include all particles involved in the collision, the internal reaction forces from the particles contacting one another sum to zero. Therefore, if there are no external forces acting on the system, then the applied external linear impulse is zero and the system's linear momentum is conserved. Note that just because the momentum of the system does not change, that doesn't mean that the momenta of the individual bodies within the system cannot change. Bodies within the system can gain or lose momentum. It is just the sum of the momenta of all the bodies that remains constant. Note that energy is generally not conserved during a collision even though the internal forces have equal magnitude and opposite direction. This is because the forces are not necessarily conservative and energy can be lost in the process of deforming inelastic members in the system during the collision. Only if the collision is completely elastic will energy be conserved. This concept will be explored further in the section on *impact*.

Conceptual Example 9.3-4

You are an astronaut floating in the vacuum of outer space and you have been stranded a distance away from your spacecraft. Which of the following options would you choose to help propel yourself back to the craft?

a) Wave your arms and kick your legs.
b) Throw a wrench toward the spacecraft.
c) Throw a wrench in the opposite direction of the spacecraft.
d) None of the above.

Conceptual Example 9.3-5

A 0.1-lb bullet is travelling at 1000 ft/s when it strikes a 5-lb block traveling at 10 ft/s on a smooth surface in the directions shown. If the bullet hits the block centrally and is embedded in the block, which of the shown paths do you think most closely approximates the direction taken by the block/bullet system following the impact?

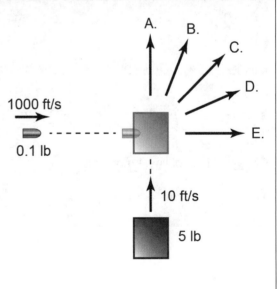

Conceptual Example 9.3-6

Consider that two bullets of equal mass have been fired with equal velocity toward a wooden block resting on a frictionless surface. One bullet is rubber and bounces off of the block, while the second bullet is made of steel and embeds in the block upon impact. In which case does the block have the larger velocity following the impact?

a) rubber bullet
b) steel bullet
c) same velocity in both cases
d) not enough information provided

Equation Summary

Abbreviated variable definition list

$\Sigma\mathbf{F}$ = external resultant force acting on all the particles in the system

$\mathbf{G}$ = linear momentum of a system of particles

$\Delta\mathbf{G}$ = change in linear momentum of the system of particles

$\mathbf{I}$ = impulse
m = mass
t = time
$\mathbf{v}$ = velocity

Linear momentum

$$\mathbf{G} = m\mathbf{v}$$

Linear momentum for a system of particles

$$\mathbf{G} = \sum m_i \mathbf{v}_i$$

Linear impulse

$$\mathbf{I} = \int_{t_1}^{t_2} \sum \mathbf{F}\, dt$$

Principle of impulse and momentum

$$\mathbf{I} = \Delta\mathbf{G} \qquad \int_{t_1}^{t_2} \sum \mathbf{F}\, dt = m\mathbf{v}_2 - m\mathbf{v}_1$$

Linear impulse-moment principle for a system of particles

$$\mathbf{I} = \Delta\mathbf{G} \qquad \int_{t_1}^{t_2} \sum \mathbf{F}\, dt = \sum_i m_i \mathbf{v}_{i,2} - \sum_i m_i \mathbf{v}_{i,1}$$

Conservations of linear momentum

$$\mathbf{G}_1 = \mathbf{G}_2 \qquad m\mathbf{v}_1 = m\mathbf{v}_2 \qquad \sum m_i \mathbf{v}_{i1} = \sum m_i \mathbf{v}_{i2}$$

Example Problem 9.3-7

Find the velocity in m/s of a 12-kg mass, initially resting against a support, after force **F** acts on the mass for 3 seconds. The time profile of the force is shown in the figure. The angle of the surface is θ = 35 degrees and the coefficient of kinetic friction is 0.4 between the mass and incline.

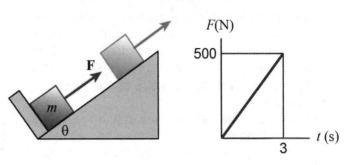

Given:

Find:

Solution:

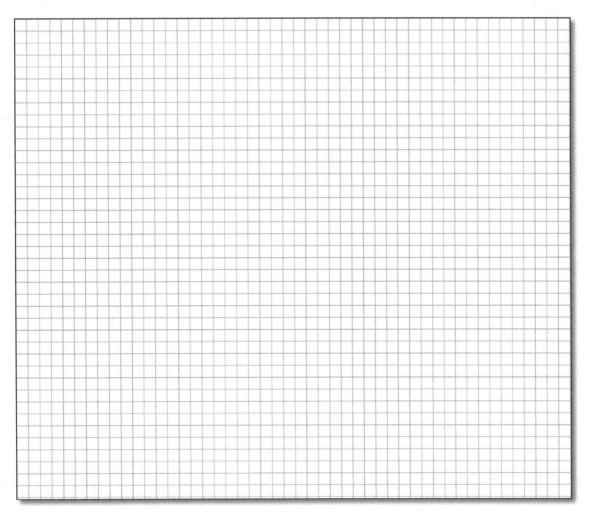

Example Problem 9.3-8

In a receiving warehouse, a 30-kg crate slides down a conveyor belt with negligible friction and is brought to rest by a hydraulic mechanism applying a constant force. If the conveyor belt is inclined at θ = 25 degrees and the package is released from rest, determine

(a) the speed of the package when it arrives at the bottom of the ramp, and

(b) the force applied by the hydraulic mechanism if the package is brought to rest in 0.10 seconds.

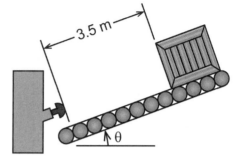

Given:

Find:

Solution:

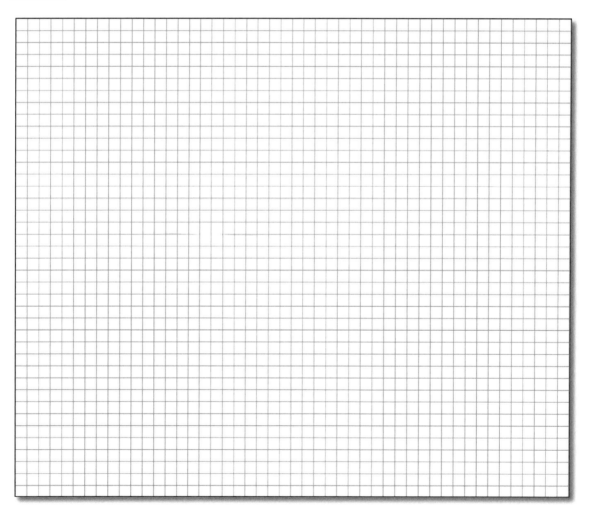

Example Problem 9.3-9

A 0.45-kg soccer ball rolls to a player with a velocity of 3 m/s, in the direction shown, before being kicked by the player. After the ball is kicked, it moves in the direction shown with a velocity of 10 m/s. If the ball is in contact with the player's foot for 0.10 seconds, determine the magnitude and direction of the average force exerted by the player during the kick.

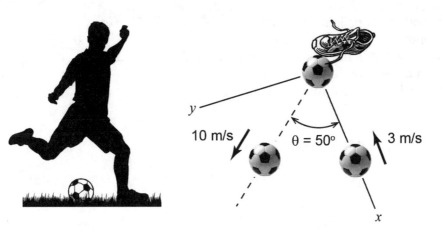

Given:

Find:

Solution:

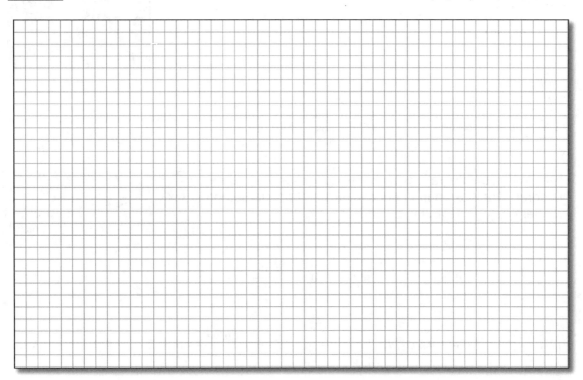

Solved Problem 9.3-10

A block of mass m rests on a frictionless horizontal table and is subjected to the exponentially decreasing force **F**. The angle θ at which the force is applied also changes with time. Both the force magnitude and the application angle are represented in the two graphs shown, where b and c are constants. If the block is stationary at time $t_o = 0$, determine its velocity **v** as a function of time.

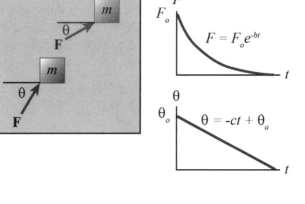

Given: $F = F_o e^{-bt}$ $\theta = -ct + \theta_o$
             $v_o = 0$ $t_o = 0$

Find: $\mathbf{v}(t)$

Solution

Principle of linear impulse-momentum

The principle of linear impulse-momentum equation shown, is a vector equation. The block does not move off the table, therefore, we will only consider motion in the plane of the table. We will apply the impulse-momentum equation in the x- and y-direction independently.

$$\int_{t_o}^{t} \sum \mathbf{F}\, dt = m\mathbf{v} - m\mathbf{v}_o$$

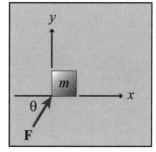

x-direction

$$\int_{t_o}^{t} \sum F_x\, dt = mv_x - mv_{x,o} \qquad\qquad \int_{t_o}^{t} F_o e^{-bt} \cos(-ct + \theta_o)\, dt = mv_x$$

We will use integration by parts to solve the above integral. The box on the right contains a shortcut method for performing integration by parts.

u		dv
$F_o e^{-bt}$	+	$\cos(-ct+\theta_o)$
$-bF_o e^{-bt}$	−	$(-\sin(-ct+\theta_o))/c$
$b^2 F_o e^{-bt}$	+	$(-\cos(-ct+\theta_o))/c^2$

$$\int_{t_o}^{t} F_o e^{-bt} \cos(-ct + \theta_o)\, dt = -\frac{F_o e^{-bt}}{c} \sin(-ct + \theta_o) - \frac{bF_o e^{-bt}}{c^2} \cos(-ct + \theta_o) - \frac{b^2}{c^2} \int_{t_o}^{t} F_o e^{-bt} \cos(-ct + \theta_o)\, dt$$

$$\left(1 + \frac{b^2}{c^2}\right) \int_{t_o}^{t} F_o e^{-bt} \cos(-ct + \theta_o)\, dt = \left[-\frac{F_o e^{-bt}}{c} \sin(-ct + \theta_o) - \frac{bF_o e^{-bt}}{c^2} \cos(-ct + \theta_o) \right]_0^t$$

$$\int_{t_o}^{t} F_o e^{-bt} \cos(-ct + \theta_o)\, dt = \left(1 + \frac{b^2}{c^2}\right)^{-1} \left(\frac{F_o}{c}\right)(e^{-bt} - 1)\left[\sin(\theta_o) - \sin(-ct + \theta_o) + \frac{b}{c}\left(\cos(\theta_o) - \cos(-ct + \theta_o)\right)\right]$$

From the impulse-momentum equation

$$v_x = \left(1 + \frac{b^2}{c^2}\right)^{-1} \left(\frac{F_o}{mc}\right)(e^{-bt} - 1)\left[\sin(\theta_o) - \sin(-ct + \theta_o) + \frac{b}{c}\left(\cos(\theta_o) - \cos(-ct + \theta_o)\right)\right]$$

y-direction

We can use a similar procedure to solve the *y*-direction velocity.

$$\int_{t_o}^{t} \sum F_y\, dt = mv_y - mv_{y,o} \qquad\qquad \int_{t_o}^{t} F_o e^{-bt} \sin(-ct + \theta_o)\, dt = mv_y$$

$$v_y = \left(1 + \frac{b^2}{c^2}\right)^{-1} \left(\frac{F_o}{mc}\right)(e^{-bt} - 1)\left[\cos(-ct + \theta_o) - \cos(\theta_o) + \frac{b}{c}\left(\sin(\theta_o) - \sin(-ct + \theta_o)\right)\right]$$

Example Problem 9.3-11

A 800-lb ram of a pile driver falls h = 3 yards from rest and strikes a 600-lb pile partially embedded in the ground. It was observed that the ram had no perceptible rebound after striking the pile. If the pile drives d = 1 yard into the ground, determine the resistive impulse of the ground on the pile.

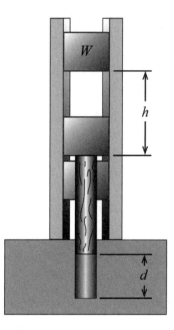

Given:

Find:

Solution:

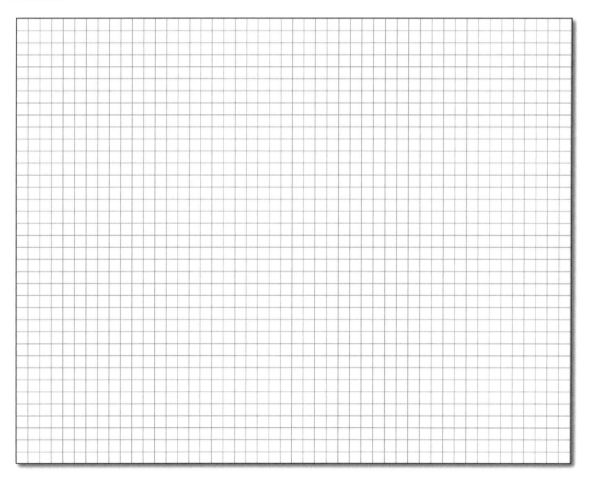

Solved Problem 9.3-12

A 1.5-oz golf ball is released from rest at the top of a ramp (h = 2 ft) as shown in the figure. It rolls down the ramp without slipping and with negligible rolling resistance. At the bottom of the ramp, it rolls into a 3-oz glass and sticks to the bottom of the glass due to the use of double sided tape. The glass and ball move as a unit and slide across the

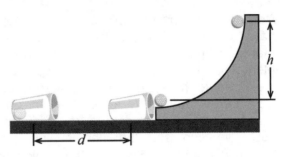

carpeted floor. The coefficient of kinetic friction between the glass and carpet is μ_k = 2. Derive an expression for the velocity as a function of time for the glass/ball system as it slides across the floor. Determine the distance d that the glass/ball system moves across the floor.

<u>Given:</u> W_{ball} = 1.5 oz W_{glass} = 3 oz
          h = 2 ft μ_k = 2
          golf ball is released from rest

<u>Find:</u> $v(t)$, d

<u>Solution</u>

Setting up the problem

Before we start solving the problem it is a good idea to identify key states. State 1 is when the golf ball is at rest at the top of the ramp (i.e. v_1 = 0). State 2 is when the ball has reached the bottom of the ramp and still has not entered the glass (i.e. $v_{glass,2}$ = 0). Between state 1 and 2 energy is conserved. State 3 is when the golf ball and glass come to rest (i.e. v_3 = 0).

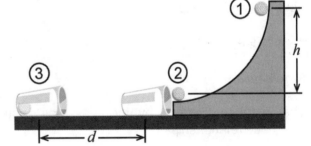

Conservation of energy

We can apply the conservation of energy balance equation between states 1 and 2 because there is negligible rolling resistance and the friction force acts at a point of zero velocity.

$$\cancel{T_1} + V_1 = T_2 + \cancel{V_2}$$ $$m_{ball}gh = \frac{1}{2}m_{ball}v_{ball,2}^2$$ $$v_{ball,2} = \sqrt{2gh}$$

Impulse-momentum equation

We can apply the principle of impulse-momentum between states 2 and 3. The impulse of the ball hitting the glass is internal to the system and we will not need to consider this interaction. However, the friction force that acts on the glass is external to the system and will need to be considered. To get a better idea of the forces that act on the system, it is always a good idea to draw a free-body diagram. From the FBD, we can see that the friction force is equal to $F_{fk} = \mu_k(W_{glass} + W_{ball})$. Since all the motion happens in the x-direction, we will apply the impulse-momentum equation in that direction. Note that the velocities are negative. This is due to our choice of axes.

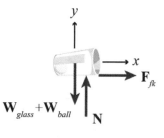

$$\int_0^t \sum F_x\, dt = \sum_i m_i v_{ix} - \sum_i m_i v_{ix,2} \qquad\qquad F_{fk}t = (m_{ball} + m_{glass})(-v) - m_{ball}(-v_{ball,2})$$

$$\mu_k(W_{glass} + W_{ball})gt = -(W_{ball} + W_{glass})v + W_{ball}\sqrt{2gh} \qquad\qquad v = \frac{W_{ball}\sqrt{2gh} - \mu_k(W_{glass} + W_{ball})gt}{(W_{ball} + W_{glass})}$$

$$\boxed{v = 3.783 - 64.4t \ \frac{\text{ft}}{\text{s}}}$$

The time at which the system comes to rest may be solved for using the velocity equation and setting $v = 0$.

$t_3 = 0.059$ s

Kinematics

The distance traveled by the glass/ball system will need to be determined using kinematic relationships.

$$v = \frac{dx}{dt} \qquad\qquad \int v\, dt = \int ds \qquad\qquad \int_0^t (3.783 - 64.4t)\, dt = \int_0^s ds$$

$$s = 3.783t - 32.2t^2 \qquad\qquad \boxed{s_3 = d = 0.11 \text{ ft}}$$

Example Problem 9.3-13

A demi-cannon weighing 3200 lb rolls, without resistance, down a hill (h = 9 ft). Once at the bottom of the hill, the cannon fires a cannon ball weighing 32 lb at a speed of 2300 ft/s relative to the cannon. What is the final velocity of the cannon?

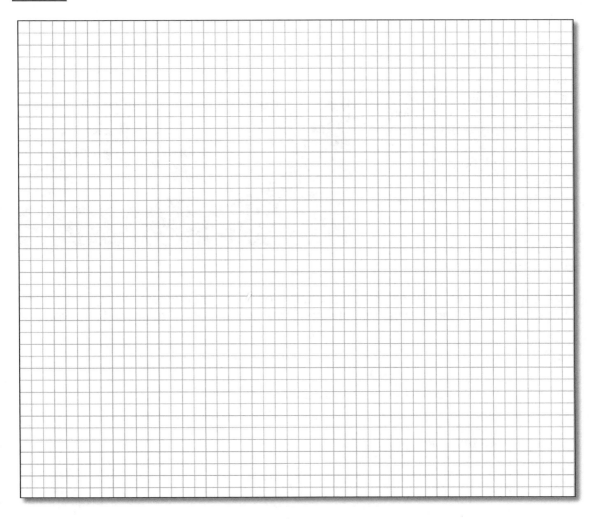

Given:

Find:

Solution:

Solved Problem 9.3-14

A boy (W_{boy} = 80 lb) runs and jumps on a stationary skateboard (W_{board} = 3 lb) with a speed of 8 mph. The boy's velocity as he lands on the skateboard is directed slightly downward at an angle of 15 degrees from the horizontal. How high will the boy travel up the half-pike before coming momentarily at rest?

What was the boy's energy before he impacted the skateboard compared to the energy of the system following his landing? Where did the difference in energy go?

Given: W_{boy} = 80 lb W_{board} = 3 lb
           $v_{boy,1}$ = 8 mph = 11.73 ft/s θ = 15°

Find: h

Solution

Setting up the problem

Let's start by labeling the important states. State 1 indicates the boy just before he lands on the skateboard. State 2 is immediately after the boy and board start to move as one. And state 3 is the system at its maximum height.

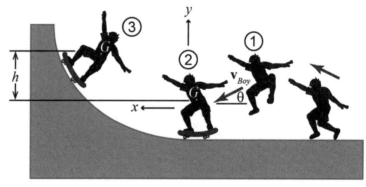

Conservation of linear momentum

We can apply the principle of conservation of linear momentum, in the x-direction, between states 1 and 2. Note that at state 2, the boy and board have the same velocity.

$$\sum m_i v_{ix,1} = \sum m_i v_{ix,2}$$

$$m_{boy} v_{boy\,x,1} + m_{board} \cancel{v_{board\,x,1}} = m_{boy} v_{boy\,x,2} + m_{board} v_{board\,x,2} = (m_{boy} + m_{board})v_{x,2}$$

$$m_{boy} v_{boy,1} \cos\theta = (m_{boy} + m_{board})v_{x,2} \qquad\qquad v_{x,2} = v_2 = 10.92\,\frac{\text{ft}}{\text{s}}$$

Conservation of energy

We will apply the conservation of energy balance between states 2 and 3.

$$T_2 + \cancel{V_2} = \cancel{T_3} + V_3 \qquad \qquad \frac{1}{2}m_{total}v_2^2 = m_{total}gh \qquad \qquad \boxed{h = 1.85 \text{ ft}}$$

We can determine the amount of energy lost by determining the change in kinetic energy between states 1 and 2.

$$\Delta T_{1-2} = \sum T_{i,1} - \sum T_{i,2} = \frac{1}{2}m_{boy}v_{boy,1}^2 + \frac{1}{2}m_{board}\cancel{v_{board,1}^2} - \frac{1}{2}(m_{boy} + m_{board})v_2^2 = 17.2 \text{ ft-lb}$$

This energy loss is mainly due to the deformation of the board and wheels. Other energy is lost with the boy's body.

9.4) IMPACT

9.4.1) IMPACT

If you play billiards, you know a lot about the subject of *impact* (see Figure 9.4-1). You know that if you hit a ball with the cue ball dead on, the impacted ball will take off in the initial direction that the cue ball was traveling (i.e. direct central impact). On the other hand, if you hit a ball off center with the cue ball, the impacted ball will take off at an angle to the initial direction that the cue ball was traveling (i.e. oblique central impact). It is also apparent that these impacts last but a short time.

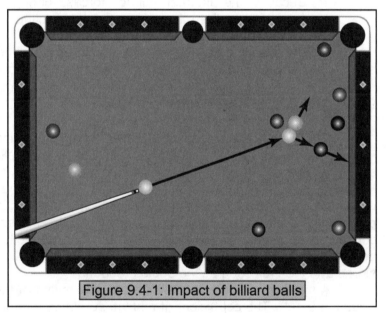

Figure 9.4-1: Impact of billiard balls

To be more precise, we will define an impact to be when two bodies collide over a very small interval of time and exert relatively large forces on one another. In the previous section on *Conservation of Linear Momentum*, we saw that the concept of conservation of linear momentum was useful in analyzing collisions. If we define the system to include both bodies involved in the collision, the internal interaction forces between the two bodies will be internal to the system and will cancel. If these were the only forces acting on the system in certain directions, the linear momentum of the system is conserved in these directions, though the energy of the system is often not conserved. By making a judicious choice of coordinate axes, we can also achieve conservation of the linear momenta of the individual particles in at least one direction.

Figure 9.4-2 shows an impact and two lines based on the impact geometry. The **line of impact** is drawn between the centers of mass of the two bodies and passes through the contact point. The **plane of contact** is perpendicular to the line of impact and resides at the contact between the two bodies. In general, the linear momentum of the system is conserved, but the momenta of the individual particles are not conserved in the direction of the line of impact. This is because the impact forces act in that direction. Assuming negligible friction, the momenta of the individual particles are conserved in the direction of the plane of contact since the impact forces have no components in this plane. Therefore, the components of the velocities of the particles in the plane of contact remain constant.

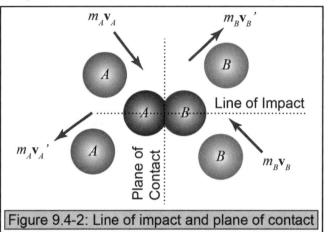

Figure 9.4-2: Line of impact and plane of contact

When analyzing impacts, the following two simplifications are often assumed:

1. Because the internal impact forces, during the collision, are so large compared to the other external forces acting on the particles, the external forces can be neglected during the collision.
2. Because the duration of the impact is so short, the displacements of the particles during the collision do not need to be considered.

9.4.2) COEFFICIENT OF RESTITUTION

Energy is not necessarily conserved during a collision. Energy is often expended in permanently deforming the bodies involved in the collision. Energy can also be expended as heat and sound. A collision in which energy is not conserved is defined as **inelastic**. A collision in which the two bodies "stick together" following the collision, that is, where the bodies have the same velocity following the collision, is said to be **perfectly inelastic** (i.e. plastic). Conversely, a collision during which energy is conserved is defined as **perfectly elastic**. In a perfectly elastic collision, the bodies behave like ideal springs. Work goes into deforming the bodies, but that energy is stored and then released as the bodies rebound.

A measure of the elasticity of a collision is its **coefficient of restitution** given in Equation 9.4-1. The coefficient of restitution relates the incoming and outgoing particle velocities along the line of impact. A perfectly inelastic collision has a coefficient of restitution of $e = 0$ and a perfectly elastic collision has a coefficient of restitution of $e = 1$.

Name examples of impacts that are nearly perfectly elastic and perfectly inelastic.

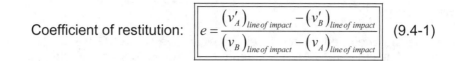
Coefficient of restitution:
$$e = \frac{\left(v'_A\right)_{line\,of\,impact} - \left(v'_B\right)_{line\,of\,impact}}{\left(v_B\right)_{line\,of\,impact} - \left(v_A\right)_{line\,of\,impact}} \qquad (9.4\text{-}1)$$

e = coefficient of restitution
$(v_A)_{line\,of\,impact}$ = speed of particle A prior to impact along the line of impact
$(v_A')_{line\,of\,impact}$ = speed of particle A after impact along the line of impact

$(v_B)_{line\,of\,impact}$ = speed of particle B prior to impact along the line of impact
$(v_B')_{line\,of\,impact}$ = speed of particle B after impact along the line of impact

Figure 9.4-3 shows the impact of two deformable bodies. As the bodies collide, they exert an impulse on one another ($\int \mathbf{D}\,dt$), where $\mathbf{D}$ is the deformation force and t is the time during the impact. At maximum deformation, the two bodies instantaneously have the same velocity ($\mathbf{u}_{max\,deformation}$). Then the material forces try to restore each body to its original shape. Each body exerts a restorative impulse on one another ($\int \mathbf{R}\,dt$), where $\mathbf{R}$ is the restorative force. The coefficient of restitution is the ratio between the restorative impulse and the deformation impulse as shown in Equation 9.4-2 and gives an indication of how much energy is being lost during the collision. The coefficient of restitution is also a measure of the elasticity of the collision. A useful form of the coefficient of restitution was given in Equation 9.4-1. This equation expresses the coefficient of restitution e as a function of the speeds before the collision and after the collision of the two particles along the line of impact.

Coefficient of restitution:
$$e = \frac{\int R\,dt}{\int D\,dt} \qquad (9.4\text{-}2)$$

e = coefficient of restitution R = restorative force
D = deformation force t = time

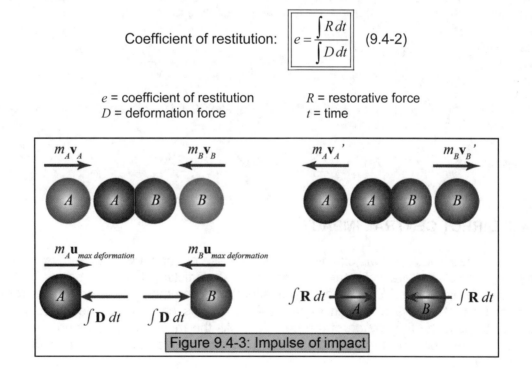

Figure 9.4-3: Impulse of impact

Equation Derivation

To derive the coefficient of restitution equation, we will apply the impulse-momentum equation along the line of impact. Applying the impulse-momentum equation to particle A during deformation and again during restoration, we get the following equations. The variables in this equation are: v_A is the component of the velocity along the line of impact just before impact, v_A' is the component of the velocity along the line of impact just after impact, and $u_{max\,deformation}$ is the speed at maximum deformation.

$$\int D\,dt = m_A u_{max\,deformation} - m_A v_A \qquad\qquad \int R\,dt = m_A v_A' - m_A u_{max\,deformation}$$

Dividing the above equations, we get the following expression.

$$e = \frac{\int R\,dt}{\int D\,dt} = \frac{m_A v_A' - m_A u_{max\,deformation}}{m_A u_{max\,deformation} - m_A v_A} = \frac{v_A' - u_{max\,deformation}}{u_{max\,deformation} - v_A}$$

We can use a similar logic to derive an equation for particle B.

$$e = \frac{\int R\,dt}{\int D\,dt} = \frac{-m_B v_B' + m_B u_{max\,deformation}}{-m_B u_{max\,deformation} + m_B v_B} = \frac{-v_B' + u_{max\,deformation}}{-u_{max\,deformation} + v_B}$$

If we take both expressions for e and add their numerators and denominators, the resulting ratio is unchanged. Therefore, we arrive at the following alternate expression for e.

$$e = \frac{(v_A' - u_{max\,deformation}) + (-v_B' + u_{max\,deformation})}{(u_{max\,deformation} - v_A) + (-u_{max\,deformation} + v_B)}$$

The final version of the coefficient of restitution (Equation 9.4-1) is obtained by simplifying the above equation and is shown below.

$$e = \frac{\left(v_A'\right)_{line\,of\,impact} - \left(v_B'\right)_{line\,of\,impact}}{\left(v_B\right)_{line\,of\,impact} - \left(v_A\right)_{line\,of\,impact}}$$

9.4.3) DIRECT CENTRAL IMPACT

To begin with, we will examine a specific case of impact where the two particles have velocities that are collinear. In the case of particles, this also means that the velocities lie along the *line of impact*. This situation is commonly referred to as a **direct central impact** and is shown in Figure 9.4-4. In this case, only one coordinate is needed to describe the motion of the particles. All the motion occurs along the line of impact. The coefficient of restitution for a direct central impact is described by Equation 9.4-1.

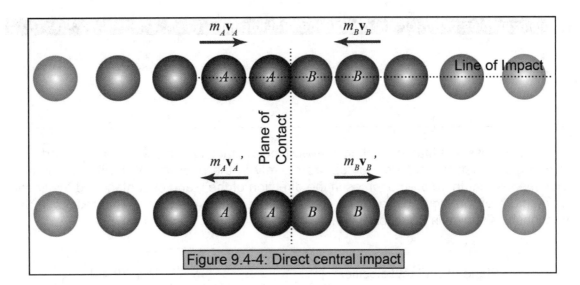

Figure 9.4-4: Direct central impact

Considering a situation of direct central impact as described above, we can apply the principle of conservation of linear momentum to generate Equation 9.4-3.

Conservation of momentum: $$m_A v_A + m_B v_B = m_A v'_A + m_B v'_B$$ (9.4-3)

v_A = speed of particle A before the impact v_B' = speed of particle B after the impact
v_A' = speed of particle A after the impact m_A = mass of particle A
v_B = speed of particle B before the impact m_B = mass of particle B

If given the initial velocities and the masses of the particles A and B, we are left with two unknowns (i.e. the two final velocities) and thus do not have sufficient information to solve the problem. If, however, we are also given, or can estimate, the coefficient of restitution, then we have two equations and two unknowns and can completely determine the behavior of the system. This, of course, presumes that the simplifications, previously stated, are appropriate.

9.6.4) OBLIQUE CENTRAL IMPACT

The more general case for the impact of two particles moving in a plane allows for the initial and final velocities to not be collinear. This situation is pictured in Figure 9.4-5 and will be referred to as an **oblique central impact**. Since momentum is a vector quantity, we can analyze this case by considering the conservation of the linear momentum of the system in two orthogonal directions (i.e. along the line of impact and along the plane of contact).

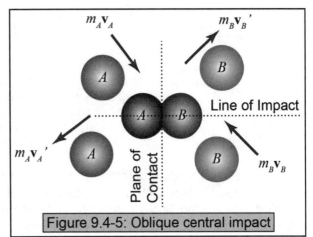

Figure 9.4-5: Oblique central impact

Name examples of direct and oblique central impacts.

If we assume that the friction between the contacting surfaces of the two particles is negligible, then the collision forces are entirely in the direction of the line of impact. This means that there are no forces in the direction of the plane of contact and the linear momentum of the individual particles is conserved in this direction. This results in the relationships shown in Equation 9.4-4.

Plane of contact velocities:
$$\left(v_A\right)_{plane\,of\,contact} = \left(v_A'\right)_{plane\,of\,contact}$$
$$\left(v_B\right)_{plane\,of\,contact} = \left(v_B'\right)_{plane\,of\,contact}$$
(9.4-4)

$(v_A)_{plane\,of\,contact}$ = speed of particle A prior to impact along the plane of contact
$(v_A')_{plane\,of\,contact}$ = speed of particle A after impact along the plane of contact
$(v_B)_{plane\,of\,contact}$ = speed of particle B prior to impact along the plane of contact
$(v_B')_{plane\,of\,contact}$ = speed of particle B after impact along the plane of contact

If we are given the initial velocities and masses of the two particles involved in the impact, the above relationships demonstrate that we are able to immediately determine the components of the final velocities in the direction of the plane of contact.

The linear momenta of the individual particles are conserved in the direction of the plane of contact, but are not conserved in the direction of the line of impact due to the reaction forces. However, the linear momentum of the system is conserved in both directions since there are no forces that are external to the system in the plane of motion, or they are negligible compared to the impact forces. Applying the conservation of momentum to the system in the direction of the line of impact, we arrive at Equation 9.4-5.

Conservation of momentum along the line of impact:
$$m_A\left(v_A\right)_{line\,of\,impact} + m_B\left(v_B\right)_{line\,of\,impact} = m_A\left(v_A'\right)_{line\,of\,impact} + m_B\left(v_B'\right)_{line\,of\,impact}$$
(9.4-5)

$(v_A)_{line\,of\,impact}$ = speed of particle A prior to impact along the line of impact
$(v_A')_{line\,of\,impact}$ = speed of particle A after impact along the line of impact
$(v_B)_{line\,of\,impact}$ = speed of particle B prior to impact along the line of impact
$(v_B')_{line\,of\,impact}$ = speed of particle B after impact along the line of impact
m_A = mass of particle A
m_B = mass of particle B

This relationship, combined with the expression for the coefficient of restitution (Equation 9.4-1), allows for the determination of the components of the final velocities in the direction of the line of impact. Therefore, we have completely determined the behavior of the system immediately following impact if we know the initial velocities and masses and the assumptions we have made are reasonable.

Conceptual Example 9.4-1

A car initially moving with speed v heads directly toward a parked car. The driver reacts by slamming on the brakes. Due to icy conditions, the car slides and is unable to decrease its speed. The car hits the parked car and the two stick together. If the cars are approximately the same weight, what is their speed after the collision?

a) v
b) $0.5v$
c) 0
d) need more information

Conceptual Example 9.4-2

A golf ball hits a billiard ball, which is initially at rest, and bounces back. After the impact, which of the following properties increase and which decrease?

	Kinetic energy	Momentum (magnitude)
Golf ball		
Billiard ball		
The system		

Equation summary

Abbreviated variable definition list

e = coefficient of restitution
m = mass
v = speed before the impact
v' = speed after the impact
$(v)_{line\ of\ impact}$ = speed prior to impact along the line of impact

$(v')_{line\ of\ impact}$ = speed after impact along the line of impact
$(v)_{plane\ of\ contact}$ = speed prior to impact along the plane of contact
$(v')_{plane\ of\ contact}$ = speed after impact along the plane of contact

Coefficient of restitution

$$e = \frac{\left(v'_A\right)_{line\ of\ impact} - \left(v'_B\right)_{line\ of\ impact}}{\left(v_B\right)_{line\ of\ impact} - \left(v_A\right)_{line\ of\ impact}}$$

Conservation of momentum

$$m_A v_A + m_B v_B = m_A v'_A + m_B v'_B$$

Plane of contact velocities

$$\left(v_A\right)_{plane\ of\ contact} = \left(v'_A\right)_{plane\ of\ contact} \qquad \left(v_B\right)_{plane\ of\ contact} = \left(v'_B\right)_{plane\ of\ contact}$$

Conservation of momentum along the line of impact

$$m_A\left(v_A\right)_{line\ of\ impact} + m_B\left(v_B\right)_{line\ of\ impact} = m_A\left(v'_A\right)_{line\ of\ impact} + m_B\left(v'_B\right)_{line\ of\ impact}$$

Example Problem 9.4-3

A steel sphere is dropped from rest from a height of 0.5 m above a marble countertop. The sphere rebounds to a height of 0.4 m following the impact. Determine the coefficient of restitution e for this situation.

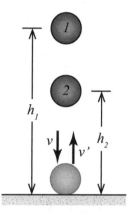

Given:

Find:

Solution:

Example Problem 9.4-4

A cue ball with an initial velocity of 5 m/s strikes another billiard ball that is initially at rest at the angle θ = 30 degrees shown. If the coefficient of restitution is $e = 0.9$, determine the motion of both balls immediately following the impact. Also, calculate the percentage of energy lost as a result of the impact.

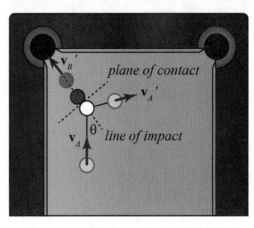

Given:

Find:

Solution:

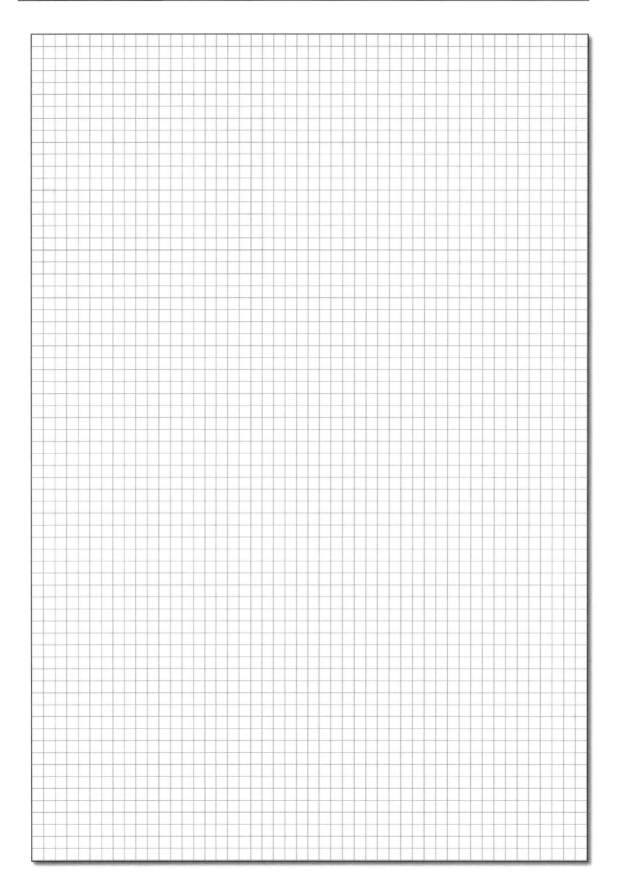

Solved Problem 9.4-5

Two identical hockey pucks moving with initial speeds of $v_A = 6$ m/s and $v_B = 10$ m/s, in the directions shown, collide. If the pucks move in the directions shown following the collision, and $v_A' = 7$ m/s, determine the coefficient of restitution.

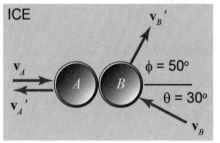

<u>Given:</u> $m_A = m_B = m$ $v_A = 6$ m/s
          $v_B = 10$ m/s $v_A' = 7$ m/s
          $\theta = 30°$ $\phi = 50°$

<u>Find:</u> e

<u>Solution</u>

Setting up the problem

When solving impact problems, it is helpful to apply the momentum equation in two special coordinate directions. These directions correspond to the *line of impact* and the *plane of contact*. Before we start our analysis, we should identify these two directions.   The system conserves momentum in the direction of the line of impact, while each particle conserves momentum in the direction of the plane of contact.

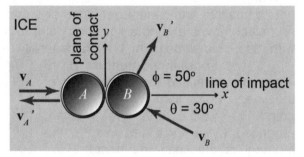

Plane of contact

The momentum of the individual particles is conserved in the direction of the plane of contact.

$$mv_{Ay} = mv'_{Ay} \qquad\qquad v'_{Ay} = 0$$

$$mv_{By} = mv'_{By} \qquad\qquad v'_{By} = v_B \sin\theta = 5\,\frac{m}{s}$$

Line of impact

The momentum of the system is conserved in the direction of the line of impact. Remember that the velocity direction is important in the following equation. Use a + or - to indicate direction.

$$mv_{Ax} + mv_{Bx} = mv'_{Ax} + mv'_{Bx} \qquad\qquad v_A - v_B \cos\theta = -v'_A + v'_B \cos\phi$$

$$6 - 10\cos 30 = -7 + v'_B \cos 50 \qquad\qquad v'_B = 6.75\,\frac{m}{s}$$

$$v_B'^2 = v_{Bx}'^2 + v_{By}'^2 \qquad\qquad v_{Bx}' = 4.54\,\frac{m}{s}$$

Coefficient of restitution

Now that we know the x-direction velocity of puck B after the impact, we can calculate the coefficient of restitution.

$$e = \frac{v_{Bx}' - v_{Ax}'}{v_{Ax} - v_{Bx}} = \frac{4.54 - 7}{6 - 10\cos 30} \qquad \boxed{e = 0.93}$$

9.5) ANGULAR MOMENTUM

Angular momentum is the rotational counterpart to *linear momentum* ($\mathbf{G} = m\mathbf{v}$). The units of angular momentum and linear momentum are different. Therefore, they are not compatible and cannot be added or subtracted. Angular momentum describes how an object moves around a specified point. All moving objects have angular momentum with respect to some point, but it is generally used to describe rotating bodies.

The **angular momentum** $\mathbf{H}_O$ of a particle is defined as the moment of the linear momentum ($\mathbf{G} = m\mathbf{v}$) about point O.   Similar to the moment produced by a force ($\mathbf{M} = \mathbf{r} \times \mathbf{F}$), the moment of the linear momentum is a function of the moment arm ($\mathbf{r}$) at which

Units of Angular Momentum
SI units:
• Kilogram-meter^2 per second [kg-m^2/s]
• Newton-meter-second [N-m-s]
US customary units:
• Slug-foot^2 per second [slug-ft^2/s]
• Foot-pound-second [ft-lb-s]

the linear momentum ($m\mathbf{v}$) is acting with respect to point O. Angular momentum is a different type of quantity than linear momentum even though they are related. This is analogous to how the moment induced by a force is a fundamentally different thing than the force itself. Particles moving about a point in a curved or straight path can have angular momentum with respect to a reference point.   In the chapter on *Rigid-Body Impulse & Momentum,* we will also see that rigid bodies that are rotating or translating may have angular momentum. Whether or not a body has angular momentum depends on the choice of reference point. Mathematically, angular momentum is defined by the cross product relationship shown in Equation 9.5-1.

Angular momentum of a particle: $\boxed{\mathbf{H}_o = \mathbf{r} \times \mathbf{G} = \mathbf{r} \times m\mathbf{v}}$ (9.5-1)

$\mathbf{H}_o$ = angular momentum with respect to O m = mass of the particle
$\mathbf{r}$ = particle's position with respect to O $\mathbf{v}$ = velocity of the particle

In Equation 9.5-1, **r** is the position vector of the particle with respect to point O as shown in Figure 9.5-1.   Recall that the result of a cross product of two vectors is a vector that is perpendicular to the plane defined by the two original vectors.   The direction of the resultant perpendicular vector can be determined by the right-hand rule.   For example, consider the two vectors **r** and *mv* in Figure 9.5-1. Visualizing both vectors as originating at the same point, take your right hand and roll your fingers from the first vector (**r**) towards the second vector (*mv*).   The direction of your thumb indicates the direction of the resultant vector, which is out of the page in this instance.   In the case of planar motion, which

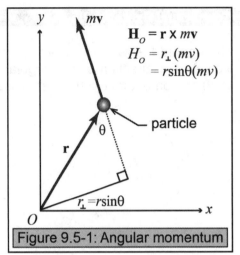

$$\mathbf{H}_o = \mathbf{r} \times m\mathbf{v}$$
$$H_o = r_\perp (mv)$$
$$= r\sin\theta(mv)$$

Figure 9.5-1: Angular momentum

we will deal with primarily, the angular momentum vector $\mathbf{H}_O$ will always point into or out of the plane of motion.   Under these conditions, the magnitude of $\mathbf{H}_O$ can be calculated by Equation 9.5-2.

In defining the angle between the two vectors, θ can be chosen as the angle that is greater than $180°$ or less than $180°$ because $\sin\theta = \sin(180 - \theta)$. The $r\sin\theta$ term can be interpreted as the moment arm from point O to the line of action of the linear momentum vector *mv*. This implies that the cross product of two parallel vectors is zero. Further details regarding the cross product operation can be found in Appendix B.

Magnitude of angular momentum of a particle: $\boxed{H_o = r_\perp mv = r\sin\theta\, mv}$ (9.5-2)

H_o = magnitude of the angular momentum with respect to O
r = particle's position with respect to O

m = mass of the particle
v = speed of the particle
θ = angle between the **r** and **v** vectors

Conceptual Example 9.5-1

The angular momentum of a particle is ...

    a) independent of the specified origin.
    b) zero when the position and momentum vectors are parallel.
    c) zero when the position and momentum vectors are perpendicular.
    d) None of the above.

Conceptual Example 9.5-2

A 3-kg ball is attached to the end of a string that is 0.5 m long. A man whirls the ball over his head in a circle with a velocity of 4 m/s. What is the angular momentum of the ball with respect to the center of the circle?

    a) 1.5 kg-m²/s   b) 2 kg-m²/s   c) 6 kg-m²/s   d) 12 kg-m²/s

Conceptual Example 9.5-3

The figures below show similar particles moving along different paths. In each case, indicate the direction of the angular momentum vector with respect to point O. If the angular momentum is zero, just indicate zero.

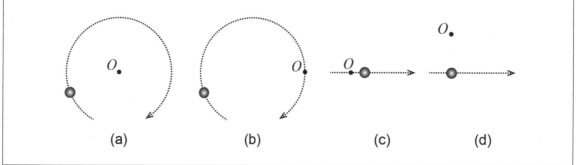

| (a) | (b) | (c) | (d) |

Conceptual Example 9.5-4

A particle of mass m is traveling with respect to a fixed point O in a variety of situations. Rank each of the given situations from greatest to least amount of angular momentum possessed by the particle with respect to the point O for the instants shown. You may assume that out of the page is considered positive.

Greatest ____   Next ____   Next ____   Next ____   Next ____   Least ____

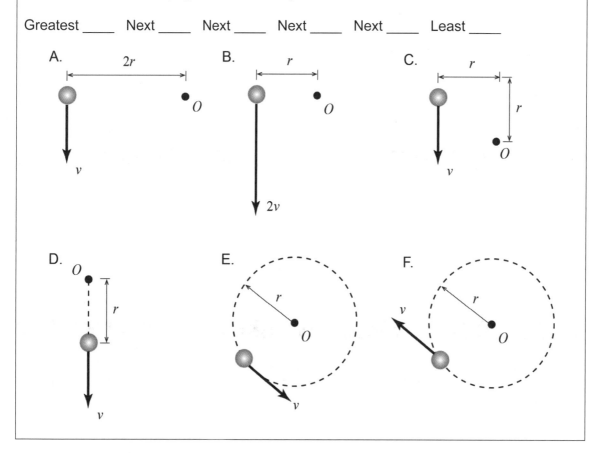

Equation Summary

<u>Abbreviated variable definition list</u>

$\mathbf{H}_o$ = angular momentum with respect to O $\mathbf{v}$ = velocity of the particle
m = mass of the particle θ = angle between the $\mathbf{r}$ and $\mathbf{v}$ vectors
$\mathbf{r}$ = particle position with respect to O

<u>Angular momentum for a particle</u>

$$\mathbf{H}_O = \mathbf{r} \times \mathbf{G} = \mathbf{r} \times m\mathbf{v} \qquad H_O = r_{\perp} mv = r\sin\theta\, mv$$

Example Problem 9.5-5

A 3-lb particle moving through space has a linear momentum of $\mathbf{G} = \mathbf{i} - 3\mathbf{j} + 5\mathbf{k}$ lb-s. Calculate the angular momentum of the particle with respect to point O if its position relative to point O is given by $\mathbf{r} = 2\mathbf{i} + 3\mathbf{j} - 4\mathbf{k}$ ft.

Given:

Find:

Solution:

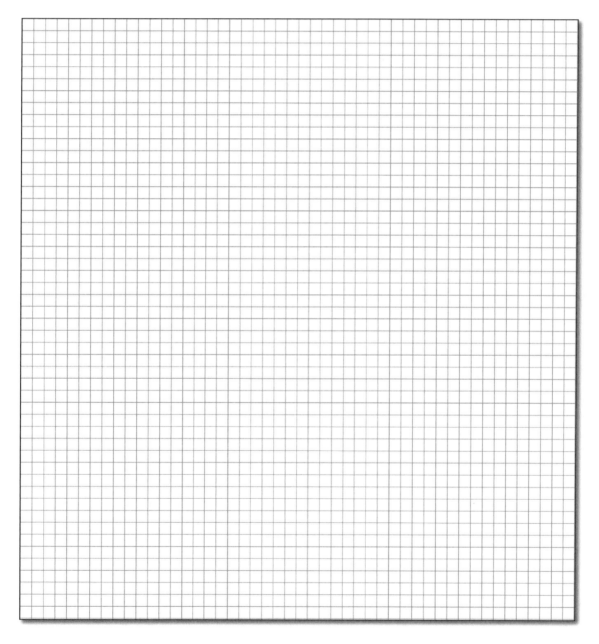

9.6) ANGULAR IMPULSE

Just as *linear impulse* ($\mathbf{I} = \int_{t_1}^{t_2} \sum \mathbf{F}\, dt$) defines the cumulative effect of a force over time, **angular impulse** ($\mathbf{J}_O = \int_{t_1}^{t_2} \sum \mathbf{M}_O\, dt$) defines the cumulative effect of a moment over time. Recall that a moment is defined as the cross product of the moment arm $\mathbf{r}$ and the applied force $\mathbf{F}$ ($\mathbf{M}_O = \mathbf{r} \times \mathbf{F}$). As with angular and linear momentum, the units for angular and linear impulse are not compatible and these two quantities cannot be added or subtracted.

Mathematically, the total *angular impulse* is defined as the integral of the sum of all the moments acting on the particle over a given interval of time as shown in Equation 9.6-1. Since a particle has no size, all the forces acting on the particle act at the same point and

Units of Angular Impulse
SI units:
• Kilogram-meter^2 per second [kg-m^2/s]
• Newton-meter-second [N-m-s]
US customary units:
• Slug-foot^2 per second [slug-ft^2/s]
• Foot-pound-second [ft-lb-s]

hence each has the same moment arm $\mathbf{r}$. As was the case with linear impulse, angular impulse is also a vector quantity and its direction derives from the direction of the resultant moment. For the case of planar motion, the angular impulse vector will always point into or out of the plane of motion. The units of angular impulse are the same as those for angular momentum.

Angular impulse:
$$\mathbf{J}_O = \int_{t_1}^{t_2} \mathbf{r} \times \sum \mathbf{F}\, dt = \int_{t_1}^{t_2} \sum \mathbf{M}_O\, dt \qquad (9.6\text{-}1)$$

$\mathbf{J}_O$ = Angular impulse with respect to O
$\mathbf{r}$ = particle's position with respect to O
$\sum \mathbf{F}$ = resultant force acting on the particle

$\mathbf{M}_O$ = moments induced by the forces acting on the particle
t = time

9.7) ANGULAR IMPULSE-MOMENTUM PRINCIPLE

9.7.1) ANGULAR IMPULSE-MOMENTUM PRINCIPLE

It is apparent that the size of the applied moment and the length of time the moment is applied to a particle affects the particle's velocity and, therefore, the angular momentum of the particle. Note that the magnitude of the moment and angular momentum has a dependency on the reference point location. The relationship between moment application and change of angular momentum is represented in the *principle of angular impulse-momentum principle* and is shown in Equation 9.7-1. The principle of angular impulse-momentum shown in Equation 9.9-1 is a vector equation. This equation may be applied independently in each orthogonal coordinate direction.

> **Principle of angular impulse and momentum:** The impulse of the resultant moment acting on a particle is equal to the change in the particle's angular momentum.

Angular impulse-momentum principle:

$$\boxed{\begin{aligned} \mathbf{J}_O &= \Delta \mathbf{H}_O \\ \int_{t_1}^{t_2} \sum \mathbf{M}_O \, dt &= \mathbf{H}_{O,2} - \mathbf{H}_{O,1} \\ &= (\mathbf{r}_2 \times m\mathbf{v}_2) - (\mathbf{r}_1 \times m\mathbf{v}_1) \end{aligned}}$$

(9.7-1)

Time rate of change of angular momentum: $\boxed{\dot{\mathbf{H}}_O = \sum \mathbf{M}_O}$ (9.7-2)

$\mathbf{J}_O$ = angular impulse applied to the particle with respect to O

$\Delta \mathbf{H}_O$ = change in angular momentum of the particle with respect to O

$\mathbf{H}_{O,1}$ = angular momentum of the particle with respect to O at state 1

$\mathbf{H}_{O,2}$ = angular momentum of the particle with respect to O at state 2

$\dot{\mathbf{H}}_O$ = time rate of change in angular momentum with respect to O

$\mathbf{M}_O$ = moments acting on the particle with respect to O

$\mathbf{r}_1$ = position of the particle with respect to O at state 1

$\mathbf{r}_2$ = position of the particle with respect to O at state 1

t = time

t_1 = time at state 1

t_2 = time at state 2

m = mass of the particle

$\mathbf{v}_1$ = velocity of the particle at state 1

$\mathbf{v}_2$ = velocity of the particle at state 2

Equation Derivation

The following derives the angular impulse-momentum principle by differentiation of the expression for angular momentum. This follows similar logic to that used in deriving the linear impulse-momentum principle. We start with the moment equation.

$$\sum \mathbf{M}_O = \mathbf{r} \times \mathbf{F}$$

From Newton's second law, we know that $\Sigma \mathbf{F} = m\mathbf{a}$, and from kinematics, we know that $\mathbf{a} = d\mathbf{v}/dt$. Inserting these relationships into the above equation, we get the following.

$$\sum \mathbf{M}_O = \mathbf{r} \times m\frac{d\mathbf{v}}{dt} = \mathbf{r} \times m\dot{\mathbf{v}}$$

Now, let's take the derivative of the angular momentum equation.

$$\mathbf{H}_O = \mathbf{r} \times m\mathbf{v}$$

$$\frac{d\mathbf{H}_O}{dt} = \frac{d\mathbf{r}}{dt} \times m\mathbf{v} + \mathbf{r} \times m\frac{d\mathbf{v}}{dt}$$

Notice in the above equation that $d\mathbf{r}/dt = \mathbf{v}$ and recall that the cross product of two parallel vectors equals zero. Therefore, we get the following equation.

$$\dot{\mathbf{H}}_O = \mathbf{r} \times m\dot{\mathbf{v}}$$

Equating the above equation with $\sum \mathbf{M}_O = \mathbf{r} \times m\dot{\mathbf{v}}$, we get an expression that relates the resultant moment to the time rate of change of angular momentum.

$$\dot{\mathbf{H}}_O = \sum \mathbf{M}_O$$

If we integrate the above expression with respect to time, we get the relationship between angular impulse and angular momentum.

$$\sum \mathbf{M}_O = \frac{d\mathbf{H}_O}{dt}$$

$$\int_{t_1}^{t_2} \sum \mathbf{M}_O \, dt = \int_{\mathbf{H}_{O,1}}^{\mathbf{H}_{O,2}} d\mathbf{H}_O$$

$$= \mathbf{H}_{O,2} - \mathbf{H}_{O,1}$$

$$= (\mathbf{r}_2 \times m\mathbf{v}_2) - (\mathbf{r}_1 \times m\mathbf{v}_1)$$

Conceptual Example 9.7-1

Imagine that you are playing a game of tetherball. There are two different tetherball poles available to use. One has a string length of L (case a), the other has a string length of $2L$ (case b). If you hold the ball out to the extent of its string length and hit the ball as hard as you can, which case has the larger resulting velocity and angular velocity?

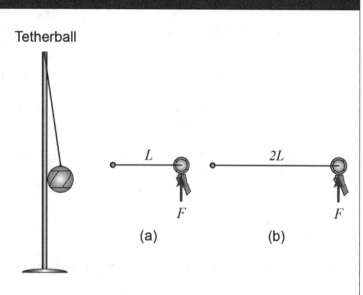

Tetherball

(a)   (b)

9.7.2) ANGULAR IMPULSE-MOMENTUM PRINCIPLE FOR A SYSTEM OF PARTICLES

Just as with linear impulse and momentum, the principle of angular impulse and momentum is not restricted to use on a single particle. The total momentum of a system of particles is simply the sum of the momenta of the individual particles as expressed in Equation 9.7-3. The total impulse applied to the system can be calculated according to Equation 9.7-4. When using Equation 9.7-4, the forces internal to the system (and the moments that they induce) do not need to be explicitly considered. This is because each pair of internal interaction forces are equal and opposite, and thus, don't induce a net moment. The overall linear impulse-momentum principle for a system of particles is given in Equation 9.7-5 and the relationship between the resultant moment and the time rate of change of angular momentum is given in Equation 9.7-6.

System of particles equation

Angular momentum: $$\boxed{\mathbf{H}_O = \sum (\mathbf{r}_i \times m_i \mathbf{v}_i)}$$ (9.7-3)

Angular impulse: $$\boxed{\mathbf{J}_O = \int_{t_1}^{t_2} \sum \mathbf{M}_O \, dt}$$ (9.7-4)

Angular impulse-momentum principle:
$$\boxed{\begin{aligned} \mathbf{J}_O &= \int_{t_1}^{t_2} \sum \mathbf{M}_O \, dt = \mathbf{H}_{O,2} - \mathbf{H}_{O,1} \\ &= \sum (\mathbf{r}_{i,2} \times m_i \mathbf{v}_{i,2}) - \sum (\mathbf{r}_{i,1} \times m_i \mathbf{v}_{i,1}) \end{aligned}}$$ (9.7-5)

Time rate of change of angular momentum: $$\boxed{\dot{\mathbf{H}}_O = \sum \mathbf{M}_O}$$ (9.7-6)

$\mathbf{J}_O$ = angular impulse applied to a system of particles with respect to O

$\mathbf{H}_{O,1}$ = angular momentum of a system of particles with respect to O at state 1

$\mathbf{H}_{O,2}$ = angular momentum of a system of particles with respect to O at state 2

$\dot{\mathbf{H}}_O$ = time rate of change in angular momentum of a system of particles with respect to O

$\mathbf{v}_{i,1}$ = velocity of particle i at state 1
$\mathbf{v}_{i,2}$ = velocity of particle i at state 2

$\mathbf{M}_O$ = moments acting on a system of particles with respect to O

$\mathbf{r}_i$ = position of particle i with respect to O
$\mathbf{r}_{i,1}$ = position of particle i with respect to O at state 1
$\mathbf{r}_{i,2}$ = position of particle i with respect to O at state 2

t = time
t_1 = time at state 1
t_2 = time at state 1
m_i = mass of particle i

9.7.3) CONSERVATION OF ANGULAR MOMENTUM

Angular momentum of a particle or a system of particles is conserved when the angular impulse is zero over a specified interval of time. In other words, when the resultant moment about a reference point O of all the forces acting on a particle (or system of particles) is zero, then angular momentum remains constant. This can be seen by looking at the principle of angular impulse-momentum equation shown below. The conservation of angular momentum expression is given in Equations 9.7-7 and 9.7-8. Note that angular momentum may be conserved about one axis and not another.

Angular impulse-momentum principle:

$$\mathbf{J}_O = \int_{t_1}^{t_2} \sum \mathbf{M}_O \, dt = \mathbf{H}_{O,2} - \mathbf{H}_{O,1} = (\mathbf{r}_2 \times m\mathbf{v}_2) - (\mathbf{r}_1 \times m\mathbf{v}_1)$$

Conservation of angular momentum: $\boxed{\mathbf{H}_{O,1} = \mathbf{H}_{O,2} \qquad \mathbf{r}_1 \times m\mathbf{v}_1 = \mathbf{r}_2 \times m\mathbf{v}_2}$ (9.7-7)

Conservation of angular impulse-momentum for a system of particles:

$$\boxed{\mathbf{H}_{O,2} - \mathbf{H}_{O,1} = \sum(\mathbf{r}_{i,2} \times m_i \mathbf{v}_{i,2}) - \sum(\mathbf{r}_{i,1} \times m_i \mathbf{v}_{i,1})} \quad (9.7\text{-}8)$$

$\mathbf{H}_{O,1}$ = angular momentum of the particle with respect to O at state 1

$\mathbf{H}_{O,2}$ = angular momentum of the particle with respect to O at state 2

$\mathbf{r}_i$ = position of particle i with respect to O

$\mathbf{r}_{i,1}$ = position of particle i with respect to O at state 1

$\mathbf{r}_{i,2}$ = position of particle i with respect to O at state 2

m = mass of the particle

m_i = mass of particle i.

$\mathbf{v}_1$ = velocity of the particle at state 1

$\mathbf{v}_2$ = velocity of the particle at state 2

$\mathbf{v}_{i,1}$ = velocity of particle i at state 1

$\mathbf{v}_{i,2}$ = velocity of particle i at state 2

t = time

Note that the resultant moment may be zero even if the position vector and resultant force are nonzero. For example, if the vectors are parallel, then their cross product is zero. This fact indicates that a judicious choice of the point O may allow conservation of angular momentum to be applied to solve a given problem. A specific type of problem where conservation of angular momentum can be applied, and is very useful for analysis, is the *central-force problem*. This type of problem is defined such that the resultant force acting on the body has a line of action that always passes through a fixed-point O. By choosing this fixed-point O as the reference, angular momentum will be conserved. Examples of central-force problems include: a car rounding a curve with constant speed, a satellite orbiting the earth, and an object tied to a string being swung around one's head.

Conceptual Example 9.7-2

A bullet is fired horizontally with a velocity v_o into a box of sand hung by a piece of rope. The box of sand is initially at rest and the bullet embeds in the box. Which of the following quantities are conserved during the impact of the bullet with the box if the system is considered to consist of both the bullet and the box?

a) energy
b) linear momentum
c) angular momentum with respect to O

Which of the following quantities are conserved, after the impact, as the box with embedded bullet swings upward?

a) energy
b) linear momentum
c) angular momentum with respect to O

Equation Summary

Abbreviated variable definition list

$\Sigma \mathbf{F}$ = resultant force acting on the particle
$\mathbf{H}_O$ = angular momentum with respect to O
$\dot{\mathbf{H}}_O$ = time rate of change in angular momentum
        with respect to O
$\mathbf{J}_O$ = angular impulse with respect to O

m = mass
$\mathbf{M}_O$ = moments with respect to O
$\mathbf{r}$ = position
t = time
$\mathbf{v}$ = velocity

Angular momentum

$$\mathbf{H}_O = \mathbf{r} \times \mathbf{G} = \mathbf{r} \times m\mathbf{v}$$

Angular momentum for a system of particles

$$\mathbf{H}_O = \sum (\mathbf{r}_i \times m_i \, \mathbf{v}_i)$$

Angular impulse

$$\mathbf{J}_O = \int_{t_1}^{t_2} \sum \mathbf{M}_O \, dt$$

Angular Impulse for a system of particles

$$\mathbf{J}_O = \int_{t_1}^{t_2} \mathbf{r} \times \sum \mathbf{F} \, dt = \int_{t_1}^{t_2} \sum \mathbf{M}_O \, dt$$

Time rate of change of angular momentum

$$\dot{\mathbf{H}}_O = \sum \mathbf{M}_O$$

Principle of angular impulse-momentum

$$\mathbf{J}_O = \Delta \mathbf{H}_O \qquad \int_{t_1}^{t_2} \sum \mathbf{M}_O \, dt = \mathbf{H}_{O,2} - \mathbf{H}_{O,1} = (\mathbf{r}_2 \times m\mathbf{v}_2) - (\mathbf{r}_1 \times m\mathbf{v}_1)$$

Principle of angular impulse-momentum for a system of particles

$$\mathbf{J}_O = \int_{t_1}^{t_2} \sum \mathbf{M}_O \, dt = \mathbf{H}_{O,2} - \mathbf{H}_{O,1} = \sum (\mathbf{r}_{i,2} \times m_i \mathbf{v}_{i,2}) - \sum (\mathbf{r}_{i,1} \times m_i \mathbf{v}_{i,1})$$

Conservation of angular momentum

$$\mathbf{H}_{O,1} = \mathbf{H}_{O,2} \qquad \mathbf{r}_1 \times m\mathbf{v}_1 = \mathbf{r}_2 \times m\mathbf{v}_2$$

$$\mathbf{H}_{O,2} - \mathbf{H}_{O,1} = \sum (\mathbf{r}_{i,2} \times m_i \, \mathbf{v}_{i,2}) - \sum (\mathbf{r}_{i,1} \times m_i \mathbf{v}_{i,1})$$

Example Problem 9.7-3

A 3-lb ball is attached to a spring and moves along a frictionless horizontal table under the influence of the constant force shown. The length of the spring and the speed of the ball in the transverse direction are represented in the graphs shown. Determine the force if ϕ equals a constant 20 degrees.

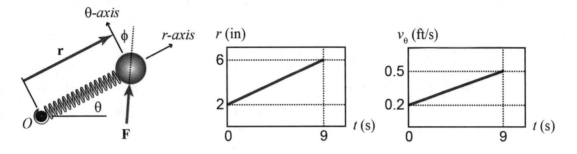

Given:

Find:

Solution:

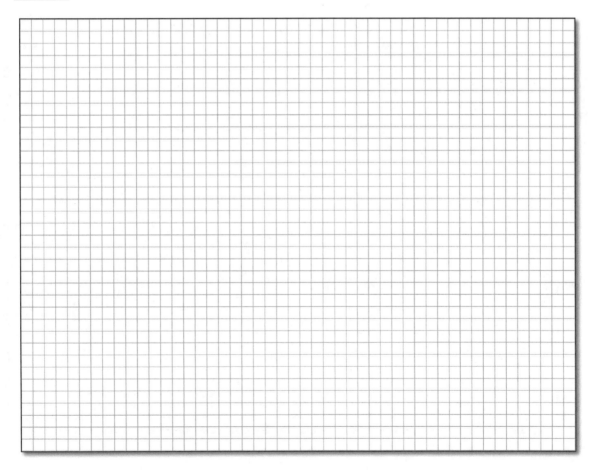

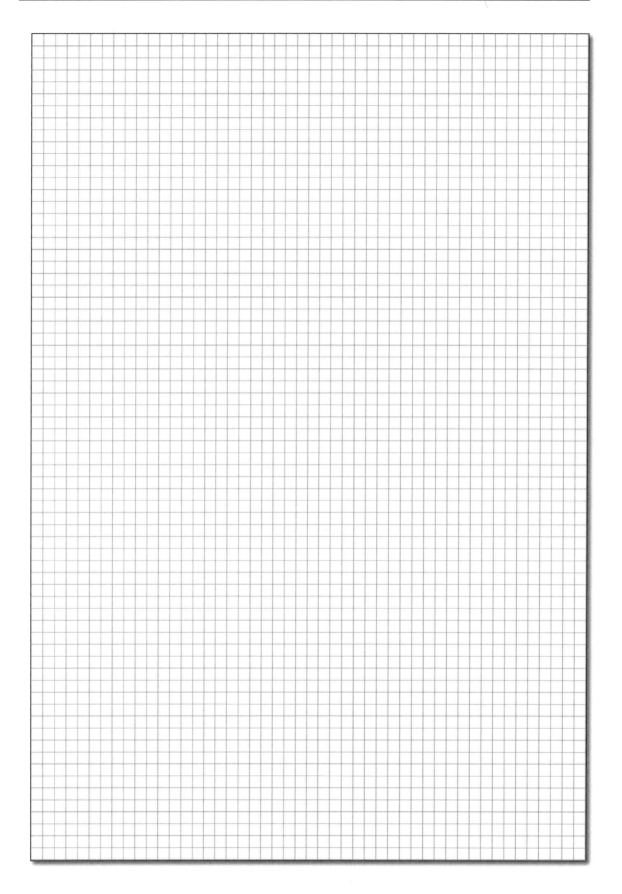

Solved Problem 9.7-4

A 2-kg sphere is attached to an inextensible cord that is being pulled through a hole in a smooth table at a constant rate of v_r = 0.1 m/s. The sphere is also being acted on in a direction perpendicular to the cord by a force $\mathbf{F}$ that has a constant magnitude of 10 N. If the sphere begins from a radius r = 1 m with zero velocity in the transverse direction, determine the sphere's velocity as a function of time.

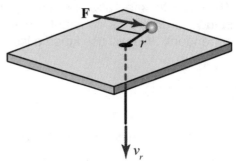

Given: m = 2 kg $\qquad$ F = 10 N = constant
$\qquad$ r_1 = 1 m $\qquad\qquad$ $r_2 = r$
$\qquad$ v_r = -0.1 m/s = constant $\quad$ $v_2 = v$
$\qquad$ no friction

Find: $\mathbf{v}(t)$

Solution

Angular impulse and momentum

We will use the principle of angular impulse and momentum to solve this problem. Remember that this is a vector equation and coordinate directions are important. Since the mass is moving around a central point, polar coordinates would be a good choice.

$$\mathbf{J}_O = \mathbf{H}_{O,2} - \mathbf{H}_{O,1}$$

$$\int_{t_1}^{t_2} \sum \mathbf{M}_O \, dt = \mathbf{r}_2 \times m\mathbf{v}_2 - \mathbf{r}_1 \times m\mathbf{v}_1$$

$$\int_{t_1}^{t_2} -Fr\,\mathbf{k}\,dt = r_2\mathbf{e}_r \times m(v_{r,2}\mathbf{e}_r + v_{\theta,2}\mathbf{e}_\theta) - r_1\mathbf{e}_r \times m(v_{r,1}\mathbf{e}_r + v_{\theta,1}\mathbf{e}_\theta)$$

In the above equation, we will make the state 2 a general state and integrate from t_1 = 0 to $t_2 = t$. Note that $v_{\theta,1}$ = 0 and that $\mathbf{e}_r \times \mathbf{e}_r = 0$.

$$\int_0^t -Fr\,\mathbf{k}\,dt = r\mathbf{e}_r \times m(v_r\mathbf{e}_r + v_\theta\mathbf{e}_\theta) - r_1\mathbf{e}_r \times m(v_r\mathbf{e}_r + \cancel{v_{\theta,1}}\mathbf{e}_\theta) = rmv_\theta\mathbf{k}$$

Kinematics

We need to integrate the left side of the angular impulse-momentum equation given below. Force F is constant, so it will come out of the integral. However, r is time dependent. We will use kinematics to determine this dependency.

$$-\int_0^t Fr\,dt = rmv_\theta$$

The speed at which the rope is pulled is constant, therefore, the distance r is equal to

$$r = r_1 + v_r t$$

Plugging the above relationship into the integral we get

$$-F\int_0^t (r_1 + v_r t)\,dt = (r_1 + v_r t)mv_\theta \qquad -F\left(r_1 t + \frac{v_r t^2}{2}\right) = (r_1 + v_r t)mv_\theta$$

$$v_\theta = -\frac{F}{m(r_1 + v_r t)}\left(r_1 t + \frac{v_r t^2}{2}\right) \qquad \mathbf{v}(t) = v_r \mathbf{e}_r - \left[\frac{F}{m(r_1 + v_r t)}\left(r_1 t + \frac{v_r t^2}{2}\right)\right]\mathbf{e}_\theta$$

$$\boxed{\mathbf{v}(t) = -0.1\mathbf{e}_r - \left[\frac{5t - 0.25t^2}{1 - 0.1t}\right]\mathbf{e}_\theta \; \frac{m}{s}}$$

Example Problem 9.7-5

Four masses (m) are attached to a light shaft. The masses are located a distance r from the axis of rotation. The shaft rotates through the action of a belt system suspended on frictionless bearings. The force **T** is applied to a belt wrapped around a pulley of radius r_p. You may assume the belt is inextensible and doesn't slip on the pulley. If the system starts from rest, determine the time it will take for the system to reach an angular velocity of ω.

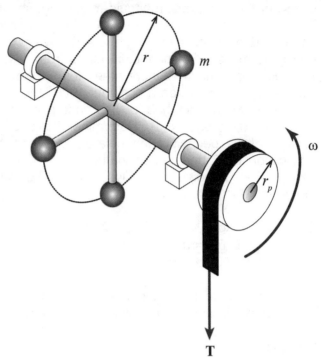

Given:

Find:

Solution:

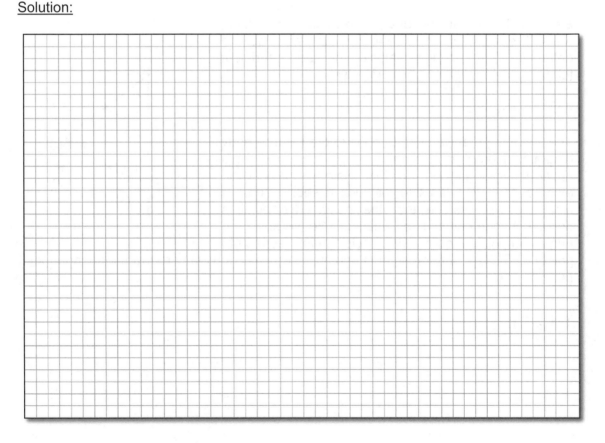

Solved Problem 9.7-6

Consider the symmetric triangular pendulum shown. The pendulum consists of two 5-kg masses connected by beams of length L = 2 meters and negligible mass. The pendulum starts from rest in the position shown where the left beam is completely vertical when a constant 1000-N force F is applied to the left bob of the pendulum for a period of 0.03 seconds. Determine the angular velocity of the pendulum immediately following the application of F and the maximum angle through which the pendulum swings before coming momentarily to rest.

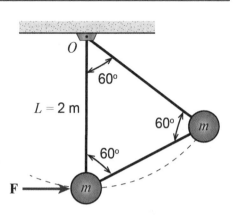

Given: m = 5 kg L = 2 m
      ω_1 = 0 @ t = 0 F = 1000 N
      Δt = 0.03 seconds

Find: angular velocity after impact, ω_2 (at state 2)
        maximum angle pendulum swings through, θ (at state 3)

Solution:

Getting familiar with the system

Since the force is applied over a known duration of time, we will employ an impulse-momentum approach to determine the angular velocity of the pendulum following the application of the impulse.

Free-body diagram

The first thing that we need to do is to draw a free-body diagram in order to determine which forces/moments apply external impulse to the system. Since the external force F is applied for such a short period of time, we will assume that the pendulum's position doesn't change during its application. We will apply the angular-impulse momentum principle with respect to

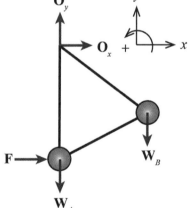

point O. The reaction forces at $\mathbf{O}$ and the weight $\mathbf{W}_A$ don't impart a moment about point O.

$$\int_{t_1}^{t_2} \mathbf{M}_O dt = \mathbf{H}_{O,2} - \mathbf{H}_{O,1}$$

Angular impulse

In calculating the angular impulse, force $\mathbf{F}$ and weight $\mathbf{W}_B$ are the two forces that impart a moment about the reference point O. These forces are constant, and based on the assumption that the pendulum remains approximately in the same location during the application of $\mathbf{F}$, we will treat the moment arms as constant also.

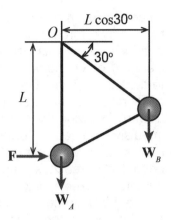

$$\int_{t_1}^{t_2} \mathbf{M}_O dt = \int_{t_1}^{t_2} (FL - W_B L\cos(30))\mathbf{k}\, dt \qquad (1)$$

$$= (FL - W_B L\cos(30))\Delta t\, \mathbf{k}$$

Angular momentum

Since we have a system of particles, the angular momentum of the system is the sum of the angular momenta of the individual particles (the two bobs). Also, the angular momentum of the system is initially zero since the system begins from rest.

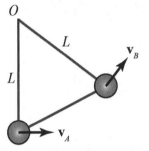

$$\mathbf{H}_{O,2} - \mathbf{H}_{O,1} = \sum \mathbf{r}_{i,2} \times m_i \mathbf{v}_{i,2} - \sum \mathbf{r}_{i,1} \times m_i \mathbf{v}_{i,1} = \mathbf{r}_{A,2} \times m_A \mathbf{v}_{A,2} + \mathbf{r}_{B,2} \times m_B \mathbf{v}_{B,2}$$

Recognizing that the pendulum rotates about a fixed axis through O, each bob will move in a circular path about O with a speed $v = r\omega$ in a direction perpendicular to the vector pointing from O toward the bob.

$$\mathbf{H}_{O,2} - \mathbf{H}_{O,1} = (L^2 m_A \omega_2 + L^2 m_B \omega_2)\mathbf{k} \qquad (2)$$

Angular impulse-momentum principle

Applying the angular impulse-momentum principle, we set Equation (1) and Equation (2) equal to one another and solve for the angular velocity ω of the system following the application of the external force F.

$$\int_{t_1}^{t_2} \mathbf{M}_O dt = \mathbf{H}_{O,2} - \mathbf{H}_{O,1} \qquad\qquad (FL - W_B L\cos(30))\Delta t = (L^2 m_A \omega_2 + L^2 m_B \omega_2)$$

$$\omega_2 = \frac{(F - W_B \cos(30))\Delta t}{L(m_A + m_B)} = \boxed{1.436\ \frac{\text{rad}}{\text{s}}\ \text{ccw}}$$

Conservation of energy

Once the external force F is released, the only forces acting on the pendulum are either conservative (the weights) or do no work (the reaction forces). Therefore, energy is conserved as the pendulum swings upward following the application of the impulsive force. Furthermore, the kinetic energy at state 3, when the pendulum reaches its maximum height, is zero since the system momentarily comes to rest. As with angular momentum, the energy of a system of particles (kinetic and potential) is simply the sum of the energies of the individual particles.

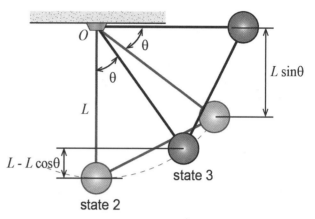

$$T_2 + \cancel{K_2} = \cancel{K_3} + V_3$$

$$\frac{1}{2}m_A v_{A,2}^2 + \frac{1}{2}m_B v_{B,2}^2 = m_A g \Delta h_{A,2-3} + m_B g \Delta h_{B,2-3} = m_A g L(1-\cos\theta) + m_B g L \sin\theta$$

Employing the kinematic relationship, $v = r\omega$ and the value for angular velocity at state 2 solved for earlier, we can solve for the maximum angle θ. To solve the resulting nonlinear equation for θ you can use numerical techniques (e.g. a graphing calculator).

$$\frac{1}{2}m_A L^2 \omega_2^2 + \frac{1}{2}m_B L^2 \omega_2^2 = m_A g L(1-\cos\theta) + m_B g L \sin\theta$$

$$\cos\theta - \sin\theta + \frac{L\omega_2^2}{g} - 1 = 0 \qquad\qquad \boxed{\theta = 0.367 \text{ rad} = 20.8^o}$$

Example Problem 9.7-7

A satellite is in an elliptic orbit about the earth having a perigee of r_P = 5000 mi and an apogee of r_A = 15,000 mi as measured from the center of the earth. The satellite's angular velocity at perigee is 0.001 rad/s. Determine the magnitude of the satellite's velocity v_A at apogee.

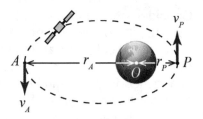

Given:

Find:

Solution:

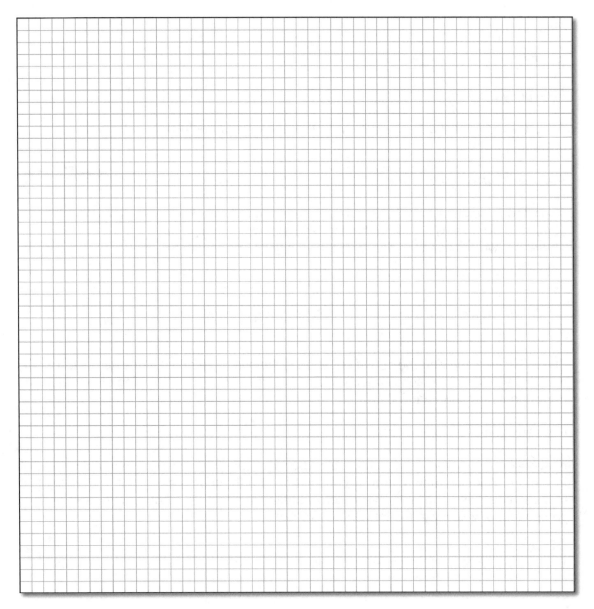

Consider the thin rod of negligible mass shown. It rotates in the horizontal xy-plane about a vertical axis passing through O. A 2-kg collar is able to slide along the smooth rod and is attached to the origin O by a spring that has an unstretched length of 1 m. If the rod initially

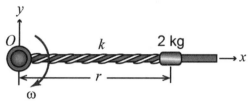

has an angular velocity of 2 rad/s and the collar is located a distance r = 0.5 m from O, determine the angular velocity of the rod when the spring is unstretched. Assume the friction in the joint at O is negligible.

Given: m = 2 kg l_o = 1 m
       ω_1 = 2 rad/s r_1 = 0.5 m
       r_2 = 1 m   no friction

Find: ω_2

Solution

Free-body diagram

To get an idea of the forces acting on the collar, let's begin by drawing a FBD.

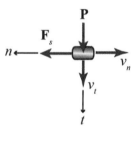

Angular impulse-momentum

Apply the principle of angular impulse-momentum to the collar to see if this will give us ω_2. Note that for fixed-axis rotation, $v_t = r\omega$. Therefore, we only need to deal with the velocity in the tangential direction.

$$\int_{t_1}^{t_2}\sum \mathbf{M}_O\, dt = \mathbf{r}_2\times m\mathbf{v}_2 - \mathbf{r}_1\times m\mathbf{v}_1 \qquad \int_{t_1}^{t_2} Pr\, dt = r_2 mv_{t,2} - r_1 mv_{t,1} = m(r_2^2\omega_2 - r_1^2\omega_1)$$

The issue with applying the above equation is that we have no way of estimating $\mathbf{P}$. We don't know if the bar is accelerated uniformly or not. Also, r is a function of time but we don't know what that function is.

Conservation of angular momentum

If we take the system to include the rod and collar, there are no moments about O. Therefore, angular momentum is conserved. Furthermore, since the rod is light, its mass is negligible and will have a zero angular momentum. Again, we need only the component of velocity in the tangential direction.

$$\mathbf{r}_2\times m\mathbf{v}_2 = \mathbf{r}_1\times m\mathbf{v}_1 \qquad\qquad r_2 mv_{t,2} = r_1 mv_{t,1} \qquad\qquad r_2^2\omega_2 = r_1^2\omega_1 \qquad\qquad \boxed{\omega_2 = 0.5\ \dfrac{\text{rad}}{\text{s}}}$$

9.8) CHOOSING A KINETIC ANALYSIS METHOD

In the course of this book, we have analyzed the motion of individual particles using three different techniques: Newtonian mechanics, work-energy, and impulse-momentum. In this section, we identify characteristics of each of these three analysis techniques to help us identify which to apply in a particular situation. In general, each technique is valid for any given situation, but often one is easier to apply based on the information that is readily available. Discussing how to choose between these three techniques will also serve as a good review of material presented earlier in the book.

9.8.1) NEWTONIAN MECHANICS METHOD

Newton's second law (Equation 9.8-1) provides an instantaneous relationship between forces and acceleration. Therefore, it is well suited to identifying a force or an acceleration at a given instant. If it is desired to determine the cumulative effect of the forces over a given time or distance, then kinematic relationships must also be employed.

Newton's second law: $\boxed{\sum \mathbf{F} = m\mathbf{a}}$ (9.8-1)

$\mathbf{F}$ = forces acting on the particle
m = mass of the particle
$\mathbf{a}$ = acceleration of the particle

9.8.2) WORK-ENERGY METHOD

In general, the work-energy principle (Equation 9.8-2) provides a relationship between forces, displacement, and velocity directly. It, in essence, captures the cumulative effect of a force over the distance that the force acts. Therefore, an indication that this approach to analysis should be used is if information is given, or desired, about the velocities and the distances over which the forces are applied.

$\boxed{U_{1-2,non} = \Delta T + \Delta V}$ (9.8-2)

$U_{1-2,non}$ = non-conservative work
ΔT = change in kinetic energy
ΔV = change in potential energy (conservative work)

One advantage of this approach is that only forces that do work must be considered. For example, consider the situation where a ball rolls down the side of a bowl. The normal force changes direction and magnitude during the ball's motion. If the impulse-momentum method were used, the cumulative effect of the normal force would be difficult to determine. However, the work done by the normal force is zero, therefore, the normal force does not need to be considered explicitly when applying the work-energy approach. The work done by the normal force is zero because at every point along the ball's trajectory, the force is perpendicular to the direction of the ball's motion. Another advantage of the work-energy approach is that when only conservative forces are acting on the system, the path taken by the system does not need to be known. Only the initial and final endpoints of the system need to be known.

A work-energy approach can also be employed for determining the instantaneous behavior of a system by taking the time derivative of Equation 9.8-2. This is an alternative to using Newton's second law for determining the instantaneous behavior of a system. This approach is especially useful in the case where energy is conserved because the derivative of the total energy is then just equal to zero.

9.8.3) IMPULSE-MOMENTUM METHOD

The linear impulse-momentum principle (Equation 9.8-3) provides a direct relationship between forces, time, and velocity. This relationship captures the cumulative effect of the forces over the time which they act. Therefore, an indication that this approach to analysis should be used is if information is given about the velocities and the time over which the forces are applied.

$$\boxed{\int_{t_1}^{t_2} \Sigma \mathbf{F}\, dt = \Delta \mathbf{G}} \quad (9.8\text{-}3)$$

$\mathbf{F}$ = forces acting on the particle
t = time
$\Delta \mathbf{G}$ = change in linear momentum

The linear impulse-momentum relationship is especially useful for analyzing collisions. In a collision, the interaction forces between the two colliding bodies are equal and opposite, resulting in them summing to zero. If there are no external forces, then the linear momentum of the system is conserved.

The angular impulse-momentum principle (Equation 9.8-4) provides a relationship between moments, time, and velocity. The angular impulse-momentum relationship is especially useful for central-force problems. Central-force problems are characterized by the resultant force acting on the body that has a line of action that always passes through a fixed-point O. By choosing this fixed-point O as the reference, with respect to which the angular impulse and momentum are calculated, the applied moments will equal zero ($\mathbf{r} \times \Sigma \mathbf{F} = 0$) and the angular momentum will be conserved with respect to O.

$$\boxed{\int_{t_1}^{t_2} \Sigma \mathbf{M}_O\, dt = \int_{t_1}^{t_2} (\mathbf{r} \times \Sigma \mathbf{F})\, dt = \Delta \mathbf{H}_O} \quad (9.8\text{-}4)$$

$\mathbf{F}$ = forces acting on the particle
$\mathbf{r}$ = position vector of the particle relative to the
  reference point O
$\mathbf{M}_O$ = moments acting on the particle

t = time
t_1 = time at state 1
t_2 = time at state 2
$\Delta \mathbf{H}_O$ = change in angular momentum

In general, there is a strong similarity between our work-energy relationships (Equation 9.8-2) and our impulse-momentum relationships (Equation 9.8-3 and 9.8-4). This follows from the fact that the work-energy principle is derived by integrating Newton's second law with respect to displacement and the impulse-momentum relationship is derived by integrating Newton's second law with respect to time. Table 9.8-1 summarizes situations and the corresponding analysis method that is most appropriate.

Givens / Need to find	Most Appropriate Method
Forces / Acceleration	Newtonian Mechanics
Forces / Displacement / Velocity	Work-Energy
Forces / Time / Velocity	Linear Impulse-Momentum
Moments / Time / Velocity	Angular Impulse-Momentum

Table 9.8-1: Appropriate method summary

CHAPTER 9 REVIEW PROBLEMS

RP9-1) The impulse-momentum principle is mostly used for solving problems involving what parameters?

    a) force, velocity, time
    b) force, acceleration, time
    c) velocity, acceleration, time
    d) force, velocity, acceleration

RP9-2) The linear impulse-momentum principle is derived by integrating Newton's second law with respect to time. (True, False)

RP9-3) If the mass of a particle doubles then its linear momentum and kinetic energy also both double. (True, False)

RP9-4) The impulse applied to a particle by a force is

    a) defined only for large forces applied over short intervals of time.
    b) equal to the change in kinetic energy of the particle.
    c) equal to the area under the force vs. time graph.

RP9-5) A particle, initially at rest, is acted upon by a time varying force $F(t)$. Rank the following force profiles from greatest to least achieved particle speed.

Greatest: _____   Next: _____   Next: _____   Least: _____

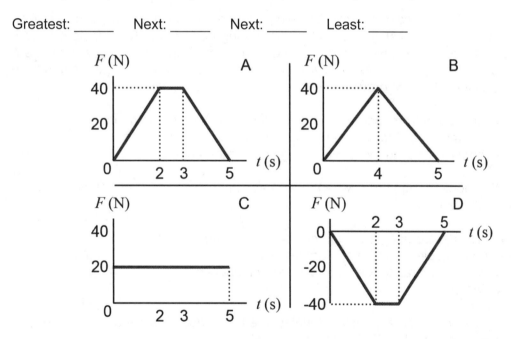

RP9-6) Conservation of linear momentum of a particle over a period of time requires that

    a) there be no external forces applied to the particle over that time interval.
    b) there be no external moments applied to the particle over that time interval.
    c) the external forces applied to the particle sum to zero over that time interval.

RP9-7) If linear momentum is conserved in one component direction then it must also be conserved in the perpendicular component direction. (True, False)

RP9-8) Which quantities are in general conserved during a collision between multiple particles?

    a) linear momentum
    b) energy
    c) both linear momentum and energy

RP9-9) Both linear momentum and energy are conserved during which type of collision?

    a) elastic
    b) inelastic
    c) all collisions

RP9-10) A rocket is propelled by expelling high-velocity gas. The resulting motion is a consequence of which of the following.

    a) conservation of linear momentum
    b) conservation of energy
    c) both
    d) neither

RP9-11) Consider two cases of the following situation. You need to push a broken down car back to town. First, you push the car starting from rest with a constant force F for 1 minute. Second, you push the car already rolling at 3 mph with a constant force F for 1 minute. In which case does the car experience a larger change in momentum? In both cases, assume that the ground is level and friction is negligible.

    a) The car starting from rest experiences a larger change in momentum.
    b) The car starting at 3 mph experiences a larger change in momentum.
    c) They both experience the same change in momentum.

RP9-12) Why does an egg crack when tossed against a brick wall but not when tossed against a taught bed sheet?

    a) The force decelerating the egg acts for a longer period of time.
    b) The force decelerating the egg decreases.
    c) The force decelerating the egg acts over a larger area.
    d) All of the above.

RP9-13) If the mass of a particle doubles then its linear momentum and angular momentum also both double. (True, False)

RP9-14) Which of the following quantities of a particle are dependent on the particle's position?

    a) linear momentum
    b) angular momentum
    c) energy
    d) all
    e) none

RP9-15) Which is the correct formula for the angular momentum of a particle?

    a) mass $\bullet$ velocity
    b) mass $\bullet$ velocity $\bullet$ radius
    c) mass $\bullet$ velocity^2 $\bullet$ radius
    d) mass $\bullet$ velocity $\bullet$ radius^2
    e) mass $\bullet$ velocity^2 $\bullet$ radius^2

RP9-16) The angular momentum of a particle can be conserved over an interval of time where the external forces applied to the particle do not sum to zero. (True, False)

RP9-17) A 1.4-kg toy rocket is initially resting on the ground when its engine is fired causing the rocket to travel vertically upward.   The thrust generated by the rocket's engine is given in the attached graph. If the engine's maximum thrust is 30 N, determine the rocket's maximum velocity. Assume that air resistance is negligible.

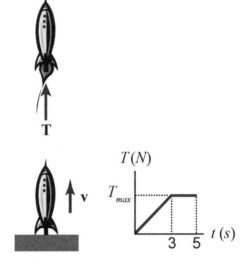

Given: $m = 1.4$ kg $T_{max} = 30$ N
$v_1 = 0$
neglect air resistance

Find: v_{max}

Solution:

Determine the thrust as a function of time.	Determine the maximum velocity

$T =$ _____

Draw a free-body diagram

$v_{max} =$ _____

RP9-18) Three train cars are rolling along a horizontal track. The velocities of car A, B and C are 3, 2 and 1 mph respectively in the directions shown. The weights of car A, B and C are 140, 30 and 100 tons respectively. After the cars collide they couple and move with a common velocity. Determine the car's final velocity and the percentage of energy that is lost in the collision.

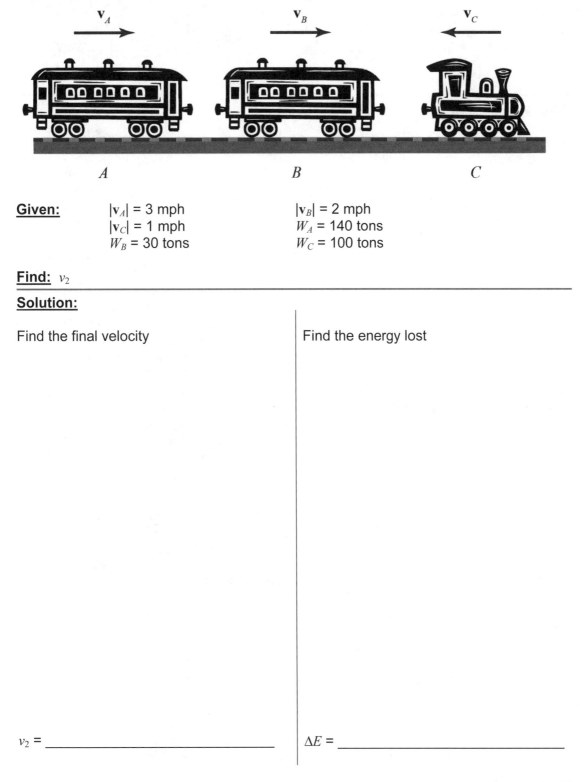

Given: $|\mathbf{v}_A|$ = 3 mph $|\mathbf{v}_B|$ = 2 mph
              $|\mathbf{v}_C|$ = 1 mph W_A = 140 tons
              W_B = 30 tons W_C = 100 tons

Find: v_2

Solution:

Find the final velocity   Find the energy lost

$v_2 =$ _____ $\Delta E =$ _____

RP9-19) Two identical hockey pucks sliding on ice, with the velocities and directions shown, collide. If the coefficient of restitution is $e = 0.8$, determine the velocities after impact and the energy lost during the collision.

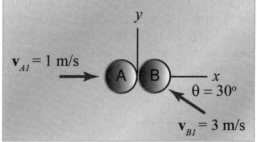

Given: $v_{A,1} = 1$ m/s $v_{B,1} = 3$ m/s
$\theta = 30°$ $e = 0.8$

Find: v_2, ΔE

Solution:

Plane of contact equations - conservation of momentum for each particles.

$v_{Ay,2} =$ _____

$v_{By,2} =$ _____

Line of impact equation - conservation of momentum for the system.

$v_{Ax,2}(v_{Bx,2}) =$ _____

Coefficient of restitution.

$v_{B,2} =$ _____

$v_{A,2} =$ _____

Change in energy.

$\Delta E =$ _____

RP9-20) Two 1-kg masses are attached to light bars of length $L = 1.2$ m which are fixed to a rotating shaft. The shaft is initially rotating at 10 rad/s when a 2-N-m moment is applied in the same direction as the initial rotation.   Determine the velocity of the masses 20 seconds after the moment is applied.

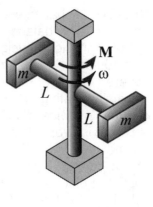

Given: $m = 1$ kg $L = 1.2$ m
                $\omega_1 = 10$ rad/s $\Delta t = 20$ s
                $M = 2$ N-m

Find: v_2

Solution:

Determine ω_2.

Determine v_2.

$\omega_2 =$ _____

$v_2 =$ _____

RP9-21) A balance pivots freely about O. The balance consists of a 300 mm long weightless arm (l_1 = 100 mm, l_2 = 200 mm), a 5-kg sphere (A) and a 1-kg cup (C) and is initially rotating at 10 rad/s in the counterclockwise direction. Block B (7 kg) is dropped from rest at a height of h = 150 mm directly above cup C. The drop is timed

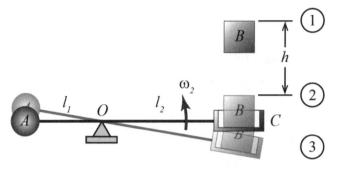

such that block B hits cup C when the balance arm is horizontal. Determine the angular velocity of the balance immediately after block B lands in cup C.

Given: m_A = 5 kg m_B = 7 kg
$\qquad\quad m_C$ = 1 kg l_1 = 100 mm
$\qquad\quad l_2$ = 200 mm h = 150 mm
$\qquad\quad \omega_2$ = 10 rad/s $v_{B,1}$ = 0

Find: ω_3

___

Solution:

Determine the velocity of block B just before it hits cup C.

Determine the angular velocity of the balance arm immediately after block B lands in cup C.

$v_{B,2}$ = _____

ω_3 = _____

CHAPTER 9 PROBLEMS

P9.1) BASIC LEVEL LINEAR IMPULSE-MOMENTUM PROBLEMS

P9.1-1) A 1-kg ball is dropped from a height of h_1 = 5 ft. The ball hits the floor and rebounds to a height of h_3 = 4.3 ft. What is the magnitude of the impulsive force that the floor imparts on the ball if the duration of the impact is 0.2 seconds?

Ans: F = 1.77 lb

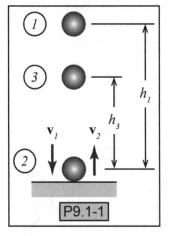

P9.1-1

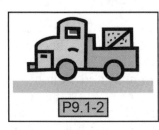

P9.1-2

P9.1-2)^{fe} A truck carrying a crate is traveling down the road at 60 mph. Determine the minimum time with which the driver could bring the truck to a complete stop without the crate sliding along the truck bed. The coefficients of static friction and kinetic friction are μ_s = 0.6 and μ_k = 0.4, respectively.

a) t = 1.05 s   b) t = 3.11 s   c) t = 4.55 s   d) t = 14.95 s

P9.1-3)^{fe} A 3000-lb car impacts a concrete barrier traveling at a speed of 4 mi/hr. If the duration of the impact is 0.3 seconds and the car rebounds from the barrier with a speed of 1 mi/hr, then determine the average horizontal force experienced by the car during the impact.

a) F = -279.5 lb   b) F = -2279 lb

c) F = -40,030 lb   d) F = -931.7 lb

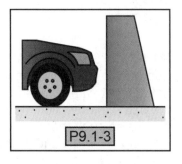

P9.1-3

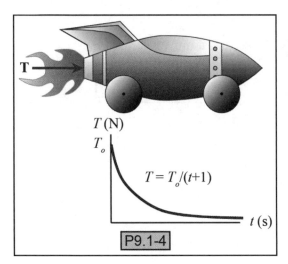

P9.1-4

P9.1-4) A 5-kg toy rocket car, starting from rest, drives along a horizontal surface once the engine is engaged.   The thrust T produced by the engine is time dependent and its magnitude is described by the attached graph, where T_o = 10 N. Neglecting rolling and air resistance, determine the speed of the car after 10 seconds.

Ans: v_2 = 4.8 m/s

P9.1-5)[fe] Ball A and ball B roll, in the same direction, with the following masses and initial speeds: $m_A = 5$ kg, $m_B = 10$ kg, $v_A = 4$ m/s and $v_B = 2$ m/s. After ball A collides with ball B it moves in the opposite direction with a speed of $v_A' = 1$ m/s. Determine the direction and speed of ball B after the collision.

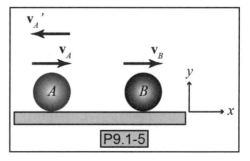

P9.1-5

a) $v_B' = 4.5$ m/s

b) $v_B' = 3.5$ m/s

c) $v_B' = 2.5$ m/s

d) $v_B' = 1.5$ m/s

P9.1-6

P9.1-6) A 150-lb man stands on a 3-lb skateboard and throws a 5-lb ball horizontally with a speed of 70 mph. If the man and skateboard are initially at rest, determine the speed of the man and skateboard after he throws the ball. If rolling resistance is considered, determine the speed of the man and skateboard 1 second after he throws the ball. Rolling resistance by be estimated by multiplying the total normal force with the rolling resistance coefficient $f_r = 0.2$.

Ans: $v_{s,2} = 1.37$ mph, $v_{s,2} = 1.24$ mph

P9.1-7)[fe] Car A has a mass of 1400 kg and is approaching the intersection shown at the same time Car B with a mass of 1100 kg is approaching the same intersection. If the two cars are involved in a collision such that they become entangled and the ensuing wreckage moves with velocity $v = 10\mathbf{i} + 7\mathbf{j}$ m/s, then determine the speed of each car prior to impact. Also determine the percentage of energy lost in the process of the collision.

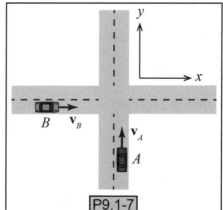

P9.1-7

a) $v_{A,1} = 22.7$ m/s, $v_{B,1} = 12.5$ m/s

b) $v_{A,1} = 12.5$ m/s, $v_{B,1} = 22.7$ m/s

c) $v_{A,1} = 10.5$ m/s, $v_{B,1} = 20.7$ m/s

d) $v_{A,1} = 20.7$ m/s, $v_{B,1} = 10.5$ m/s

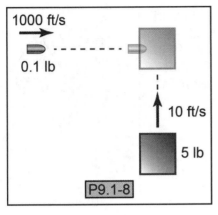

P9.1-8) A 0.1-lb bullet is travelling at 1000 ft/s when it strikes a 5-lb block travelling at 10 ft/s on a smooth surface in the directions shown. If the bullet hits the block centrally and is embedded in the block, determine the speed and direction of the bullet/block system following the collision.

a) $v = 21.92$ ft/s, $\theta = 63.4°$

b) $v = 21.92$ ft/s, $\theta = 26.6°$

c) $v = 29.41$ ft/s, $\theta = 26.6°$

d) $v = 29.41$ ft/s, $\theta = 63.4°$

P9.1-9)[fe] A 3750-lb automobile travels down a 10-degree hill at 50 mph when the driver applies the brakes. If the braking system supplies an average 1700-lb braking force, determine the time required to bring the vehicle to rest once the brakes are applied.

a) $t = 8.14$ s   b) $t = 4.16$ s   c) $t = 5.55$ s   d) $t = 26.7$ s

P9.2) INTERMEDIATE LINEAR IMPULSE-MOMENTUM PROBLEMS

P9.2-1) A 4-oz baseball traveling at $v_1 = 100$ mph gets struck by a bat. The baseball leaves the bat traveling $v_2 = 140$ mph at an angle of $\theta = 35°$ as shown in the figure. Determine the magnitude and direction of the average force that the bat applies to the ball if the impact lasts for 0.02 seconds.

Ans: $F = 130.5$ lb, $\phi = 20.5°$

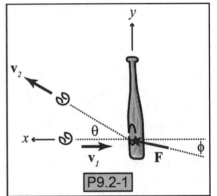

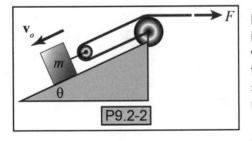

P9.2-2) A 50-kg mass attached to a pulley system is initially traveling down a 20° incline at 2 m/s when a 100-N force is applied as shown in the figure. Determine the time it takes the mass to start traveling up the incline and the speed of the mass 10 seconds after the application of the force.

Ans: $t = 3.1$ s, $v = 4.45$ m/s

P9.2-3)[fe] A very light 3-in radius flywheel is attached to and rotates about a bearing at O. The bearing is poorly lubricated and creates a constant frictional moment of 1 lb-ft. A 15-lb weight is attached via a string to the outer rim of the flywheel. The weight is released from rest and allowed to fall. Determine the time it takes the weight to reach a speed of 10 ft/s.

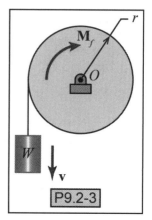

P9.2-3

a) $t = 13.6$ s   b) $t = 5.8$ s

c) $t = 1.12$ s   d) $t = 0.42$ s

P9.2-4

P9.2-4) Consider the accompanying figure of a man pushing a 100-lb wood crate along a painted concrete floor. If he wishes to accelerate the crate to at least 3 mph starting from rest in the span of 5 seconds, what constant pushing force (**P**) is needed if it is applied at an angle of $\theta = 35°$? The kinetic and static coefficients of friction are 0.2 and 0.28, respectively. Use the principle of impulse and momentum to solve this problem. Use one other method to solve this problem and compare and contrast the methods.

Ans: $P = 42.5$ lb

P9.2-5) A cue ball ($m_{cue} = 0.17$ kg) hits the 8 ball ($m_8 = 0.16$ kg) with a velocity of 3 m/s in the direction shown ($\theta = 20°$). After hitting the 8 ball, the cue ball's velocity is 2 m/s in the direction shown ($\phi = 10°$). Find the direction and magnitude of the impulsive force that the cue ball applies to the 8 ball and the velocity of the 8 ball after it has been struck. The impact lasts for 0.1 seconds. Assume that the impulsive force is constant for the duration of the impact.

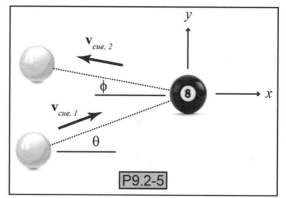

P9.2-5

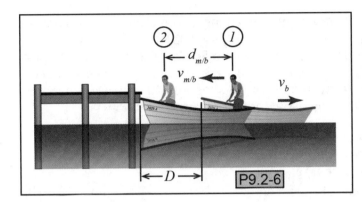

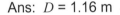

P9.2-6

P9.2-6) A man (m_{man} = 90 kg) stands in the aft of a small boat (M_{boat} = 400 kg). The boat is stationary and the bow of the boat just touches the dock. The man walks, at a constant rate, a distance of 4 meters toward the dock relative to the boat in a span of 3 seconds. Determine the distance the bow of the boat moves away from the dock during these 3 seconds.

Ans: D = 1.16 m

P9.2-7) A worker throws a 20-lb package onto a 100-lb stationary cart. The package strikes the cart with a speed of 3 ft/s at an angle of 25° as shown in the figure. After the package lands on the cart, the cart and package move as one. Determine the carts speed after the package lands on it and the energy lost during the impact.

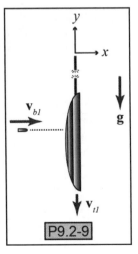

P9.2-7

Ans: v_2 = 0.453 ft/s, ΔE = 2.41 lb-ft

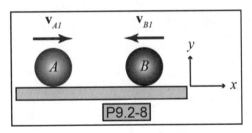

P9.2-8

P9.2-8) Two 3-kg balls roll on a horizontal table and eventually collide. The balls are initially rolling at $v_{A,1}$ = 7 m/s and $v_{B,1}$ = 15 m/s in the directions shown. If the coefficient of restitution is 0.7, determine the velocity of ball B after the collision. Also, determine the amount of energy lost in the collision.

Ans: $v_{B,2}$ = 3.7 m/s, ΔE = 185 J

P9.2-9) A 3-kg rifle target is suspended by a rope that breaks. At the exact instant that the rope breaks, a shooter fires a 30-g bullet from his gun. The bullet embeds in the target 1 second after being fired and arrives with a speed of 200 m/s entirely in the horizontal direction. Neglecting air resistance and the target size, determine the horizontal distance that the target travels before it hits the ground. The target center is 3 m above the ground when the bullet strikes it.

Ans: Δx = 2.15 m

P9.2-9

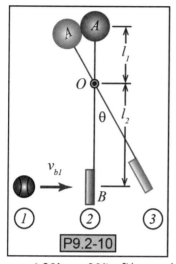

P9.2-10) A weighted throwing target consists of a light rod with a counter weight (W_A = 3 lb) and target (W_B = 2 lb) attached to it as shown in the figure. The rod is free to pivot about O. A tennis ball (W_b = 3 oz) is thrown, hits and sticks to the Velcro target with a speed of 40 ft/s. Determine the maximum angle that the target swings through.

Ans: $\theta = 20.1°$

P9.2-11) Two smooth spheres of identical mass collide as shown. The velocities of the two spheres prior to impact are $v_A = (-20i - 20j)$ ft/s and $v_B = (15i + 5j)$ ft/s. Assuming the coefficient of restitution is 0.7, determine the velocities of the spheres following the impact.

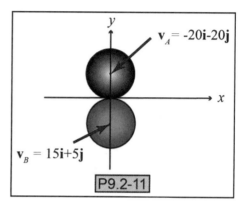

Ans: $v_A = -20i + 1.25j$ ft/s, $v_B = 15i - 16.25j$ ft/s

P9.3) ADVANCED LEVEL LINEAR IMPULSE-MOMENTUM PROBLEMS

P9.3-1) A 2-kg box is initially at rest when it is pushed along a rough floor (μ_k = 0.4) by a force **F** whose magnitude as a function of time is given in the attached figure. If the force **F** is applied to the box at position 1 and is removed 3 seconds later at position 2, determine the total distance, D, that the box travels before coming to rest at position 3. In the given force profile you may take F_o to equal 30 N.

Ans: $D = 42$ m

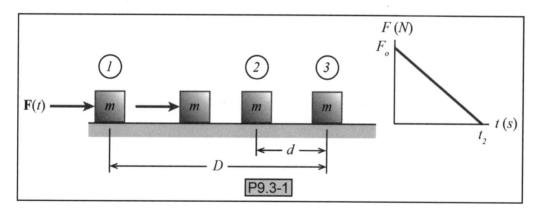

P9.3-2) A 5-kg particle, initially at rest, is acted upon by force **F** that acts in a straight line and has a magnitude represented in the figure. Neglecting friction, determine the speed of the particle 15 seconds after the force is applied. Also, determine the maximum displacement of the particle assuming that force F continues to decrease at the same constant rate shown after t = 15 s.

Ans: v = 30 m/s, Δs = 425 m

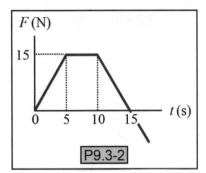

P9.3-3) Consider the 1000-kg satellite shown. It is inserted into orbit with an angular velocity of 1.5 rad/sec about its z-axis and an initial velocity of 5000 m/s entirely in the x-direction. The primary satellite booster generates 500 N of constant thrust. The thrust vector initially points in the y-direction. If the booster is fired starting at the instant shown and remains on until the satellite has rotated $\pi/4$ radians, determine the resulting component of the satellite's velocity in the y-direction. Assume that the satellite's mass remains constant and that the thrust is directed through the satellite's mass center (allows us to model the satellite as a particle).

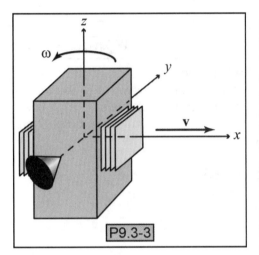

Ans: $v_{y,2}$ = 0.236 m/s

P9.3-4) Two small spheres of identical mass hang from strings of length L. Sphere A is released from rest from an angle of α and strikes sphere B that is initially at rest in the position shown. As a result of the collision, sphere B swings through an angle β. Determine the coefficient of restitution.

Ans: $e = 2\sqrt{\dfrac{1-\cos\beta}{1-\cos\alpha}} - 1$

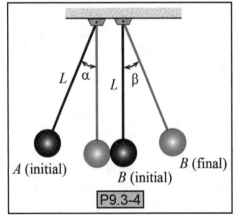

P9.4) BASIC LEVEL ANGULAR IMPULSE-MOMENTUM PROBLEMS

P9.4-1) A particle moves through space with a velocity of 5 ft/s in the direction shown. What is the particle's angular momentum about O when it is located at the coordinate position (7,6) ft?

Ans: $\mathbf{H}_O$ = -1.24 **k** slug-ft^2/s

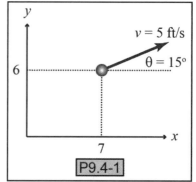

P9.4-1

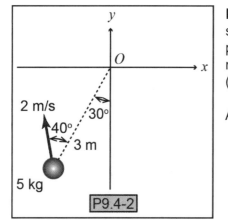

P9.4-2

P9.4-2) A 5-kg particle moves in the x-y plane with a speed of 2 m/s in the direction shown. For the instant pictured, determine the particle's (a) linear momentum, (b) angular momentum about point O and (c) kinetic energy.

Ans: $\mathbf{G}$ = -1.74 **i** + 9.85 **j** kg-m/s,
  $\mathbf{H}_O$ = -19.3 **k** kg-m^2/s, T = 10 J

P9.4-3) A pendulum consists of a 5-kg bob attached to the end of 2-m long bar of negligible mass. At the instant shown, the bob has a velocity of 0.2 m/s and θ equals 20 degrees. For this instant, determine in Cartesian and polar coordinates (a) the linear momentum **G** of the bob and (b) the angular momentum $\mathbf{H}_o$ of the bob.

Ans: $\mathbf{G}$ = 1 $\mathbf{e}_\theta$ kg-m/s = 0.94 **i** + 0.342 **j** kg-m/s,
  $\mathbf{H}_O$ = -2 **k** kg-m^2/s

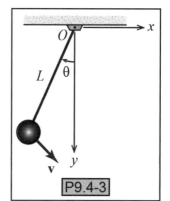

P9.4-3

P9.4-4) A 10-lb particle's position is given by the position vector $\mathbf{r} = t\mathbf{i} - 3\mathbf{j} + 7t^2\mathbf{k}$ ft. Determine the particle's angular momentum with respect to the origin at t = 5 seconds.

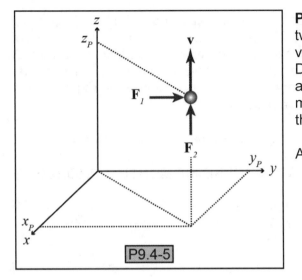

P9.4-5) A particle in space is acted on by two forces ($\mathbf{F}_1$, $\mathbf{F}_2$) and moves with a velocity $\mathbf{v}$ as shown in the figure. Determine the particle's angular momentum and time rate of change of angular momentum with respect to the origin when the particle is at position (x_P, y_P, z_P).

Ans: $\mathbf{H}_O = mv(y_P\mathbf{i} - x_P\mathbf{j})$,

$\dot{\mathbf{H}}_O = y_P F_2\mathbf{i} + (z_P F_1 - x_P F_2)\mathbf{j} - y_P F_1\mathbf{k}$

P9.4-5

P9.4-6)[fe] A light rod is free to pivot about its center (O). The rod is initially resting horizontally when two masses simultaneously strike and stick to the rod. Determine the angular velocity of the rod after the impact assuming the masses and velocities of the particles are as shown in the figure.

a) $\omega = v/mL$ rad/s ccw

b) $\omega = v/L$ rad/s ccw

c) $\omega = v$ rad/s ccw

d) $\omega = 4v/L$ rad/s ccw

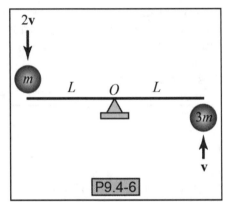

P9.4-6

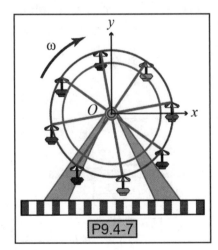

P9.4-7

P9.4-7)[fe] The Ferris wheel shown consists of 8 cars that each have a mass of approximately 500 kg when loaded with riders. The cars are located about the point O at a radius of 8 m. If the Ferris wheel is spinning clockwise with an angular speed of 0.25 rad/sec, determine the system's angular momentum with respect to the wheel's center O. You may treat the mass of the Ferris wheel structure as negligible.

a) $\mathbf{H}_O = -64{,}000\ \mathbf{k}$ kg-m^2/s

b) $\mathbf{H}_O = 64{,}000\ \mathbf{k}$ kg-m^2/s

c) $\mathbf{H}_O = -8{,}000\ \mathbf{k}$ kg-m^2/s

d) $\mathbf{H}_O = 8{,}000\ \mathbf{k}$ kg-m^2/s

9.4-8)[fe] Two men of mass 100 kg and mass 80 kg sit at the edge of a merry-go-round with a 2 meter radius. The merry-go-round is initially rotating with an angular speed of 10 rpm. If the 80-kg man walks inward on the merry-go-round and stops a distance of 1 meter from the center O about which the ride is spinning, estimate the new angular speed of the system. You may assume that the merry-go-round rotates with negligible friction and has a much smaller mass than the men.

a) ω = 12.9 rpm   b) ω = 13.5 rpm   c) ω = 14.3 rpm   d) ω = 15 rpm

P9.5) INTERMEDIATE LEVEL ANGULAR IMPULSE-MOMENTUM PROBLEMS

P9.5-1) A 2-kg sphere is attached to an inextensible cord that is being pulled through a hole in a smooth table by a force F. The sphere is initially rotating about the hole in the table with a velocity of 5 m/s when the radius is 1 m.   If the force F is increased until the radial distance has been halved, determine the speed of the particle at this new radial position. Also determine the magnitude of the force F in this state.

Ans: v = 10 m/s, F = 400 N

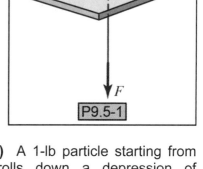

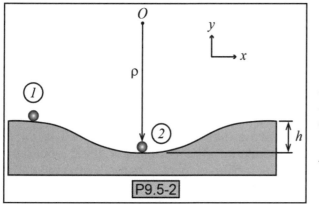

P9.5-2) A 1-lb particle starting from rest, rolls down a depression of height h = 2 ft and a radius of curvature of ρ = 5 ft. What is the angular momentum of the particle about O when it reaches the bottom of the depression?

Ans: $\mathbf{H}_o$ = 1.76 k slug-ft^2/s

P9.5-3) Consider the thin rod of negligible mass shown rotating in the horizontal xy-plane about the vertical axis through O with an angular speed of 3 rad/s. A 2-kg collar is sliding along the rod toward the origin O with a speed of 2 m/s. Determine the torque that must be applied at this instant about the vertical z-axis to maintain the rod's angular speed at 3 rad/s. Assume the friction in the joint at O is negligible.

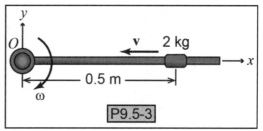

Ans: T = 12 N-m

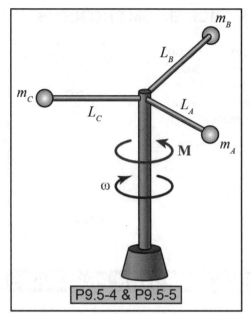

P9.5-4 & P9.5-5

9.5-4) Three spheres lying in a horizontal plane are rigidly attached to a vertical shaft via light rods as shown. The system is initially rotating about the vertical shaft with an angular speed of 20 rad/sec when a moment of magnitude $M = 2t$ N-m, where t is in seconds, is applied in the opposite direction. Determine the time required to bring the system to rest if $m_A = 5$ kg, $m_B = 5$ kg, $m_C = 5$ kg, $L_A = 2$ m, $L_B = 3$ m, and $L_C = 4$ m.

Ans: $t = 53.85$ s

9.5-5) Three spheres lying in a horizontal plane are rigidly attached to a vertical shaft via light rods as shown. The system is initially at rest when a moment of magnitude $M = t^2$ N-m, where t is in seconds, is applied in the direction shown (note that the resulting ω will be in the opposite direction from that shown in the figure). Determine the angular speed of the system as a function of time if $m_A = 5$ kg, $m_B = 4$ kg, $m_C = 2$ kg, $L_A = 2$ m, $L_B = 3$ m, and $L_C = 4$ m.

Ans: $\omega = 0.004t^3$ rad/s

9.5-6) A flyball governor, such as the one depicted, was employed by James Watt to provide feedback to mechanically control the speed of his steam engine.   Consider that the apparatus shown consists of two spheres of equal mass that are mounted to the ends of long rods of length L and negligible mass.   The other ends of the rods are pinned to a vertical shaft at O. The rods make an angle of $\theta = 45$ degrees with the vertical shaft when the shaft is rotating with an angular speed of 90 rpm.   Determine the angle the rods make with the vertical shaft when the vertical shaft is rotating with an angular speed of 60 rpm.

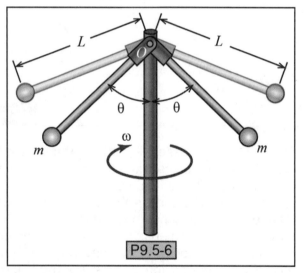

P9.5-6

Ans: $\theta = 60°$

P9.6) ADVANCED LEVEL ANGULAR IMPULSE-MOMENTUM PROBLEMS

P9.6-1) Consider the simple pendulum shown. If the pendulum is released from rest from an angle of $\theta = 90$ degrees, determine the pendulum's velocity when it reaches its lowest position ($\theta = 0$ degrees) using the angular impulse-momentum principle. Compare this approach to employing a work-energy approach.

Ans: $v = \sqrt{2gL}$

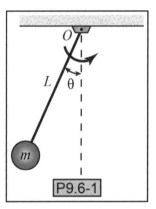

CHAPTER 9 COMPUTER PROBLEMS

C9-1) A 5-kg particle is initially at rest when it is acted upon by force **F**. The force acts in a straight line and has the magnitude represented in the figure. Neglecting friction, plot the speed and displacement of the particle from $t = 0$ to $t = 20$ seconds.

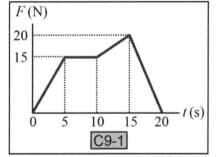

CHAPTER 9 DESIGN PROBLEMS

D9-1) Design a baseball bat to maximize the velocity with which a baseball leaves the bat. Research and consider the following: standard major league baseball mass and the coefficient of restitution between the ball and bat. Remember the coefficient of restitution will depend on the bat material. Start your design with the standard major league bat dimensions, mass, and material. From there, optimize the bat design. You may change its dimensions, mass, and material. Justify all your design choices and calculate the velocity of the ball when it leaves the bat as a function of a constant moment (simulating the batter's input) applied to the bat. Model the bat as a particle during the impact, but model it as a rigid body for any calculations performed before the impact.

CHAPTER 9 ACTIVITIES

A9-1) This activity demonstrates the concept of conservation of linear momentum.

Group size: 3 students (It is best if each student is different in weight.)

Supplies

- two identical chairs with wheels
- masking tape

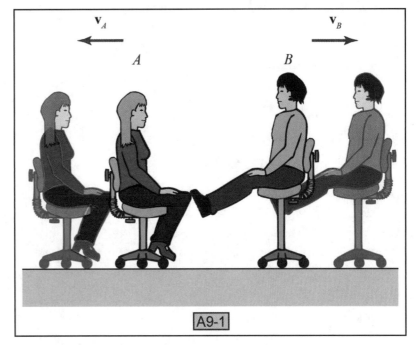

A9-1

Procedure

1. Measure the weight of each student.
2. Place the wheeled chairs on a smooth floor facing each other.
3. Choose two people and have them sit one in each chair.
4. Mark the initial location of each chair with the masking tape.
5. Have one subject push the other subject with his/her feet, causing both chairs to move away from each other as shown in the figure.
6. The third student should measure the distance each chair moved.
7. Repeat steps 5 and 6 but this time have the other subject push.
8. Repeat the entire experiment two more times so that each student has a chance to measure the chair distance.

Calculations and Discussion

- Calculate the average speed of each chair for every experimental run.
- Calculate the linear momentum of the system immediately after the push.
- List all of the assumptions that were applied in your calculations.
- Discuss your results
- List possible sources of experimental error.

NOTES

PART V: KINETICS - IMPULSE & MOMENTUM

CHAPTER 10: RIGID-BODY IMPULSE & MOMENTUM

CHAPTER OUTLINE

CHAPTER SUMMARY

In the chapter on *Particle Impulse and Momentum*, the impulse-momentum method was introduced as the third and final kinetic analysis technique that will be covered in this book. In this chapter, we will build on the particle impulse-momentum method and apply it to systems of particles and rigid bodies. As was the case with our previous techniques, rigid-body analysis introduces complications. Unlike a particle, rigid bodies have size and can rotate as well as translate. In other words, if an unbalanced force is applied at a distance from the center of mass of a  rigid body, it will cause the body to translate and rotate. As was the case with Newtonian mechanics and the work-energy method, we will consider angular impulse-momentum relationships with respect to different types of motion and different reference points. In particular, we will consider pure translation, pure rotation, and general planar motion.

10.1) IMPULSE-MOMENTUM FOR SYSTEMS OF PARTICLES

10.1.1) LINEAR IMPULSE AND MOMENTUM FOR SYSTEMS OF PARTICLES

In prior chapters, we have seen that the total energy of a system of particles is simply the sum of the energies of the individual particles. This type of relationship is also true for linear and angular momentum. Furthermore, by defining a system to include multiple particles, advantages can be derived from the fact that the internal reaction forces cancel.

The concepts of linear impulse and linear momentum for systems of particles were introduced in the chapter on *Particle Impulse and Momentum*. These relationships are repeated here in Equations 10.1-1 through 10.1-3. The linear momentum of the system is simply the sum of the linear momenta of the individual particles making up that system. The linear momentum of the system may also be determined by using the total mass of the system (m) and the velocity of the system's mass center ($\mathbf{v}_G$) as described in Equation 10.1-1. The total impulse applied to the system is equal to the impulse applied by an equivalent resultant force $\Sigma\mathbf{F}$ made up of all the external forces applied to the system as shown in Equation 10.1-2. Any internal reaction forces acting between the particles do not need to be explicitly considered because each pair of interaction forces cancel. Finally, the impulse-momentum principle for a system of particles is given in Equation 10.1-3 and, just as in the case of an individual particle, the total impulse applied to the system is equal to the change in linear momentum of the system.

> Units of Linear Impulse & Momentum
> SI units:
> - kilogram–meter per second [kg-m/s]
> - Newton–second [N-s]
>
> US customary units:
> - pound–second [lb-s]

System of particles equations

Linear momentum:
$$\mathbf{G} = \sum m_i \mathbf{v}_i = m\mathbf{v}_G \quad (10.1\text{-}1)$$

Linear impulse:
$$\mathbf{I} = \int_{t_1}^{t_2} \sum \mathbf{F}\, dt \quad (10.1\text{-}2)$$

Linear impulse-momentum principle:
$$\int_{t_1}^{t_2} \sum \mathbf{F}\, dt = \Delta\mathbf{G} = \sum m_i \mathbf{v}_{i,2} - \sum m_i \mathbf{v}_{i,1} = m\mathbf{v}_{G,2} - m\mathbf{v}_{G,1} \quad (10.1\text{-}3)$$

$\Sigma\mathbf{F}$ = resultant forces acting on all the particles in the system
$\mathbf{G}$ = linear momentum of the system of particles
$\Delta\mathbf{G}$ = change in linear momentum of the system of particles
$\mathbf{I}$ = external impulse applied to the system
m_i = mass of the ith particle

m = total mass of the system of particles
t = time
$\mathbf{v}_{i,1}$, $\mathbf{v}_{i,2}$ = velocity of the ith particle at state 1 and state 2, respectively
$\mathbf{v}_{G,1}$, $\mathbf{v}_{G,2}$ = velocity of the system's mass center at state 1 and state 2, respectively

Equation Derivation

Linear momentum for a system of particles may be calculated by summing the individual momenta for each particle or by looking at the system as a whole. In the later case, the total mass of the system and the velocity of the system's center of mass is used to calculate the system's linear momentum as shown below.

$$\mathbf{G} = \sum m_i \mathbf{v}_i = m\mathbf{v}_G$$

To see how the two forms of the linear momentum equation are related, let's consider the velocity of a single particle in terms of the velocity of the system's center of mass. We can rewrite $\mathbf{G} = \sum m_i \mathbf{v}_i$ as follows.

$$\mathbf{G} = \sum m_i (\mathbf{v}_G + \mathbf{v}_{i/G}) = \left(\sum m_i \right) \mathbf{v}_G + \sum m_i \mathbf{v}_{i/G}$$

Recalling, from the previous discussions of systems of particles, that $\sum m_i \mathbf{v}_{i/G} = 0$ and letting m be the total mass of the system, the above becomes the alternative expression for the linear momentum as shown below. This expression for the linear momentum of a system of particles is the same as for an individual particle except that the velocity is specifically for the system's mass center.

$$\mathbf{G} = m\mathbf{v}_G$$

10.1.2) ANGULAR IMPULSE AND MOMENTUM FOR SYSTEM OF PARTICLES

As described in the previous chapter on *Particle Impulse and Momentum*, angular momentum of a particle is equal to the moment of its linear momentum with respect to some reference point, as shown in

Units of Angular Impulse & Momentum
SI units:
• kilogram- meter^2 per second [kg-m^2/s]
• Newton-meter-second [N-m-s]
US customary units:
• foot-pound-second [ft-lb-s]

Equation 10.1-4. The angular momentum of a system of particles is equal to the sum of the angular momenta of the individual particles as shown in Equation 10.1-5. It may also be calculated by considering the system as a whole as shown in Equation 10.1-6. In this case, the distance (r_G), speed (v_G) and mass moment of inertia (I_G) are with respect to the system's mass center G. Note that the center of mass version of the angular momentum definition assumes that the particles within the system are rigidly connected and moving in a plane. The direction of H_O is either in or out of the plane. This alternate expression for the system's angular momentum can be understood as the sum of the angular momentum due to the system's translation relative to O and the angular momentum due to the system's rotation about its mass center G.

Angular momentum for a particle: $\boxed{\mathbf{H}_O = \mathbf{r} \times \mathbf{G} = \mathbf{r} \times m\mathbf{v}}$ (10.1-4)

Angular momentum for a system of particles:

$$\boxed{\mathbf{H}_O = \sum (\mathbf{r}_i \times \mathbf{G}_i) = \sum (\mathbf{r}_i \times m_i \mathbf{v}_i)}\quad (10.1\text{-}5)$$

Angular momentum for a system of particles moving in a plane:

$$\boxed{\mathbf{H}_O = \mathbf{r}_G \times m\mathbf{v}_G + I_G \boldsymbol{\omega}}\quad (10.1\text{-}6)$$

$\mathbf{G}$ = linear momentum of a particle
$\mathbf{G}_i$ = linear momentum of the ith particle
$\mathbf{H}_O$ = angular momentum of the particle/system with respect to O
I_G = mass moment of inertia of the system
m = mass of the particle
m_i = mass of the the ith particle

$\mathbf{r}, \mathbf{r}_i, \mathbf{r}_G$ = position of the particle, the ith particle, and the system's mass center G with respect to O, respectively
$\mathbf{v}, \mathbf{v}_i, \mathbf{v}_G$ = velocity of the particle, the ith particle, and the system's mass center G, respectively
$\boldsymbol{\omega}$ = angular velocity of the system

Equation Derivation

As was done for linear momentum, if we assume that the individual particles in a system are rigidly connected, we can derive another expression for the angular momentum of the system in terms of the motion of the system's center of mass. We will begin with the equation for the angular momentum of a system of individual particles ($\mathbf{H}_O = \Sigma(\mathbf{r}_i \times m_i \mathbf{v}_i)$). The position of each particle $\mathbf{r}_i$ can be related to the position of the system's mass center $\mathbf{r}_G$ and the position of the particle relative to the mass center $\mathbf{r}_{i/G}$ through vector addition. Making this substitution, we get the following.

$$\mathbf{H}_O = \sum (\mathbf{r}_i \times m_i \mathbf{v}_i) = \sum ((\mathbf{r}_G + \mathbf{r}_{i/G}) \times m_i \mathbf{v}_i) = \sum (\mathbf{r}_G \times m_i \mathbf{v}_i) + \sum (\mathbf{r}_{i/G} \times m_i \mathbf{v}_i)$$

$$= \mathbf{r}_G \times \sum (m_i \mathbf{v}_i) + \sum (\mathbf{r}_{i/G} \times m_i \mathbf{v}_i)$$

Similarly, the velocity of each particle $\mathbf{v}_i$ can be expressed in terms of the velocity of the system's mass center $\mathbf{v}_G$ and the particle's velocity relative to the mass center $\mathbf{v}_{i/G}$.

$$\mathbf{H}_O = \mathbf{r}_G \times \sum (m_i (\mathbf{v}_G + \mathbf{v}_{i/G})) + \sum (\mathbf{r}_{i/G} \times m_i (\mathbf{v}_G + \mathbf{v}_{i/G}))$$

$$= \mathbf{r}_G \times \mathbf{v}_G \sum (m_i) + \mathbf{r}_G \times \sum (m_i \mathbf{v}_{i/G}) + \sum (\mathbf{r}_{i/G} \times m_i \mathbf{v}_G) + \sum (\mathbf{r}_{i/G} \times m_i \mathbf{v}_{i/G})$$

Since m_i is a scalar and switching the order of a cross product changes the sign of the resultant vector, we can rewrite the third term in the above equation as follows.

$$\sum (\mathbf{r}_{i/G} \times m_i \mathbf{v}_G) = -\sum (\mathbf{v}_G \times m_i \mathbf{r}_{i/G}) = -\mathbf{v}_G \times \sum (m_i \mathbf{r}_{i/G})$$

Assuming that the particles are rigidly connected, we can write the velocity of each particle relative to the mass center in terms of the system's angular velocity ($\mathbf{v}_{i/G} = \boldsymbol{\omega} \times \mathbf{r}_{i/G}$) to arrive at the following.

$$\mathbf{H}_O = \mathbf{r}_G \times m\mathbf{v}_G + \mathbf{r}_G \times \sum (m_i \mathbf{v}_{i/G}) - \mathbf{v}_G \times \sum (m_i \mathbf{r}_{i/G}) + \sum (\mathbf{r}_{i/G} \times m_i (\boldsymbol{\omega} \times \mathbf{r}_{i/G}))$$

Recall that $\Sigma m_i r_{i/G} = 0$, and consequently that $\Sigma m_i v_{i/G} = 0$. Using these results and taking advantage of the fact that we generally perform our analysis for planar motion, the previous expression reduces to the following.

$$\mathbf{H}_O = \mathbf{r}_G \times m\mathbf{v}_G + \boldsymbol{\omega} \sum \left(m_i r_{i/G}^2 \right)$$

The definition of the mass moment of inertia of a system of particles with respect to an axis through G is $I_G = \Sigma m_i r_{i/G}^2$. Therefore, we can rewrite the above as the expression given below.

$$\mathbf{H}_O = \mathbf{r}_G \times m\mathbf{v}_G + I_G \boldsymbol{\omega}$$

The expression for the angular impulse applied to a system of particles is slightly different from the expression for a single particle (Equation 10.1-7) because now the forces can act at different locations. This is shown in Equation 10.1-8 where each radius $\mathbf{r}_i$ is the vector from the reference point O to the point of application of the corresponding resultant force $\Sigma \mathbf{F}$. A pair of internal interaction forces still cancel though, since they both act at the same point of contact.

Angular Impulse for a single particle: $$\mathbf{J}_O = \int_{t_1}^{t_2} \mathbf{r} \times \sum \mathbf{F} \, dt = \int_{t_1}^{t_2} \sum \mathbf{M}_O \, dt \qquad (10.1\text{-}7)$$

Angular impulse for a system of particles: $$\mathbf{J}_O = \int_{t_1}^{t_2} \sum \left(\mathbf{r}_i \times \mathbf{F}_i \right) dt = \int_{t_1}^{t_2} \sum \mathbf{M}_O \, dt \qquad (10.1\text{-}8)$$

$\mathbf{F}$ = force acting on the particle or system
$\mathbf{J}_O$ = angular impulse of the particle/system
   with respect to O
$\mathbf{M}_O$ = resultant moment acting on the particle or
   system of particles

$\mathbf{r}, \mathbf{r}_i$ = position of the particle and the ith particle
   with respect to O respectively
t = time

As was the case with an individual particle, the angular impulse applied to the system is equal to the change in angular momentum of the system as shown in Equation 10.1-9. Note that even though the angular impulse and angular momentum may be determined for a moving reference frame (i.e. O is arbitrary), the angular impulse-momentum principle only applies for an inertial reference frame (i.e. O is fixed).

Angular impulse-momentum principle for a system of particles:

$$\mathbf{J}_O = \Delta \mathbf{H}_O$$

$$\int_{t_1}^{t_2} \sum \mathbf{M}_O \, dt = \sum \mathbf{H}_{Oi,2} - \sum \mathbf{H}_{Oi,1} = \sum (\mathbf{r}_i \times m_i \mathbf{v}_{i,2}) - \sum (\mathbf{r}_i \times m_i \mathbf{v}_{i,1})$$

(10.1-9)

$\mathbf{H}_{Oi,1}$, $\mathbf{H}_{Oi,2}$ = angular momentum at state 1 and state 2 of the ith particle, respectively

$\Delta \mathbf{H}_O$ = total change in angular momentum for the system with respect to O

$\mathbf{J}_O$ = external angular impulse applied to the system with respect to O

m_i = mass of the ith particle

$\Sigma \mathbf{M}_O$ = resultant moment acting on the system

$\mathbf{r}_i$ = position of the ith particle with respect to O

t = time

$\mathbf{v}_{i,1}$, $\mathbf{v}_{i,2}$ = velocity at state 1 and state 2 of the ith particle, respectively

Conceptual Example 10.1-1

A horizontal force is applied over a small interval of time to the system of particles in the situations shown. Rank the magnitude of the change in linear velocity of each body's mass center from before the application of the impulse to after the removal of the impulse. Then rank the magnitude of the change in angular velocity of each body from before the application of the impulse to after the removal of the impulse.

Linear velocity

Greatest _____ Next _____ Next _____ Least _____

Angular velocity

Greatest _____ Next _____ Next _____ Least _____

Conceptual Example 10.1-2

A large force in the x-direction is applied to the system of particles shown over a small interval of time. Following the removal of the force, how do you expect the system to move?

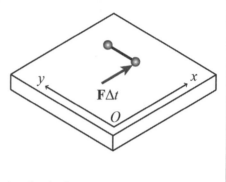

a) Translate up (pos y) and to the right (pos x) without rotating.
b) Translate to the right (pos x) without rotating.
c) Translate up (pos y) and to the right (pos x) while rotating counterclockwise.
d) Translate to the right (pos x) while rotating counterclockwise.

Equation Summary

Abbreviated variable definition list

F = force
G = linear momentum
H_O = angular momentum with respect to O
I = linear impulse
I_G = mass moment of inertia with respect to
    the mass center
J_O = angular impulse with respect to O

m = mass
M_O = moment with respect to O
r = position
t = time
v = velocity
ω = angular velocity

System of particles

Linear momentum

$$G = \sum m_i v_i = mv_G$$

Angular momentum

$$H_O = \sum (r_i \times m_i v_i) \quad \text{or} \quad H_O = r_G \times mv_G + I_G\omega$$

Linear impulse

$$I = \int_{t_1}^{t_2} \sum F \, dt$$

Angular impulse

$$J_O = \int_{t_1}^{t_2} \sum M_O \, dt$$

Linear impulse-momentum principle

$$I = G_2 - G_1$$

Angular impulse-momentum principle

$$J_O = H_{O,2} - H_{O,1}$$

Example Problem 10.1-3

Consider a dumbbell consisting of two 2-kg spheres rigidly connected by a 3-m bar of negligible mass. The dumbbell slides across a smooth table. If the center of mass of the system is translating with velocity $\mathbf{v}_G$ = 4$\mathbf{i}$ m/s and the system is rotating with angular velocity ω = 1$\mathbf{k}$ rad/sec, determine a) the linear momentum of the system and b) the angular momentum of the system with respect to O for the instant shown in the figure.

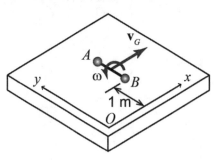

Given:

Find:

Solution:

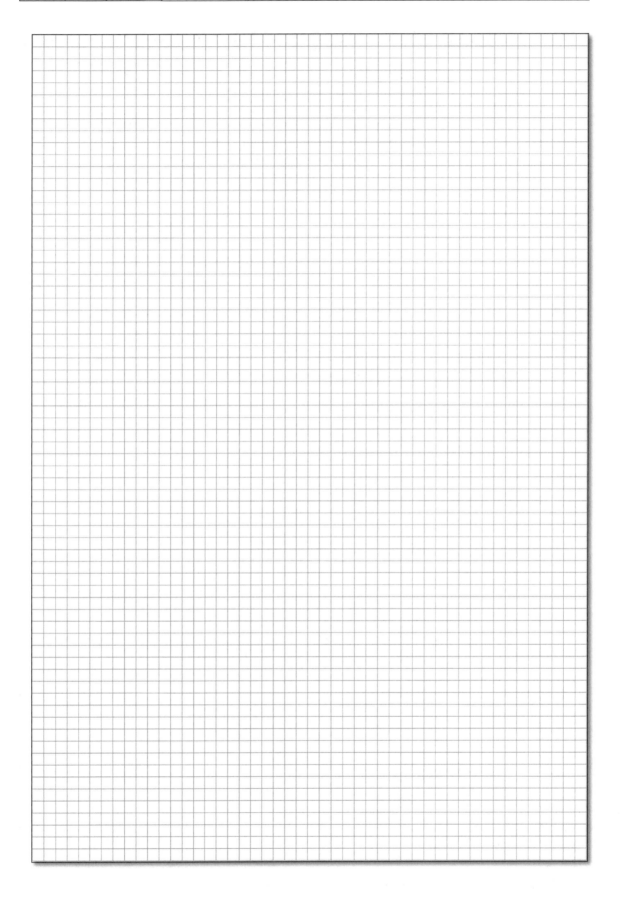

Solved Problem 10.1-4

Consider a dumbbell consisting of two 2-kg spheres rigidly connected by a 3-m bar of negligible mass. The dumbbell is initially at rest on a smooth, horizontal table when a 20**i** N-s impulse is applied to sphere A. If the impulse is applied over a short duration of time, determine the linear velocity of both spheres immediately after the application of the impulse.

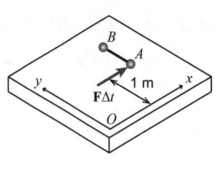

Given: $m_A = m_B = m = 2$ kg $L = 3$ m

      $v_{A,1} = v_{B,1} = 0$ $\mathbf{I} = 20\ \mathbf{i}$ N-s

Find: $v_{A,2},\ v_{B,2}$

Solution:

Linear impulse-momentum

We will apply the linear impulse-momentum principle to the system of particles. Using our physical understanding of the situation, we can deduce that $v_{Ay} = v_{By} = 0$.

$$\mathbf{I} = \mathbf{G}_2 - \mathbf{G}_1 \qquad\qquad \int \sum F\, dt = \sum m_i v_{ix,2} - \sum m_i v_{ix,1}$$

$$\int \sum F\, dt = m_A v_{Ax,2} + m_B v_{Bx,2} = m(v_{Ax,2} + v_{Bx,2}) \qquad v_{Ax,2} + v_{Bx,2} = 10\ \frac{m}{s} \qquad (1)$$

Angular impulse-momentum

We need another equation to solve for the two velocities. We will apply the angular impulse-momentum principle to obtain another equation. We will use the origin (O) as the reference point.

$$\mathbf{J}_O = \mathbf{H}_{O,2} - \mathbf{H}_{O,1} \qquad\qquad \int \sum \mathbf{M}_O\, dt = \mathbf{r} \times \int \sum \mathbf{F}_O\, dt = \sum \mathbf{r}_{i/O,2} \times m\mathbf{v}_{i,2} - \sum \mathbf{r}_{i/O,1} \times m\mathbf{v}_{i,1}$$

$$r\int \sum F_O\, dt(-\mathbf{k}) = (r_{y,A/O} m_A v_{Ax,2} + r_{y,B/O} m_B v_{Bx,2})(-\mathbf{k}) = m(r_{y,A/O} v_{Ax,2} + r_{y,B/O} v_{Bx,2})(-\mathbf{k})$$

$$1(20) = 2(v_{Ax,2} + 4v_{Bx,2}) \qquad\qquad v_{Ax,2} = 10 - 4v_{Bx,2} \qquad (2)$$

Substitute equation (2) into (1).

$$10 - 4v_{Bx,2} + v_{Bx,2} = 10 \qquad\qquad \boxed{v_{Bx,2} = 0} \qquad\qquad \boxed{v_{Ax,2} = 10\ \frac{m}{s}}$$

10.2) LINEAR MOMENTUM

10.2.1) PARTICLE LINEAR MOMENTUM

In the previous chapter, we learned about impulse and momentum for particles. The equation for a particle's linear momentum is given in Equation 10.2-1. The linear momentum of a particle depends on its mass and velocity. Think about how the particle linear momentum equation might change for a rigid body. Remember that a rigid body has size and can rotate.

Linear momentum of a particle: $\boxed{\mathbf{G} = m\mathbf{v}}$ (10.2-1)

$\mathbf{G}$ = linear momentum
m = mass
$\mathbf{v}$ = velocity

Example 10.2-1

Answer the following questions.

a) Write down the equation for the linear momentum of a particle.

b) If a rigid body is rotating as well as translating, do all points on the rigid body have the same velocity?

c) If we apply the particle linear momentum equation to a rigid body, what point do you think we should use for the velocity?

d) Based on your answers to the above questions, write down what you think is the equation for the linear momentum of a rigid body.

10.2.2) RIGID-BODY LINEAR MOMENTUM

Since rigid bodies have size, they can rotate as well as translate. In this section, we will restrict ourselves to the case of planar motion where the body is constrained to translate and rotate within a two-dimensional plane. To gain some

Units of Linear Momentum
SI units:
- kilogram–meter per second [kg-m/s]
- Newton–second [N-s]
US customary units:
- pound–second [lb-s]

insight into the impulse and momentum equations for a rigid body, we will start by considering a rigid body as a collection of particles. Imagine that the rigid body consists of a system of infinitesimally small pieces or particles of mass m_i as shown in Figure 10.2-1. Each particle is located at some position $\mathbf{r}_i$ relative to a reference point.

Remember that a rigid body can rotate, therefore, the velocity of each piece of the rigid body may have a different velocity as shown in Figure 10.2-2. Each of the small pieces making up the rigid body has linear momentum ($\mathbf{G}_i = m_i\mathbf{v}_i$). If we want to determine the momentum of the rigid body as a whole, we would have to sum all of the individual momenta as we did with systems of particles, see Equation 10.2-2. It can then be shown, as was done for systems of particles, that the linear momentum of a rigid body can be expressed in terms of the body's overall mass and the velocity of its mass center (Equation 10.2-3), rather than in terms of the momenta of the individual particles. One interesting implication of Equation 10.2-3 is that the linear momentum of a rigid body does not depend on the body's rotation.

Linear momentum of a rigid body as a system of particles:

$$\mathbf{G} = \sum m_i\mathbf{v}_i \quad (10.2\text{-}2)$$

Linear momentum of a rigid body: $\boxed{\mathbf{G} = m\mathbf{v}_G}$ (10.2-3)

$\mathbf{G}$ = linear momentum
m = mass of the rigid body
m_i = mass of the individual particles
    comprising the rigid body

$\mathbf{v}_G$ = velocity of the body's mass center
$\mathbf{v}_i$ = velocity of the individual particles
    comprising the rigid body

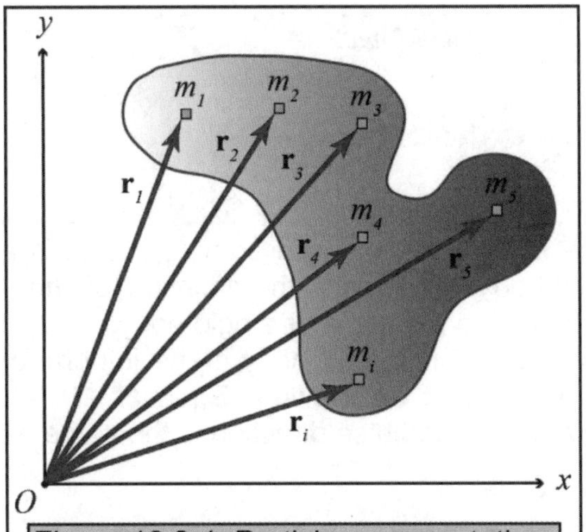

Figure 10.2-1: Particle representation of a rigid body

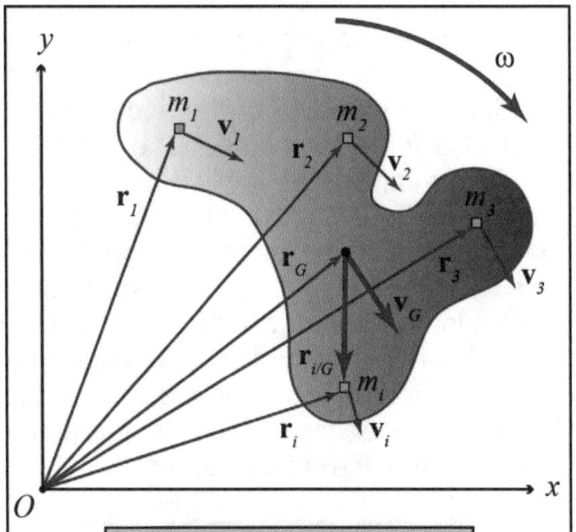

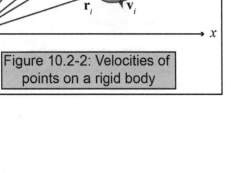

Figure 10.2-2: Velocities of points on a rigid body

Conceptual Example 10.2-2

A sphere moves down a smooth frictionless incline. Which of the following quantities increase?

a) angular velocity
b) total energy
c) potential energy
d) linear momentum

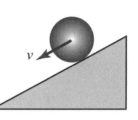

What if the incline is rough?

a) angular velocity
b) total energy
c) potential energy
d) linear momentum

10.3) LINEAR IMPULSE

In the chapter on *Particle Impulse and Momentum*, we learned that **linear impulse** is a measure of the accumulated effect of a force over time. The equation for a linear impulse as applied to a particle is given in Equation 10.3-1. Think about

Units of Linear Impulse
SI units:
- kilogram–meter per second [kg-m/s]
- Newton–second [N-s]

US customary units:
- pound–second [lb-s]

how Equation 10.3-1 might change if the force is applied to a rigid body. A force is a force whether it is applied to a particle or a rigid body. Therefore, the definition of linear impulse is unchanged whether the impulse is applied to a particle or a rigid body.

Mathematically, the total linear impulse acting on a rigid body is the integral of the sum of forces acting on the body with respect to time as shown in Equation 10.3-1. Recall that linear impulse is a vector quantity where the impulse direction is the same as the direction of the resultant force.

Linear impulse: The accumulated effect of forces acting over time.

Linear impulse: $$\mathbf{I} = \int_{t_1}^{t_2} \sum \mathbf{F}\, dt \quad (10.3\text{-}1)$$

$\mathbf{I}$ = impulse
$\sum \mathbf{F}$ = resultant of all the external forces acting on the particle
t = time

10.4) LINEAR IMPULSE-MOMENTUM PRINCIPLE

10.4.1) RIGID BODIES

In the previous chapter, we built some intuition about how the linear impulse applied to a particle relates to the particle's change in linear momentum. The larger the applied force and the longer its duration of application, the larger the resulting change in the particle's velocity. The exact relationship between linear impulse and linear momentum from the chapter on *Particle Impulse and Momentum* still holds for rigid bodies. You could demonstrate this fact by again considering a rigid body as a collection of infinitesimally small particles.

Take your pencil and place it on a desk. Tap it with your finger at the midpoint of the pencil so that it rolls along the desk as shown in Figure 10.4-1. What

> *Note: In actuality the linear momentum in the two cases of pencil motion are not equal due to the action of friction. However, we are idealizing the situation.

happened? You applied a force to the pencil and its momentum ($m\mathbf{v}$) changed. Repeat what you just did except this time tap the pencil near one of its ends. What was different about the motion of the pencil this time? You should have noticed that instead of rolling in a straight line, the pencil rotated as well. Assuming that you applied the same force in both cases, does this mean that the change in linear momentum was different? In one case it only rolled and in the other it rolled and rotated. It turns out that the change in linear momentum for both cases is the same[*]. The linear momentum of the pencil only relies on the velocity of the pencil's center of mass and not on the fact that it rotated.

It is interesting to note that the linear impulse applied to a rigid body does not depend on the point of application of the underlying force. Therefore, whether a force is applied at a body's mass center or at a distance offset from the body's mass center, the resulting change in linear momentum is the same as long as the force was applied for the same duration in the two cases.

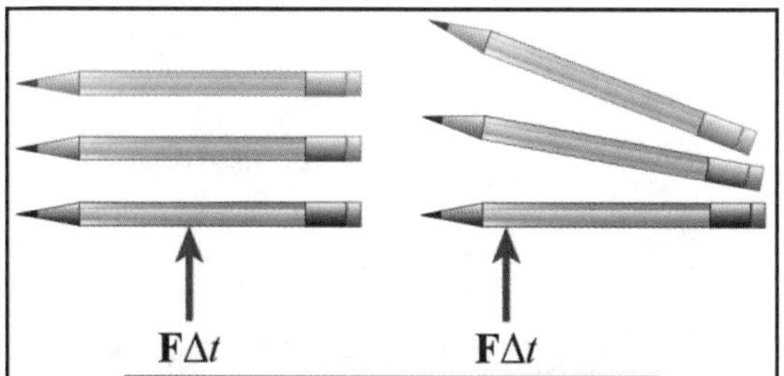

$$F\Delta t \qquad F\Delta t$$

Figure 10.4-1: Rolling pencil

The quantity that affects a body's change in linear momentum is the product between force and the duration of application or the *impulse*. This is reflected in the linear impulse-momentum principle given in Equation 10.4-1. The difference between applying the principle of linear impulse-momentum to a particle and to a rigid body is the location of the velocity. Since particles are defined as having negligible size, they move with a single velocity. On the other hand, each point on a rigid body may have a different velocity depending on whether or not the body is rotating. The linear momentum of a rigid body is specifically a function of the velocity of the body's mass center. Also, recall that the vector nature of the linear impulse-momentum principle allows the relationship to be broken down into components.

> **Principle of linear impulse and momentum:** The impulse of the resultant force acting on a rigid body is equal to the change in the body's linear momentum.

Linear impulse-momentum principle for a rigid body:

$$\mathbf{I} = \Delta\mathbf{G} \qquad \int_{t_1}^{t_2} \sum \mathbf{F}\, dt = m\mathbf{v}_{G,2} - m\mathbf{v}_{G,1} \qquad (10.4\text{-}1)$$

$\mathbf{I}$ = linear impulse applied to the body
$\Delta\mathbf{G}$ = change in linear momentum
$\sum\mathbf{F}$ = resultant force acting on the body

t = time
m = mass of the rigid body
$\mathbf{v}_{G,1}$, $\mathbf{v}_{G,2}$ = velocity of the body's mass center at state 1 and state 2, respectively

10.4.2) SYSTEMS OF RIGID BODIES

The total momentum of a system of rigid bodies is simply the sum of the momenta of the individual bodies as expressed in Equation 10.4-2. The total impulse applied to the system can be calculated according to Equation 10.4-3. When using Equation 10.4-3, the forces internal to the system do not need to be explicitly considered. This is because each pair of internal interaction forces sum to zero. Each internal force has equal magnitude but opposite direction to its interaction partner and is applied over the same interval of time. This is one of the great advantages of the impulse-momentum approach. The overall linear impulse-momentum principle for a system of rigid bodies is given in Equation 10.4-4.

Systems of rigid bodies equations

Linear momentum: $\boxed{\mathbf{G} = \sum m_i \mathbf{v}_{Gi}}$ (10.4-2)

Linear impulse: $\boxed{\mathbf{I} = \int_{t_1}^{t_2} \sum \mathbf{F}\, dt}$ (10.4-3)

Linear impulse-momentum principle:

$$\mathbf{I} = \Delta\mathbf{G} \qquad \int_{t_1}^{t_2} \sum \mathbf{F}\, dt = \sum m_i \mathbf{v}_{Gi,2} - \sum m_i \mathbf{v}_{Gi,1} \qquad (10.4\text{-}4)$$

$\sum\mathbf{F}$ = external resultant force acting on all the rigid bodies in the system
$\mathbf{G}$ = linear momentum of the system of rigid bodies
$\Delta\mathbf{G}$ = change in linear momentum of the system of rigid bodies

$\mathbf{I}$ = external linear impulse applied to the system
m_i = mass of the ith body
t_1, t_2 = time at state 1, and time at state 2, respectively
$\mathbf{v}_{Gi,1}$, $\mathbf{v}_{Gi,2}$ = mass center velocity of the ith body at state 1 and state 2, respectively

10.4.3) CONSERVATION OF LINEAR MOMENTUM

Since the linear impulse-momentum principle is the same for particles and rigid bodies, it makes sense that the conditions under which linear momentum is conserved would also be the same. If the external forces acting on a rigid body sum to zero over a period of time, then the linear momentum of the body is constant over that interval as expressed in Equations 10.4-5 and 10.4-6. More generally, if the linear impulse applied to a body (or system) over an interval of time is equal to zero, then the initial and final linear momenta of that body (or system) are equal. Recall that linear momentum can be conserved in one direction but may not be conserved in another.

Conservation of linear momentum for a rigid body:

$$\boxed{\mathbf{G}_1 = \mathbf{G}_2 \qquad m\mathbf{v}_{G,1} = m\mathbf{v}_{G,2}} \quad (10.4\text{-}5)$$

Conservation of linear momentum for a system of rigid bodies:

$$\boxed{\mathbf{G}_1 = \mathbf{G}_2 \qquad \sum m_i \mathbf{v}_{Gi,1} = \sum m_i \mathbf{v}_{Gi,2}} \quad (10.4\text{-}6)$$

$\mathbf{G}_1, \mathbf{G}_2 =$ linear momentum at state 1 and state 2, respectively
$m =$ mass of the rigid body
$m_i =$ mass of the ith rigid body

$\mathbf{v}_{G,1}, \mathbf{v}_{G,2} =$ velocity of the body's mass center at state 1 and state 2
$\mathbf{v}_{Gi,1}, \mathbf{v}_{Gi,2} =$ velocity of the ith body's mass center at state 1 and state 2, respectively

---

Conceptual Example 10.4-1

A particle traveling horizontally strikes a rigid body at rest and adheres to the body such that the particle and body move together following the collision. Rank the magnitude of the velocity of the center of mass of each system following the collision from greatest to least for the situations shown.

Greatest _____   Next _____   Next _____   Least _____

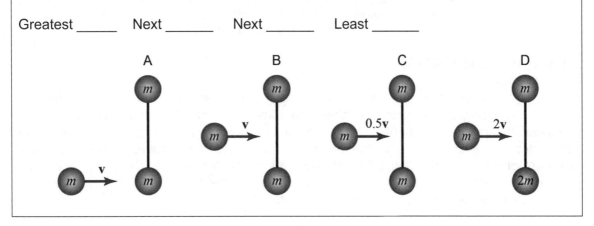

10.5) ANGULAR MOMENTUM

In this section, we will address the principles of angular momentum for rigid bodies. The exact form of the angular momentum equation depends on the selected reference point. Any rotating body has angular momentum

Units of Angular Momentum
SI units:
• kilogram-meter^2 per second [kg-m^2/s]
• Newton-meter-second [N-m-s]
US customary units:
• foot-pound- second [ft-lb-s]

about its center of mass G. However, the exact amount of momentum depends on the reference point chosen. The three specific cases we will examine are: (i) pure translation (ii) pure rotation, and (iii) general planar motion. A precise derivation will be presented for each of the cases mentioned.

10.5.1) PURE TRANSLATION

We will begin by examining the case where the body is only translating, then we will examine the case of pure rotation, and finally the case of general planar motion. In the case where the rigid body is undergoing pure translation, the angular momentum relative to a point O takes the form given in Equation 10.5-1. Remember that the velocity of every point on a rigid body undergoing pure translation is the same. Note that in this case, the equation for angular momentum is almost the same as the angular momentum of a particle. The only difference is that the position vector now measures the location of the body's mass center with respect to O. This relationship for a rigid body can be derived by again considering the body as consisting of a system of particles. Note that angular momentum can be defined with respect to a fixed or a moving reference frame (attached to O).

Pure translation angular momentum: $\boxed{\mathbf{H}_O = \mathbf{r}_{G/O} \times m\mathbf{v}}$ (10.5-1)

$\mathbf{H}_O$ = angular momentum of the rigid body with respect to O
$\mathbf{r}_{G/O}$ = position of mass center G relative to O
$\mathbf{v}$ = velocity of the rigid body

10.5.2) PURE ROTATION

In the case where the rigid body is rotating in a plane about a fixed-axis through point O, the angular momentum takes the form given in Equation 10.5-2. In this equation the angular momentum and angular velocity are written as vectors, but their directions are limited to be in the positive or negative $\mathbf{k}$ direction, where the mass moment of inertia is with respect to an axis that is perpendicular to the plane of motion through O (parallel to the z-axis). The derivation of this relationship follows the logic employed in the section on systems of particles.

Pure rotation angular momentum: $\boxed{\mathbf{H}_O = I_O \boldsymbol{\omega}}$ (10.5-2)

$\mathbf{H}_O$ = angular momentum of the rigid body with respect to O
I_O = mass moment of inertia of the rigid body about the axis through O
$\boldsymbol{\omega}$ = angular velocity of the rigid body

Equation Derivation

We will start the derivation with the angular momentum equation for a system of particles moving in a plane ($\mathbf{H}_O = \Sigma(\mathbf{r}_{i/O} \times m_i \mathbf{v}_i)$). In the instance where the rigid body is undergoing pure rotation, the speed of any point on the body is equal to $\omega \times \mathbf{r}$. Assuming that the rigid body is a collection of small particles, we may substitute the velocity relationship $\mathbf{v}_i = \omega \times \mathbf{r}_{i/O}$ to obtain the following.

$$\mathbf{H}_O = \sum \left(\mathbf{r}_{i/O} \times m_i \mathbf{v}_i \right) = \sum \left(\mathbf{r}_{i/O} \times m_i (\omega \times \mathbf{r}_{i/O}) \right) = \sum m_i r_{i/O}^2 \omega$$

The summation of elements $\Sigma m_i r_{i/O}^2$ corresponds to the body's mass moment of inertia with respect to the point O. Therefore, the above equation becomes the following, which is the formal definition of angular momentum with respect to O of a rigid body undergoing planar motion.

$$\mathbf{H}_O = I_O \omega$$

10.5.3) GENERAL PLANAR MOTION WITH A CENTER OF MASS REFERENCE

We will now consider the angular momentum for general planar motion of rigid bodies, that is, bodies that can translate as well as rotate in a two-dimensional plane. We will first examine the situation where the reference point is the body's mass center G. In the next section, we look at the situation where the angular momentum is with respect to an arbitrary point O. If we use the body's mass center G as our reference point, the angular momentum equation is very similar to the pure rotation case. The angular momentum of a rigid body with a center of mass reference is given by Equation 10.5-3. Again, this equation presumes that the angular momentum and angular velocity are in the positive or negative $\mathbf{k}$ direction, even though they are written as vectors.

General planar motion angular momentum: $\boxed{\mathbf{H}_G = I_G \omega}$ (10.5-3)

$\mathbf{H}_G$ = angular momentum of the rigid body with respect to G
I_G = mass moment of inertia of the rigid body about the axis through G
ω = angular velocity of the rigid body

Equation Derivation

Consider a rigid body to be constructed of many particles moving in a plane. In the case of a system of particles, the angular momentum with a center of mass reference is given by the following equation.

$$\mathbf{H}_G = \sum \left(\mathbf{r}_{i/G} \times m_i \mathbf{v}_i \right)$$

Each particle that makes up the rigid body will have its own velocity. The individual velocities can be related to the body's mass center velocity through the relative velocity relationship ($\mathbf{v}_i = \mathbf{v}_G + \mathbf{v}_{i/G}$). Substituting this relationship into the above equation we obtain the following.

$$\mathbf{H}_G = \sum \left(\mathbf{r}_{i/G} \times m_i \mathbf{v}_i \right) = \sum \left(\mathbf{r}_{i/G} \times m_i (\mathbf{v}_G + \mathbf{v}_{i/G}) \right) = \sum \left((\mathbf{r}_{i/G} \times m_i \mathbf{v}_G) + (\mathbf{r}_{i/G} \times m_i \mathbf{v}_{i/G}) \right)$$

The first term of the above equation ($\Sigma(\mathbf{r}_{i/G} \times m_i \mathbf{v}_G)$) can be rewritten as $\Sigma(m_i \mathbf{r}_{i/G} \times \mathbf{v}_G)$ since the mass is a scalar. The term $\Sigma m_i \mathbf{r}_{i/G}$ equates to zero since G is the mass center. The second term of the above equation ($\Sigma(\mathbf{r}_{i/G} \times m_i \mathbf{v}_{i/G})$) can be rewritten by recognizing that each relative velocity is equal to $r_{i/G}\omega$ in a direction that is perpendicular to the radius from the mass center which is equivalent to $\boldsymbol{\omega} \times \mathbf{r}_{i/G}$. Using these substitutions, we get the following equation.

$$\mathbf{H}_G = \sum\left(\mathbf{r}_{i/G} \times m_i(\boldsymbol{\omega} \times \mathbf{r}_{i/G})\right) = \sum m_i r_{i/G}^2 \boldsymbol{\omega}$$

The summation of elements $\Sigma m_i r_{i/G}^2$ corresponds to the body's mass moment of inertia with respect to the axis through G and the above equation becomes the equation for general planar angular momentum with a mass center reference.

$$\mathbf{H}_G = I_G \boldsymbol{\omega}$$

10.5.4) GENERAL PLANAR MOTION

The angular momentum for general planar motion with respect to an arbitrary point O is given in Equation 10.5-4. This is the most general form of the angular momentum equation and is consistent with the form found for systems of particles. All other forms (i.e. for pure translation, pure rotation and with respect to the body's mass center) may be derived from this equation. Equation 10.5-4 consists of a term that is equal to the angular momentum of a particle of mass m located at the center of mass of the body and a term that is equal to the angular momentum of the body with respect to its center of mass. In other words, the first term is the angular momentum due to the body's translation and the second term is due to the body's rotation. Note that Equation 10.5-4 is only valid for planar motion.

General planar motion angular momentum: $\boxed{\mathbf{H}_O = \mathbf{r}_{G/O} \times m\mathbf{v}_G + I_G \boldsymbol{\omega}}$ (10.5-4)

$\mathbf{H}_O$ = angular momentum with respect to O
I_G = mass moment of inertia of the rigid body G
m = mass of the rigid body

$\mathbf{r}_{G/O}$ = position of the body's mass center G relative to O
$\mathbf{v}_G$ = velocity of the body's mass center G
ω = angular velocity of the rigid body

To better understand the angular momentum of a rigid body, let's take a closer look at Equation 10.5-4. Imagine that you are on a tea cup amusement ride as shown in Figure 10.5-1. All of the tea cups rotate about a common center, and each teacup individually rotates about its own center. For each tea

Figure 10.5-1: Tea cup amusement park ride

cup, the first part of Equation 10.5-4 ($\mathbf{r}_{G/O} \times m\mathbf{v}_G$) represents the angular momentum due to the tea cups translating with respect to some point O (the common center for example), and the second part of the equation represents the rotation of the tea cup about its own mass center ($I_G\omega$). More precisely, the first part of the equation represents the moment of the linear momentum of the teacup with respect to point O, while the second part of the equation represents the angular momentum due to the rotation of the tea cup.

Conceptual Example 10.5-1

Imagine you are riding in the tea cup ride shown in Figure 10.5-1.

 a) If a tea cup is not spinning about its center, but the base holding all the tea cups is still moving, does it still have angular momentum? Write down the angular momentum equation of the tea cup for this case.

 b) If the platform stops rotating, but a tea cup is still spinning, does it still have angular momentum? Write down the angular momentum equation for this case.

Conceptual Example 10.5-2

A bar of mass m and length l moves differently in each of the situations shown. Rank the angular momentum of the bar with respect to the point O from greatest to least for the given situations.

Greatest _____ Next _____ Next _____ Least _____

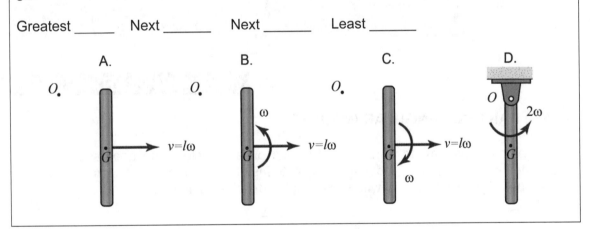

Equation Derivation

Starting again with the expression for the angular momentum of a system of particles, we can equate the individual particle positions to the position of the body's mass center using the relative position expression ($\mathbf{r}_i = \mathbf{r}_{G/O} + \mathbf{r}_{i/G}$). Substituting this expression into the momentum equation we get the following equation.

$$\mathbf{H}_O = \sum \left(\mathbf{r}_i \times m_i \mathbf{v}_i \right) = \sum \left((\mathbf{r}_{G/O} + \mathbf{r}_{i/G}) \times m_i \mathbf{v}_i \right) = \sum \left((\mathbf{r}_{G/O} \times m_i \mathbf{v}_i) + (\mathbf{r}_{i/G} \times m_i \mathbf{v}_i) \right)$$

Using the same logic employed in examining the linear momentum of a rigid body, the first term $\sum(\mathbf{r}_{G/O} \times m_i \mathbf{v}_i)$ equates to $\mathbf{r}_{G/O} \times m\mathbf{v}_G$. The second term $\sum(\mathbf{r}_{i/o} \times m_i \mathbf{v}_i)$ corresponds to $I_G\omega$. Therefore, the angular momentum with respect to an arbitrary point O is formally defined in the following equation.

$$\mathbf{H}_O = \mathbf{r}_{G/O} \times m\mathbf{v}_G + I_G\omega$$

Equation Derivation

Equation 10.5-4 is the most general form of angular momentum. The forms for pure rotation and rotation about the body's mass center may be derived from this equation. To obtain the pure rotation form, first substitute $\mathbf{v}_G = \omega \times \mathbf{r}_{G/O}$. and then apply the parallel axis theorem ($I = I_G + md^2$).

$$\mathbf{H}_O = \mathbf{r}_{G/O} \times m\mathbf{v}_G + I_G\omega = \mathbf{r}_{G/O} \times m(\omega \times \mathbf{r}_{G/O}) + I_G\omega = mr_{G/O}{}^2\omega + I_G\omega = \left(mr_{G/O}{}^2 + I_G \right)\omega = I_O\omega$$

To obtain the general planar motion form with a mass center, G, reference, substitute G for O.

$$\mathbf{H}_O = \mathbf{r}_{G/O} \times m\mathbf{v}_G + I_G\omega = \mathbf{r}_{G/G} \times m\mathbf{v}_G + I_G\omega = I_G\omega$$

10.6) ANGULAR IMPULSE

10.6.1) PARTICLE ANGULAR IMPULSE

A force may impart an angular impulse on a particle. The angular impulse is often defined using the induced moment instead of the force, as shown in Equation 10.6-1. This quantity was introduced in the preceding chapter.

$$\text{Angular impulse:} \quad \boxed{\mathbf{J}_O = \int_{t_1}^{t_2} \mathbf{r} \times \sum \mathbf{F} \, dt = \int_{t_1}^{t_2} \sum \mathbf{M}_O \, dt} \quad (10.6\text{-}1)$$

$\sum \mathbf{F}$ = resultant of all the external forces acting on the particle

$\mathbf{J}_O$ = angular impulse with respect to O

$\sum \mathbf{M}_O$ = resultant moment acting on the particle

$\mathbf{r}$ = particle position with respect to O

t = time

10.6.2) RIGID-BODY ANGULAR IMPULSE

As was the case with particles, *angular impulse* defines the cumulative effect of a moment over time, where the moment induced by a force is defined as the cross product of the moment arm **r** and the force **F** ($\mathbf{M}_O = \mathbf{r} \times \mathbf{F}$). One slight difference with rigid bodies is that each individual force can act at a different location since rigid bodies have size. Therefore, each moment arm **r** defines the position of the point of application of the corresponding force. Mathematically, angular impulse for a rigid body is defined in Equation 10.6-2. For the case of planar motion, the angular impulse vector will always point into or out of the plane of motion.

Angular Impulse for a rigid body:
$$\mathbf{J}_O = \int_{t_1}^{t_2} \sum (\mathbf{r} \times \mathbf{F})\, dt = \int_{t_1}^{t_2} \sum \mathbf{M}_O \, dt \qquad (10.6\text{-}2)$$

F = external forces acting on the body
$\mathbf{J}_O$ = angular impulse with respect to O

$\Sigma \mathbf{M}_O$ = resultant moment acting on the body
r = force application position with respect to O
t = time

10.7) ANGULAR IMPULSE-MOMENTUM PRINCIPLE

Consider the fan blade shown in Figure 10.7-1. If the blade is at rest, it has no angular momentum with respect to a fixed point. Imagine that you take your finger, apply a force to the fan blade and spin it. What happens? The force applied by your finger induces a moment and in turn applies an angular impulse to the fan. After the force is removed, the fan blades continue to rotate. The blades now have angular momentum. As you can see with this simple example, there is a relationship between applied angular impulse and angular momentum. This relationship is analogous to the one we have between linear impulse and linear momentum.

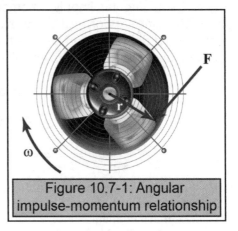

Figure 10.7-1: Angular
impulse-momentum relationship

Principle of angular impulse-momentum: The impulse of the resultant moment acting on a body is equal to the change in the body's angular momentum.

As seen in the *Angular Momentum of Rigid Bodies* section, the angular momentum depends on the reference point choice. Therefore, the angular impulse-momentum principle will depend on the type of motion and the chosen reference point. The three specific cases we will examine are: (i) pure translation (ii) pure rotation, and (iii) general planar motion.

10.7.1) PURE TRANSLATION

We will begin by examining the case where the body is only translating, then we will examine the case of pure rotation, and finally the case for general planar motion. The angular impulse-momentum principle for pure translation with a fixed-point O reference is given by Equation 10.7-1. This relationship has the same form as the particle case, except the position vector corresponds specifically to the location of the body's mass center G. This relationship follows from the consideration of a rigid body as a collection of infinitesimally small particles.

Pure translation angular impulse-momentum principle:

$$\mathbf{J}_O = \Delta\mathbf{H}_O \qquad \int_{t_1}^{t_2} \sum \mathbf{M}_O \, dt = \mathbf{r}_{G/O,2} \times m\mathbf{v}_2 - \mathbf{r}_{G/O,1} \times m\mathbf{v}_1 \qquad (10.7\text{-}1)$$

$\sum\mathbf{M}_O$ = resultant moment acting on the rigid body

$\Delta\mathbf{H}_O$ = change in angular momentum of the rigid body with respect to fixed-point O

$\mathbf{J}_O$ = applied angular impulse with respect to fixed-point O

$\mathbf{r}_{G/O,1}$, $\mathbf{r}_{G/O,2}$ = position of the body's mass center G with respect to fixed-point O at state 1 and state 2 respectively

t_1, t_2 = time at state 1 and time at state 2, respectively

$\mathbf{v}_1$, $\mathbf{v}_2$ = velocity of the rigid body at state 1 and state 2 respectively

10.7.2) PURE ROTATION

The angular impulse-momentum principle for pure rotation about a fixed-axis through point O is given by Equation 10.7-2. This relationship is derived in the same manner we did for systems of particles and presumes the moment, angular momentum, and angular velocity vectors are all in the positive or negative $\mathbf{k}$ directions.

Pure rotation angular impulse-momentum principle for planar motion:

$$\mathbf{J}_O = \Delta\mathbf{H}_O \qquad \int_{t_1}^{t_2} \sum \mathbf{M}_O \, dt = I_{O,2}\boldsymbol{\omega}_2 - I_{O,1}\boldsymbol{\omega}_1 \qquad (10.7\text{-}2)$$

$\Delta\mathbf{H}_O$ = change in angular momentum of the rigid body with respect to O

$I_{O,1}$, $I_{O,2}$ = mass moment of inertia of the rigid body about fixed-point O at state 1 and state 2, respectively

$\mathbf{J}_O$ = applied angular impulse with respect to fixed-point O

$\sum\mathbf{M}_O$ = resultant moments with respect to O acting on the system

t_1, t_2 = time at state 1 and time at state 2, respectively

$\boldsymbol{\omega}_1$, $\boldsymbol{\omega}_2$ = angular velocity of the rigid body at state 1 and state 2, respectively

10.7.3) GENERAL PLANAR MOTION WITH CENTER OF MASS G REFERENCE

The angular impulse-momentum relationship in the case of the general planar motion of a rigid body has the same form as the pure rotation case if the body's mass center G is taken as the reference point as shown in Equation 10.7-3. This relationship can be derived by imagining the rigid body as consisting of a system of particles.

General planar motion angular impulse-momentum principle:

$$\mathbf{J}_G = \Delta \mathbf{H}_G \qquad \int_{t_1}^{t_2} \sum \mathbf{M}_G \, dt = I_{G,2} \omega_2 - I_{G,1} \omega_1 \qquad (10.7\text{-}3)$$

$\Delta \mathbf{H}_G$ = change in angular momentum of the rigid body with respect to G

$I_{G,1}, I_{G,2}$ = mass moment of inertia of the rigid body about its mass center at state 1 and state 2, respectively

$\mathbf{J}_G$ = applied angular impulse with respect to the body's mass center G

$\sum \mathbf{M}_G$ = resultant moment with respect to G acting on the system

t_1, t_2 = time at state 1 and time at state 2, respectively

ω_1, ω_2 = angular velocity of the rigid body at state 1 and state 2, respectively

10.7.4) GENERAL PLANAR MOTION

The principle of angular impulse-momentum given in Equation 10.7-4 is stated with respect to an arbitrary fixed-point O and matches the form derived for systems of rigidly connected particles. This equation may be used for a body that is purely translating, purely rotating, or rotating and translating. A judicious choice of the reference point O can greatly simplify the calculations necessary for analyzing a body by making the moment induced by a force zero, or by simplifying the calculation of the angular momentum (e.g. by choosing G as the reference). The angular impulse-momentum principle given in Equation 10.7-4 is written as a vector equation, but in the case of planar motion, I_G is a scalar and all of the resulting vectors will be parallel (i.e into or out of the plane of motion). All of the other forms of the angular impulse-momentum principle for rigid bodies follow directly from Equation 10.7-4.

General planar motion angular impulse-momentum principle:

$$\mathbf{J}_O = \Delta \mathbf{H}_O \qquad \int_{t_1}^{t_2} \sum \mathbf{M}_O \, dt = (\mathbf{r}_{G/O,2} \times m\mathbf{v}_{G,2} + I_{G,2} \omega_2) - (\mathbf{r}_{G/O,1} \times m\mathbf{v}_{G,1} + I_{G,1} \omega_1) \qquad (10.7\text{-}4)$$

$\sum \mathbf{M}_O$ = resultant moment acting on the rigid body

$\Delta \mathbf{H}_O$ = change in angular momentum of the rigid body with respect to O

$I_{G,1}, I_{G,2}$ = mass moment of inertia of the rigid body about its mass center at state 1 and state 2, respectively

$\mathbf{J}_O$ = applied angular impulse with respect to fixed-point O

m_i = mass of the ith body

$\mathbf{r}_{G/O,1}, \mathbf{r}_{G/O,2}$ = position of the body's mass center G with respect to point O at state 1 and state 2, respectively

t_1, t_2 = time at state 1 and time at state 2, respectively

$\mathbf{v}_{G,1}, \mathbf{v}_{G,2}$ = velocity of the rigid body at state 1 and state 2, respectively

ω_1, ω_2 = angular velocity of the rigid body at state 1 and state 2, respectively

Conceptual Example 10-7.1

Where should a force be applied on an object in order to cause the greatest impact on the object's angular momentum?

- a) At the geometric center of the object.
- b) At the object's center of mass.
- c) At the edge of the object.
- d) The placement of the force makes no difference.

10.7.5) SYSTEMS OF RIGID BODIES

The impulse-momentum principle equation for a system of rigid bodies is very similar to that for a single rigid body. The difference is that the total angular momentum of a system of rigid bodies is the sum of the momenta of the individual bodies. The angular impulse-momentum principle for a system of rigid bodies undergoing general planar motion is given by Equations 10.7-5 and 10.7-6. As was the case with systems of particles, the moments induced by forces internal to the system will cancel and do not need to be explicitly considered.

Angular impulse-momentum principle for a system of rigid bodies moving in a plane:

$$\mathbf{J}_G = \Delta \mathbf{H}_G \qquad \int_{t_1}^{t_2} \sum \mathbf{M}_G \, dt = \sum (I_{Gi,2} \boldsymbol{\omega}_{i,2}) - \sum (I_{Gi,1} \boldsymbol{\omega}_{i,1}) \qquad (10.7\text{-}5)$$

$$\mathbf{J}_O = \Delta \mathbf{H}_O \qquad \int_{t_1}^{t_2} \sum \mathbf{M}_O \, dt = \sum (\mathbf{r}_{Gi/O,2} \times m_i \mathbf{v}_{Gi,2} + I_{Gi,2} \boldsymbol{\omega}_{i,2}) - \sum (\mathbf{r}_{Gi/O,1} \times m_i \mathbf{v}_{Gi,1} + I_{Gi,1} \boldsymbol{\omega}_{i,1})$$

(10.7-6)

$\sum \mathbf{M}_O, \sum \mathbf{M}_G$ = external resultant moment acting on all the rigid bodies in the system

$\Delta \mathbf{H}_O, \Delta \mathbf{H}_G$ = change in angular momentum of the system

$I_{G,1}, I_{G,2}$ = mass moment of inertia of the rigid body about its mass center at state 1 and state 2, respectively

$\mathbf{J}_O, \mathbf{J}_G$ = applied angular impulse with respect to fixed-point O and the body's mass center G, respectively

m_i = mass of the ith body

$\mathbf{r}_{Gi/O,1}, \mathbf{r}_{Gi/O,2}$ = position of the ith body at state 1, and state 2 from the body's mass center G to point O, respectively

t_1, t_2 = time at state 1 and time at state 2, respectively

$\mathbf{v}_{Gi,1}, \mathbf{v}_{Gi,2}$ = mass center velocity of the ith body at state 1, and state, 2 respectively

10.7.6) CONSERVATION OF ANGULAR MOMENTUM

Angular momentum is conserved if the externally applied moments sum to zero. More generally, if the angular impulse $\int \sum \mathbf{M} dt$ applied to a body over an interval of time is equal to zero, then the initial and final angular momenta of that body are equal. Recall that angular momentum may be conserved with respect to one point, but not another. Therefore, care should be taken to choose the reference point in an intelligent manner.

The conservation of angular momentum for a rigid body undergoing general planar motion is given in Equations 10.7-7 and 10.7-8. As was seen in the *Impulse-Momentum for a System of Particles* section, we were able to define a system to consist of multiple particles so that internal reaction forces/moments between the particles canceled leading momentum to be conserved. We can do the same thing for a system of rigid bodies. The momentum of a system of rigid bodies is equal to the sum of the momenta of each of the bodies in the system. The conservation of angular momentum for a system of bodies is given by Equations 10.7-9 and 10.7-10.

Conservation of angular momentum for a rigid body:

$$\boxed{\mathbf{H}_{G,1} = \mathbf{H}_{G,2} \qquad I_{G,1}\boldsymbol{\omega}_1 = I_{G,2}\boldsymbol{\omega}_2} \quad (10.7\text{-}7)$$

$$\boxed{\mathbf{H}_{O,1} = \mathbf{H}_{O,2} \qquad \mathbf{r}_{G/O,1} \times m\mathbf{v}_{G,1} + I_{G,1}\boldsymbol{\omega}_1 = \mathbf{r}_{G/O,2} \times m\mathbf{v}_{G,2} + I_{G,2}\boldsymbol{\omega}_2} \quad (10.7\text{-}8)$$

Conservation of angular momentum for a system of rigid bodies:

$$\boxed{\mathbf{H}_{G,1} = \mathbf{H}_{G,2} \qquad \sum I_{Gi,1}\boldsymbol{\omega}_{i,1} = \sum I_{Gi,2}\boldsymbol{\omega}_{i,2}} \quad (10.7\text{-}9)$$

$$\boxed{\mathbf{H}_{O,1} = \mathbf{H}_{O,2} \qquad \sum \left(\mathbf{r}_{Gi/O,1} \times m_i \mathbf{v}_{Gi,1} + I_{Gi,2}\boldsymbol{\omega}_{i,1} \right) = \sum \left(\mathbf{r}_{Gi/O,2} \times m_i \mathbf{v}_{Gi,2} + I_{Gi,2}\boldsymbol{\omega}_{i,2} \right)} \quad (10.7\text{-}10)$$

$\mathbf{H}_{O,1}, \mathbf{H}_{O,2}$ = angular momentum of body with respect to O at state 1 and state 2, respectively.

$\mathbf{H}_{G,1}, \mathbf{H}_{G,2}$ = angular momentum of body with respect to G at state 1 and state 2, respectively

$I_{G,1}, I_{G,2}$ = mass moment of inertia of the rigid body about its mass center at state 1 and state 2, respectively

m = mass of the rigid body

$\mathbf{r}_{G/O,1}, \mathbf{r}_{G/O,2}$ = position of the body's mass center with respect to O at state 1 and state 2, respectively

$\mathbf{v}_{G,1}, \mathbf{v}_{G,2}$ = velocity of mass center at state 1 and state 2, respectively

$\boldsymbol{\omega}_1, \boldsymbol{\omega}_2$ = angular velocity of the rigid body at state 1 and state 2, respectively

Conceptual Example 10.7-2

An Olympic platform diver is initially rotating in a tucked position during a dive before unfolding himself prior to entering the water. If we neglect air resistance, as the diver extends his body his angular velocity ...

    a) increases.
    b) decreases.
    c) remains the same.
    d) not enough information provided

For the above situation, what happens to the diver's kinetic energy?

    a) It stays the same as he releases from the tuck.
    b) It becomes larger when he releases from the tuck.
    c) It becomes smaller when he releases from the tuck.

Equation Summary

Abbreviated variable definition list

$\Sigma\mathbf{F}$ = resultant force
$\mathbf{G}$ = linear momentum
$\mathbf{H}_O$ = angular momentum with respect to O
$\mathbf{I}$ = linear impulse
I = mass moment of inertia
$\mathbf{J}_O$ = angular impulse with respect to O

m = mass
$\Sigma\mathbf{M}_O$ = resultant moment
$\mathbf{r}$ = position
t = time
$\mathbf{v}$ = velocity
ω = angular velocity

Linear momentum

Rigid body

$$\mathbf{G} = m\mathbf{v}_G$$

System of rigid bodies

$$\mathbf{G} = \sum m_i \mathbf{v}_{Gi}$$

Linear impulse

$$\mathbf{I} = \int_{t_1}^{t_2} \sum \mathbf{F}\, dt$$

Linear impulse-momentum principle

Rigid body

$$\int_{t_1}^{t_2} \sum \mathbf{F}\, dt = m\mathbf{v}_{G,2} - m\mathbf{v}_{G,1}$$

System of rigid bodies

$$\int_{t_1}^{t_2} \sum \mathbf{F}\, dt = \sum m_i \mathbf{v}_{Gi,2} - \sum m_i \mathbf{v}_{Gi,1}$$

Angular momentum

Pure translation

$$\mathbf{H}_O = \mathbf{r}_{G/O} \times m\mathbf{v}$$

General planar motion

$$\mathbf{H}_G = I_G \omega$$

$$\mathbf{H}_O = \mathbf{r}_{G/O} \times m\mathbf{v}_G + I_G \omega$$

Pure rotation

$$\mathbf{H}_O = I_O \omega$$

Angular impulse

$$\mathbf{J}_O = \int_{t_1}^{t_2} \sum (\mathbf{r}_{force/Oi} \times \mathbf{F}_i)\, dt = \int_{t_1}^{t_2} \sum \mathbf{M}_O\, dt$$

$$\mathbf{J}_G = \int_{t_1}^{t_2} \sum (\mathbf{r}_{force/Gi} \times \mathbf{F}_i)\, dt = \int_{t_1}^{t_2} \sum \mathbf{M}_G\, dt$$

Angular impulse-momentum principle

Pure translation

$$\int_{t_1}^{t_2} \sum \mathbf{M}_O\, dt = \mathbf{r}_{G/O,2} \times m\mathbf{v}_2 - \mathbf{r}_{G/O,1} \times m\mathbf{v}_1$$

Pure rotation in a plane

$$\int_{t_1}^{t_2} \sum \mathbf{M}_O\, dt = I_{O,2}\omega_2 - I_{O,1}\omega_1$$

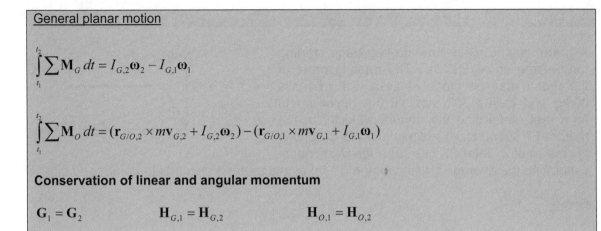

General planar motion

$$\int_{t_1}^{t_2} \sum \mathbf{M}_G \, dt = I_{G,2}\boldsymbol{\omega}_2 - I_{G,1}\boldsymbol{\omega}_1$$

$$\int_{t_1}^{t_2} \sum \mathbf{M}_O \, dt = (\mathbf{r}_{G/O,2} \times m\mathbf{v}_{G,2} + I_{G,2}\boldsymbol{\omega}_2) - (\mathbf{r}_{G/O,1} \times m\mathbf{v}_{G,1} + I_{G,1}\boldsymbol{\omega}_1)$$

Conservation of linear and angular momentum

$$\mathbf{G}_1 = \mathbf{G}_2 \qquad\qquad \mathbf{H}_{G,1} = \mathbf{H}_{G,2} \qquad\qquad \mathbf{H}_{O,1} = \mathbf{H}_{O,2}$$

Example Problem 10.7-3

A given electric motor has the two-stage starting torque-time curve shown. The rotating parts of the motor may be modeled as a disk of mass 30 kg and radius 300 mm. If the motor starts from rest, determine the angular velocity of the motor at 7 seconds. A magnetic brake is applied to the motor after it has run for 7 seconds. Determine the average braking torque if it takes the motor 4 seconds to come to rest.

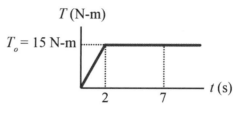

Given:

Find:

Solution:

Example Problem 10.7-4

A 1000-kg satellite is tethered to a 200-kg satellite by a light cable. For the situation shown, the satellites and cable are rotating with an angular velocity of 0.5 rpm. If the larger satellite slowly releases another 10 m of cable, determine the final angular velocity of the system. Treat the satellites as particles. Reconsider the situation, but now take into account the mass moments of inertia of the two satellites. Assume that the larger satellite has a mass moment of inertia about an axis through its mass center that is parallel to the axis of rotation of 500 kg-m^2. Likewise, assume the smaller satellite has a mass moment inertia about a parallel axis through its mass center of 10 kg-m^2.

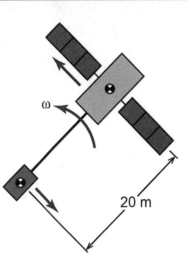

Given:

Find:

Solution:

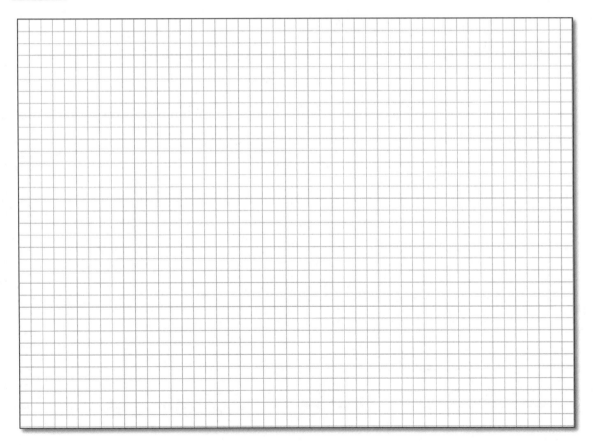

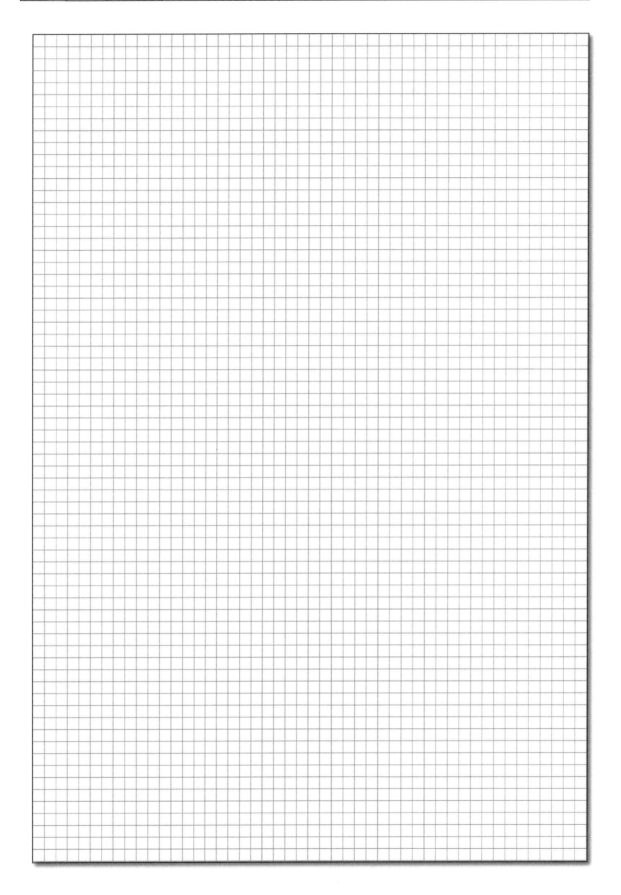

Solved Problem 10.7-5

The center G of the 5-kg wheel, with radius of gyration of 1.2 m about G, has velocity 3 m/s down the incline when a force $F = 20$ N is applied to the cord wrapped around its inner hub. If the wheel rolls without slipping, calculate the velocity v of the center G when F has been applied for 4 seconds. The wheel has an inner radius of 1 m and an outer radius of 2 m.

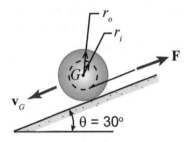

Given: $v_{G,1} = 3$ m/s $\qquad k_G = 1.2$ m
$\qquad\quad r_i = 1$ m $\qquad\qquad r_o = 2$ m
$\qquad\quad \Delta t = 4$ s $\qquad\qquad m = 5$ kg
$\qquad\quad F = 20$ N $\qquad\qquad$ No slip

Find: $v_{G,2}$

Solution:

Free-body diagram

We will use both the linear and angular impulse-momentum principle to solve this problem. Free-body diagrams are very useful when applying the impulse-momentum principles. They allow you to visually see the forces and moments that are being applied to the body.

Note that the friction force is static (no slip), however, we do not know the magnitude of this force.

Linear impulse-momentum principle

$$\mathbf{I} = \mathbf{G}_2 - \mathbf{G}_1 \qquad\qquad \int \sum \mathbf{F} dt = m\mathbf{v}_{G,2} - m\mathbf{v}_{G,1}$$

The linear impulse-momentum principle is a vector equation that may be applied in each coordinate direction.

x-direction: $\qquad \displaystyle\int_0^{\Delta t} (-F_{fs} - mg\sin\theta + F)\,dt = m(-v_{Gx,2} + v_{Gx,1})$

The mg and F forces are constant and may come out of the integral. Note that the directions of the velocities are indicated with a positive or negative sign.

$$\int_0^{\Delta t} -F_{fs}\,dt = (mg\sin\theta - F)\Delta t + m(-v_{Gx,2} + v_{Gx,1}) \qquad\qquad (1)$$

Angular impulse-momentum principle

The linear impulse-momentum principle gave us an equation with 2 unknowns. We will use the angular impulse-momentum principle to get the second equation that will allow us to solve for $v_{G,2}$.

$$\mathbf{J}_G = \mathbf{H}_{G,2} - \mathbf{H}_{G,1} \qquad \int \sum M_G \, dt = I_G \omega_2 - I_G \omega_1 \qquad \int_0^{\Delta t} (Fr_i - F_{fs} r_o) \, dt = I_G (\omega_2 - \omega_1)$$

$$\int_0^{\Delta t} -F_{fs} \, dt = \frac{mk_G^2 (\omega_2 - \omega_1) - Fr_i \Delta t}{r_o} \qquad (2)$$

Equate Equations (1) and (2).

$$\frac{mk_G^2 (\omega_2 - \omega_1) - Fr_i \Delta t}{r_o} = (mg \sin \theta - F) \Delta t + m(-v_{Gx,2} + v_{Gx,1})$$

No slip implies that $v = r\omega$.

$$\frac{mk_G^2 (v_{Gx,2}/r_o - v_{Gx,1}/r_o) - Fr_i \Delta t}{r_o} = (mg \sin \theta - F) \Delta t + m(-v_{Gx,2} + v_{Gx,1})$$

$$mk_G^2 (v_{Gx,2} - v_{Gx,1}) - Fr_i r_o \Delta t = r_o^2 (mg \sin \theta - F) \Delta t + mr_o^2 (-v_{Gx,2} + v_{Gx,1})$$

$$v_{Gx,2} = \frac{r_o^2 mg \sin \theta \Delta t + F \Delta t (r_i r_o - r_o^2)}{(mk_G^2 + mr_o^2)} + v_{Gx,1} \qquad \boxed{v_{Gx,2} = 11.54 \frac{\text{m}}{\text{s}}} \text{ down the incline}$$

10.8) ECCENTRIC IMPACT

In the preceding chapter we analyzed the situation of two colliding particles. We learned that by choosing two special coordinate directions the analysis of impact was greatly simplified. We defined a coordinate frame consisting of a direction lying in the *plane of contact* of the two bodies and an orthogonal direction along the *line of impact*. Since particles have no size, it was always the case that the centers of mass of the colliding bodies were located along the line of impact. We did not have to consider that the collision may cause the bodies to "spin." When dealing with rigid bodies, an impact force that is offset from the mass center of a colliding body, will alter the body's rotation as well as its translation. **Eccentric impact**

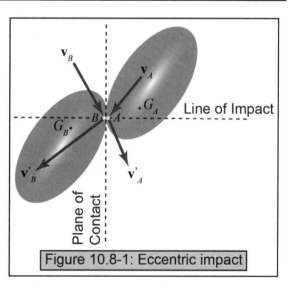

Figure 10.8-1: Eccentric impact

occurs when the line of impact of the collision is offset from one or both of the colliding body's mass centers.

If there are no external forces or moments applied to the bodies during the impact, or if they are small compared to the impact forces and moments, we can assume that linear momentum and angular momentum are conserved. Assuming planar motion, the conservation of linear momentum is represented by Equation 10.9-1 and the conservation of angular momentum is represented by Equation 10.9-2 . Note, Equation 10.9-2 assumes angular momentum is calculated with respect to the bodies' mass centers, but it can just as easily be done about other points.

Conservation of linear momentum: $$\boxed{m_A \mathbf{v}_{GA} + m_B \mathbf{v}_{GB} = m_A \mathbf{v}'_{GA} + m_B \mathbf{v}'_{GB}}$$ (10.9-1)

Conservation of angular momentum: $$\boxed{I_{GA}\boldsymbol{\omega}_A + I_{GB}\boldsymbol{\omega}_B = I_{GA}\boldsymbol{\omega}'_A + I_{GB}\boldsymbol{\omega}'_B}$$ (10.9-2)

m_A, m_B = mass of rigid body A and rigid body B, respectively

I_{GA}, I_{GB} = mass moment of inertia about its mass center of rigid body A and rigid body B, respectively

$\mathbf{v}_{GA}$, $\mathbf{v}_{GB}$ = velocity prior to impact of the mass center of body A and body B, respectively

$\mathbf{v}'_{GA}$, $\mathbf{v}'_{GB}$ = velocity following impact of the mass center of body A and body B, respectively

$\boldsymbol{\omega}_A$, $\boldsymbol{\omega}_B$ = angular velocity prior to impact of body A and body B, respectively

$\boldsymbol{\omega}'_A$, $\boldsymbol{\omega}'_B$ = angular velocity following impact of body A and body B, respectively

Assuming planar motion, Equation 10.9-1 can be decomposed into orthogonal components (i.e perpendicular coordinate directions). If we assume as we did in the preceding chapter that the friction between the two surfaces is negligible, the components of the velocities along the plane of contact are unchanged by the impact as shown in Equation 10.9-3. Along the line of impact linear momentum is conserved as shown in Equation 10.9-4.

Plane of contact velocities: $$\boxed{\begin{aligned} \left(v_{GA}\right)_{plane\,of\,contact} &= \left(v'_{GA}\right)_{plane\,of\,contact} \\ \left(v_{GB}\right)_{plane\,of\,contact} &= \left(v'_{GB}\right)_{plane\,of\,contact} \end{aligned}}$$ (10.9-3)

$(v_{GA})_{plane\,of\,contact}$ = speed of body A prior to impact along the plane of contact
$(v'_{GA})_{plane\,of\,contact}$ = speed of body A after impact along the plane of contact
$(v_{GB})_{plane\,of\,contact}$ = speed of body B prior to impact along the plane of contact
$(v'_{GB})_{plane\,of\,contact}$ = speed of body B after impact along the plane of contact

Conservation of linear momentum along the line of impact:

$$\boxed{m_A \left(v_{GA}\right)_{line\,of\,impact} + m_B \left(v_{GB}\right)_{line\,of\,impact} = m_A \left(v'_{GA}\right)_{line\,of\,impact} + m_B \left(v'_{GB}\right)_{line\,of\,impact}}$$ (10.9-4)

$(v_{GA})_{line\,of\,impact}$ = speed of body A prior to impact along the line of impact
$(v'_{GA})_{line\,of\,impact}$ = speed of body A after impact along the line of impact
$(v_{GB})_{line\,of\,impact}$ = speed of body B prior to impact along the line of impact
$(v'_{GB})_{line\,of\,impact}$ = speed of body B after impact along the line of impact
m_A = mass of body A
m_B = mass of body B

If we are given the initial motion of the two colliding bodies, we are then left with 4 unknowns between Equations 10.9-2 and 10.9-4. As was the case with central impact, an additional relationship is needed. Available to help us determine the behavior of the system following impact is the coefficient of restitution (e). In the case of eccentric impact, the **coefficient of restitution** is still defined as the ratio between the relative velocity of separation and the relative velocity of approach. For rigid bodies, it is more specifically for the points of contact (point A and point B in Figure 10.9-1). This is expressed in Equation 10.9-5. The coefficient of restitution is a very complicated parameter that depends on the body's material, the geometry, and the impact velocities. This means that rigid-body impact problems may be solved, and have meaning, for only the most simplified cases.

Coefficient of restitution: $$e = \frac{\left(v'_B\right)_{line\ of\ impact} - \left(v'_A\right)_{line\ of\ impact}}{\left(v_A\right)_{line\ of\ impact} - \left(v_B\right)_{line\ of\ impact}}$$ (10.9-5)

e = coefficient of restitution
$(v_A)_{line\ of\ impact}$ = velocity of contact point A along the line of impact prior to impact
$(v'_A)_{line\ of\ impact}$ = velocity of contact point A along the line of impact following impact
$(v_B)_{line\ of\ impact}$ = velocity of contact point B along the line of impact prior to impact
$(v'_B)_{line\ of\ impact}$ = velocity of contact point B along the line of impact following impact

Using our knowledge of kinematics, the velocities of the points of contact can be expressed in terms of the velocities of the mass centers and the angular velocities of the bodies.

CHAPTER 10 REVIEW PROBLEMS

RP10-1) Consider an ice hockey puck and a street hockey puck. They both have equal mass and their mass centers are traveling with equal linear velocities as shown in the figure. Which one has the greater linear momentum?

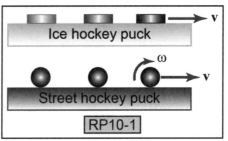

a) The ice hockey puck has the greater linear momentum.
b) The street hockey puck has the greater linear momentum.
c) They both have equal linear momentum.

Which one has the greater angular momentum with respect to its mass center?

a) The ice hockey puck has the greater angular momentum.
b) The street hockey puck has the greater angular momentum.
c) They both have equal angular momentum.

RP10-2) An externally applied force cannot change the angular momentum of a spinning object. (True, False)

RP10-3) No matter where a force is applied, it will have the same effect on the angular momentum. (True, False)

RP10-4) A bar and particle of the same mass m and translating perpendicular to each other collide. Immediately preceding the collision they are traveling with the same speed v. The particle strikes the bar near the edge and adheres to the bar. Which kinetic analysis approach would you employ to determine the linear and angular velocities of the system immediately after the collision?

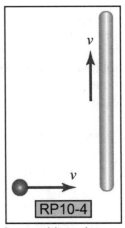

a) Newtonian mechanics
b) Work-energy
c) Impulse-momentum

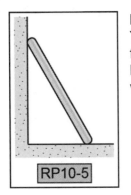

RP10-5) A 24-kg ladder is released from rest in the position shown. You may model the ladder as a slender uniform bar and neglect friction. If you wish to determine the angular acceleration of the ladder at the instant of release, which kinetic analysis approach would you employ?

a) Newtonian mechanics
b) Work-energy
c) Impulse-momentum

RP10-6) The 1-kg uniform slender bar and 2-kg disk shown are released from rest when the bar is horizontal. The disk rolls without slip on the curved surface shown. If you wish to determine the bar's angular velocity when the bar is vertical, which kinetic analysis approach would you employ?

a) Newtonian mechanics
b) Work-energy
c) Impulse-momentum

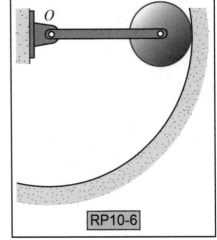

RP10-7)   The 5-kg rectangular plate shown rests on a smooth horizontal surface when the two large impulsive forces displayed, F_A = 1000 N and F_B = 500 N, are applied to the plate at the angles of θ = 45° and ϕ = 30°. Determine the angular velocity of the plate and the velocity of its mass center after the forces have been applied for 0.02 seconds. You may assume the plate moves a negligible distance during the application of the forces.

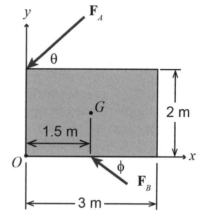

Given: F_A = 1000 N $\theta = 45°$
        F_B = 500 N $\phi = 30°$
        m = 5 kg $\Delta t = 0.02$ s
        $v_{G,1} = 0$ $\omega_1 = 0$

Find: $\mathbf{v}_{G,2}$, ω_2

Solution:

Linear impulse and momentum	Angular impulse and momentum

$\mathbf{v}_{G,2}$ = _____ **i** + _____ **j** m/s ω_2 = _____ **k** rad/s

RP10-8) An alternative to using thrusters for controlling satellite orientation is to use reaction wheels. Each reaction wheel is firmly attached to the satellite and essentially consists of a motor driving a heavy flywheel. Reaction wheels are advantageous in that they don't require propellant (they are powered from the solar panels) and they can position a satellite with high accuracy. Consider a 1200-kg satellite with a radius of gyration with respect to the mass center of 0.85 m about the z-axis with a reaction wheel actuator that includes a 10-kg flywheel of radius 0.2 m. If the system is initially at rest and the flywheel is spun up to 10,000 rpm, determine the resulting angular velocity of the satellite. You may neglect the mass of the rest of the reaction wheel assembly.

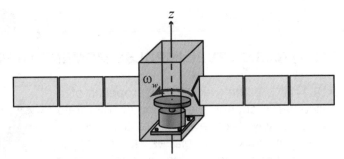

Given: m_s = 1200 kg m_w = 10 kg
       k_G = 0.85 m r = 0.2 m
       ω_1 = 0 $\omega_{w,2}$ = 10,000 **k** rpm

Find: $\omega_{s,2}$

Solution:

Calculate the mass moments of inertia for the wheel and satellite about the mass center.

$I_{wG} =$ _____

$I_{sG} =$ _____

Does this system conserve angular momentum? Yes   No

Calculate the final angular velocity of the satellite.

$\omega_{s,2} =$ _____

CHAPTER 10 PROBLEMS

P10.1) BASIC LEVEL IMPULSE-MOMENTUM PROBLEMS

P10.1-1)[fe]   If the vertical tail propeller of a helicopter is absent, the helicopter will rotate about its yaw axis every time the angular speed of its main blades is changed. Assuming that each of the helicopter's four main blades are L = 6 m long with a mass of 30 kg, estimate the final angular momentum of the helicopter cabin if it is initially at rest, the vertical tail propeller is broken, and the angular speed of the main blades is increased from 200 rpm to 400 rpm. The mass moment of inertia with respect to the mass center of the helicopter cabin is 250 kg-m^2 and the mass center is on the same axis as the center of the main rotor's blades. Neglect air drag on the main blades.

a) ω = 1002 rpm   b) ω = 1102 rpm   c) ω = 1122 rpm   d) ω = 1152 rpm

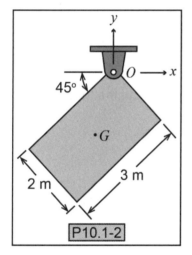

P10.1-2) A 10-kg thin plate is pinned at fixed point O and released from rest such that it reaches an angular velocity of 4 rad/sec at the position shown. Determine the plate's linear momentum and the plate's angular momentum (a) about its center of mass G, and (b) about the point O.

Ans: $G = 70.7\,\mathbf{i} - 14.2\,\mathbf{j}$ kg-m/s, $H_G = 43.33\,\mathbf{k}$ kg-m^2/s, $H_O = 173.32\,\mathbf{k}$ kg-m^2/s

P10.1-3) Three small identical 5-kg spheres lying in a horizontal plane are rigidly attached to a vertical shaft via rods as shown of lengths L_A = 2 m, L_B = 4 m, and L_C = 3 m. The bars have a mass of 1 kg per meter. The system is initially rotating about the vertical shaft with an angular speed of 20 rad/s when a moment of magnitude $M = 2t$ N-m, where t is in seconds, is applied in the opposite direction. Determine the time required to bring the system to rest.

Ans: t = 59.7 s

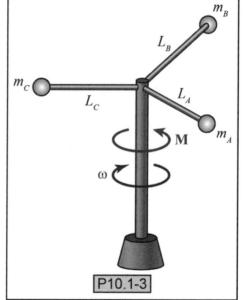

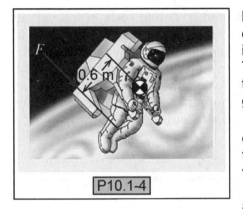

P10.1-4

P10.1-4)[fe] An astronaut with thruster pack has a combined mass of 130 kg and a mass moment of inertia about the combined mass center of 50 kg·m². The astronaut is initially not rotating when the thruster pack is fired. If the thruster initially generates a force according to the equation $F = 150\sqrt{t}$ N, where t is in seconds, determine the duration of time the thruster needs to fire to provide the astronaut with a counterclockwise angular velocity of 0.4 rad/s.

a) $t = 0.48$ s   b) $t = 0.38$ s

c) $t = 0.28$ s   d) $t = 0.18$ s

P10.1-5)[fe]   A   1200-kg satellite is inserted into its orbit with its solar panels collapsed rotating about its z-axis   at   a   rate   of 1.5 rev/sec. Determine the satellite's angular velocity after   its   solar   panels deploy. You may assume that   the   satellite has a radius   of   gyration   with

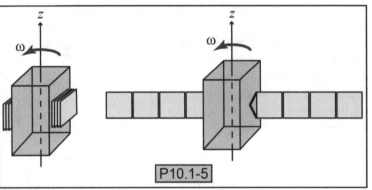

P10.1-5

respect to the mass center about the z-axis of 0.5 m when its solar panels are collapsed and a radius of gyration of 0.85 m when its solar panels are deployed.

a) $\omega = 0.52$ rev/s   b) $\omega = 0.62$ rev/s   c) $\omega = 0.72$ rev/s   d) $\omega = 0.82$ rev/s

P10.2) INTERMEDIATE LEVEL IMPULSE-MOMENTUM PROBLEMS

P10.2-1) A 2-kg sphere is moving at 3 m/s when it strikes the end of a 5-kg slender bar that is pinned at O. Determine the angular velocity of the bar after impact if the bar has length 2 m and the coefficient of restitution between the bar and sphere is $e = 0.7$.

Ans: $\omega = 1.39$ rad/s

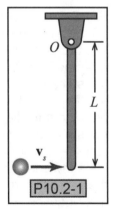

P10.2-1

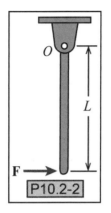

P10.2-2) A 25-kg slender bar is pinned at O and rests in the vertical position shown when a large impulsive force $F = 1000$ N is applied to its end for a time interval of 0.005 s. Determine the angular speed of the bar after the impulse has been applied. Also, find the pin reaction at O during the application of the force. You may assume that the bar moves a negligible distance during the period of the application of the impulsive force and the length of the bar is 3 m.

P10.2-2

P10.2-3)   The 5-slug pulley ($r = 0.5$ ft, $k_O = 0.35$ ft) shown in the figure is initially at rest when a a) 10-lb force and b) 10-lb weight is attached to the end of the rope that is wrapped tightly about the pulley's outer radius. Determine the angular velocity of the pulley 1 second later for both cases.

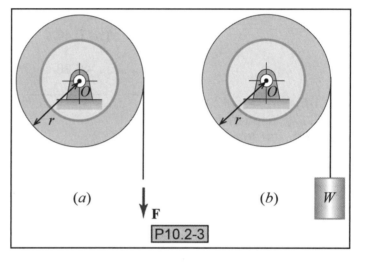

(a)

F

(b) W

P10.2-3

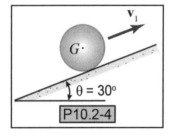

$\mathbf{v}_1$

$G\cdot$

$\theta = 30°$

P10.2-4

P10.2-4) The center G of the uniform disk shown has an initial velocity $v_1 = 1$ m/s up the 30°-incline at time $t = 0$ s. If the disk rolls without slipping, determine its velocity when $t = 5$ s.

Ans: $v = 15.35$ m/s down the incline

P10.2-5) A 1200-kg satellite is in a stationary orbit above the earth. The principle inertias of the satellite are I_{xx} = 570 kg-m^2, $I_{yy} = 650$ kg-m^2, and I_{zz} =340 kg-m^2.   If thrusters 1 and 2 fire for a short period of time and impart a total linear impulse of -30 N-s **j** onto the satellite, determine the change in the translational and angular velocity of the satellite. You may assume that the origin of the shown x-y-z coordinate frame is located at the satellite's mass center G, and thruster 1 has location (-0.5 m, 0.5 m, 1 m) while thruster 2 has location (0.5 m, 0.5 m, 1 m).

Ans: $\Delta\mathbf{v}_G = -0.025$ **j** m/s, $\Delta\omega = 0.53$ **i** rad/s

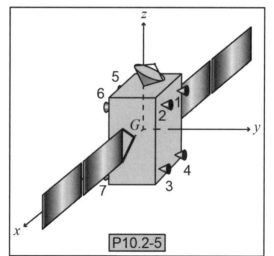

P10.2-5

P10.2-6) An alternative to using thrusters for controlling satellite orientation is to use reaction wheels. Each reaction wheel is firmly attached to the satellite and essentially consists of a motor driving a heavy flywheel. Reaction wheels are advantageous in that they don't require propellant (they are powered from the solar panels) and they can position a satellite with high accuracy. Consider a 1200-kg satellite with a radius of gyration with respect to the mass center of 0.85 m about the z-axis with a reaction wheel actuator that includes a 10-kg flywheel of radius 0.2 m. If the system is initially at rest and the flywheel is spun up to ω_p = 10,000 rpm according to the profile shown, determine the final angular velocity of the satellite and its overall change in angular position between the beginning and end of the 10 second period shown. You may neglect the mass of the rest of the reaction wheel assembly.

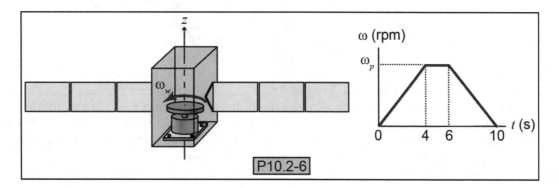

P10.2-7) A bar and particle of the same mass m and translating perpendicular to each other collide with the same initial speed v. If the particle strikes the bar near the edge and adheres to the bar following the collision, determine the resulting translational and angular velocity of the combined system. Also, specify the location of the above solved tranlational velocity.

Ans: $\mathbf{v}_G = \dfrac{v}{2}(\mathbf{i}+\mathbf{j})$, $\bar{y} = \dfrac{L}{4}$, $\omega = 1.2v/L$

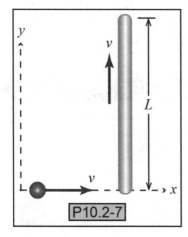

P10.3) ADVANCED LEVEL IMPULSE-MOMENTUM PROBLEMS

P10.3-1) A satellite is in orbit above the earth and it needs to be able to maintain its attitude about its z-axis within ± 5 degrees of desired. The satellite is equipped with thrusters that each generate 10 N of force and have a minimum impulse bit of 30 mN·sec. The minimum impulse bit defines the smallest amount of time the thrusters can be turned on for. Because the thrusters can only be turned on or off, they can't be modulated, the satellite attitude will bounce back and forth across the allowed attitude band as alternating thrusters are fired. For example, thrusters 1 and 6 will fire to cause the satellite to rotate in the positive **k** direction, then when

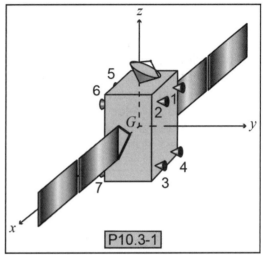

P10.3-1

the attitude reaches the limits of its allowed band, thrusters 2 and 5 will fire to rotate the satellite in the opposite direction. If the satellite's mass moment of inertia about its z-axis is 340 kg-m^2 and the specific impulse of each thruster is 2000 N-s/kg (means 1 kg of fuel is used to generate 2000 N-s of impulse), estimate the minimum amount of fuel needed to maintain the satellite's attitude for 1 year. You may assume that the origin of the shown x-y-z coordinate frame is located at the satellite's mass center G, and thruster 1 has location (0.5 m, -0.5 m, 1 m), thruster 2 has location (0.5 m, 0.5 m, 1 m), thruster 5 has location (-0.5 m, 0.5 m, 1 m), and thruster 6 has location (-0.5 m, -0.5 m, 1 m).

CHAPTER 10 COMPUTER PROBLEMS

C10-1) Consider the symmetric triangular pendulum shown consisting of two 10-kg masses connected by beams of length $L = 2$ meters and negligible mass. The pendulum starts from rest in the position shown where the left beam is completely vertical when a constant 100-N force **F** is applied to the left bob of the pendulum for a period of 1 second. Assuming the direction of the applied force remains constant, plot the angular position of the pendulum over the full period of the external force's application.

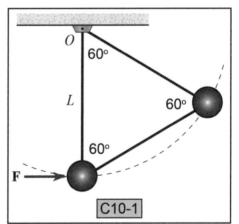

C10-1

CHAPTER 10 DESIGN PROBLEMS

D10-1)   Consider the "tea-cup" amusement park ride shown.   Each cup has an occupancy limit of three riders. The cups are rotated with respect to the platform by the application of torque from the riders.   The platform itself is turned by a large electric motor. Use your knowledge of dynamics to address the following issues. It is desired that the platform be driven with the angular velocity profile shown below.

- Recommend the peak torque that the electric motor must supply (for short bursts of time).
- Recommend the sustained torque that the electric motor must supply.

It is known that the platform and cups are made from steel. Also, an experiment was performed where the platform was brought to a speed of 1 rad/sec and then allowed to come to rest on its own. Repeated experiments showed that the coast-down time took approximately 30 seconds. If there is further information needed to finish your design, explain how you would obtained estimates of the necessary data.

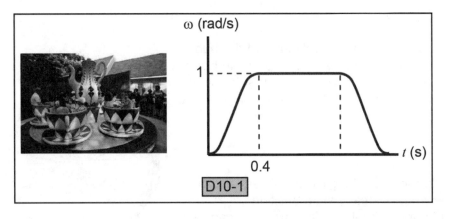

CHAPTER 10 ACTIVITIES

A10-1) This activity gives students a physical understanding of angular momentum.

Group size: 1-2 students

Supplies

- Dumbbells
- Swivel chair with minimal rotational friction

Procedure

1. Sit in a swivel chair.
2. Hold a dumbbell close to your body.
3. Spin yourself in the chair as fast as you can.
4. Extend the dumbbell out away from your body and note the affect.
5. Bring the dumbbell back into your body and note the effect.

Calculations and Discussion

- List all of your assumptions.
- Explain in words and calculations why your angular speed changed when you changed the position of the dumbbell.
- List possible sources of experimental error.

A10-2) This activity gives students a physical understanding of angular momentum.

Group size: 1-2 students

Supplies

- Bicycle wheel
- Rope
- Swivel chair

Experimental setup

Take the rope and tie it securely around the hub of the wheel. Make sure it will not slip off and does not interfere with the rotational motion of the wheel. Find a swivel chair that has as little rotational friction as possible.

Procedure

1. Hold the rope dangling the wheel and note the wheel's steady state position.
2. Spin the wheel as fast as you can and move the wheel into the vertical position. Does the wheel stay in this position?
3. Stop the wheel and remove the rope.
4. Sit on the swivel chair holding the wheel.
5. Spin the wheel as fast as you can in the horizontal position.
6. Move the wheel into the vertical position. Note what happens.
7. Flip the wheel 180°. Note what happens.

Calculations and Discussion

- List all of your assumptions.
- Explain in words and calculations why the wheel, hanging by the rope, stayed in the vertical position.
- Explain in words and calculations why the swivel chair rotated when changing the wheel's position.
- List possible sources of experimental error.

APPENDIX A: MOMENTS OF INERTIA

APPENDIX A OUTLINE

A.1) MASS MOMENTS OF INERTIA TABLES

A.1.1) MASS MOMENT OF INERTIA OF HOMOGENEOUS SOLIDS

Figure	Volume and Center of Gravity Location	Mass Moment of Inertia
Slender Rod	$x_G = \dfrac{L}{2}$	$I_{xG} = 0$ $\qquad I_{y'} = \dfrac{mL^2}{3}$ $I_{yG} = \dfrac{mL^2}{12}$ $I_{zG} = \dfrac{mL^2}{12}$ $\qquad I_{z'} = \dfrac{mL^2}{3}$
Cylinder	$V = \pi r^2 h$ $y_G = \dfrac{h}{2}$	$I_{xG} = \dfrac{m(3r^2 + h^2)}{12}$ $I_{yG} = \dfrac{mr^2}{2}$ $I_{zG} = \dfrac{m(3r^2 + h^2)}{12}$ $I_{x'} = \dfrac{m(3r^2 + 4h^2)}{12}$ $I_{z'} = \dfrac{m(3r^2 + 4h^2)}{12}$
Tube	$V = \pi h (r_o^2 - r_i^2)$ $y_G = \dfrac{h}{2}$	$I_{xG} = \dfrac{m(3r_o^2 + 3r_i^2 + h^2)}{12}$ $I_{yG} = \dfrac{m(r_o^2 + r_i^2)}{2}$ $I_{zG} = \dfrac{m(3r_o^2 + 3r_i^2 + h^2)}{12}$ $I_{x'} = \dfrac{m(3r_o^2 + 3r_i^2 + 4h^2)}{12}$ $I_{z'} = \dfrac{m(3r_o^2 + 3r_i^2 + 4h^2)}{12}$

Figure	Volume and Center of Gravity Location	Mass Moment of Inertia
Disk		$I_{xG} = \dfrac{mr^2}{4}$ $I_{yG} = \dfrac{mr^2}{4}$ $\qquad$ $I_{z'} = \dfrac{3mr^2}{2}$ $I_{zG} = \dfrac{mr^2}{2}$
Thin Ring		$I_{xG} = \dfrac{mr^2}{2}$ $\qquad$ $I_{x'} = \dfrac{3mr^2}{2}$ $I_{yG} = \dfrac{mr^2}{2}$ $\qquad$ $I_{y'} = \dfrac{3mr^2}{2}$ $I_{zG} = mr^2$ $\qquad$ $I_{z'} = 3mr^2$
Thin Plate	$x_G = \dfrac{a}{2}$ $y_G = \dfrac{b}{2}$	$I_{xG} = \dfrac{mb^2}{12}$ $I_{yG} = \dfrac{ma^2}{12}$ $I_{zG} = \dfrac{m(a^2 + b^2)}{12}$
Block	$V = abc$ $x_G = \dfrac{a}{2}$ $y_G = \dfrac{b}{2}$ $z_G = \dfrac{c}{2}$	$I_{x'} = \dfrac{m(b^2 + c^2)}{3}$ $\qquad$ $I_{xG} = \dfrac{m(b^2 + c^2)}{12}$ $I_{y'} = \dfrac{m(a^2 + c^2)}{3}$ $\qquad$ $I_{yG} = \dfrac{m(a^2 + c^2)}{12}$ $I_{z'} = \dfrac{m(a^2 + b^2)}{3}$ $\qquad$ $I_{zG} = \dfrac{m(a^2 + b^2)}{12}$

Figure	Volume and Center of Gravity Location	Mass Moment of Inertia
Sphere	$V = \dfrac{4\pi r^3}{3}$	$I_{xG} = \dfrac{2mr^2}{5}$ $I_{yG} = \dfrac{2mr^2}{5}$ $I_{zG} = \dfrac{2mr^2}{5}$
Hemisphere	$V = \dfrac{2\pi r^3}{3}$ $z_G = \dfrac{3r}{8}$	$I_{xG} = \dfrac{83mr^2}{320}$ $I_{yG} = \dfrac{83mr^2}{320}$ $I_{zG} = \dfrac{2mr^2}{5}$
Cone	$V = \dfrac{\pi r^2 h}{3}$ $z_G = \dfrac{h}{4}$	$I_{xG} = \dfrac{3m(4r^2 + h^2)}{80}$ $I_{yG} = \dfrac{3m(4r^2 + h^2)}{80}$ $I_{zG} = \dfrac{3mr^2}{10}$

A.2) MASS MOMENTS OF INERTIA

In order to analyze the angular motion of a rigid body, a property that accounts for the distribution of the body's mass with respect to a given axis is needed. This property is called the body's *mass moment of inertia*. The **mass moment of inertia** (I) is a measure of a body's resistance to angular acceleration about a given axis. A body's mass moment is its rotational inertia. Understanding the mass moment of inertia of a rigid body is an important part of analyzing its motion. A quantity that uses the same variable (I) as its nomenclature, is the area moment of inertia. The *area moment of inertia* is not the same as the *mass moment of inertia*. The **area moment of inertia** is a measure of a body's resistance to bending about a given axis. The area moment of inertia is used extensively in analyzing the strength of materials.

A.2.1) MASS MOMENT OF INERTIA FOR A SYSTEM OF PARTICLES

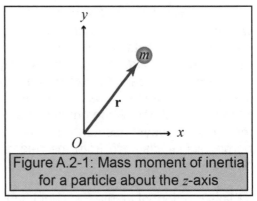

The mass moment of inertia for a single particle with respect to an axis perpendicular to its plane of motion is given by Equation A.2-1. A representative particle is shown in Figure A.2-1. Equation A.2-1 states that the moment of inertia is proportional to the particle's mass m, and the squared distance r^2, where r is the perpendicular distance from the reference axis through O to the particle.   Interpreting Equation A.2-1 shows that the further the particle is away from O or the more massive the particle is, the

Figure A.2-1: Mass moment of inertia for a particle about the z-axis

larger I will become. The larger I becomes, the harder it will be to rotate the particle about the axis passing through O.

Mass moment of inertia for a particle: $\boxed{I_O = mr^2}$ (A.2-1)

I_O = mass moment of inertia about the axis through O
r = perpendicular distance from the axis passing through O to the particle
m = mass of the particle

For a system of particles like that shown in Figure A.2-2, the overall mass moment of inertia with respect to an axis perpendicular to the plane of motion is simply the sum of the mass moments of the individual particles, as defined by Equation A.2-2. Some further intuition for where this equation comes from will be provided when we introduce impulse and momentum.

Figure A.2-2: Mass moment of inertia for a system of particle about the z-axis

Mass moment of inertia for a system of particles: $\boxed{I_O = \sum m_i r_i^2}$ (A.2-2)

I_O = mass moment of inertia about the axis through O
r_i = perpendicular distance from the axis passing through O to the ith particle
m_i = mass of the ith particle

A.2.2) MASS MOMENT OF INERTIA FOR A RIGID BODY

As we did with the center of mass, we can consider a rigid body as a collection of infinitesimally small particles as shown in Figure A.2-3. In this case, Equation A.2-2 can be expressed using integral notation as shown in Equation A.2-3. Each incremental piece of the rigid body has mass dm and perpendicular distance from the reference axis r_i.

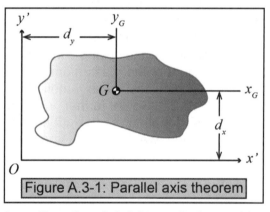

Figure A.2-3: Rigid body as a collection of infinitesimal particles

Mass moment of inertia for a rigid body:

$$I_O = \int r^2 dm \qquad \text{(A.2-3)}$$

I_O = mass moment of inertia about the axis through O
r = perpendicular distance of each incremental mass element dm from an axis through O

\A.3) PARALLEL-AXIS THEOREM

Mass moments of inertia for many simple rigid body shapes are given in Section A.1. Many of these inertias are given with respect to the axis that passes through the body's center of mass (G). In order to analyze rigid-body motion, we may need the mass moment of inertia of the body with respect to an arbitrary axis as shown in Figure A.3-1. The parallel-axis theorem (Equation A.3-1) may be used to calculate the mass moment of inertia of the body with respect to an axis parallel to the axis passing through the body's mass center. To illustrate how Equation A.3-1 is applied, consider the body shown in Figure A.3-1. If we want to know the mass moment of inertia about the x'-axis, but are only given the mass moment of inertia about the x_G axis (the axis through the body's center of mass) and the distance between the two axes, the equation would be $I_{x'} = I_{x_G} + md_x^2$.

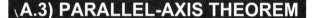

Figure A.3-1: Parallel axis theorem

Parallel-axis theorem: $\quad I' = I_G + md^2 \qquad$ (A.3-1)

I' = mass moment of inertia with respect to an axis parallel to the axis passing through G
I_G = mass moment of inertia with respect to an axis passing through G

m = mass of the body
d = the perpendicular distance from the parallel axis to mass center G

A.4) RADIUS OF GYRATION

It is often the case that a body's mass moment of inertia with respect to an axis is expressed in terms of its radius of gyration (k) with respect to that axis as shown in Equation A.4-1. This radius of gyration can be thought of as the weighted average of the distances from each point of the body to the reference axis.

Mass moment of inertia/radius of gyration relationship: $\boxed{I = mk^2}$ (A.4-1)

I = mass moment of inertia
k = radius of gyration
m = mass of the body

APPENDIX A REFERENCES

[1] www.efunda.com/math/solids/index solid.cfm

APPENDIX B: MATH FOR DYNAMICS

APPENDIX B OUTLINE

B.1) GEOMETRY

B.1.1) WHERE GEOMETRY IS USED IN DYNAMICS

Geometry is important to the study of dynamics is many ways. Geometry gives us a basis for communication, allows us to calculate lengths, angles, areas and volumes. Whether you realize it or not, without the subject of geometry, we wouldn't know how to communicate the concept of a Line or the fact the two objects move in a parallel manner. Geometry also give use the rules and formulas to calculate lengths and angles necessary to analyze rigid body systems. Finally, the ability to calculate areas and volumes becomes important when we start dealing with moments of inertia.

B.1.2) LINES

Point: A point is an exact location and has no size. The distance between two points in space may be calculated using Equation B.1-1.

Distance between two points in space: $\sqrt{(x_2^2 - x_1^2) + (y_2^2 - y_1^2) + (z_2^2 - z_1^2)}$ (B.1-1)

(x_1, y_1, z_1), (x_2, y_2, z_2) = Coordinates of the two points

Line: A line is a one dimensional figure that has infinite length but no width or height.

Line segment: A line segment is a line with finite length, see Figure B.1-1. The midpoint of a line segment may be calculated using Equation B.1-2.

Midpoint of a line segment:

$$\left(\left(\frac{x_1 + x_2}{2}\right), \left(\frac{y_1 + y_2}{2}\right), \left(\frac{z_1 + z_2}{2}\right)\right)$$ (B.1-2)

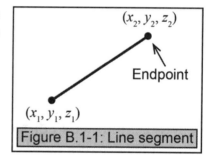

Endpoint

(x_1, y_1, z_1)

Figure B.1-1: Line segment

(x_1, y_1, z_1), (x_2, y_2, z_2) = Coordinate points of the line segment endpoints

Ray: A line that starts at a point and goes off in a particular direction to infinity.

Plane: A plane is a two dimension figure that has infinite length and width but no height.

Space: The collection of all points. It consists of an infinite number of planes.

B.1.3) ANGLES

Angle: Two rays that share the same endpoint form an angle. The point where the rays intersect is called the vertex of the angle. The two rays are called the sides of the angle. We measure the size of an angle using the units of *degrees* or *radians*. Various types of angles are shown in Figure B.1-2.

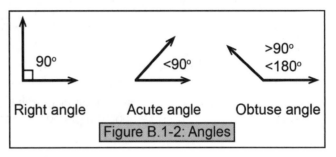

Figure B.1-2: Angles

- Right angle: An angle that is exactly 90° or $\pi/2$ rad.
- Acute angle: Any angle that is less than 90° or $\pi/2$ rad.
- Obtuse angle: Any angle that is larger than 90° or $\pi/2$ rad and less than 180° or π rad.
- Complementary angles: Two angles that add up to be 90° or $\pi/2$ rad.
- Supplementary angles: Two angles that add up to be 180° or π rad.
- Zero angle: An angle that is exactly 0°.
- Straight angle: An angle that is exactly 180°.

Vertical angles: Vertical angles are created when two lines intersect as shown in Figure B.1-3. Vertical angles have the same degree measurement.

Alternate Interior, alternate exterior, and corresponding angles: If two parallel lines are intersected by a third line several alternate interior (θ), alternate exterior (ϕ), and corresponding angles (α, β) are created as shown in Figure B.1-3. These alternate interior, alternate exterior, and corresponding angle pairs have the same degree measurement.

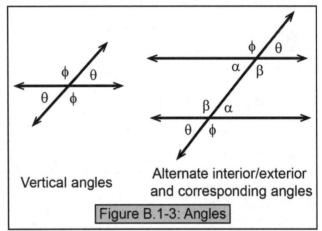

Vertical angles

Alternate interior/exterior and corresponding angles

Figure B.1-3: Angles

B.1.4) POLYGONS

Polygon: A polygon is made by joining lines segments together to create a closed figure. Each line segment only touches two other line segments at their endpoints.

Regular polygon: A regular polygon is a polygon whose sides are all the same length. The interior angles of a regular polygon are all equal. The sum of all the interior angles of a regular polygon with n sides, if $n > 2$, is equal to 180*(n - 2) degrees. Examples of regular polygons are a square and an equilateral triangle. The area of a regular polygon with n sides can be calculated using Equation B.1-3. The perimeter or distance around the outside of a regular polygon can be calculated using equation B.1-4.

Area of a regular polygon: $A = \dfrac{n}{4}a^2 \cot\left(\dfrac{\pi}{n}\right) = \dfrac{n}{2}r^2 \sin\left(\dfrac{2\pi}{n}\right) = nd^2 \tan\left(\dfrac{\pi}{n}\right)$ (B.1-3)

Perimeter of a regular polygon: $P = na$ (B.1-4)

n = number of sides
a = side length

r = radius (distance from the center to a vertex)
d = apothem (perpendicular distance from the center to one of the sides)

Triangle: A triangle is a three-sided polygon as shown in Figure B.1-4. The sum of its interior angles is equal to 180°. The area of a triangle may be calculated using Equation B.1-5.

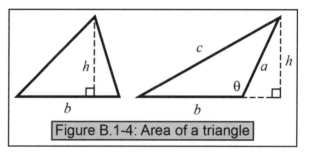

Figure B.1-4: Area of a triangle

Equilateral triangle: A triangle with the length of all sides being equal as shown in Figure B.1-5. Each interior angle equals 60°. The area of an equilateral triangle may be calculated using Equation B.1-6.

Right triangle: A triangle having one right angle as shown in Figure B.1-5. The side opposite the right angle is called the hypotenuse. The two sides that form the right angle are called the legs. The lengths of the sides of a right triangle are related by Pythagorean's Theorem given in Equation B.1-7.

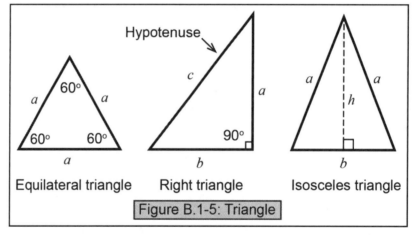

Equilateral triangle   Right triangle   Isosceles triangle

Figure B.1-5: Triangle

Isosceles triangle: A triangle with two sides of equal length as shown in Figure B.1-5. The base, side and height dimensions may be calculated by manipulating Pythagorean's Theorem as given in Equation B.1-8.

Area of a Triangle: $$A = \frac{1}{2}bh = \frac{1}{2}ab\sin(\theta) = \frac{1}{4}\sqrt{P(P-a)(P-b)(P-c)}$$ (B.1-5)

Area of an equilateral triangle: $$A = \frac{\sqrt{3}}{4}a^2$$ (B.1-6)

Pythagorean's theorem for a right triangle: $$a^2 + b^2 = c^2$$ (B.1-7)

Pythagorean's theorem for an isosceles triangle: $$a^2 = h^2 + \left(\frac{b}{2}\right)^2$$ (B.1-8)

a = side length $\qquad\qquad$ h = perpendicular height
b = base length $\qquad\qquad$ θ = angle enclosed by sides a and b
c = length of the hypotenuse $\qquad$ P = perimeter of the triangle = $a + b + c$

Scalene triangle: A triangle with all side having different lengths.

Acute triangle: A triangle having three acute interior angles.

Obtuse triangle: A triangle having one obtuse interior angle.

Quadrilaterals: A four sided polygon. The sum of the interior angles is equal to 360°.

Rectangle: A four-sided polygon where all of the interior angles are equal to 90° as shown in Figure B.1-6. The sum of the interior angles is equal to 360°. The area of a rectangle may be calculated using Equation B.1-9. The length of the diagonal may be calculated using Pythagorean's Theorem given in Equation B.1-10.

Square: A four-sided regular polygon where all of the interior angles are equal to 90°. A square is a rectangle with equal sides. The sum of the interior angles is equal to 360°.

Parallelogram: A four-sided polygon where two opposing sides are equal in length and parallel as shown in Figure B.1-6. This property of course makes the other two sides equal in length and parallel to each other. The sum of the interior angles is equal to 360°. The area of a parallelogram may be calculated using Equation B.1-11.

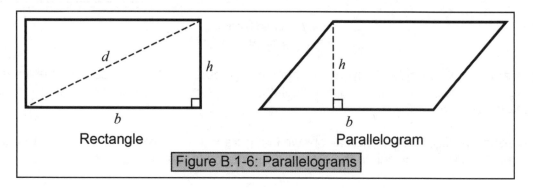

Figure B.1-6: Parallelograms

Rectangle area: $\boxed{A = bh}$ (B.1-9)   Rectangle diagonal: $\boxed{d^2 = b^2 + h^2}$ (B.1-10)

Parallelogram area: $\boxed{A = bh}$ (B.1-11)

b = base
h = height
d = diagonal

Rhombus: A four-sided polygon. A rhombus is a parallelogram that has all sides the same length as shown in Figure B.1-7. The sum of the interior angles is equal to 360°. To calculate the area of a rhombus, use Equation B.1-12.

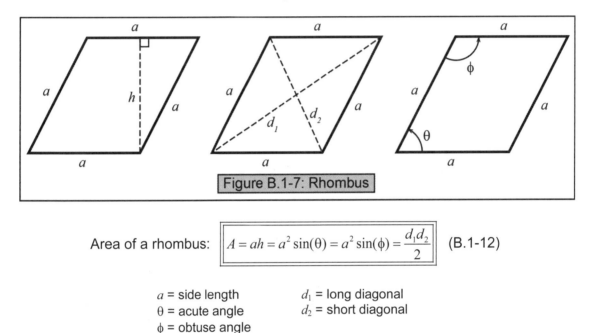

Figure B.1-7: Rhombus

Area of a rhombus: $\boxed{A = ah = a^2 \sin(\theta) = a^2 \sin(\phi) = \dfrac{d_1 d_2}{2}}$ (B.1-12)

a = side length d_1 = long diagonal
θ = acute angle d_2 = short diagonal
ϕ = obtuse angle

Trapezoid: A four-sided polygon where only two sides are parallel as shown in Figure B.1-8. The two sides that are parallel are called the bases of the trapezoid. The sum of the interior angles is equal to 360°. To calculate the area of a trapezoid, use Equation B.1-13.

Area of a trapezoid: $\boxed{\dfrac{1}{2}h(a + b)}$ (B.1-13)

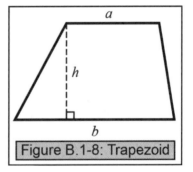

Figure B.1-8: Trapezoid

a, b = base lengths
h = perpendicular height

Pentagon: A five-sided polygon as shown in Figure B.1-9. The sum of the interior angles is equal to 540°. To calculate the area of a regular pentagon, use Equation B.1-14.

Hexagon: A six-sided polygon as shown in Figure B.1-9. The sum of the interior angles is equal to 720°. To calculate the area of a regular hexagon, use Equation B.1-15.

Heptagon: A seven-sided polygon as shown in Figure B.1-9. The sum of the interior angles is equal to 900°. To calculate the area of a regular heptagon, use Equation B.1-16.

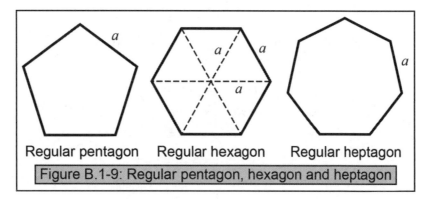

Figure B.1-9: Regular pentagon, hexagon and heptagon

Octagon: An eight-sided polygon as shown in Figure B.1-10. The sum of the interior angles is equal to 1080°. To calculate the area of a regular octagon, use Equation B.1-17.

Nonagon: A nine-sided polygon as shown in Figure B.1-10. The sum of the interior angles is equal to 1260°. To calculate the area of a regular nonagon, use Equation B.1-18.

Decagon: A ten-sided polygon as shown in Figure B.1-10. The sum of the interior angles is equal to 1440°. To calculate the area of a regular decagon, use Equation B.1-19.

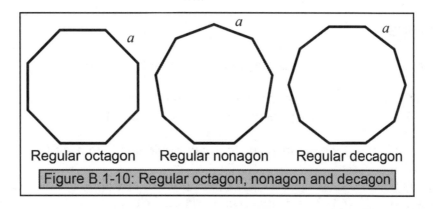

Figure B.1-10: Regular octagon, nonagon and decagon

Regular pentagon area: $$A = \frac{5}{4}a^2 \cot\left(\frac{\pi}{5}\right)$$ (B.1-14)

Regular hexagon area: $$A = \frac{3\sqrt{3}}{2}a^2$$ (B.1-15)

Regular heptagon area: $$A = \frac{7}{4}a^2 \cot\left(\frac{\pi}{7}\right)$$ (B.1-16)

Regular pentagon area: $$A = 2a^2 \cot\left(\frac{\pi}{8}\right)$$ (B.1-17)

Regular nonagon area: $$A = \frac{9}{4}a^2 \cot\left(\frac{\pi}{9}\right)$$ (B.1-18)

Regular decagon area: $$A = \frac{5}{2}a^2 \cot\left(\frac{\pi}{10}\right)$$ (B.1-19)

A = area
a = side length

B.1.5) CONICS

Conic: Shapes that can be created by a plane intersecting a cone.

Circle: A set of points in a plane that are equidistant from a common point as shown in Figure B.1-11. Various circle parameters may be calculated using Equations B.1-20 through B.1-25.
- Center point: A common point that lies equidistant from all the points on the circle.
- Radius (r): The distance from the center point to any point on the circle.
- Circumference (C): The distance around the circle.
- Arc length (S): A distance along the curved line whose beginning and end is defined by a central angle (θ).
- Sector area (A_s): A pie shaped area within the circle defined by a central angle (θ).
- Chord Length (l): The line that links the beginning and end of an arc along the circle.
- Segment area (A_1): The area between an associated arc and chord.

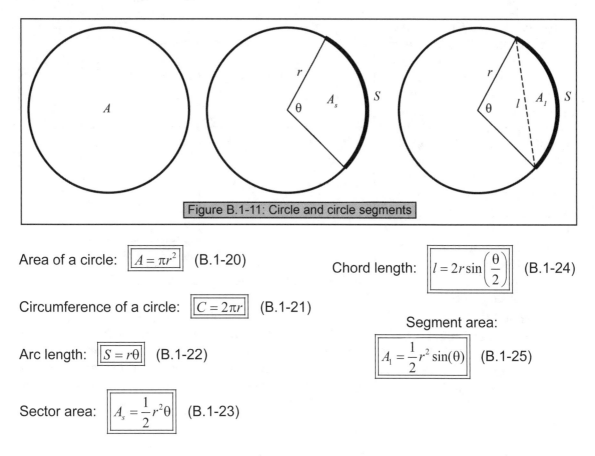

Figure B.1-11: Circle and circle segments

Area of a circle: $A = \pi r^2$ (B.1-20)

Circumference of a circle: $C = 2\pi r$ (B.1-21)

Arc length: $S = r\theta$ (B.1-22)

Sector area: $A_s = \dfrac{1}{2}r^2\theta$ (B.1-23)

Chord length: $l = 2r\sin\left(\dfrac{\theta}{2}\right)$ (B.1-24)

Segment area:

$A_1 = \dfrac{1}{2}r^2\sin(\theta)$ (B.1-25)

r = radius of the circle
θ = central angle in radians

Equation of a circle: relationships for a circle in the x-y coordinate system are given by Equation B.1-26 through B.1-28 and illustrated in Figure B.1-12.

x-coordinate: $x = r\cos\theta + x_C$ (B.1-26)

y-coordinate: $y = r\sin\theta + y_C$ (B.1-27)

$(x - x_C)^2 + (y - y_C)^2 = r^2$ (B.1-28)

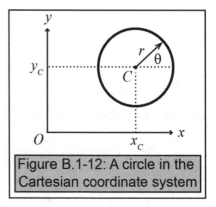

Figure B.1-12: A circle in the Cartesian coordinate system

r = radius of the circle
C = circle center

x_C, y_C = x- and y-location of C
θ = angle between a line drawn from C to a location on the circle and the x-axis

Ellipse: A set of points in a plane whose distance from two points in a plane have a constant sum as shown in Figure B.1-13. The area of an ellipse may be calculated using Equation B.1-29. The equation for an ellipse where the minor and major axes correspond to the Cartesian coordinate axes with the center at the origin is given by Equation B.1-30.

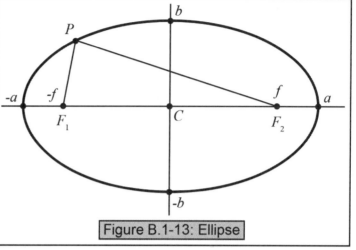

Figure B.1-13: Ellipse

- **Major and minor axes:** The major and minor axes of the ellipse are the lines that cut the ellipse in half. The major axis contains the two foci (F_1, F_2) and is longer than the minor axis.
- **Major diameter ($2a$):** The length of the major axis interior to the ellipse.
- **Minor diameter ($2b$):** The length of the minor axis interior to the ellipse.
- **Center point (C):** A point at the intersection of the major and minor axes.
- **Foci (F_1, F_2):** Two special points that lie on the major axis and are equidistance from the center point. The sum of the distances between the two foci and any point on the ellipse is equal to the major diameter ($PF_1 + PF_2 = 2a$).

Area of an ellipse: $A = \pi ab$ (B.1-29)

$$\frac{x}{a^2} + \frac{y}{b^2} = 1$$ (B.1-30)

a = major radius
b = minor radius

B.2) ALGEBRA

B.2.1) WHERE ALGEBRA IS USED IN DYNAMICS

Ideal dynamics problems should be solved in variable form. There are two advantages to variable solutions. First, it increases the accuracy of the final answer. Second, it leads to generality. This means that the solution may be used to get numerical solutions for many situations and not just for the problem at hand. Solving problems using variables leads us into equation manipulation and the subject of algebra.

B.2.2) EQUALITIES

Many times we deal with equations that have equal signs. I am not going to spend time listing all the mathematical properties and theorems. Just remember the following.

What is done to one side of the equation has to be done to the other side of the equation to maintain equality.

This means that if you add a number to one side of the equation, you must add the same number to the other side of the equation. If you multiply a number to one side of the equation, you must multiply every term of the other side of the equation by that same number.

Solved Problem B.2-1

The following are examples where the equation $y = x + z$ is manipulated by the number a and still maintains equality.

Addition: $y + a = x + z + a$

Multiplication: $ay = ax + az$

Division: $\dfrac{y}{a} = \dfrac{x}{a} + \dfrac{z}{a}$

Power: $y^a = (x + z)^a$

B.2.3) INEQUALITIES

Dealing with an equation that has an inequality in it is very similar to dealing with an equation that has an equality in it. The guiding rule is still the same. What you do to one side of the equation, you must do to the other side of the equation in order to maintain the direction of the inequality. However, the difference comes in when you multiply or divide by a negative number. **If you multiply or divide by a negative number you must reverse the inequality to maintain the equations original intent.** It is as if you subtracted every term to the other side of the inequality.

Solved Problem B.2-2

Solve the equations $y^2 + x > 5 + 2x$ and $-3y + x^2 \le 8 + y$ for y in terms of x.

Solution:

$y^2 + x > 5 + 2x$

$y^2 + x - x > 5 + 2x - x$

$y^2 > 5 + x$

$\sqrt{y^2} > \sqrt{5 + x}$

$y > \sqrt{5 + x}$

$-3y + x^2 \le 8 + y$

$-3y - y + x^2 - x^2 \le 8 + y - y + x^2$

$-4y \le 8 + x^2$

$\dfrac{-4y}{-4} \ge \dfrac{8 + x^2}{-4}$

$y \ge \dfrac{8 + x^2}{-4}$

B.2.4) SOLVING POLYNOMIAL EQUATIONS

A polynomial function consists of a linear combination of a variable and its powers with exponents of non-negative integers. An example is shown in Equation B.2-1. The solutions to a polynomial equation are called the roots of the equation. If the order of the polynomial equation is two, the equation is easy to solve. A second-order polynomial equation is called a quadratic equation, shown in Equation B.2-2, and it can be solved using the quadratic formula given in Equation B.2-3. If the order is higher than two, then solving the equation becomes more complex. Methods of solution include factoring out roots, graphing the equation and numerical methods. Fortunately, many calculator and computer programs solve higher-order polynomial equations.

Polynomial equation: $$a_n x^n + a_{n-1} x^{n-1} + a_{n-2} x^{n-2} + ... + a_2 x^2 + a_1 x + a_0 = 0 \qquad \text{(B.2-1)}$$

Quadratic equation: $$ax^2 + bx + c = 0 \qquad \text{(B.2-2)}$$

Quadratic formula: $$x = \frac{-b \pm \sqrt{b^2 - 4ac}}{2a} \qquad \text{(B.2-3)}$$

n = the degree of the polynomial
a, b, c = constants
x = variable

B.2.5) SOLVING NONLINEAR EQUATIONS

B.2.5.1) Natural log and the exponential function (e)

The **natural logarithm** is the logarithm to the base e. e is approximately the number 2.718281828. The **exponential function** is the function e^x. The function is sometimes written as $\exp(x)$, especially when the exponent is complex. The natural logarithm is generally written as $\ln(x)$ or $\log_e(x)$. The natural logarithm of a number x is the power to which e would have to be raised to equal x. The natural log of e itself ($\ln(e)$) is 1 because $e^1 = e$, while the natural logarithm of 1 ($\ln(1)$) is 0, since $e^0 = 1$. The natural logarithm can be defined for any positive real number a as the area under the curve $y = 1/x$ from 1 to a. The natural logarithm function ($\ln(x)$), if considered as a real-valued function of a real variable, is the inverse function of the exponential function (e^x), leading to the identities given in Equations B.2-4 and B.2-5.

$$e^{\ln(x)} = x \qquad \text{if } x > 0 \qquad \text{(B.2-4)}$$

$$\ln(e^x) = x \qquad \text{if } x > 0 \qquad \text{(B.2-5)}$$

Like all logarithms, the natural logarithm maps multiplication into addition and division into subtraction as shown in Equations B.2-6 through B.2-9.

$$\ln(xy) = \ln(x) + \ln(y) \qquad \text{(B.2-6)}$$

$$\exp(x + y) = \exp(x)\exp(y) \qquad \text{(B.2-8)}$$

$$\ln\left(\frac{x}{y}\right) = \ln(x) - \ln(y) \qquad \text{(B.2-7)}$$

$$\exp(x - y) = \frac{\exp(x)}{\exp(y)} \qquad \text{(B.2-9)}$$

Solved Problem B.2-3

Solve the equations $ae^x = b$ and $a\ln(x) = b$.

Solution:

$ae^x = b$

$\ln(ae^x) = \ln(b)$

$\ln(a) + \ln(e^x) = \ln(b)$

$x = \ln(b) - \ln(a)$

$x = \ln\left(\frac{b}{a}\right)$

$a\ln(x) = b$

$\ln(x) = \frac{b}{a}$

$\exp(\ln(x)) = \exp\left(\frac{b}{a}\right)$

$x = e^{b/a}$

B.3) TRIGONOMETRY

B.3.1) LAW OF SINES AND COSINES

Referring to the Figure B.3-1, the law of sines and cosines are given by Equation B.3-1 and B.3-2 respectively. These relations are especially useful for finding unknown angles and side lengths for triangles that don't have a right angle.

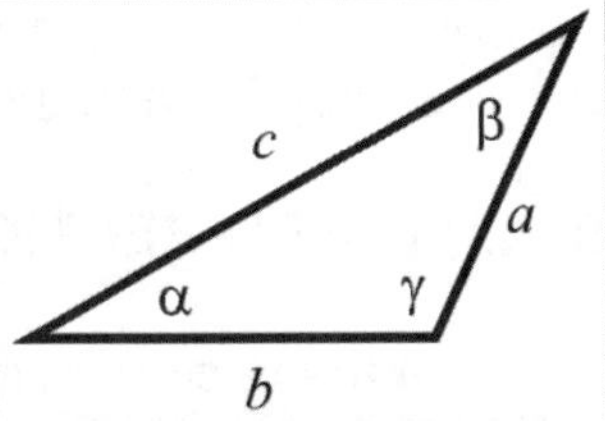

Law of sines: $$\frac{a}{\sin\alpha} = \frac{b}{\sin\beta} = \frac{c}{\sin\gamma} \qquad \text{(B.3-1)}$$

Figure B.3-1: Triangle

Law of cosines: $$b^2 = a^2 + c^2 - 2ac\cos\beta \qquad \text{(B.3-2)}$$

B.3.2) TRIGONOMETRIC IDENTITIES

The Table B.3-1 gives a list of useful trigonometric identities.

$\sin\theta = \cos(90^\circ - \theta)$	$\cos\theta = \sin(90^\circ - \theta)$
$\tan\theta = \dfrac{\sin\theta}{\cos\theta}$	$\sin^2\theta + \cos^2\theta = 1$

$\sin^2\theta = \dfrac{1-\cos(2\theta)}{2}$	$\cos^2\theta = \dfrac{1+\cos(2\theta)}{2}$	$\tan^2\theta = \dfrac{1-\cos(2\theta)}{1+\cos(2\theta)}$

$\sin(\theta+\phi) = \sin\theta\cos\phi + \cos\theta\sin\phi$	$\cos(\theta+\phi) = \cos\theta\cos\phi - \sin\theta\sin\phi$
$\sin(\theta-\phi) = \sin\theta\cos\phi - \cos\theta\sin\phi$	$\cos(\theta-\phi) = \cos\theta\cos\phi + \sin\theta\sin\phi$
$\tan(\theta+\phi) = \dfrac{\tan\theta + \tan\phi}{1 - \tan\theta\tan\phi}$	$\tan(\theta-\phi) = \dfrac{\tan\theta - \tan\phi}{1 + \tan\theta\tan\phi}$
$\sin\theta + \sin\phi = 2\sin\dfrac{1}{2}(\theta+\phi)\cos\dfrac{1}{2}(\theta-\phi)$	$\sin\theta - \sin\phi = 2\cos\dfrac{1}{2}(\theta+\phi)\sin\dfrac{1}{2}(\theta-\phi)$
$\cos\theta + \cos\phi = 2\cos\dfrac{1}{2}(\theta+\phi)\cos\dfrac{1}{2}(\theta-\phi)$	$\cos\theta - \cos\phi = -2\sin\dfrac{1}{2}(\theta+\phi)\sin\dfrac{1}{2}(\theta-\phi)$

Table B.3-1: Trigonometric identities

B.4) VECTORS

A vector is used to represent a physical quantity that has both a direction and magnitude. A vector is generally represented by the symbols $\underline{U}, \vec{U}, \bar{U},$ or $\mathbf{U}$. On a diagram, a vector is drawn as an arrow. The direction of the arrow represents the vector's direction and the length of the arrow represents the vecotr's magnitude.

B.4.1) UNIT DIRECTION VECTOR

If the vector resides in the x-y-z coordinate system, then the unit direction vectors **i**, **j**, and **k** are used to indicate the vector components in the x-, y- and z-directions respectively. Unit direction vectors have a magnitude of one and, therefore, do not increase the size of any scalar that they are assigning direction to.

What is the magnitude of a unit direction vector?

If I multiply a scalar with a unit direction vector, will the magnitude of the resulting vector be *less than*, *equal to*, or *greater than* the original scalar?

B.4.2) VECTOR ALGEBRA

B.4.2.1) Vector operations and laws

If $\mathbf{P}$ and $\mathbf{Q}$ are vectors given by $\mathbf{P} = P_x\mathbf{i} + P_y\mathbf{j} + P_z\mathbf{k}$, $\mathbf{Q} = Q_x\mathbf{i} + Q_y\mathbf{j} + Q_z\mathbf{k}$, then you may perform the mathematical operations and employ the laws given in Table B.4-1.

Addition $\mathbf{P} + \mathbf{Q} = (P_x + Q_x)\mathbf{i} + (P_y + Q_y)\mathbf{j} + (P_z + Q_z)\mathbf{k}$	Subtraction $\mathbf{P} - \mathbf{Q} = (P_x - Q_x)\mathbf{i} + (P_y - Q_y)\mathbf{j} + (P_z - Q_z)\mathbf{k}$
Multiplication by a scalar $a\mathbf{P} = \mathbf{P}a = aP_x\mathbf{i} + aP_y\mathbf{j} + aP_z\mathbf{k}$	Commutative $\mathbf{P} + \mathbf{Q} = \mathbf{Q} + \mathbf{P}$
Distributive $a(b\mathbf{P}) = (ab)\mathbf{P},$ $(a+b)\mathbf{P} = a\mathbf{P} + b\mathbf{P},$ $a(\mathbf{P} + \mathbf{Q}) = a\mathbf{P} + a\mathbf{Q}$	Associative $(\mathbf{P} + \mathbf{Q}) + \mathbf{V} = \mathbf{Q} + (\mathbf{P} + \mathbf{V})$

Table B.4-1: Vector operations and laws

B.4.2.2) Magnitude

Consider the vector $\mathbf{P} = P_x\mathbf{i} + P_y\mathbf{j} + P_z\mathbf{k}$. The magnitude of $\mathbf{P}$ is given by Equation B.4-1

Vector magnitude: $$P = \sqrt{P_x^2 + P_y^2 + P_z^2}$$ (B.4-1)

B.4.2.2) Dot product

Consider the vectors $\mathbf{P} = P_x\mathbf{i} + P_y\mathbf{j} + P_z\mathbf{k}$ and $\mathbf{Q} = Q_x\mathbf{i} + Q_y\mathbf{j} + Q_z\mathbf{k}$, the dot product is given by Equation B.4-2. The dot product satisfies the properties given in Table B.4-2.

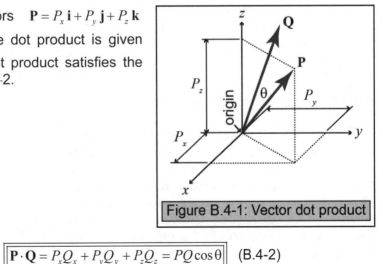

Figure B.4-1: Vector dot product

Dot product: $$\mathbf{P} \cdot \mathbf{Q} = P_x Q_x + P_y Q_y + P_z Q_z = PQ\cos\theta$$ (B.4-2)

Commutative	Associative
$\mathbf{P} \cdot \mathbf{Q} = \mathbf{Q} \cdot \mathbf{P}$	$a(\mathbf{P} \cdot \mathbf{Q}) = (a\mathbf{P}) \cdot \mathbf{Q} = \mathbf{P} \cdot (a\mathbf{Q})$
Distributive	
$\mathbf{P} \cdot (\mathbf{Q} + \mathbf{V}) = \mathbf{P} \cdot \mathbf{Q} + \mathbf{P} \cdot \mathbf{V}$	

Table B.4-2: Dot product properties

B.4.3) CROSS PRODUCT

Rigid bodies have size and, therefore, if a force is applied in a location other than the center of mass, the force will create a moment that rotates the body. The moment created by a force about an arbitrary point P is given by $\mathbf{M}_P = \mathbf{r} \times \mathbf{F}$. Therefore, being able to perform cross products is important in the analysis of rigid-body kinetics. Cross products also help us find the directions of the velocity and acceleration of point on a rigid body in these rigid body kinematics equations $\mathbf{v} = \boldsymbol{\omega} \times \mathbf{r}$ and $\mathbf{a} = \boldsymbol{\alpha} \times \mathbf{r} + \boldsymbol{\omega} \times (\boldsymbol{\omega} \times \mathbf{r})$.

B.4.3.1) Calculating the cross product

Consider the cross product between vectors $\mathbf{P} = P_x \mathbf{i} + P_y \mathbf{j} + P_z \mathbf{k}$ and $\mathbf{Q} = Q_x \mathbf{i} + Q_y \mathbf{j} + Q_z \mathbf{k}$, the cross product $\mathbf{P} \times \mathbf{Q}$ is calculated using Equation B.4-3.

Cross product:

$$\mathbf{P} \times \mathbf{Q} = \begin{vmatrix} \mathbf{i} & \mathbf{j} & \mathbf{k} \\ P_x & P_y & P_z \\ Q_x & Q_y & Q_z \end{vmatrix} = \mathbf{i}(P_y Q_z - Q_y P_z) - \mathbf{j}(P_x Q_z - Q_x P_z) + \mathbf{k}(P_x Q_y - Q_x P_y) \quad \text{(B.4-3)}$$

Answer the following questions.

The result of a dot product $\mathbf{P} \cdot \mathbf{Q}$ is a (vector, scalar).

The result of a dot product $\mathbf{P} \times \mathbf{Q}$ is a (vector, scalar).

True or False? $\mathbf{P} \cdot \mathbf{Q} = \mathbf{Q} \cdot \mathbf{P}$

True or False? $\mathbf{P} \times \mathbf{Q} = \mathbf{Q} \times \mathbf{P}$

B.4.3.2) The direction of a cross product

Unlike the dot product which produces a scalar, the cross product between two vectors results in a vector that is perpendicular to both the original vectors as shown in Figure B.4-2. The calculation of the cross product of two vectors was discussed previously. More generally, we can compute the cross product by the results for the cross product of the unit vectors $\mathbf{i}$, $\mathbf{j}$, and $\mathbf{k}$. The direction of the vector resulting from a cross product also may be determined using the right hand rule*.

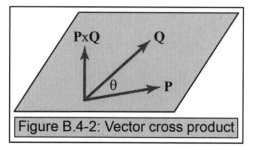

Figure B.4-2: Vector cross product

Use the right hand rule to find the direction of the following cross products.

i x j =

j x i =

k x i =

k x j =

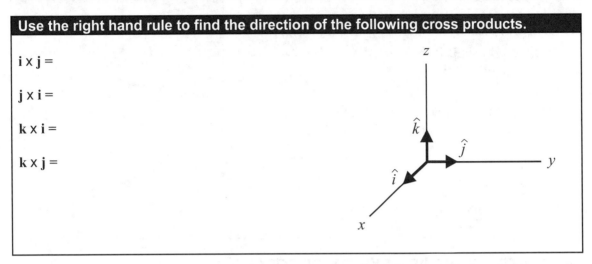

Another helpful tool is the diagram shown in Figure B.4-3. The implication of this diagram is that following the direction of the arrows, the product of two successive elements is the third unit vector. For example, $j \times k = i$. If the cross product of the two elements is being taken in the reverse order indicated by the arrows, then the result is the negative of the third element. For example, $k \times j = -i$.

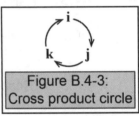

Figure B.4-3:
Cross product circle

B.4.4) RESOLVING A 2-D FORCE INTO COMPONENTS

For two dimensions, we can use Equations B.4-4 and B.4-5 to resolve a vector $\mathbf{F} = F_x\mathbf{i} + F_y\mathbf{j}$ into vector components F_x and F_y, which are parallel to the x- and y-axis.

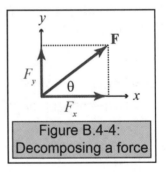

Figure B.4-4:
Decomposing a force

x-direction force component: $\boxed{F_x = F\cos\theta}$ (B.4-4)

y-direction force component: $\boxed{F_y = F\sin\theta}$ (B.4-5)

Solved Problem B.4-1

What are the x- and y-components of F, if its magnitude is 50 N and $\theta = 35°$?

$$\mathbf{F} = F(\cos\theta\,\mathbf{i} + \sin\theta\,\mathbf{j})$$
$$= 50(\cos 35\,\mathbf{i} + \sin 35\,\mathbf{j}) \quad \text{N}$$
$$= 40.96\,\mathbf{i} + 28.68\,\mathbf{j} \quad \text{N}$$

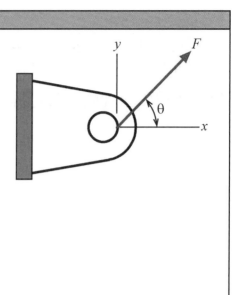

What are the x- and y-components of F, if its magnitude is 50 N and $\theta = -35°$?

$$\mathbf{F} = F(\cos\theta\,\mathbf{i} + \sin\theta\,\mathbf{j})$$
$$= 50(\cos(-35)\,\mathbf{i} + \sin(-35)\,\mathbf{j}) \quad \text{N}$$
$$= 40.96\,\mathbf{i} - 28.68\,\mathbf{j} \quad \text{N}$$

B.4.5) RESULTANT OF A SYSTEM OF FORCES

The resultant of a system of n forces is given by Equation B.4-6.

Resultant force:
$$\mathbf{F} = \sum_{i=1}^{n} F_{xi}\,\mathbf{i} + \sum_{i=1}^{n} F_{yi}\,\mathbf{j} + \sum_{i=1}^{n} F_{zi}\,\mathbf{k} \qquad \text{(B.4-6)}$$

Solved Problem B.4-2

What is the resultant force of the following system of forces?

$A = 50\mathbf{i}$
$B = 200(\cos 45\,\mathbf{i} + \sin 45\,\mathbf{j})$
$C = 100(-\cos 30\,\mathbf{i} + \sin 30\,\mathbf{j})$

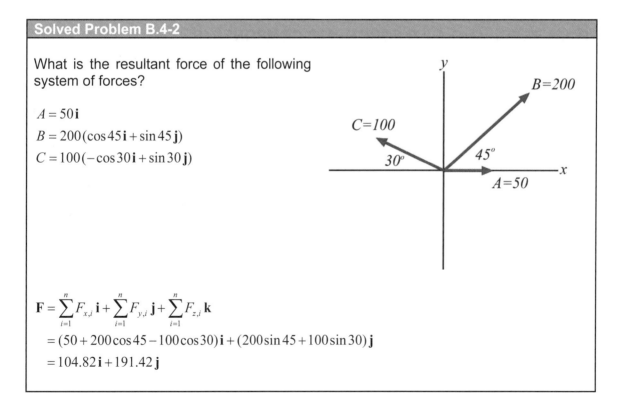

$$\mathbf{F} = \sum_{i=1}^{n} F_{x,i}\,\mathbf{i} + \sum_{i=1}^{n} F_{y,i}\,\mathbf{j} + \sum_{i=1}^{n} F_{z,i}\,\mathbf{k}$$
$$= (50 + 200\cos 45 - 100\cos 30)\,\mathbf{i} + (200\sin 45 + 100\sin 30)\,\mathbf{j}$$
$$= 104.82\,\mathbf{i} + 191.42\,\mathbf{j}$$

B.4.6) PROJECTION OF A FORCE ON A LINE

The projection of **F** onto line AB is given by the dot product of **F** with the unit vector in the direction of line AB is given by Equation B.4-7.

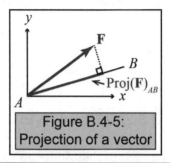

Figure B.4-5:
Projection of a vector

Projection of a vector: $\boxed{\text{Proj}(\mathbf{F})_{AB} = \mathbf{F} \cdot \mathbf{u}_{AB}}$ (B.4-7)

Solved Example B.4-3

A plane contains the vectors **A** and **B**. Determine the projection of **A** onto **B**.

$$\mathbf{A} = 4\mathbf{i} - 2\mathbf{j} + 3\mathbf{k} \qquad \mathbf{B} = -2\mathbf{i} + 6\mathbf{j} - 5\mathbf{k}$$

$$\text{Proj}(\mathbf{A})_B = \mathbf{A} \cdot \mathbf{u}_B = (4\mathbf{i} - 2\mathbf{j} + 3\mathbf{k}) \cdot \left(\frac{-2\mathbf{i} + 6\mathbf{j} - 5\mathbf{k}}{\sqrt{2^2 + 6^2 + 5^2}} \right)$$

$$= (4\mathbf{i} - 2\mathbf{j} + 3\mathbf{k}) \cdot (4\mathbf{i} - 2\mathbf{j} + 3\mathbf{k})$$

$$= -0.248\mathbf{i} + 0.744\mathbf{j} - 0.62\mathbf{k}$$

B.5) CALCULUS - DIFFERENTIATION

Calculus can be used to help solve most engineering problems, particularly those that have rate of change concepts. Another common use is to find the maximum or minimum values of a function.

B.5.1) THE SLOPE OF A LINE AND A FUNCTION

B.5.1.1) Slope of a line

The slope (m) of the line shown in Figure B.5-1 is given by Equation B.5-1 and the equation of the line is given by Equation B.5-2.

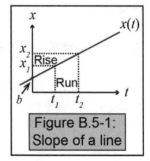

Figure B.5-1:
Slope of a line

Slope: $\boxed{m = \dfrac{\text{rise}}{\text{run}} = \dfrac{(x_2 - x_1)}{(t_2 - t_1)}}$ (B.5-1)

Equation of a line: $\boxed{x = mt + b}$ (B.5-2)

m = slope of the line
b = y-intercept
t = time

B.5.1-2) Slope of a function

Consider a particle who's speed is given by the function $v(t)$ shown in Figure B.5-2. How do we determine the slope of the tangent line at Point 1? We take the derivative of $v(t)$ and evaluate it at t_1.

B.5.2) DERIVATIVES

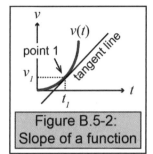

Figure B.5-2:
Slope of a function

Taking derivatives of continuous functions is an essential part of dynamics. For example, if you know a particle's velocity as a function of time ($v(t)$), you can determine its acceleration by taking the derivative of $v(t)$ with respect to time. The position (x), velocity (v), and acceleration (a) of a particle are related through the following equations.

$$v = \frac{dx}{dt} = \dot{x} \qquad\qquad a = \frac{dv}{dt} = \dot{v} = \ddot{x}$$

Solved Problem B.5-1

A particle's displacement is given by $x(t) = x_o + c_1 t + c_2 t^3$, where x_o is the particle's initial displacement at time $t_o = 0$, and c_1 and c_2 are constants. Find the particle's acceleration.

Solution:

Notice that $x(t)$ is a function of time. To get the acceleration, we need to differentiate $x(t)$ with respect to time to get the velocity $v(t)$. Then we differentiate $v(t)$ to get the acceleration $a(t)$.

$$v = \frac{dx}{dt} = c_1 + 3c_2 t^2 \qquad\qquad a = \frac{dv}{dt} = 6c_2 t$$

A particle's position, velocity and acceleration may come in a variety of forms. Therefore, it is necessary to know how to differentiate many different types of functions.

B.5.3) COMMON DIFFERENTIATION RULES

Let $x(t)$, $f(t)$ and $g(t)$ be functions of time and $\dot{x}$, $\dot{f}$, and $\dot{g}$ be their respective time derivatives. A list of common differentiation rules are given in Table B.5-1, where n, a and b are constants.

Rule 1) $x = a$ $\Rightarrow$ $\dot{x} = 0$		Rule 2) $x = at$ $\Rightarrow$ $\dot{x} = a$	
Rule 3) $x = at^n$ $\Rightarrow$ $\dot{x} = nat^{n-1}$		Rule 4) $x = \dfrac{1}{t^n} = t^{-n}$ $\Rightarrow$ $\dot{x} = -nt^{-n-1} = \dfrac{-n}{t^{n+1}}$	
Rule 5) $x = f + g$ $\Rightarrow$ $\dot{x} = \dot{f} + \dot{g}$		Rule 6) $x = \sin t$ $\Rightarrow$ $\dot{x} = \cos t$	
Rule 7) $x = \cos t$ $\Rightarrow$ $\dot{x} = -\sin t$		Rule 8) $x = \ln(t)$ $\Rightarrow$ $\dot{x} = \dfrac{1}{t}$	
Rule 9) $x = e^{at}$ $\Rightarrow$ $\dot{x} = ae^{at}$		Rule 10) <u>The product rule:</u> $x = fg \Rightarrow \dot{x} = \dot{f}g + \dot{g}f$ Memory phrase: The derivative of the first times the second plus the derivative of the second times the first.	
Rule 11) <u>Composite functions – the chain rule:</u> $u = g(t)$ $x = f(u)$ $\Rightarrow$ $\dfrac{dx}{dt} = \dfrac{dx}{du}\dfrac{du}{dt}$ Memory phrase: The derivative of the outside times the derivative of the inside.		Rule 12) <u>The quotient rule:</u> $x = \dfrac{f}{g} = fg^{-1} \Rightarrow$ $\dot{x} = \dot{f}g^{-1} + (-1g^{-2}\dot{g})f = \dfrac{\dot{f}g - \dot{g}f}{g^2}$	

Table B.5-1: Common derivatives

Solved Problem B.5-2

Common derivatives

Rule 2: $x = 2t$, $\dot{x} = 2$

Rule 3: $x = t^4$, $\dot{x} = 4t^3$

Rule 4: $x = \dfrac{1}{t^2}$, $\dot{x} = \dfrac{-2}{t^3}$

Rule 5: $x = at + bt^3$, $\dot{x} = a + 3bt^2$

Rule 10: $x = (1 + t^2)(2 - t^4)$, $\dot{x} = 2t(2 - t^4) - 4t^3(1 + t^2)$

Rule 11: $x = \left(f(t)\right)^n$, $\dot{x} = n\left(f(t)\right)^{n-1}\dot{f}$

Rule 11: $x = (2t + t^3)^5$,   **Let** $u = 2t + t^3$ then $x = u^5$ $\dot{x} = 5u^4\dot{u} = 5(2t + t^3)^4(2 + 3t^2)$

Rule 12: $x = \sin\left(\theta(t)\right)$, $\dot{x} = \dot{\theta}\cos\theta$

Rule 12: $x = \cos^3\left(\theta(t)\right)$, $\dot{x} = 3\cos^2\theta(-\sin\theta)\dot{\theta}$

B.5.4) LOCAL MAXIMA AND MINIMA

Many times in dynamics we would like to maximize or minimize the displacement, velocity or acceleration. Consider the velocity curve shown in Figure B.5-3. Point 1 is a local maximum and point 2 is a local minimum. The slope or derivative at both of these points is zero. This means that the tangent line is horizontal. To determine at what t_1 and t_2 these maximum and minimum points occur, we need to: take the time derivative of $v(t)$, set the derivative (the slope) equal to zero, and solve for t as shown in Equation B.5-3.

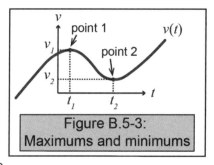

Figure B.5-3:
Maximums and minimums

Finding maxima and minima: $\boxed{\dfrac{dv(t)}{dt} = 0}$ (B.5-3)

Solved Problem B.5-4

What are the maximum and minimum velocities that occur between 0 and 3 seconds?

$$v = \frac{t^3}{3} - \frac{5t^2}{2} + 6t + 3$$

Solution:

$$\dot{v} = a = t^2 - 5t + 6 = 0 \qquad t = 2 \text{ or } 3$$

A local maximum occurs at t = 2 and v = 7.6.
A local minimum occurs at t = 3 and v = 7.5.

B.6) CALCULUS - INTEGRATION

Integration is the inverse operation of differentiation. There are two types of integrals: definite integrals, which are restricted to a specific range, and indefinite integrals, which have an unrestricted range. The solution to any indefinite integral will contain a constant (usually denoted by C). This constant cancels out in a definite integral.

Integrals are important in dynamics if you need to calculate the velocity of an object given the acceleration, or you need to calculate the position given the velocity. The position (x), velocity (v), and acceleration (a) of a particle are related through the following equations.

$$v = \frac{dx}{dt} = \dot{x} \qquad\qquad a = \frac{dv}{dt} = \dot{v} = \ddot{x}$$

These equations may be rearranged to give the integral form.

$$\int v\,dt = \int dx \qquad\qquad \int a\,dt = \int dv$$

Solved Problem B.6-1

A particle's acceleration in m/s^2 is given by $a = 2t^3 - 5$. What is its velocity at $t = 2$ seconds if it starts from rest at $t = 0$?

Solution:

$$\int a\,dt = \int dv$$

$$\int_0^t (2t^3 - 5)\,dt = \int_0^v dv$$

$$2\frac{t^4}{4} - 5t = v$$

$$v_{t=2} = 6 \ \text{m/s}$$

B.6.1) BASIC INTEGRALS

A list of basic integrals are given in Table B.5-2.

Rule 1) $\int a\,f(x)\,dx = a\int f(x)\,dx$	Rule 2) $\int \big(f(x)+g(x)\big)\,dx = \int f(x)\,dx + \int g(x)\,dx$
Rule 3) $\int dx = x + C$	Rule 4) $\int x^n\,dx = \dfrac{x^{n+1}}{n+1} + C \qquad n \neq -1$
Rule 5) $\int \dfrac{1}{x}\,dx = \ln|x| + C$	Rule 6) $\int e^{ax}\,dx = \dfrac{1}{a}e^{ax} + C$
Rule 7) $\int \sin(ax)\,dx = -\dfrac{1}{a}\cos(ax) + C$	Rule 8) $\int \cos(ax)\,dx = \dfrac{1}{a}\sin(ax) + C$

Table B.5-2: Basic integrals

Solved Problem B.6-2

Basic Integrals

Rule 4: $\int (2x^4 - 3x + 2)\,dx = \dfrac{2x^5}{5} - \dfrac{3x^2}{2} + 2x + C$

Rule 8: $\int (3\cos(x) - 4\sin(2x))\,dx = 3\sin(x) + 2\cos(2x) + C$

B.6.2) INTEGRATION TECHNIQUES

B.6.2-1) Integration by substitution

This is a method that changes the form of the integrand to, hopefully, a simpler form.

$$\int f(g(t))\,dt \qquad \text{let} \qquad u = g(t) \qquad \text{then} \qquad du = \dot{g}\,dt \qquad \text{and} \qquad dt = \frac{du}{\dot{g}}$$

$$\int f(g(t))\,dt = \int \frac{f(u)}{\dot{g}}\,du$$

Solved Problem B.6-3

Integration by substitution

$$\int 2x\sin(x^2)\,dx \qquad\qquad u = x^2 \qquad\qquad du = 2x\,dx$$

$$\int 2x\sin(x^2)\,dx = \int \sin(u)\,du = -\cos(u) + C = -\cos(x^2) + C$$

$$\int \sin^2\theta\cos\theta\,d\theta \qquad\qquad u = \sin\theta \qquad\qquad du = \cos\theta\,d\theta$$

$$\int \sin^2\theta\cos\theta\,d\theta = \int u^2\,du = \frac{u^3}{3} + C = \frac{\sin^3\theta}{3} + C$$

$$\int \frac{4x^3}{x^4+1}\,dx \qquad\qquad u = x^4+1 \qquad\qquad du = 4x^3$$

$$\int \frac{4x^3}{x^4+1}\,dx = \int \frac{1}{u}\,du = \ln|u| + C = \ln\left|x^4+1\right| + C$$

B.6.2-2) Integration by parts

$$\int u\,dv = uv - \int v\,du$$

Integration by parts

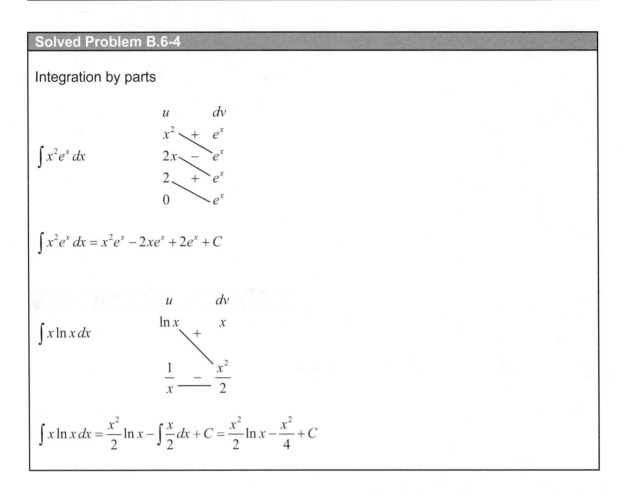

$$\int x^2 e^x \, dx = x^2 e^x - 2xe^x + 2e^x + C$$

$$\int x \ln x \, dx = \frac{x^2}{2} \ln x - \int \frac{x}{2} dx + C = \frac{x^2}{2} \ln x - \frac{x^2}{4} + C$$

APPENDIX B PROBLEMS

PB-1) Find the derivative $\dot{x}$ for the following functions.

a) $x = 5t^3$

b) $x = \sqrt{3t}$

c) $x = (2t+1)^2$

d) $x = \dfrac{1}{(t+3)}$

e) $x = \cos(3t)$

f) $x = \sin^2(\theta(t))$

g) $x = \sin(\theta(t))\cos(\theta(t))$

h) $x = t^2(1+2t)$

i) $x(t) = 3t^4 + 2t + 5$

j) $x(t) = \ln 3t^2$

k) $x(t) = 3t^2 \cos t$

l) $x = \left(\dfrac{3t+1}{2t+1}\right)^4$

PB-2) Find the slope of the curve $v = 3t^3 - 5t^2 + 6t + 3$. Find the (acceleration) slope of the curve at $t = 2.5$ seconds.

PB-3) The velocity of an object is $v = t^3 - 2t^2 + 10$, what are the maximum and minimum velocities of the object between 0 and 2 seconds.

PB-4) Find the integral of the following functions.

a) $\int \sqrt{x}\,dx$

b) $\int \dfrac{1}{x^3}\,dx$

c) $\int \dfrac{x}{1+x^2}\,dx$

d) $\int e^{x^5}(5x^4)\,dx$

e) $\int (1+x^3)^5 x^2\,dx$

f) $\int x\sin x\,dx$

g) $\int e^x \cos x\,dx$

h) $\int (3t^4 + 2t + 5)\,dt$

i) $\int_3^4 \sqrt{x}\,dx$

j) $\int_1^3 e^{3t}\,dt$

APPENDIX B REFERENCES

[1] www.mathleague.com

[2] http://library.thinkquest.org/2647/geometry/geometry.htm

[3] www.mathopenref.com

APPENDIX C: UNITS & PROPERTIES

APPENDIX C OUTLINE

C.1) UNITS

C.1.1) SI UNITS

The *International System of Units (SI)* is founded on seven SI base units that are assumed to be mutually independent. These base units are given in Table C.1-1. There are also 22 derived units that are obtained algebraically from the base units. The derived units relevant to mechanics are given in Table C.1-2. There are also units outside the SI that are accepted for use with the SI. The units outside the SI and relevant to mechanics are given in Table C.1-3.

SI Base Unit	Symbol	What it measures
meter	m	distance
kilogram	kg	mass
second	s	time
Ampere	A	electric current
Kelvin	K	temperature
mole	mol	amount of substance
candela	cd	Intensity of light

Table C.1-1: SI base units

SI Derived Unit	Symbol	What it measures
square meter	m^2	area
cubic meter	m^3	volume
meter per second	m/s	velocity, speed
meter per second squared	m/s^2	acceleration
kilogram per cubic meter	kg/m^3	mass density
radian	$rad = m*m^{-1}$	plane angle
Newton	$N = kg\text{-}m/s^2$	force
Pascal	$Pa = N/m^2$	pressure
Joule	J = N-m	work, energy
Watt	W = J/s	power
Newton-meter	N-m	moment, torque
radian per second	rad/s	angular velocity
radian per second squared	rad/s^2	angular acceleration

Table C.1-2: SI derived units

SI Base Unit	Symbol	What it measures
minute	min	time
hour	h	time
day	d	time
degree	°	angle
minute	'	angle
second	"	angle
liter	$L = 10^{-3} \ m^3$	volume
metric ton	$t = 10^3 \ kg$	mass

Table C.1-3: SI accepted units

C.1.2) METRIC UNITS

The units of measure used in the Metric system are described in section on *SI Units*. In the *Metric* system, designations of any derived unit within the same measurement class may be arrived at by adding the prefixes deka, hecto, and kilo meaning, respectively, 10, 100, and 1000, and deci, centi, and milli, meaning, respectively, one-tenth, one-hundredth, and one-thousandth. In certain cases, it becomes convenient to provide for multiples larger than 1000 and for subdivisions smaller than one-thousandth. All prefixes are given in Table C.1-4.

Prefix	Symbol	Meaning	Prefix	Symbol	Meaning
yotta	Y	10^{24}	deci	d	10^{-1}
zetta	Z	10^{21}	centi	c	10^{-2}
exa	E	10^{18}	milli	m	10^{-3}
peta	P	10^{15}	micro	μ	10^{-6}
tera	T	10^{12}	nano	n	10^{-9}
giga	G	10^{9}	pico	p	10^{-12}
mega	M	10^{6}	femto	f	10^{-15}
kilo	k	10^{3}	atto	a	10^{-18}
hecto	h	10^{2}	zepto	z	10^{-21}
deka	da	10^{1}	yocto	y	10^{-24}

Table C.1-4: Metric prefixes

The conversions between the Metric system units of measure and the US Customary units of measure and other Metric units are given in Table C.1-5.

Measured unit	Equivalent
Length (1 meter equals)	0.001 kilometer (km)
	100 centimeters (cm)
	1000 millimeters (mm)
	39.37008 inches (in)
	3.28084 feet (ft)
	1.09361 yards (yd)
	0.00062 miles (mi)
	0.00054 nautical miles (nmi)
Angle (1 radian equals)	57.29578 degrees ($^\circ$)
Angle (1 degree equals)	0.01745 radians (rad) (2π rad = 360°)
Mass (1 kilogram equals)	1000 grams (g)
	0.001 metric tons (t)
	2.20462 avoirdupois pounds (lb_m)
	35.27396 avoirdupois ounces (oz_m)
	0.06847 slugs
Time (1 second equals)	0.01667 minutes (1 min = 60 s)
	0.00028 hours (1 h = 60 min)
Force (1 Newton equals)	0.22481 pounds (lb_f)
Energy (1 Joule equals)	0.73756 foot-pounds (ft-lb_f)
	0.00095 British Thermal Units (BTU)
	0.00024 kilocalories (kcal)
	10^7 ergs
Power (1 Watt equals)	0.001 kilowatts (kW)
	0.00134 horsepower (hp)
	0.73756 foot-pounds per second (ft-lb_f/s)
	3.41442 BTU/hour (BTU/h)
	0.00024 kilocalories per second (kcal/s)
	10^7 ergs per second (erg/s)

Table C.1-5: Metric unit conversions

C.1.3) US CUSTOMARY UNITS

The US Customary system contains many units, several within the same unit class. For example, mass can be measured using slugs, avoirdupois pounds, grains, the pennyweight and the dram. However, there are preferred units and units that make dynamic calculations more straight forward and simple. Table C.1-6 lists the preferred units.

US Customary Unit	Symbol	What it measures
foot	ft	distance
slug	slug = lb_f-s^2 / ft	mass
second	s	time
pound	lb_f = slug-ft / s^2	force
British thermal units	BTU	energy
horsepower	hp	power

Table C.1-6: US customary preferred units

The conversions between the US Customary system units of measure and the Metric units of measure and other US Customary units are given in Table C.1-7.

Measured unit	Equivalent
Length (1 foot equals)	0.0003 kilometers (km)
	0.3048 meters (m)
	30.48 centimeters (cm)
	304.8 millimeters (mm)
	12 inches (in)
	0.33333 yards (yd) (1 yd = 3 ft)
	0.0001894 miles (mi) (1 mi = 5280 ft)
	0.00016 nautical miles (nmi) (1 nmi = 6076.11549 ft)
Angle (1 radian equals)	57.29578 degrees ($^{\circ}$)
Angle (1 degree equals)	0.01745 radians (rad) (2π rad = 360°)
Mass (1 slug equals)	14,605.67431 grams (g)
	14.60567 kilograms (kg)
	0.01461 metric tons (t)
	32.2 avoirdupois pounds (lb_m)
	515.2 avoirdupois ounces (oz_m)
Mass (1 avoirdupois pound equals)	453.59237 grams (g)
	0.45359 kilograms (kg)
	0.00045 metric tons (t)
	16 avoirdupois ounces (oz_m)
	0.03106 slugs
Time (1 second equals)	0.01667 minutes (1 min = 60 s)
	0.00028 hours (1 h = 60 min)
Force (1 pound equals)	4.44822 Newtons (N)
Energy (1 BTU equals)	778.16934 foot-pounds ($ft\text{-}lb_f$)
	1055.056 Joules (J)
	0.25216 kilocalories (kcal)
Power (1 horsepower equals)	0.746 kilowatts (kW)
	746 watts (W)
	550.22134 foot-pounds per second ($ft\text{-}lb_f/s$)
	2547.16089 BTU/hour (BTU/h)
	0.1783 kilocalories per second (kcal/s)

Table C.1-7: US customary unit conversions

C.2) PROPERTIES

C.2.1) CONSTANTS

Constant	Variable	Value
Acceleration due to gravity on the Earth	g	$9.80665 \ m/s^2$ $32.1740 \ ft/s^2$
Acceleration due to gravity on the Moon	g_{moon}	$1.67 \ m/s^2$ $5.47 \ ft/s^2$
Gravitational constant	G	$6.67 \times 10^{-11} \ N\text{-}m^2/kg^2$ $3.434 \times 10^{-8} \ lb\text{-}ft^2/slug^2$
Average radius of the Earth	r_{earth}	3956.6 mi 6371 km Pole/Equator = 6356.8/6378.1 km
Mass of the Earth	m_{earth}	5.9721986×10^{24} kg
Average radius of the Moon	r_{moon}	1737.1 km
Mass of the Moon	m_{moon}	7.347×10^{22} kg
Distance between the Earth and the Moon	$d_{earth\text{-}moon}$	363,104 km - 405,696 km
Distance between the Earth and the Sun	$d_{earth\text{-}sun}$	147,098,074 km - 152,097,701 km
Mass of the Sun	m_{sun}	1.98892×10^{30} kg

Table C.2-1: Physical constants

C.2.2) DENSITY OF SELECTED MATERIAL

Material	Density (kg/m^3)	Material	Density (kg/m^3)
Air at standard conditions	1.225	Lead	11342
Aluminum 2011	2830	Nickel 20	8090
Aluminum 2024	2770	Nickel 200	8890
Aluminum 6061	2720	Nickel 330	7940
Aluminum 7075	2810	Platinum	21460
Brass	8216	Silver	10501
Bronze	8860	Stainless 300	8090
Cast Iron	7150	Stainless 400	7930
Chromium	7150	Steel 1020	7870
Copper	8933	Tin	7287
Gold	19282	Titanium	4540
Iron	7874		

Table C.2-2: Density of selected materials

C.2.3) FRICTION BETWEEN SURFACES

The coefficients of friction given in Table C.2-3 were obtained under very specific conditions[6]. For any application other than solving textbook problems and estimations, the ideal method of determining the coefficient of friction is by trials.

Material 1	Material 2	Coefficient of Friction			
		Dry		Greasy	
		Static	Sliding	Static	Sliding
Aluminum	Aluminum	1.05 - 1.35	1.4	0.3	
Aluminum	Mild Steel	0.61	0.47		
Brass	Cast Iron		0.3		
Bronze	Cast Iron		0.22		
Cast Iron	Cast Iron	1.1	0.15		0.07
Cast Iron	Oak		0.49		0.075
Copper	Cast Iron	1.05	0.29		
Copper	Mild Steel	0.53	0.36		0.18
Glass	Glass	0.9 - 1.0	0.4	0.1 - 0.6	0.09 - 0.12
Glass	Nickel	0.78	0.56		
Leather	Oak (Parallel grain)	0.61	0.52		
Nickel	Nickel	0.7 - 1.1	0.53	0.28	0.12
Nickel	Mild Steel		0.64		0.178
Oak	Oak (parallel grain)	0.62	0.48		
Oak	Oak (cross grain)	0.54	0.32		0.072
Rubber	Asphalt (Dry)	0.9	0.5-0.8		
Rubber	Asphalt (Wet)		0.25 - 0.75		
Rubber	Concrete (Dry)		0.6-0.85		
Rubber	Concrete (Wet)		0.45 - 0.75		
Steel(Mild)	Brass	0.51	0.44		
Steel (Mild)	Cast Iron		0.23	0.183	0.133
Steel (Mild)	Phos. Bros		0.34		0.173
Steel (Mild)	Steel (Mild)	0.74	0.57		0.09 - 0.19
Steel(Hard)	Steel (Hard)	0.78	0.42	0.05 - 0.11	0.029 - 0.12
Steel	Zinc (Plated on steel)	0.5	0.45		
Tin	Cast Iron		0.32		
Wood - waxed	Dry snow		0.04		
Wood	Brick	0.6			
Wood	Concrete	0.62			
Zinc	Cast Iron	0.85	0.21		

Table C.2-3: Coefficients of friction

APPENDIX C REFERENCES

[1] "NIST Handbook 44, Appendix C, Specifications, Tolerances, and Other Technical Requirements for Weighing and Measuring Devices, General Tables of Units of Measurement" http://ts.nist.gov/weightsandmeasures/publications/appxc.cfm

[2] "The NIST Reference on Constants, Units, and Uncertainty" http://physics.nist.gov/cuu/index.html

[3] "The Engineer's Handbook", http://www.engineershandbook.com/Tables/frictioncoefficients.htm

[4] WolframAlpha, http://www.wolframalpha.com

[5] Universe Today, http://www.universetoday.com

[6] The Engineering Toolbox, http://www.engineeringtoolbox.com/metal-alloys-densities-d_50.html

[7] Avlan design, http://www.avlandesign.com/density_metal.htm

APPENDIX D: EQUATIONS

APPENDIX D OUTLINE

D.1.1) GRAVITATION

$$\mathbf{W} = m\mathbf{g}$$

$$F = G\frac{m_A m_B}{d^2}$$

D.1.2) KINEMATICS - RECTILINEAR MOTION

Rectilinear motion

$$v = \frac{ds}{dt}$$

$$a = \frac{dv}{dt}$$

$$a\,ds = v\,dv$$

$$\Delta s = s_{final} - s_{initial}$$

$$v_{ave} = \frac{\Delta s}{\Delta t}$$

$$a_{ave} = \frac{\Delta v}{\Delta t}$$

Constant acceleration

$$a = a_0 \qquad v = a_0(t - t_0) + v_0 \qquad s = \frac{1}{2}a_0(t - t_0)^2 + v_0(t - t_0) + s_0$$

$$v^2 = v_0^2 + 2a_0(s - s_0)$$

Constant velocity

$$s = v_o(t - t_o) + s_o$$

Non-constant acceleration

$$\int_{t_0}^{t} v(t)\,dt = \int_{s_0}^{s} ds$$

$$\int_{t_o}^{t} dt = \int_{s_o}^{s} \frac{1}{v(s)}\,ds$$

$$\int_{t_0}^{t} a(t)\,dt = \int_{v_0}^{v} dv$$

$$\int_{s_0}^{s} a(s)\,ds = \int_{v_0}^{v} v\,dv$$

$$\int_{s_o}^{s} ds = \int_{v_o}^{v} \frac{v}{a(v)}\,dv$$

$$\int_{t_o}^{t} dt = \int_{v_o}^{v} \frac{1}{a(v)}\,dv$$

D.1.3) KINEMATICS - CURVILINEAR MOTION

Curvilinear motion

$$\mathbf{r} = x\mathbf{i} + y\mathbf{j}$$

$$\mathbf{v} = \frac{d\mathbf{r}}{dt}$$

$$\mathbf{a} = \frac{d\mathbf{v}}{dt}$$

$$\mathbf{a}\,d\mathbf{r} = \mathbf{v}\,d\mathbf{v}$$

$$\Delta\mathbf{r} = \mathbf{r}_{final} - \mathbf{r}_{initial}$$

$$\mathbf{v}_{ave} = \frac{\Delta\mathbf{r}}{\Delta t}$$

$$\mathbf{a}_{ave} = \frac{\Delta\mathbf{v}}{\Delta t}$$

Non-constant acceleration

$$\int_{t_o}^{t} \mathbf{v}\, dt = \int_{\mathbf{r}_o}^{\mathbf{r}} d\mathbf{r} \qquad\qquad \int_{t_o}^{t} \mathbf{a}\, dt = \int_{\mathbf{v}_o}^{\mathbf{v}} d\mathbf{v} \qquad\qquad \int_{\mathbf{r}_o}^{\mathbf{r}} \mathbf{a}\, d\mathbf{r} = \int_{\mathbf{v}_o}^{\mathbf{v}} \mathbf{v}\, d\mathbf{v}$$

D.1.4) KINEMATICS - PROJECTILE MOTION

Velocity

$$\mathbf{v}_o = v_{xo}\mathbf{i} + v_{yo}\mathbf{j} = v_o(\cos\theta\,\mathbf{i} + \sin\theta\,\mathbf{j})$$

x-direction

$$a_x = 0 \qquad\qquad v_x = \frac{dx}{dt} = \text{constant} \qquad\qquad x = x_o + v_{xo}t = x_o + (v_o\cos\theta)t$$

y-direction

$$a_y = \frac{dv_y}{dt} = -g \qquad\qquad v_y = \frac{dy}{dt} = v_{yo} - gt = v_o\sin\theta - gt \qquad\qquad v_{y(apex)} = 0$$

$$y = y_o + v_{yo}t - \frac{gt^2}{2} = y_o + (v_o\sin\theta)t - \frac{gt^2}{2} \qquad\qquad a_y\, dy = v_y\, dv_y$$

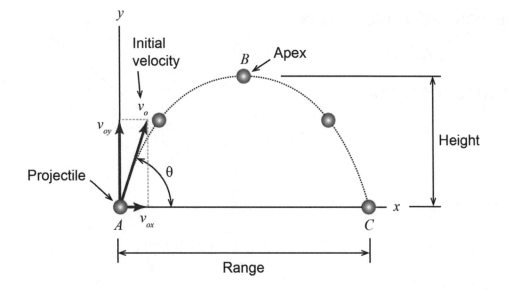

D.1.5) KINEMATICS - NORMAL AND TANGENTIAL COORDINATES

$$\mathbf{v} = v\,\mathbf{e}_t \qquad v = \frac{ds}{dt} = \rho\dot{\theta} = r\omega \qquad \mathbf{a} = \dot{v}\mathbf{e}_t + \frac{v^2}{\rho}\mathbf{e}_n = r\alpha\,\mathbf{e}_t + r\omega^2\,\mathbf{e}_n$$

$$\rho = \frac{\left[1 + \left(\dfrac{dy}{dx}\right)^2\right]^{3/2}}{\left|\dfrac{d^2 y}{dx^2}\right|}$$

Circular motion

$$v = r\dot{\theta} \qquad a_t = r\ddot{\theta} \qquad a_n = \frac{v^2}{r} = r\dot{\theta}^2$$

D.1.6) KINEMATICS - POLAR COORDINATES

$$\mathbf{r} = r\,\mathbf{e}_r \qquad \mathbf{v} = \dot{r}\mathbf{e}_r + r\dot{\theta}\mathbf{e}_\theta \qquad \mathbf{a} = (\ddot{r} - r\dot{\theta}^2)\mathbf{e}_r + (r\ddot{\theta} + 2\dot{r}\dot{\theta})\mathbf{e}_\theta$$

Circular motion

$$\mathbf{v} = r\dot{\theta}\mathbf{e}_\theta \qquad \mathbf{a} = -r\dot{\theta}^2\,\mathbf{e}_r + r\ddot{\theta}\mathbf{e}_\theta$$

D.1.7) KINEMATICS - RELATIVE MOTION

$$\mathbf{r}_B = \mathbf{r}_A + \mathbf{r}_{B/A} \qquad \mathbf{v}_B = \mathbf{v}_A + \mathbf{v}_{B/A} \qquad \mathbf{a}_B = \mathbf{a}_A + \mathbf{a}_{B/A}$$

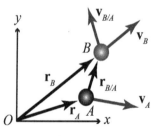

D.1.8) KINEMATICS - ROTATION

$$\omega = \frac{d\theta}{dt} \qquad \alpha = \frac{d\omega}{dt} \qquad \alpha\,d\theta = \omega\,d\omega$$

Constant acceleration

$$\alpha = \alpha_o \qquad \omega = \alpha_o(t - t_o) + \omega_o \qquad \theta = \frac{1}{2}\alpha_o(t - t_o)^2 + \omega_o(t - t_o) + \theta_o$$

$$\omega^2 = \omega_o^2 + 2\alpha_o(\theta - \theta_o)$$

Non-constant acceleration

$$\int_{t_o}^{t} \omega(t)\,dt = \int_{\theta_o}^{\theta} d\theta \qquad \int_{t_o}^{t} \alpha(t)\,dt = \int_{\omega_o}^{\omega} d\omega \qquad \int_{\theta_o}^{\theta} \alpha(\theta)\,d\theta = \int_{\omega_o}^{\omega} \omega\,d\omega$$

Pure rotation

$$s_{A/O} = r_{A/O}\,\theta \qquad\qquad \mathbf{v}_{A/O} = r_{A/O}\omega\,\mathbf{e}_t = \boldsymbol{\omega}\times\mathbf{r}_{A/O}$$

$$\begin{aligned}
\mathbf{a}_{A/O} &= r_{A/O}\alpha\,\mathbf{e}_t + r_{A/O}\omega^2\,\mathbf{e}_n \\
&= \boldsymbol{\alpha}\times\mathbf{r}_{A/O} - \omega^2\mathbf{r}_{A/O} \\
&= \boldsymbol{\alpha}\times\mathbf{r}_{A/O} + \boldsymbol{\omega}\times(\boldsymbol{\omega}\times\mathbf{r}_{A/O})
\end{aligned}$$

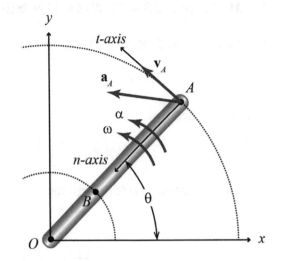

Gears and belts

$$v_{contact} = r_1\omega_1 = r_2\omega_2 \qquad\qquad N_1\omega_1 = N_2\omega_2 \qquad\qquad a_t = r_1\alpha_1 = r_2\alpha_2$$

D.1.9) RIGID-BODY GENERAL PLANAR MOTION

$$\begin{aligned}
\mathbf{v}_B &= \mathbf{v}_A + \mathbf{v}_{B/A} \\
&= \mathbf{v}_A + \boldsymbol{\omega}\times\mathbf{r}_{B/A}
\end{aligned}$$

$$\begin{aligned}
\mathbf{a}_B &= \mathbf{a}_A + (\mathbf{a}_{B/A})_t + (\mathbf{a}_{B/A})_n \\
&= \mathbf{a}_A + \boldsymbol{\alpha}\times\mathbf{r}_{B/A} - \omega^2\mathbf{r}_{B/A} \\
&= \mathbf{a}_A + \boldsymbol{\alpha}\times\mathbf{r}_{B/A} + \boldsymbol{\omega}\times(\boldsymbol{\omega}\times\mathbf{r}_{B/A})
\end{aligned}$$

Rolling without slip

$$s_O = r\theta \qquad\qquad v_O = r\omega \qquad\qquad a_O = r\alpha$$

$$\mathbf{v}_O = \boldsymbol{\omega}\times\mathbf{r}_{O/IC} \qquad\qquad \mathbf{a}_O = \boldsymbol{\alpha}\times\mathbf{r}_{O/IC}$$

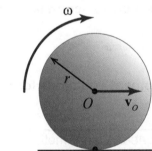

Rotating reference frame

$$\mathbf{v}_A = \mathbf{v}_B + \boldsymbol{\omega}\times\mathbf{r}_{A/B} + \mathbf{v}_{A,rel} \qquad\qquad \mathbf{a}_A = \mathbf{a}_B + \mathbf{a}_{A/B} = \mathbf{a}_B + 2\boldsymbol{\omega}\times\mathbf{v}_{A,rel} + \boldsymbol{\alpha}\times\mathbf{r}_{A/B} - \omega^2\mathbf{r}_{A/B} + \mathbf{a}_{A,rel}$$

$$\mathbf{a}_A = \mathbf{a}_B + \mathbf{a}_{A/B} = \mathbf{a}_B + 2\boldsymbol{\omega}\times\mathbf{v}_{A,rel} + \boldsymbol{\alpha}\times\mathbf{r}_{A/B} + \boldsymbol{\omega}\times(\boldsymbol{\omega}\times\mathbf{r}_{A/B}) + \mathbf{a}_{A,rel}$$

D.1.10) KINETICS – NEWTONIAN MECHANICS

Spring and damping forces

$$F_s = kx \qquad F_d = c\dot{x}$$

Friction forces

$$F_{fs,\max} = \mu_s N \qquad F_{fs} \leq F_{fs,\max} \qquad F_{fk} = \mu_k N$$

Newton's second law for particles

$$\sum \mathbf{F} = m\mathbf{a}$$

	Coordinate Directions		
Cartesian coordinates:	$\sum F_x = ma_x$	$\sum F_y = ma_y$	$\sum F_z = ma_z$
Normal–tangential coordinates:	$\sum F_t = ma_t$	$\sum F_n = ma_n = m\dfrac{v^2}{\rho}$	$\sum F_b = 0$
Polar coordinates:	$\sum F_r = ma_r = m(\ddot{r} - r\dot{\theta}^2)$	$\sum F_\theta = ma_\theta = m(r\ddot{\theta} + 2\dot{r}\dot{\theta})$	

Newton's second law for systems of particles

$$\sum \mathbf{F}_i = \sum m_j \mathbf{a}_j \qquad \sum \mathbf{F} = m\mathbf{a}_G$$

Newton's second law for rigid bodies

$$\sum \mathbf{F} = m\mathbf{a}_G \qquad \mathbf{a}_G = \mathbf{a}_P + \boldsymbol{\alpha} \times \mathbf{r}_{G/P} - \omega^2 \mathbf{r}_{G/P}$$

Moment equations for rigid bodies

	Moments	
General planar motion (rotation about a moving point P)	$\sum \mathbf{M}_P = I_P \boldsymbol{\alpha} + \mathbf{r}_{G/P} \times m\mathbf{a}_P$	$\sum \mathbf{M}_P = I_G \boldsymbol{\alpha} + \mathbf{r}_{G/P} \times m\mathbf{a}_G$
Translation	$\sum \mathbf{M}_P = \mathbf{r}_{G/P} \times m\mathbf{a}_P$	$\sum \mathbf{M}_G = 0$
Rotation about a fixed point (O)	$\sum \mathbf{M}_O = I_O \boldsymbol{\alpha}$	
Rotation about G	$\sum \mathbf{M}_G = I_G \boldsymbol{\alpha}$	

Parallel Axis Theorem: $\quad I = I_G + md^2$ Radius of Gyration: $\quad I = mk^2$

D.1.11) KINETICS – WORK AND ENERGY

Work

	Work	Conditions
Particle	$U_{1-2} = \int_{\mathbf{r_1}}^{\mathbf{r_2}} \mathbf{F} \cdot d\mathbf{r}$	
	$U_{1-2} = \mathbf{F} \cdot \Delta\mathbf{r}_{1-2}$	$\mathbf{F}$ cannot be a function of $\mathbf{r}$
	$U_{1-2} = F d_{1-2}$	F and d are parallel
	$U_{1-2} = (F \cos\theta)\, d_{1-2}$	θ = angle between F and d
Rigid body	$U_{1-2} = \int_{\mathbf{r_1}}^{\mathbf{r_2}} \mathbf{F} \cdot d\mathbf{r}$	
	$U_{1-2} = \int_{\theta_1}^{\theta_2} F_{\perp}\, r\, d\theta = \int_{\theta_1}^{\theta_2} M\, d\theta$	Work done by a force induced moment
	$U_{1-2} = \int_{\theta_1}^{\theta_2} \mathbf{M} \cdot d\boldsymbol{\theta}$	Work done by a moment
System of particles	$U_{1-2} = \int_{\mathbf{r_1}}^{\mathbf{r_2}} \mathbf{F} \cdot d\mathbf{r}$	

Kinetic energy

	Kinetic Energy	
Particle/Translation	$T = \dfrac{1}{2}mv^2$	
Rotation about a fixed point (O)	$T = \dfrac{1}{2}I_O\omega^2$	
Rotation about G	$T = \dfrac{1}{2}mv_G^2 + \dfrac{1}{2}I_G\omega^2$	
System of particles	$T = \sum \dfrac{1}{2}m_i v_i^2$	$T = \dfrac{1}{2}mv_G^2 + \dfrac{1}{2}I_G\omega^2$
System of rigid bodies	$T = \sum T_i$	

Potential energy

	Potential Energy		
Gravitational	$V_g = m g h_G$	$V_g = -\dfrac{mgR_e^2}{r}$	
Elastic	$V_e = \dfrac{1}{2}k x^2$		
System of particles	$V_g = \sum m_i g h_i$	$V_g = m g h_G$	$V_e = \sum \dfrac{1}{2}k_j x_j^2$
System of rigid bodies	$V = \sum V_i$		

Work-energy balance

$$T_1 + U_{1-2} = T_2 \qquad\qquad T_1 + V_1 + U_{1-2,non} = T_2 + V_2$$

Conservation of energy

$$T_1 + V_1 = T_2 + V_2$$

Power and efficiency

$$P = \frac{dU}{dt} \qquad\qquad P = \mathbf{F} \cdot \mathbf{v} \qquad\qquad \varepsilon = \frac{P_{out}}{P_{in}}$$

D.1.12) KINETICS – IMPULSE AND MOMENTUM

Linear and angular momentum

	Linear momentum	Angular momentum
Particle / translation	$\mathbf{G} = m\mathbf{v}$	$\mathbf{H}_O = \mathbf{r}_{G/O} \times m\mathbf{v} \qquad \mathbf{H}_G = 0$
System of particles	$\mathbf{G} = \sum m_i \mathbf{v}_i$	$\mathbf{H}_O = \sum (\mathbf{r}_i \times m_i \mathbf{v}_i)$
	$\mathbf{G} = m\mathbf{v}_G$	$\mathbf{H}_G = I_G \boldsymbol{\omega}$
Pure rotation		$\mathbf{H}_O = I_O \boldsymbol{\omega}$
General planar motion	$\mathbf{G} = m\mathbf{v}_G$	$\mathbf{H}_G = I_G \boldsymbol{\omega} \qquad \mathbf{H}_O = \mathbf{r}_{G/O} \times m\mathbf{v}_G + I_G \boldsymbol{\omega}$
System of rigid bodies	$\mathbf{G} = \sum m_i \mathbf{v}_{Gi}$	$\mathbf{H}_O = \sum \mathbf{H}_{Oi}$

Linear and angular impulse

Linear impulse: $\mathbf{I} = \int_{t_1}^{t_2} \sum \mathbf{F}\, dt$

Angular impulse: $\mathbf{J}_O = \int_{t_1}^{t_2} \sum (\mathbf{r}_{force/Oi} \times \mathbf{F}_i)\, dt = \int_{t_1}^{t_2} \sum \mathbf{M}_O\, dt$

$$\mathbf{J}_O = \int_{t_1}^{t_2} \sum (\mathbf{r}_{force/Gi} \times \mathbf{F}_i)\, dt = \int_{t_1}^{t_2} \sum \mathbf{M}_G\, dt$$

Principle of impulse-momentum

Linear impulse-momentum: $\mathbf{I} = \mathbf{G}_2 - \mathbf{G}_1$ \qquad Conservation of linear momentum: $\mathbf{G}_1 = \mathbf{G}_2$

Principle of angular impulse-momentum: $\mathbf{J}_O = \mathbf{H}_{O,2} - \mathbf{H}_{O,1}$ \qquad $\mathbf{J}_G = \mathbf{H}_{G,2} - \mathbf{H}_{G,1}$

Conservation of angular momentum: $\mathbf{H}_{O,2} = \mathbf{H}_{O,1}$

Time rate of change of momentum

$$\dot{\mathbf{H}}_O = \sum \mathbf{M}_O \qquad \dot{\mathbf{G}} = \sum \mathbf{F}$$

Impact

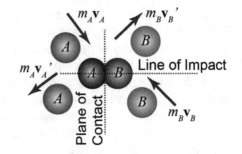

$$e = \frac{\left(v_A'\right)_{line\,of\,impact} - \left(v_B'\right)_{line\,of\,impact}}{\left(v_B\right)_{line\,of\,impact} - \left(v_A\right)_{line\,of\,impact}}$$

$$\left(v_A\right)_{plane\,of\,contact} = \left(v_A'\right)_{plane\,of\,contact}$$

$$\left(v_B\right)_{plane\,of\,contact} = \left(v_B'\right)_{plane\,of\,contact}$$

$$m_A \left(v_A\right)_{line\,of\,impact} + m_B \left(v_B\right)_{line\,of\,impact} = m_A \left(v_A'\right)_{line\,of\,impact} + m_B \left(v_B'\right)_{line\,of\,impact}$$

D.1.5) FE REFERENCE HANDBOOK EQUATIONS

What is the FE Exam?

The *Fundamentals of Engineering* (FE) exam is a measure of minimum competency to enter the engineering profession[1]. Employers hiring recent engineering graduates often look to see if the applicant has successfully passed the FE exam. Passing the exam also demonstrates how serious you are about your engineering career.

The *National Council of Examiners for Engineering and Surveying* (NCEES) is the organization that oversees the development, distribution, and grading of the FE Exam, along with its counterpart, the *Principles and Practice of Engineering* (PE) exam. The FE Exam is a "limited reference exam," which means that the only reference material an examinee can use during the exam is the NCEES FE Reference Handbook[2]. This is a 258-page publication containing equations and data needed during the exam.

The *FE Reference Handbook* is broken up by subject and discipline. Within the handbook there is a section devoted to dynamics equations. These pages, as far as they relate to the contents of this book, are reproduced in the following section. One key ingredient to passing the FE exam is familiarity with the FE Reference Handbook. This is gives you an opportunity to see and get familiar with these equations. It should be noted that minor modifications have been to the FE Reference Handbook equation notation for the sake of consistency.

What is Engineering Licensure?

According to the NCEES (National Council of Examiners for Engineering and Surveying), "Licensure is the mark of a professional". Becoming licensed as a professional engineer (P.E.), shows that your engineering knowledge is up to a recognized standard, and it provides career options and opportunities that might not be available otherwise[3]. Only licensed professionals are allowed to offer their services to the public and sign and seal plans for the public. These requirements and high standards help protect the public's safety and welfare.

The licensure procedure and requirements varies slightly from state to state, but in general, the following four steps must be completed before you may apply for a professional engineering license (as listed by the NCEES[3]).

Step 1) FE Exam: An applicant must take and pass the Fundamentals of Engineering (FE) exam.

Step 2) Graduation: The applicant must graduate from an ABET-accredited engineering program at a college or university.

Step 3) Work Experience: After passing the FE exam and graduating, actual work experience must be gained. The duration and type of work experience required depends on the state in which the license is going to be acquired. The requirement usually includes a four year internship.

Step 4) PE Exam: Take and pass the Principles and Practice of Engineering (PE) exam.

After successfully completing the four steps, the engineer is eligible for licensure by their state's licensing board. Once licensure is granted, the engineer is known as a "professional engineer", or P.E. A professional engineering license (P.E.) allows engineers to legally represent themselves to the public as an engineer, offer consulting engineering services to private and public entities, and perform engineering design or construction on public works.

DYNAMICS

KINEMATICS

Kinematics is the study of motion without consideration of the mass of, or the forces acting on, the system. For particle motion, let $r(t)$ be the position vector of the particle in an inertial reference frame. The velocity and acceleration of the particle are respectively defined as

$v = dr/dt$
$a = dv/dt$, where
v = the instantaneous velocity,
a = the instantaneous acceleration, and
t = time

Cartesian Coordinates

$\mathbf{r} = x\mathbf{i} + y\mathbf{j} + z\mathbf{k}$
$\mathbf{v} = \dot{x}\mathbf{i} + \dot{y}\mathbf{j} + \dot{z}\mathbf{k}$
$\mathbf{a} = \ddot{x}\mathbf{i} + \ddot{y}\mathbf{j} + \ddot{z}\mathbf{k}$, where

$\dot{x} = dx / dt = v_x$, etc.

$\ddot{x} = d^2x / dt^2 = a_x$, etc.

Radial and Transverse Components for Planar Motion

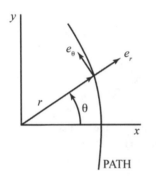

Unit vectors $\mathbf{e}_\theta$ and $\mathbf{e}_r$ are, respectively, normal to and collinear with the position vector $\mathbf{r}$. Thus:

$\mathbf{r} = r\,\mathbf{e}_r$

$\mathbf{v} = \dot{r}\mathbf{e}_r + r\dot{\theta}\mathbf{e}_\theta$

$\mathbf{a} = (\ddot{r} - r\dot{\theta}^2)\mathbf{e}_r + (r\ddot{\theta} + 2\dot{r}\dot{\theta})\mathbf{e}_\theta$, where

r = the radial distance
θ = the angle between the x axis and $\mathbf{e}_r$
$r = dr/dt$, etc.
$\ddot{r} = d^2r / dt^2$, etc.

Unit vectors $\mathbf{e}_t$ and $\mathbf{e}_n$ are, respectively, tangent and normal to the path with $\mathbf{e}_n$ pointing to the center of curvature. Thus

Plane Circular Motion

A special case of transverse and radial components is for constant radius rotation about the origin, or plane circular motion.

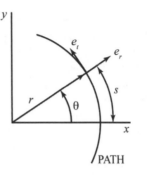

Here the vector quantities are defined as

$\mathbf{r} = r\,\mathbf{e}_r$
$\mathbf{v} = r\omega\,\mathbf{e}_t$
$\alpha = -r\omega^2\,\mathbf{e}_r + r\alpha\,\mathbf{e}_t$, where
r = the radius of the circle, and
θ = the angle between the x and $\mathbf{e}_r$ axes

The magnitudes of the angular velocity and acceleration, respectively, are defined as

$\omega = \dot{\theta}$, and

$\alpha = \dot{\omega} = \ddot{\theta}$

Arc length, tangential velocity and tangential acceleration, respectively, are

$s = r\,\theta$
$v_t = r\,\omega$
$a_t = r\,\alpha$

The normal acceleration is given by

$a_n = r\,\omega^2$

Normal and Tangential Components

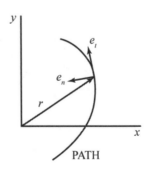

$\mathbf{v} = v(t)\,\mathbf{e}_t$

$\mathbf{a} = a(t)\mathbf{e}_t + (v_t^2 / \rho)\mathbf{e}_n$, where

ρ = instantaneous radius of curvature

Constant Acceleration

The equations for the velocity and displacement when acceleration is a constant are given as

$a(t) = a_o$

$v(t) = a_o(t - t_o) + v_o$

$s(t) = a_o(t - t_o)^2/2 + v_o(t - t_o) + s_o$, where

s = distance along the line of travel

s_o = displacement at time t_o

v = velocity along the direction of travel

v_o = velocity at time t_o

a_o = constant acceleration

t = time, and

t_o = some initial time

For a free-falling body $a_o = g$ (downward).
An additional equation for velocity as a function of position may be written as

$v^2 = v_o^2 + 2a_o(s - s_o)$

For constant angular acceleration the equations for angular velocity and displacement are

$\alpha(t) = \alpha_o$

$\omega(t) = \alpha_o(t - t_o) + \omega_o$

$\theta(t) = \alpha_o(t - t_o)^2/2 + \omega_o(t - t_o) + \theta_o$, where

θ = angular displacement

θ_o = angular displacement at time t_o

ω = angular velocity

ω_o = angular velocity at time t_o

α_o = constant angular acceleration

t = time, and

t_o = some initial time

An additional equation for angular velocity as a function of angular position may be written as

$\omega^2 = \omega_o^2 + 2\alpha_o(\theta - \theta_o)$

Non-constant Acceleration

When non-constant acceleration, $a(t)$, is considered, the equations for the velocity and displacement may be obtained from

$$v(t) = \int_{t_o}^{t} a(t)dt + v_{t_o}$$

$$s(t) = \int_{t_o}^{t} v(t)dt + s_{t_o}$$

For variable angular acceleration

$$\omega(t) = \int_{t_o}^{t} \alpha(t)dt + \omega_{t_o}$$

$$\theta(t) = \int_{t_o}^{t} \omega(t)dt + \theta_{t_o}$$

Projectile Motion

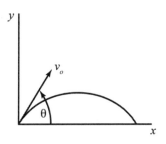

The equations for common projectile motion may be obtained for the constant acceleration equations as

$a_x = 0$

$v_x = v_o \cos(\theta)$

$x = v_o \cos(\theta) t + x_o$

$a_y = -g$

$v_y = -gt + v_o \sin(\theta)$

$y = -gt^2/2 + v_o \sin(\theta) t + y_o$

CONCEPT OF WEIGHT

$W = mg$, where

W = weight, N (lb$_f$),

m = mass, kg (lbg-sec^2/ft), and

g = local acceleration of gravity, m/sec^2 (ft/sec^2)

KINETICS

Newton's second law for a particle is

$$\Sigma F = d(mv)/dt, \text{ where}$$

ΣF = the sum of the applied forces acting on the particle,

m = the mass of the particle

v = the velocity of the particle

For constant mass

$$\Sigma F = m \, dv/dt = ma$$

One-Dimensional Motion of a Particle (Constant Mass)

When motion only exists in a single dimension then, without loss of generality, it may be assumed to be in the x direction, and

$$a_x = F_x/m, \text{ where}$$

F_x = the resultant of the applied forces which in general can depend on t, x, and v_x.

If F_x only depends on t, then

$$a_x(t) = F_x(t)/m$$

$$v_x(t) = \int_{t_o}^{t} a_x(t)dt + v_{x_o}$$

$$x(t) = \int_{t_o}^{t} v_x(t)dt + x_{t_o}$$

If the force is constant (i.e. independent of time, displacement, and velocity) then

$$a_x = F_x/m$$
$$v(t) = a_o(t - t_o) + v_o$$
$$s(t) = a_o(t - t_o)^2/2 + v_o(t - t_o) + s_o$$

Normal and Tangential Kinetics for Planar Problems

When working with normal and tangential directions, the scalar equations may be written as

$$\Sigma F_t = ma_t = mdv_t/dt \text{ and}$$

$$\Sigma F_n = ma_n = m(v_t^2/\rho)$$

Impulse and Momentum

Linear

Assuming constant mass, the equation of motion of a particle may be written as

$$mdv/dt = F$$

$$mdv = Fdt$$

For a system of particles, by integrating and summing over the number of particles, this may be expanded to

$$\sum m_i(v_i)_{t_2} = \sum m_i(v_i)_{t_1} + \sum \int_{t_1}^{t_2} F_i dt$$

The term on the left side of the equation is the linear momentum of a system of particles at time t_2. The first term on the right side of the equation is the linear momentum of a system of particles at time t_1. The second term on the right side of the equations the impulse of the force F from time t_1 to t_2. It should be noted that the above equation is a vector equation. Component scalar equations may be obtained by considering the momentum and force in a set of orthogonal directions.

Angular Momentum or Moment of Momentum

The angular momentum or the moment of momentum about point O of a particle is defined as

$$\mathbf{H}_O = \mathbf{r} \times m\mathbf{v}, \text{ or}$$

$$\mathbf{H}_O = I_O \omega$$

Taking the time derivative of the above, the equation of motion may be written as

$$\dot{\mathbf{H}}_o = d(I_o \omega)/dt = \mathbf{M}, \text{ where}$$

M is the moment applied to the particle. Now by integrating and summing over a system of any number of particles, this may be expanded to

$$\sum (\mathbf{H}_{Oi})_{t_2} = \sum (\mathbf{H}_{Oi})_{t_1} + \sum \int_{t_1}^{t_2} \mathbf{M}_{Oi} dt$$

The term on the left side of the equation is the angular momentum of a system of particles at time t. The first term on the right side of the equation is the angular momentum of a system of particles at time t_1. The second term on the right side of the equation is the angular impulse of the moment M from time t_1 to t_2.

Work and Energy

Work U is defined as

$$U = \int \mathbf{F} \cdot d\mathbf{r}$$

Kinetic Energy

The kinetic energy of a particle is the work done by an external agent in accelerating the particle from rest to a velocity v. Thus

$$T = mv^2/2$$

In changing the velocity from v_1 to v_2, the change in kinetic energy is

$$T_2 - T_1 = m(v_2^2 - v_1^2)/2$$

Potential Energy
The work done by an external agent in the presence of a conservative field is termed the change in potential energy.
Potential Energy in Gravity Field
$$V = mgh, \text{ where}$$
h = the elevation above some specified datum

Elastic Potential Energy
For a linear elastic spring with modulus, stiffness, or spring constant, the force in the spring is
$$F_s = kx, \text{ where}$$
x = the change in length of the spring from the undeformed length of the spring.

The potential energy stored in the spring when compressed or extended by an amount x is
$$V = kx^2/2$$

In changing the deformation in the spring from position x_1 to x_2, the change in the potential energy stored in the spring is
$$V_2 - V_1 = k(x_2^2 - x_1^2)/2$$

Principle of Work and Energy
If T_i and V_i are, respectively, the kinetic and potential energy of a particle at state i, then for conservative systems (no energy dissipation or gain), the law of conservation of energy is
$$T_2 + V_2 = T_1 + V_1$$

If nonconservative forces are present, then the work done by these forces must be accounted for. Hence
$$T_2 + V_2 = T_1 + V_1 + U_{1\text{-}2}, \text{ where}$$

$U_{1\text{-}2}$ = the work done by the nonconservative forces in moving between state 1 and state 2. Care must be exercised during computations to correctly compute the algebraic sign of the work term. If the forces serve to increase the energy of the system, $U_{1\rightarrow2}$ is positive. If the forces, such as friction , serve to dissipate energy, $U_{1\text{-}2}$ is negative.

Impact
During an impact, momentum is conserved while energy may or may not be conserved. For direct central impact with no external forces
$$m_1\mathbf{v}_1 + m_2\mathbf{v}_2 = m_1\mathbf{v}_1' + m_2\mathbf{v}_2', \text{ where}$$
m_1, m_2 = the masses of the two bodies,
v_1, v_2 = the velocities of the bodies just before impact,
v_1', v_2' = the velocities of the bodies just after impact.

From impacts, the relative velocity expression is
$$e = \frac{(v_2')_n - (v_1')_n}{(v_1)_n - (v_2)_n}, \text{ where}$$
e = coefficient of restitution,
$(v_i)_n$ = the velocity normal to the plane of impact just before impact, and
$(v_i')_n$ = the velocity normal to the plane of impact just after impact
The value of e is such that
$$0 \le e \le 1$$
$e = 1$, perfectly elastic (energy conserved), and
$e = 0$, perfectly plastic (no rebound)

Knowing the value of e, the velocities after the impact are given as
$$(v_1')_n = \frac{m_2(v_2)_n(1+e) + (m_1 - em_2)(v_1)_n}{m_1 + m_2}$$
$$(v_2')_n = \frac{m_2(v_1)_n(1+e) + (em_1 - m_2)(v_2)_n}{m_1 + m_2}$$

Friction
The Laws of Friction are
1. The total friction force F the can be developed is independent of the magnitude of the area of contact.
2. The total friction force F that can be developed is proportional to the normal force N.
3. For low velocities of sliding, the total frictional force that can be developed is practically independent of the velocity, although experiments show that the force F necessary to initiate slip is greater than that necessary to maintain the motion.

The formula expressing the Laws of Friction is
$$\mathbf{F} \le \mu\mathbf{N}, \text{ where}$$
μ = the coefficient of friction.

In general
$F < \mu_s N$, no slip occurring
$F = \mu_s N$, at the point of impending slip, and
$F = \mu_k N$, when slip is occurring
Here,
μ_s = the coefficient of static friction, and
μ_k = the coefficient of kinetic friction

The coefficient of kinetic friction is often approximate as 75% of the coefficient of static friction.

Mass Moment of Inertia

The definitions for the mass moments of inertia are

$$I_x = \int (y^2 + z^2)dm,$$

$$I_y = \int (x^2 + z^2)dm, \text{ and}$$

$$I_z = \int (x^2 + y^2)dm$$

A table listing moment of inertia formulas for some standard shapes is given in Appendix A.

Parallel-Axis Theorem

The mass moments of inertia may be calculated about any axis through the application of the above definitions. However, once the moments of inertia have been determined about an axis passing through a body's mass center, it may be transformed ot another parallel axis. The transformation equation is

$$I_{new} = I_c + md^2, \text{ where}$$

I_{new} = the mass moment of inertia about any specified axis

I_c = the mass moment of inertia about an axis that is parallel to the above specified axis but passes through the body's mass center

m = the mass of the body

d = the normal distance from the body's mass center to the above-specified axis

Radius of Gyration

The radius of gyration is defined as

$$r = \sqrt{I/m}$$

PLANE MOTION OF A RIGID BODY

Kinematics

Instantaneous Center of Rotation (Instant Centers)

An instantaneous center of rotation (instant center) is a point, common to two bodies, at which each has the same velocity (magnitude and direction) at a given instant. It is also a point on one body about which another body rotates, instantaneously.

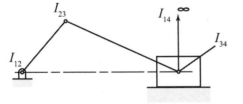

1 GROUND

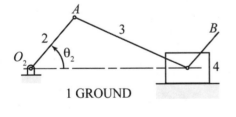

The figure shows a fourbar slider-crank. Link 2 (the crank) rotates about the fixed center, O_2. Link 3 couples the crank to the slider (link 4), whick slides against ground (link 1). Using the definition of an instant center (IC), we see that the pins at O_2, A, and B are ICs that are designated I_{12}, I_{23}, and I_{34}. The easily observable IC is I_{14}, which is located at infinity with its direction perpendicular to the interface between links 1 and 4 (the direction of sliding). To locate the remaining two ICs (for a fourbar) we must make use of Kennedy's rule.

Kennedy's Rule: When three bodies move relative to one another they have three instantaneous centers, all of which lie on the same straight line.

To apply this rule to the slider-crank mechanism, consider links 1, 2, and 3 whose ICs are I_{12}, I_{23}, and I_{13}, all of which lie on a straight line. Consider also links 1, 3, and 4 whose ICs are I_{13}, I_{34}, and I_{14}, all of which lie on a straight line. Extending the line through I_{12} and I_{23} and the line through I_{34} and I_{14} to their intersection locates I_{13}, which is common to the two groups of links that were considered.

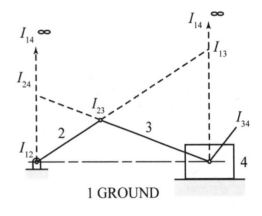

1 GROUND

Similarly, if body groups 1, 2, 4 and 2, 3, 4 are considered, a line drawn through known ICs I_{12} and I_{14} to the intersection of a line drawn through known ICs I_{23} and I_{34} locates I_{24}.

The number of ICs, c, for a given mechanism is related to the number of links, n, by

$$c = \frac{n(n-1)}{2}$$

Relative Motion

The equations for the relative position, velocity, and acceleration my by written as

Translating Axis

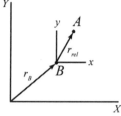

$$\mathbf{r}_A = \mathbf{r}_B + \mathbf{r}_{rel}$$
$$\mathbf{v}_A = \mathbf{v}_B + \omega \times \mathbf{r}_{rel}$$
$$\mathbf{a}_A = \mathbf{a}_B + \alpha \times \mathbf{r}_{rel} + \omega \times (\omega \times \mathbf{r}_{rel}),$$ where, ω and α are, respectively, the angular velocity and angular acceleration of the relative position vector $\mathbf{r}_{rel}$.

Rotating Axis

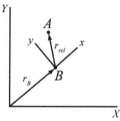

$$\mathbf{r}_A = \mathbf{r}_B + \mathbf{r}_{rel}$$
$$\mathbf{v}_A = \mathbf{v}_B + \omega \times \mathbf{r}_{rel} + \mathbf{v}_{rel}$$
$$\mathbf{a}_A = \mathbf{a}_B + \alpha \times \mathbf{r}_{rel} + \omega \times (\omega \times \mathbf{r}_{rel}) + 2\omega \times \mathbf{v}_{rel} + \mathbf{a}_{rel}$$
where, ω and α are, respectively, the total angular velocity and acceleration of the relative position vector $\mathbf{r}_{rel}$.

Rigid Body Rotation

For rigid body rotation q
$$\omega = d\theta/dt$$
$$\alpha = d\omega/dt,$$ and
$$\alpha d\theta = \omega d\omega$$

Kinetics

In general, Newton's second laws for a rigid body, with constant mass and mass moment of inertia, in plane motion may be written in vector form as

$$\Sigma\mathbf{F} = m\mathbf{a}_c$$
$$\Sigma\mathbf{M}_c = I_c\alpha$$
$$\sum\mathbf{M}_p = I_c\alpha + \rho_{pc} \times m\mathbf{a}_c$$

F are forces and $\mathbf{a}_c$ is the acceleration of the body's mass center both in the plane of motion, $\mathbf{M}_c$ are moments and α is the angular acceleration both about an axis normal to the plane of motion, I_c is the mass moment of inertia about the normal axis through the mass center, and ρ_{pc} is a vector from point p to point c.

Without loss of generality, the body may be assumed to be in the x-y plane. The scalar equations of motion may then be written as

$$\Sigma F_x = ma_{xc}$$
$$\Sigma F_y = ma_{yc}$$
$$\Sigma M_{zc} = I_{zc}\alpha,$$ where

zc indicates the z axis passing through the body's mass center, a_{xc} and a_{yc} are the acceleration of the body's mass center in the x and y directions, respectively, and α is the angular acceleration of the body about the z axis.

Rotation about an Arbitrary Fixed Axis

For rotation about some arbitrary fixed axis q
$$\Sigma M_q = I_q\alpha$$

If the applied moment acting about the fixed axis is constant then integrating with respect to time, from $t = 0$ yields

$$\alpha = M_q/I_q$$
$$\omega = \omega_o + \alpha t$$
$$\theta = \theta_o + \omega_o t + \alpha t^2/2$$

Where ω_o and θ_o are the values of angular velocity and angular displacement at time $t = 0$, respectively.

The change in kinetic energy is the work done in accelerating the rigid body from ω_o to ω

$$I_q\omega^2/2 = I_q\omega_o^2/2 + \int_{\theta_o}^{\theta} M_q d\theta$$

Kinetic Energy

In general the kinetic energy for a rigid body may be written as

$$T = mv^2/2 + I_c\omega^2/2$$

For motion in the xy plane this reduces to

$$T = m(v_{cx}^2 + v_{cy}^2)/2 + I_c\omega_z^2/2$$

For motion about an instant center,

$$T = I_{IC}\omega^2/2$$

[1] National Council of Examiners for Engineering and Surveying (NCEES)
    http://www.engineeringlicense.com/feexam/

[2] National Council of Examiners for Engineering and Surveying (NCEES), "Fundamentals
    of Engineering Supplied-Reference Handbook 8th Ed.", 2008, ISBN: 1-932613-30-8

[3] National Council of Examiners for Engineering and Surveying (NCEES), www.ncees.org

NOTES

NOTES

NOTES